Soil Mechanics—Transient and Cyclic Loads

Soil Mechanics—Transient and Cyclic Loads

Constitutive relations and numerical treatment

Edited by
G. N. Pande
O. C. Zienkiewicz

Department of Civil Engineering
University of Wales, Swansea

A Wiley–Interscience Publication

1807 1982

JOHN WILEY & SONS
Chichester · New York · Brisbane · Toronto · Singapore

TA
710
.S593
1982

Library of Congress Cataloging in Publication Data:
Main entry under title:

Soil mechanics—transient and cyclic loads.

  (Wiley series in numerical methods in engineering)
  A Wiley–Interscience publication.
  Includes index.
  1. Soil mechanics—Addresses, essays, lectures.
I. Pande, G. N. II. Zienkiewicz, O. C.
III. Series.
TA710.S593   624.1'5136   81-16485
ISBN 0 471 10046 3   AACR2

British Library Cataloguing in Publication Data:

Soil mechanics—transient and cyclic loads.—
  (Wiley series in numerical methods in engineering)
  1. Soil dynamics
  I. Pande, G. N. II. Zienkiewicz, O. C.
  624.1'513 TA710

ISBN 0 471 10046 3

Printed in Northern Ireland at The Universities Press (Belfast) Ltd., and bound at the Pitman Press Ltd., Bath.

Contributing Authors

A. M. ANSAL   *Macka Cival Engineering Faculty, Istanbul Technical University, Istanbul, Turkey*

Z. P. BAŽANT   *The Technological Institute, Northwestern University, Evanston, Illinois, U.S.A.*

P. BETTESS   *Department of Civil Engineering, University College of Swansea, Swansea, U.K.*

S. K. BHATIA   *Department of Civil Engineering, University of British Columbia, Vancouver, Canada*

J. R. BOOKER   *Department of Civil Engineering, University of Sydney, Sydney, Australia*

J. P. CARTER   *Department of Civil Engineering, University of Queensland, Brisbane, Australia*

C. T. CHANG   *ECI–SECI Engineering Consulting Team, Lampung, Sumatra, Indonesia Formerly at Department of Civil Engineering, University College of Swansea, Swansea, U.K.*

Y. F. DAFALIAS   *Department of Civil Engineering, University of California, Davis, California, U.S.A.*

H. A. M. VAN EEKELEN   *Shell Research Centre, Rijswijk, The Netherlands*

W. D. L. FINN   *Department of Civil Engineering, University of British Columbia, Vancouver, Canada*

M. A. FODA   *Parsons Laboratory, Department of Civil Engineering, Massachusetts Institute of Technology, Cambridge, Massachusetts, U.S.A.*

J. GHABOUSSI   *Department of Civil Engineering, University of Illinois at Urbana-Champaign, Urbana, Illinois, U.S.A.*

L. R. HERRMANN — *Department of Civil Engineering, University of California, Davis, California, U.S.A.*

E. HINTON — *Department of Civil Engineering, University College of Swansea, Swansea, U.K.*

I. M. IDRISS — *Woodward-Clyde Consultants, San Francisco, California, U.S.A.*

K. ISHIHARA — *Department of Civil Engineering, University of Tokyo, Bankyo–Ku, Tokyo, Japan*

R. J. KRIZEK — *The Technological Institute, Northwestern University, Evanston, Illinois, U.S.A.*

K. H. LEUNG — *Department of Civil Engineering, University College of Swansea, Swansea, U.K.*

C. C. MEI — *Parsons Laboratory, Department of Civil Engineering, Massachusetts Institute of Technology, Cambridge, Massachusetts, U.S.A.*

H. MOMEN — *Department of Civil Engineering, University of Illinois at Urbana-Champaign, Urbana, Illinois, U.S.A.*

Z. MROZ — *Institute of Fundamental Technological Research, Warsaw, Poland*

S. NEMAT-NASSER — *The Technological Institute, Northwestern University, Evanston, Illinois, U.S.A.*

V. A. NORRIS — *Department of Civil Engineering, University College of Swansea, Swansea, U.K.*

R. NOVA — *Instituto di Scienze e Technica delle Costruzioni, Technical University (Politechnics) of Milan, Italy*

G. N. PANDE — *Department of Civil Engineering, University College of Swansea, Swansea, U.K.*

M. J. PENDER — *Department of Civil Engineering, University of Auckland, New Zealand*

D. J. PICKERING — *Cook, Pickering & Doyle Limited, Vancouver, Canada*

H. E. READ — *System Science and Software, San Diego, California, U.S.A.*

P. W. ROWE — *Simon Engineering Laboratories, University of Manchester, Manchester, U.K.*

H. B. SEED
Department of Civil Engineering, University of California, Berkeley, California, U.S.A.

I. M. SMITH
Simon Engineering Laboratories, University of Manchester, Manchester, U.K.

I. TOWHATA
Department of Civil Engineering, University of Tokyo, Bunkyo–Ku, Tokyo, Japan

K. C. VALANIS
College of Engineering, University of Cincinnati, Cincinnati, Ohio, U.S.A.

A. VERRUIJT
Department of Civil Engineering, University of Delft, The Netherlands

D. M. WOOD
Engineering Department, University of Cambridge, Cambridge, U.K.

C. P. WROTH
Department of Engineering Science, University of Oxford, Oxford, U.K.

O. C. ZIENCKIEWICZ,
Department of Civil Engineering, University College of Swansea, Swansea, U.K.

Preface

The behaviour of soils even under static loads is a subject of some complexity which has attracted much research effort over the past decades and is still continuing to do so. Part of the difficulty is, of course, the two-phase nature of the material. In recent years the response of foundations of structures to dynamic loading such as that caused by earthquakes, or to slower cyclic loading experienced by marine and off-shore structures subject to wave action had to be dealt with. This increases the complexity of the relevant formulation and constitutive laws. The experimental and theoretical research currently going on in this area is enormous so that the practising engineer finds it difficult to follow developments and indeed has little guidance as to the work in progress. An International Conference on Soils under Cyclic and Transient Loads (Proceedings of the International Symposium on Soils under Cyclic and Transient Loading, Swansea, 7–11 January 1980, edited G. N. Pande and O. C. Zienkiewicz, A. A. Balkema, Rotterdam 1980) was held to bring together the experts and throw some light in the directions of on-going research and applications in practice. From this Conference the idea of grouping further the theories of particularly important work and elaborating these on a more permanent platform arose. This is the origin of this book.

The presentation includes some 22 chapters written on various aspects of the problem. The reader will find that the various chapters fall into categories and the book indeed can be considered as one of several parts.

The first part, Chapters 1–5, deals with the essential aspect of formulating the mechanics of transient dynamic behaviour for loading of various frequencies. Here coupling between the water or other fluid in the pores with the behaviour of the solid skeleton is considered in general terms. This formulation is essential for understanding and indeed development of practical procedures for dealing with the behaviour of real soils under the action of dynamic loads, whatever the constitutive relation used. The subsequent part of the book consisting of Chapters 6–18 deals with the crux of the matter. This is undoubtedly the specific form of constitutive model which has to be applied to the soil in order to model its performance generally. Here the most important models presented are those utilizing plastic or viscoplastic formulation of one kind or another. Chapters 7–13 are indeed devoted to various elaborations of critical state soil mechanics required to deal with reversed, cyclic, and transient loading conditions. In Chapters 14 and 15 attention is given to the so-called endochronic model which combines some aspects of plasticity and viscoplasticity in a novel formulation. Finally, in Chapters 17 and 18, modelling and indeed computer solutions of the degradation and shake-down problems are considered. Chapters 19, 20, and 21 deal with diverse aspects of the problem ranging from an assessment of the practical use of analytical models and their comparison with physical situations and to some aspects of soil behaviour which are associated not so much with the particular model, but with the manner in which tests are performed. Finally, Chapter 22 dealing with the liquefaction studies in the People's Republic of China contains much information hitherto inaccessible to engineers in the Western hemisphere.

We hope that this volume will serve to present a complete state-of-the-art picture of this important field at the beginning of the nineteen eighties and as a guide to the further research work and practical application which undoubtedly will evolve.

G. N. PANDE
O. C. ZIENKIEWICZ

Contents

Preface   ix

1 Soils and Other Saturated Media under Transient, Dynamic
   Conditions; General Formulation and the Validity of Vari-
   ous Simplifying Assumptions   1
   O. C. Zienkiewicz and P. Bettess

2 Boundary Layer Theory of Waves in Poro-elastic Sea Bed   .   17
   C. C. Mei and M. A. Foda

3 Approximations of Cyclic Pore Pressures Caused by Sea
   Waves in a Poro-elastic Half-plane   37
   A. Verruijt

4 On the Importance of Dissipation Effects in Evaluating Pore
   Pressure Changes due to Cyclic Loading   53
   H. B. Seed and I. M. Idriss

5 Liquefaction and Permanent Deformation under Dynamic
   Conditions — Numerical Solution and Constitutive Rela-
   tions   71
   O. C. Zienkiewicz, K. H. Leung, E. Hinton and C. T.
   Chang

6 Dynamic Response Analyses of Saturated Sands   105
   W. D. L. Finn

7 Dynamic Response Analysis of Level Ground Based on the
   Effective Stress Method   133
   K. Ishihara and I. Towhata

8 Elastoplastic and Viscoplastic Constitutive Models for Soils
   with Application to Cyclic Loading   173
   Z. Mroz and V. A. Norris

9 **A Critical State Soil Model for Cyclic Loading** **219**
  J. P. Carter, J. R. Booker and C. P. Wroth

10 **Bounding Surface Formulation of Soil Plasticity** **253**
  Y. F. Dafalias and L. R. Herrmann

11 **A Model for the Cyclic Loading of Overconsolidated Soil** . **283**
  M. J. Pender

12 **Modelling and Analysis of Cyclic Behaviour of Sands** . . . **313**
  J. Ghaboussi and H. Momen

13 **A Constitutive Model for Soil under Monotonic and Cyclic Loading** **343**
  R. Nova

14 **A New Endochronic Plasticity Model for Soils** **375**
  K. C. Valanis and H. E. Read

15 **Endochronic Models for Soils** . . . .
  Z. P. Bažant, A. M. Ansal and R. J. Krizek

16 **On Dynamic and Static Behaviour of Granular Materials** . . **439**
  S. Nemat-Nasser

17 **Fatigue Models for Cyclic Degradation of Soils** **459**
  H. A. M. van Eekelen

18 **Shakedown of Foundations Subjected to Cyclic Loads** . . . **469**
  G. N. Pande

19 **Comments on the Use of Physical and Analytical Models** . **491**
  P. W. Rowe and I. M. Smith

20 **Laboratory Investigations of the Behaviour of Soils under Cyclic Loading: a Review** **513**
  D. M. Wood

21 **The Cyclic Simple Shear Test** **583**
  W. D. L. Finn, S. K. Bhatia and D. J. Pickering

22 **Soil Liquefaction Studies in the People's Republic of China** . **609**
  W. D. L. Finn

Subject Index **629**

Soil Mechanics—Transient and Cyclic Loads
Edited by G. N. Pande and O. C. Zienkiewicz
© 1982 John Wiley & Sons Ltd

Chapter 1

Soils and other Saturated Media under Transient, Dynamic Conditions; General Formulation and the Validity of Various Simplifying Assumptions

O. C. Zienkiewicz and P. Bettess

SUMMARY

The behaviour of soil, rock or other porous materials under saturated conditions is discussed. A physical derivation of the governing equations is given, generalizing the formulation of Biot.[1-4] As the full formulation presents certain difficulties for computation and is unnecessary under some conditions, various approximations are presented ranging from so-called consolidation equations (with dynamic terms removed) through a new intermediate dynamic approximation to fully undrained behaviour. Approximate limits of validity of the respective ranges of such simplifications are given.

1.1 INTRODUCTION

The behaviour of soils (or indeed of other porous materials, such as rock, concrete, etc.) under conditions of transient loading is of obvious interest to engineers designing foundations of nuclear reactors and dams subject to earthquakes, oil exploration problems, etc. In all such cases the two-phase nature of the material has to be considered with the interstitial fluid displacing relative to the solid matrix. Similar phenomena exist in the quasi-static behaviour of consolidating foundations and the so-called "undrained behaviour" represents but a limiting case of very small relative moment.

Much attention has been given to the problems encountered and their formulation. Clearly here, the constitutive equations of the material are of paramount importance if the deformation pattern, liquefaction, etc. need to be determined. The major thrust of research is directed at such problems

1

(with numerical computer-based solutions being possible today for most complex formulations) and the later chapters of this book will address themselves to this aspect. Nevertheless, the essence of the problem, irrespective of the detail of constitutive relations, lies in the proper description of the interaction of the fluid and solid phases. With incorrect modelling of this interaction the basic features of the physics governing the problem can be missed and erroneous assumptions obtained. The classical, early work of Terzaghi[16,17] and Rendulic[12] on consolidation problems was followed by the more explicit work of Biot,[1-4] Florin[7] and others,[15,19] valid for all quasi-static situations. The derivation of Biot, limited to linear elastic behaviour, has been extended by him to the dynamic range using mechanistic modelling approach.[4] Other work, using 'mixture theory' approaches has recently become fashionable and added a plethora of papers on the subject, compounding its complexity.[5,6,9,10,14,15] On the engineering front, various approximations had to be introduced *ad hoc* to enable analysis to be carried out, often neglecting some essential features or using simplifying assumptions without their verification. It is therefore the object of this chapter to re-derive and extend the Biot equations to deal with non-linear situations and to explain some limits of validity to the various approximations. The formulation presented establishes the basic model into which detailed constitutive relationships can be inserted when full analysis is to be carried out.

1.2 THE BASIC RELATIONSHIPS FOR A POROUS SATURATED MATERIAL

1.2.1 State variables

Consider a porous material idealized in Figure 1.1. The state of this material can be fully described by:

total stress*

$$\sigma_{ij};\tag{1.1}$$

pore pressure

$$p;\tag{1.2}$$

the displacements (average) of the solid matrix

$$u_i;\tag{1.3}$$

---

* Tensorial notation is adopted in this chapter with i as the cartesian co-ordinate direction. In Appendix A the same equations are also given in a matrix type notation more suitable for numerical analysis. The sign convention of positive tension is used in this chapter to conform with the general usage in mechanics.

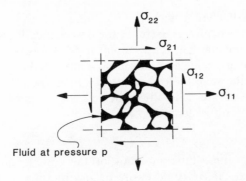

Figure 1.1 Total stress σ and pore pressure p
in an element of a porous material

and the average relative displacement of the fluid *vis-à-vis* the solid skeleton

$$w_i. \qquad (1.4a)$$

In the last quantity we shall define w_i by the ratio of the quantity of the fluid displaced over the total cross-sectional area thus making the actual pore fluid displacement

$$\frac{w_i}{n}, \qquad (1.4b)$$

where n is the porosity of the solid phase. In the following we shall further define ρ as the density of the solid plus fluid 'ensemble' and ρ_f as the density of the fluid alone.

The variables defined in (1.1)–(1.4) allow the (Lagrangian) velocities and accelerations to be evaluated as well as such derived quantities as strain. This is given for small deformations by

$$\varepsilon_{ij} = (u_{i,j} + u_{j,i})/2, \qquad (1.5)$$

where

$$u_{i,j} = \partial u_i / \partial x_j, \text{ etc.}$$

Obviously other non-linear strain measures may be defined by more general operators with equal ease.

1.2.2 Effective stress and constitutive relations

Pure statics allows us to divide the total stress state into two parts, one of these being the hydrostatic pressure p acting externally and internally on the solid skeleton (Figure 1.1) Thus

$$\sigma_{ij} = \sigma'_{ij} - \delta_{ij} p. \qquad (1.6)$$

The stress σ'_{ij} is known as the *effective stress,* and this separation, well established in soil mechanics, is useful in the description of the stress effects. We note the experimental fact that a uniform pore-pressure change cannot cause any appreciable strains and indeed no general failure of the skeleton. In soil mechanics, the slight deformations of volumetric nature caused by a pore-pressure increase are generally neglected, but in less porous materials these deformations can be computed as

$$d\bar{\varepsilon}_{ij} = -\delta_{ij} \, dp/3K_s \qquad (1.7)$$

where K_s is the average bulk modulus of the solid grains forming the skeleton. Non-linear effects could easily be incorporated if K_s were made dependent on p. Generally, however the strain $\bar{\varepsilon}_{ij}$ represents a minor effect and its inclusion is necessary only in rock-like materials.

Having postulated the virtually negligible effect of the pressure p on total strain ε_{ij}, we imply that most of the deformations are due to the effective stresses or other extraneous causes such as for instance temperature, past history of straining, etc. We can thus write quite generally.

$$d\varepsilon_{ij} = d\varepsilon^{\sigma}_{ij} + d\bar{\varepsilon}_{ij} + d\varepsilon^{0}_{ij}, \qquad (1.8)$$

where the last term denotes strains due to creep, temperature, etc. The rate-independent constitutive law relates $d\sigma_{ij}$ to $d\varepsilon^{0}_{ij}$ by

$$d\sigma'_{ij} = D_{ijkl} \, d\varepsilon^{\sigma}_{kl}. \qquad (1.9)$$

In the above the tangent modulus matrix D_{ijkl} is defined by the form of the constitutive relationships used and generally will depend on σ_{ij}, ε_{ij}, etc. Anisotropy of the solid skeleton may easily be taken into account at this stage as well as any of the elasto-plastic relations developed in later chapters.

1.2.3 Overall equilibrium

Clearly overall equilibrium between the total stress-tensor gradients, body forces and inertia forces must exist. Thus we can write

$$\sigma_{ij,j} + \rho g_i = \rho \ddot{u}_i + \rho_f \ddot{w}_i, \qquad \text{with} \quad \frac{\partial}{\partial t} u_i \equiv \dot{u}_i, \text{etc.} \qquad (1.10)$$

In the above, g_i represents the vector of gravity accelerations in the reference frame and the acceleration terms on the right-hand side come from the overall and relative fluid displacement respectively. The reader will note that the actual fluid mass in a unit of total volume is $n\rho_f$, but the disappearance of the porosity n is accounted for by the definition of the 'average' displacement.

1.2.4 Equilibrium of fluid flow

In quasi-static flow through a porous medium we note that 'resistive' forces due to the fluid viscosity balance the pressure gradient, thus giving

$$-p_{,i} = k_{ij}^{-1} \dot{w}_j. \qquad (1.11)$$

In the above, the matrix k_{ij} is known as the permeability matrix, which in general depends on $\dot{w}_i$ and n for a given fluid. In slow flow through an isotropic medium it is simply defined by a single value of permeability coefficient k, i.e.

$$k_{ij} = \delta_{ij}k. \qquad (1.12)$$

In the general dynamic case, the inertia forces need to be added, resulting in the fluid equilibrium equation:

$$-p_{,i} + \rho_f g_i = k_{ij}^{-1} \dot{w}_j + \rho_f \ddot{u}_i + \rho_f \ddot{w}_i/n. \qquad (1.13)$$

1.2.5 Mass balance of flow

The final relationship required is that defining the mass balance of flow. We note here that the divergence of the flow velocity vector must be equal to the rate of decrease of the pore space and the fluid expansion rate. Thus we can write the volumetric balance as

$$\dot{w}_{i,i} = -\dot{\varepsilon}_{ii} - (1-n)\dot{p}/K_s + \dot{\sigma}'_{ii}/3K_s - \dot{p}n/K_f. \qquad (1.14)$$

The meaning of the first and last terms of the right-hand-side of equation (1.14) is self-evident; the first of these represents simply the rate of pore volume decrease when all solid grains of the skeleton are incompressible, while the last, with K_f defining the bulk modulus of the fluid, represents the rate of fluid expansion. The two middle terms are less obvious. These represent the rate of pore volume increase associated with the compression of the solid grains. The first of these is due to the increase in hydrostatic pressure while the second is associated with the mean effective stress change.

1.2.6 Summary of governing equations

Equations (1.5)–(1.10), (1.13) and (1.14) define the problem of dynamic fluid–solid interaction completely. Below we give a summary of these equations eliminating specifically $\mathrm{d}\bar{\varepsilon}_{ij}$ and $\mathrm{d}\varepsilon^\sigma_{ij}$, by the use of relationships (1.7) and (1.8). On rearranging these equations we can thus rewrite the generalized Biot equations in the form given below:
Effective stress definition

$$\bar{\sigma}_{ij} = \sigma'_{ij} - \delta_{ij}p; \qquad (1.15a)$$

Strain definition

$$d\varepsilon_{ij} = (du_{i,j} + du_{j,i})/2; \tag{1.15b}$$

Constitutive law

$$d\sigma'_{ij} = D_{ijkl}(d\varepsilon_{kl} - d\varepsilon^0_{kl} + \underline{\delta_{kl}\, dp/3K_s}); \tag{1.15c}$$

Equilibrium (overall)

$$\sigma_{ij,j} + \rho g_i = \rho \ddot{u}_i + \rho_f \ddot{w}_i; \tag{1.15d}$$

Equilibrium of fluid flow

$$-p_{,i} + \rho_f g_i = k_{ij}^{-1} \dot{w}_j + \rho_f \ddot{u}_i + \rho_f \ddot{w}_i/n; \tag{1.15e}$$

and

Mass balance (in integrated form)

$$w_{i,i} + \varepsilon_{ii} + \frac{np}{K_f} + \underline{(1-n)p/K_s - \sigma'_{ii}/3K_s} = F(x_i). \tag{1.15f}$$

In equation (1.15f) the function $F(x_i)$ can be determined from the initial condition at $t = 0$ and for convenience we identify this simply as $F = 0$. On occasion it will be useful to use the incremental form of this equation. In Chapter 2 the reader will find an alternative form of these equations (equations 2.21, 2.22) with velocities rather than displacements as the primary variables. The form given above is generally more convenient for numerical analysis (see Chapter 4)

We note that equations (1.15a, b, c and f) can be used to eliminate σ_{ij}, σ'_{ij}, ε_{ij} and p, thus leaving two sets of equations, (1.15d, e), for the determination of the primary set of variables u_i and w_i. Full solution of such discretized equations in the linear case has been adopted as the basic formulation used by Ghaboussi and Wilson.[8]

In principle, the solution of the full set of equations (1.15) in the basic variables u_i and w_i should yield the various limiting approximations directly. We shall find that as such limits are approached we may encounter computational difficulties if discretized solution approaches are used (e.g. finite-element methods). Further, in such limits a simplified formulation is useful and more economical processes of solution can be employed.

In writing equations (1.15f and c), we have underlined the terms which have the modulus K_s in the denominator and which can be omitted in soil-mechanics problems but not to problems relating in rock-like materials where K_s and the terms contained in D_{ijkl} are of the same order.

1.3 SOME APPROXIMATIONS TO EQUATION (15)

1.3.1 The *u–p* formulation (medium-speed phenomena)

The use of the full solution of equation (1.15) in terms of the u/w variables represents six variables in three dimensions (or four in two dimensions) and it is often preferable to reduce the problem by retaining u_i and p as basic variables. The elimination of the variable w_i is simple if the $\ddot{w}_i$ terms are dropped from equation 1.15(d and c) on the assumption that the ratio

$$\ddot{w}_i/\ddot{u}_i \to 0.$$

Now solving equation (1.15e) for $\dot{w}_i$, using the differentiated form of (1.15f) and eliminating $\dot{w}_i$, we can write the more compact system as

$$\sigma_{ij} = \sigma'_{ij} - \delta_{ij} p, \tag{1.16a}$$

$$d\varepsilon_{ij} = (du_{i,j} + du_{j,i})/2, \tag{1.16b}$$

$$d\sigma'_{ij} = D_{ijkl}(d\varepsilon_{kl} - d\varepsilon^0_{kl} + \delta_{kl}\, dp/3K_s), \tag{1.16c}$$

$$\sigma_{ij,j} + \rho g_i = \rho \ddot{u}_i, \tag{1.16d}$$

$$(k_{ij}p_{,j})_{,i} - \dot{\varepsilon}_{ii} - (k_{ij}\rho_f g_j)_{,i} = -(k_{ij}\rho \ddot{u}_i)_{,i} + n\dot{p}/K_f + (1-n)\dot{p}/K_s - \dot{\sigma}'_{ii}/3K_s. \tag{1.16e}$$

The system of equations given above is more familiar. The first four equations represent simple dynamic behaviour and the fifth (1.16e) is an augmented form of the transient seepage equation.

1.3.2 Consolidation equations (very slow phenomena)

If the problem is so slow that all acceleration forces are found to be negligible, i.e.

$$\ddot{w}_i \to 0, \qquad \ddot{u}_i \to 0,$$

the well-known equations of consolidation are obtained. These are identical to the set of equations (1.16) but with the terms containing $\ddot{u}_i$ omitted in equations (1.16d and e). We shall thus number this system, for easy reference, (17a–e).

$$\sigma_{ij} = \sigma'_{ij} - p\delta_{ij}, \tag{1.17a}$$

$$d\varepsilon_{ij} = (du_{i,j} + du_{j,i})/2, \tag{1.17b}$$

$$d\sigma'_{ij} = D_{ijkl}(d\varepsilon_{kl} - d\varepsilon^0_{kl} + \delta_{kl}\, dp/3K_s), \tag{1.17c}$$

$$\sigma_{ij,j} + \rho g_i = 0, \tag{1.17d}$$

$$(k_{ij}p_{,j})_{,i} - \dot{\varepsilon}_{ii} - (k_{ij}\rho_f g_j)_{,i} = n\dot{p}/K_f + (1-n)\dot{p}/K_s - \dot{\sigma}'_{ii}/3K_s. \tag{1.17e}$$

1.3.3 Undrained behaviour (very rapid phenomena)

As the permeability k_{ij} becomes very small (or when the speed of the phenomena is such that w_i, $\dot{w}_i$ and $\ddot{w}_i$ never reach significant values) we obtain the limiting case of undrained behaviour.

Examining either the system (1.15) or the system (1.16), we find that the last equation gives (on making $k_{ij} = 0$)

$$\dot{\varepsilon}_{ii} = -\dot{p}n/K_f - \underline{(1-n)\dot{p}/K_s} + \dot{\sigma}_{ii}/3K_s, \qquad (1.18a)$$

or incrementally,

$$d\varepsilon_{ii} = -dp(n/K_f + \underline{(1-n)/K_s}) + d\dot{\sigma}_{ii}/3K_s \qquad (1.18b)$$

The total equation system can now be written omitting the flow momentum equation as

$$\sigma_{ij} = \sigma'_{ij} - \delta_{ij}p, \qquad (1.19a)$$

$$d\varepsilon_{ij} = (du_{i,j} + du_{j,i})/2, \qquad (1.19b)$$

$$d\sigma'_{ij} = D_{ijkl}(d\varepsilon_{kl} - d\varepsilon^0_{kl} + \delta_{kl}\, dp/3K_s), \qquad (1.19c)$$

$$dp = (n/K_f + \underline{(1-n)/K_s})^{-1}(-d\varepsilon_{ii} + \underline{d\sigma'_{ii}/3K_s}), \qquad (1.19d)$$

$$\sigma_{ij,j} + \rho g_i = \rho \ddot{u}_i. \qquad (1.19e)$$

In the above the primary variable is simply u_i, thus the system is in principle much easier to solve. For soils in which K_s tends to infinity we can readily express the constitutive relationships in terms of total stress. The reader can verify that the above system can be written as

$$d\sigma_{ij} = \bar{D}_{ijkl}\, d\varepsilon_{kl} - D_{ijkl}\, d\varepsilon^0_{kl}, \qquad (1.20a)$$

$$d\varepsilon_{ij} = (du_{i,j} + du_{j,i})/2, \qquad (1.20b)$$

$$\sigma_{ij,j} + \rho g = \rho \ddot{u}_i, \qquad (1.20c)$$

with

$$\bar{D}_{ijkl} = D_{ijkl} + \delta_{ij}\delta_{kl}K_f/n, \qquad (1.20d)$$

which is identical to the behaviour of a single-phase system with a redefined total stress tangent modulus. Equation (1.20d) shows the relation between drained and undrained tangent elasticity matrices which must be observed.

1.3.4 Drained behaviour

Finally we note that if all transient behaviour has ceased, i.e. we can put $\dot{w}_i = 0$, $\dot{u}_i = 0$, $\ddot{w}_i = 0$, $\ddot{u}_i = 0$, complete uncoupling of equations occurs, and starting from equation (1.16e) we can write

$$-(k_{ij}p_{,j})_{,i} + (k_{ij}\rho_f g_j)_{,i} = 0, \qquad (1.21a)$$

which can be solved independently to determine the pressures p. With p now known, the remaining equation yields

$$d\sigma'_{ij} = D_{ijkl}(d\varepsilon_{kl} - d\varepsilon^0_{kl}), \qquad (1.21b)$$

$$\sigma_{ij,j} - p_{,i} + \rho g_i = 0, \tag{1.21c}$$

$$d\varepsilon_{ij} = (du_{i,j} + du_{j,i})/2. \tag{1.21d}$$

Now we can obtain a solution and independently determine u_i using the known values of p.

Clearly that type of fully drained behaviour does not occur ever with dynamic effects.

1.4 VALIDITY OF THE VARIOUS APPROXIMATIONS

The question of the range of applicability of the various approximations is of great practical importance. If for instance, we can determine *a priori* that the behaviour of the material will be sufficiently well modelled by totally undrained assumptions, equations (1.20) or (1.21), then only one set of variables u_i remains and a relatively simple solution technique will suffice. In another context we can decide that the transient is sufficiently slow for all acceleration terms to be omitted and use pure consolidation equations to give an adequate solution.

To decide the limits of applicability of the various approximations a simple linear problem was recently solved for a periodic load input.[21] This problem involved a soil layer (Figure 1.2) of a depth L and a periodic surface loading. Dimensional analysis indicates that the solution depends

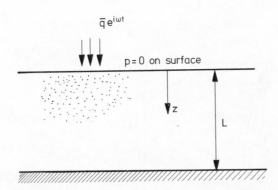

Figure 1.2 A soil layer subject to periodic loading. The free surface is drained; that is,

$$p = 0 \quad z = 0$$

$$\begin{array}{ll} u = 0 & u = 0 \\ dp/dz = 0 & w = 0 \end{array} \bigg\} \quad z = L$$

or

primarily on two non-dimensional parameters (assuming that K_s tends to infinity, i.e. that grain compressibility is not important):

$$\Pi_1 = \frac{2}{\pi} k\rho \frac{T}{\hat{T}^2} = \frac{2}{\beta\pi} \frac{\bar{k}}{g} \frac{T}{\hat{T}^2}, \qquad \Pi_2 = \pi^2 \left(\frac{T}{\hat{T}}\right)^2, \tag{1.22a}$$

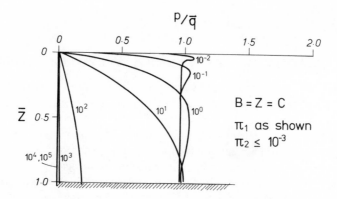

Figure 1.3(a) Variation of pore pressure with depth for various values of π_1 and π_2.

———— B   Biot theory
– – – – Z   (u–p approximation theory)
- - - - - C   (consolidation theory)
(Solution (C) is independent of π_2)

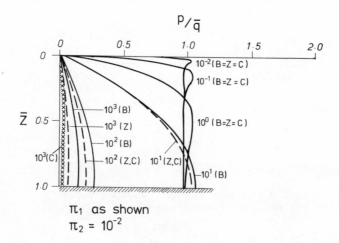

Figure 1.3(b)

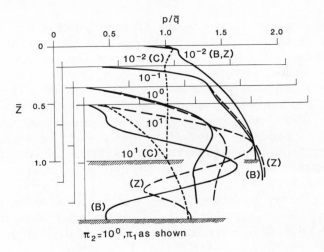

Figure 1.3(c)

where $\bar{k} = k\rho_f g$ is the kinematic permeability coefficient generally used in soil mechanics. In addition to the above parameters, there are others of minor sensitivity which we shall consider as constants.

$$\Pi_3 \equiv \beta = \rho_f/\rho \qquad \Pi_4 = n$$

$$\Pi_5 = \frac{K_f/n}{(K + K_f/n)}.$$

(1.22b)

In the above

$$v_c^2 = (K + K_f/n)/\rho, \qquad \hat{T} = 2L/V_c$$

(1.23)

represent respectively a compression wave velocity and a natural period of the layer for undrained material behaviour,

$$T = 2\pi/\omega$$

(1.24)

is the period of applied loads, and

$$K = E(1 - \nu)/(1 + \nu)(1 - 2\nu)$$

(1.25)

is the isotropic modulus of compressibility of the soil skeleton.

With typical values taken for the secondary parameters

$$\Pi_3 = \Pi_4 = \tfrac{1}{3}, \qquad \Pi_5 = 1$$

(1.26)

in Figure 1.3(a)–(c) we show the pore pressure amplitudes for various values of Π_1 and Π_2 and we note that the solution of the simplified consolidation equations does not depend on the parameter Π_2.

When the parameter Π_2 is very small, dynamic effects are totally insignificant and the full Biot solution (B), the new approximate theory (Z) and the consolidation equations (C) result in identical answers, indicating that the phenomena are 'slow'.

Further, for such 'slow' phenomena at low Π_1 values, the permeability is unimportant and a totally undrained solution is applicable throughout most of the layer. Clearly as $p = 0$ at the surface, such an undrained assumption would not be applicable in its immediate vicinity, where rapid gradients of pressure exist. Such a boundary-layer effect is discussed in Chapter 2 for 'slow' phenomena and it is shown there how this can be superposed on the undrained solution.

Figures 1.3(b) and (c) indicate further that the consolidation theory is totally inapplicable at high values of Π_2 (or high frequency values) but shows that at such values of Π_2 and relatively low values of Π_1, both (B) and (Z) solutions give comparable answers.

In Figure 1.4 we show the summary of the basic conclusions taken from the examination of many computational results,[21] in which zones of applicability of various assumptions hold for determination of pressure fluctuations (excluding for the undrained assumption the boundary-layer effects).

Extrapolating the conclusions of this simple analysis to other real problems we now have a useful guide as to the applicability of the various assumptions. Thus for any analysis we can specify a typical length L, a typical permeability coefficient, etc., and decide the type of approximation applicable with some quantitative basis (without unguarded use of such terms as 'obviously'...).

To reduce these approximate considerations for easy practical use, we write

$$\Pi_1 = \frac{2}{\beta\pi}\frac{\bar{k}}{g}\frac{T}{\hat{T}^2} \approx \frac{2\bar{k}}{g}\frac{T}{\hat{T}^2}, \tag{1.27a}$$

$$\Pi_2 \approx 10\left(\frac{\hat{T}}{T}\right)^2, \tag{1.27b}$$

with

$$\hat{T} = 2L/V_c \quad \text{and} \quad V_c \approx 1000 \text{ m/s}.$$

Thus, for example, if we consider a typical earth dam response to earthquake motion in which the range of periods is within 0.01–10 s, a typical dimension of $L = 50$ m gives $\hat{T} = 0.1$ s. We see that the range of the parameters is (taking $g \approx 10$ m/sec)

$$\frac{2k \times 10^{-2}}{10 \times (10^{-1})^2} = 0.2k < \Pi_1 < \frac{2k}{10}\frac{10}{(10^{-1})^2} = 200k,$$

$$10\left(\frac{10^{-1}}{10}\right)^2 = 10^{-3} < \Pi_2 < 10\left(\frac{10^{-1}}{10^{-2}}\right)^2 = 10^3.$$

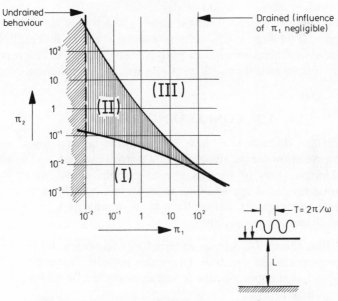

Zone I $B = Z = C$ Slow phenomena ($\ddot{w}$ and $\ddot{u}$ can be neglected)
Zone II $B = Z \neq C$ Moderate speed ($\ddot{w}$ can be neglected)
Zone III $B \neq Z \neq C$ Fast phenomena ($\ddot{w}$ can not be neglected) only
                       full Biot equation valid

Figure 1.4   Zones of applicability of various assumptions

$$\pi_1 = \frac{k\rho V_c^2}{\omega L^2} = \frac{2k\rho T}{\pi \hat{T}^2} = \left(\frac{2}{\beta\pi}\right) \frac{\bar{k}}{g} \frac{T}{\hat{T}^2}$$

$$\pi_2 = \frac{\omega^2 L^2}{V_c^2} = \pi^2 \left(\frac{\hat{T}}{T}\right)^2$$

$$\bar{k} = \rho_f g k \qquad \bar{k}\text{—kinematic permeability}$$

$$\hat{T} = 2L/V_c \qquad V_c^2 = (D + K_f/n)/\rho$$

$$\approx \beta K_f/\rho_f n$$

$$\approx K_f/\rho_f \text{ (speed of sound in water)}$$

$$\beta = \rho_f/\rho \qquad n \approx 0.33 \qquad \beta = 0.33$$

We note that only when $\bar{k}$ is of the order of 10^{-4} m/s can fully undrained behaviour be assumed and thus in a realistic mean of the expected frequencies either of the B and Z formulations is applicable.

Considering on the other hand a typical seabed problem on which waves of a period no shorter than 10 s are acting and in which the length of interest, (say the depth below surface), is of the order of 10 m, we note that $\hat{T} = 10^{-2}$ s. Hence

$$\Pi_2 < 10\left(\frac{10^{-2}}{10}\right)^2 = 10^{-5},$$

and now dynamic effects can be entirely neglected. Whether such problems fall into drained or undrained categories we leave the reader to determine. In Chapters 2^{11} and 3^{18} some detailed consideration is given to such problems, with *a priori* omission of dynamic terms.

It will be found in practice that the zone in which only the Biot equations apply (Zone III of Fig. 1.4) is seldom encountered. It generally concerns high-frequency oscillations of very porous media and is of interest in such subjects as certain mud flows, flow of vibrated liquid concrete etc.

1.5 CONCLUDING REMARKS

In the foregoing discussion we have concentrated on deriving a reasonably rigorous formulation for the dynamics of a porous medium and to show how conscious introduction of various approximations allows us to formulate more familiar ranges of applications. Indeed some guidelines to the applicability of each such approximation have been summarized.

The approximation ranges are:

(B)   Full Biot theory [u_i and w_i as variables] (equation 1.15);
(Z)   Approximation in which $\ddot{w}_i$ terms are negligible (equation 1.16);
(C)   Consolidation equation with $\ddot{u}_i$ and $\ddot{w}_i$ neglected [u_i and p as variables] (equation 1.17);
(U)   Undrained behaviour [u_i as the only variable] (equations (1.20), (1.21).

Each of these has a reasonably well-defined limit of applicability, and each allows a different finite-element discretization to be used. We shall leave the details of finite-element or other discretization processes to be discussed later, but we will remark now that formulation Z which readily includes C and U as subclasses, leads to a convenient formulation capable of dealing with phenomena ranging from rapid earthquake transients to the following slow consolidation process. In most situations we find such a formulation to be very viable and applicable to practical engineering problems as reported elsewhere[20,22] and we shall discuss its application in Chapter 4.

APPENDIX A   BASIC EQUATIONS IN MATRIX FORM

Defining displacements as $\mathbf{u}^T = [u_1, u_2, u_3]$, stresses and strains as $\boldsymbol{\sigma}^T = [\sigma_{11}, \sigma_{22}, \sigma_{33}, \sigma_{12}, \sigma_{23}, \sigma_{31}]$, $\mathbf{D}$ as a 6×6 matrix of coefficients, $\mathbf{m} = [1, 1, 1, 0, 0, 0]$, and $\mathbf{L}$ as a 6×3 matrix operator equivalent to the strain definition, we can write the basic equations in a form more suitable for finite-element discretization.

Thus the equivalent of the generalized Biot's equations (1.15) is

$$\boldsymbol{\sigma} = \boldsymbol{\sigma}' - \mathbf{m}p, \tag{A1.a}$$

$$d\boldsymbol{\varepsilon} = \mathbf{L}\,d\mathbf{u}, \tag{A1.b}$$

$$d\boldsymbol{\sigma}' = \mathbf{D}(d\boldsymbol{\varepsilon} - d\boldsymbol{\varepsilon}^0 + \underline{\mathbf{m}\,dp/3K_s}), \tag{A1.c}$$

$$\mathbf{L}^T\boldsymbol{\sigma}' + \rho\mathbf{g} = \rho\ddot{\mathbf{u}} + \rho_f\ddot{\mathbf{w}}, \tag{A1.d}$$

$$-\boldsymbol{\nabla}p + \rho_f g = \mathbf{k}^{-1}\dot{\mathbf{w}} + \rho_f\ddot{\mathbf{u}} + \rho_f\ddot{\mathbf{w}}, \tag{A1.e}$$

$$\boldsymbol{\nabla}^T\mathbf{w} + \mathbf{m}^T\boldsymbol{\varepsilon} + pn/K_f + \underline{(1-n)p/K_s - \mathbf{m}^T\boldsymbol{\sigma}'/3K_s} = 0. \tag{A1.f}$$

The approximate dynamic equations 1.16 become

$$\boldsymbol{\sigma} = \boldsymbol{\sigma}' - \mathbf{m}p, \tag{A2.a}$$

$$d\boldsymbol{\varepsilon} = \mathbf{L}\,d\mathbf{u}, \tag{A2.b}$$

$$d\boldsymbol{\sigma}' = \mathbf{D}(d\boldsymbol{\varepsilon} - d\boldsymbol{\varepsilon}^0 + \underline{\mathbf{m}\,dp/3K_s}), \tag{A2.c}$$

$$\mathbf{L}^T\boldsymbol{\sigma} + \rho\mathbf{g} = \rho\ddot{\mathbf{u}}, \tag{A2.d}$$

$$-\boldsymbol{\nabla}^T\mathbf{k}\boldsymbol{\nabla}p + \mathbf{m}^T\dot{\boldsymbol{\varepsilon}} + \boldsymbol{\nabla}^T(\mathbf{k}\rho_f\mathbf{g}) = +\boldsymbol{\nabla}^T(\mathbf{k}\rho\ddot{\mathbf{u}}) - n\dot{p}/K_f$$
$$-\underline{(1-n)\dot{p}/K_s + \mathbf{m}^T\dot{\boldsymbol{\sigma}}'/3K_s}. \tag{A2.e}$$

Finally the undrained system of equations becomes (1.20):

$$d\boldsymbol{\sigma} = \bar{\mathbf{D}}\,d\boldsymbol{\varepsilon} - \mathbf{D}\,d\boldsymbol{\varepsilon}^0, \tag{A3.a}$$

$$d\boldsymbol{\varepsilon} = \mathbf{L}\,d\mathbf{u}, \tag{A3.b}$$

$$\mathbf{L}^T\boldsymbol{\sigma} + \rho\mathbf{g} = \rho\ddot{\mathbf{u}}, \tag{A3.c}$$

with

$$\bar{\mathbf{D}} = \mathbf{D} + \mathbf{m}(K_f/n)\mathbf{m}^T. \tag{A3.d}$$

REFERENCES

1. Biot, M. A. Theory of elasticity and consolidation for a porous anisotropic solid, *J. Appl. Phys.*, **26**, 182–5, 1955.
2. Biot, M. A. General theory of three-dimensional consolidation, *J. Appl. Phys.*, **12**, 155–64, 1941.
3. Biot, M. A. Theory of stability and consolidation of a porous medium under initial stress, *J. Math. and Mech.*, **12**, 521–41, 1963.
4. Biot, M. A. Mechanics of deformation and acoustic propagation in porous medial, *J. Appl. Phys.*, **33**, 1483–98, 1962.
5. Bowen, R. M. Theory of mixtures, in A. C. Eringen, (ed.), *Continuum Physics*, Vol. III, Academic Press, New York, 1976.
6. Crochet, J., and Naghdi, P. M. A dynamic theory of interacting continua, *Int. J. Eng. Sci.*, **3**, 1965.
7. Florin, V. A. *Theory of Soil Consolidation* (in Russian), Stroyiziet, Moscow, 1942.
8. Ghaboussi, J., and Wilson, E. L. "Variational formulation of dynamics of fluid saturated porous elastic solids" *Proc. A.S.C.E.*, **98** (EM 4), 947–63, 1972.

9. Green, A. E., and Naghdi, P. M. A dynamical theory of interacting continua, *Int. J. Eng. Sci.* **3**, 231–41, 1965.
10. Green, A. E., and Steel, T. R. Constitutive equations for interacting continua, *Int. J. Eng. Sci.* **4** 483–500, 1966.
11. Mei, C. C., and Foda, M. A. Chapter 2, this book.
12. Rendulic, L. "Porenziffer und Porenwasserdruck in Tonen" *Bauingenieur*, **17** 559–64, 1936.
13. Runesson, K. *On Nonlinear Consolidation of Soft Clay*, Publ. 78. 1, Dept. of Struct. Mech., Chalmers Univ. of Techn. Goteborg, 1978.
14. Sandhu, R. S. *Fluid Flow in Saturated Porous Elastic Media*, Ph.D. thesis, University of California, Berkeley, 1968.
15. Schiffman, R., Chen, A., and Jorean, J. An analysis of consolidation theories, Proc. A.S.C.E., *J. Soil Mech. and Found. Div.*, **95** (SM 1), 285–312, 1964.
16. Terzaghi, K. Die Berechung der Durchlassigkeitsziffer des Tones aus dem Verlauf der hydro-dynamischen Spannunserscheinungen, *Akad. Wissenschaft Wein, Sitz-ungber.* **132** (H1), Mat.-Naturwissensch. Klasse.
17. Terzaghi, K. *Theoretical Soil Mechanics*, Wiley, New York, 1943.
18. Verruijt, A. Chapter 3, this book.
19. Zienkiewicz, O. C., Humpheson, C., and Lewis, R. W. *A Unified Approach to Soil Mechanics Problems Including Plasticity and Viscoplasticity*, University of Wales, Dep. of Civil Eng., C/R/250/75, Swansea, 1975, also Chapter 4 in Gudhus, G. (ed.), *Finite Elements in Geomechanics*, Wiley, pp 157–78, 1977.
20. Zienkiewicz, O. C., Chang, C. T., and Hinton, E. Non-linear seismic response and liquefaction, *Int. J. Num. Anal. Meth. in Geomech.*, **2**, 381–401, 1978.
21. Zienkiewicz, O. C., Chang, C. T., and Bettess, P. Drained, undrained, consolidating and dynamic behaviour assumptions in soils. Limits of validity, *Geotechnique*, **30**, 385–95, 1980.
22. Zienkiewicz, O. C., Leung, K. H., Hinton, E., and Chang, C. T. Earth dam analysis for earthquakes: Numerical Solution and constitutive relations for non-linear analysis. Conf. on '*Design of Dams to resist earthquake*' Inst. of Civil Eng. London Oct. 1980.
23. Zienkiewicz, O. C. Chapter 5, this book.

Soil Mechanics—Transient and Cyclic Loads
Edited by G. N. Pande and O. C. Zienkiewicz
© 1982 John Wiley & Sons Ltd

Chapter 2

Boundary Layer Theory of Waves in a Poro-elastic Sea Bed

C. C. Mei and M. A. Foda

SUMMARY

This chapter discusses the dynamics of a model sea bed under periodic loading. The sea bed is assumed to consist of a porous elastic skeleton saturated with water. Some general theoretical results obtained recently by the present authors are sketched. In particular, an effective analytical method is presented for solving the poro-elastic equations of Biot under various practical boundary conditions. A new example of forced heave of caisson is presented. The central point of this paper is that for low permeability, there is a boundary layer near the mud line where one-dimensional analysis is sufficient, while away from the boundary layer the soil behaves as a single phase. This composite picture suggests simplifications in future development of the constitutive equation and effective numerical methods. The examples treated explicitly omit the dynamic (acceleration) effects and are valid for relatively low frequencies and low permeabilities only.

2.1 INTRODUCTION

Pipelines for transporting petroleum from an offshore terminal to the shore are usually laid directly on top of the sea bed if the water depth is large, or buried in the sea bed if the water depth is small. For sufficiently shallow water or strong waves, the varying wave pressure can induce considerable stress and strain on the sea bed, thereby causing soil failure and breakage of pipelines. Many offshore platforms are supported on piles, while many of the gravity type rest directly on the sea bed. Under persistent attack of sea waves, not only the strength of the structure itself, but also the dynamic stability of the sea bed is of concern. The latter is of course a major topic in soil dynamics and has also been studied in the context of machine vibrations and earthquake engineering for some time. For foundations on land it is often possible to model the land as a dry solid only. However, it is well known that to model the non-linear dynamic behaviour of a soil realistically is an extremely difficult and so far unfinished task. To provide some reference for design or perhaps only for physical understanding, simplified models (hence not always realistic) have been used in recent theoretical studies. In fact a large amount of existing work is based on the assumption that the soil is a single-phased linearly elastic medium.

Now in the sea bed, the presence of fluid in the pores renders the single-phase theory inadequate. Recourse to a genuine two-phase theory such as the one by Terzaghi for one-dimensional consolidation is necessary. Important generalizations of Terzaghi's work to three dimensions were subsequently made by Biot,[1,2] and these have been further extended to allow for different constitutive relations (see Chapter 1). On the simplest basis of a linearly elastic soil, certain one-dimensional wave problems of interest in seismology have been investigated.[3-5]. Recently stresses induced by water waves in a porous sea bed have drawn some attention.[6,11,12] Further applications of Biot's equations can be found in hydraulic fracture, which is an important subject in petroleum extraction.

As will be shown shortly, the linearized equations which govern the coupled motions of the pore fluid and the elastic solid skeleton are much more complicated than those governing the usual elastodynamics; exact and explicit solutions are consequently hard to obtain. In a recent paper, Mei and Foda[7] have shown that a boundary-layer approximation, well known in fluid dynamics, greatly facilitates the solution of Biot's equations for a variety of boundary conditions of practical interest. The approximation also elucidates the physical mechanism involved. The purposes of this chapter are to summarize the general results in the paper quoted, and to work out a new example relevant to caisson vibration. Although the use of linear elasticity is surely a severe limitation of this theory, it is hoped that the idea herein may be fruitfully extended to soils which must be described by non-linear constitutive equations.

2.2 BIOT'S EQUATIONS OF LINEAR PORO-ELASTICITY

We shall imagine the porous solid to be a continuum and let $\bar{n}(\mathbf{x}, t)$ be the porosity of the composite medium, i.e. the volume fraction of the pores. Let $\bar{\rho}_s$ denote the density and $\mathbf{v}$ the eulerian velocity of the solid skeleton. Conservation of the solid mass requires that

$$\frac{\partial}{\partial t}[(1-\bar{n})\bar{\rho}_s] + \nabla \cdot [\bar{\rho}_s(1-\bar{n})\mathbf{v}] = 0. \tag{2.1}$$

Let $\bar{\rho}_w$ denote the density and $\mathbf{u}$ the velocity of the pore fluid; conservation of the fluid mass requires that

$$\frac{\partial}{\partial t}(\bar{n}\bar{\rho}_w) + \nabla \cdot (\bar{\rho}_w \bar{n}\mathbf{u}) = 0. \tag{2.2}$$

Since the solid grains are typically twenty-five times less compressible than pure water, it is adequate to assume that $\bar{\rho}_s$ does not change with time.

Assuming for simplicity that the soil is homogeneous, we may take

$$\bar{\rho}_s = \rho_s^0 = \text{constant.} \tag{2.3}$$

Pure water is also highly incompressible except at high frequencies (in the acoustic range). However, in natural soils, tiny air pockets are entrained in the pores, and the mixture of air and water has a relatively high compressibility when treated as a single substance. We define the effective bulk modulus of elasticity K_f by the relation

$$\frac{d\bar{\rho}_w}{\bar{\rho}_w} = \frac{dp}{K_f}. \tag{2.4}$$

It is known that K_f is related to the bulk modulus of pure water K_{f_0} and the degree of saturation S by[10]

$$\frac{1}{K_f} = \frac{1}{K_{f_0}} + \frac{1-S}{p}, \qquad 1-S \ll 1, \tag{2.5}$$

where p is the absolute pressure. For small variations p may be approximated by a constant p_0. Assuming further that S is a constant, then so is K_f. Note that $K_{f_0} = 2 \times 10^9 \, \text{N/m}^2$, but if $S = 95\%$ and $p_0 = 1 \, \text{atm.} = 10^5 \, \text{N/m}^2$, then K_f drops drastically to $10^6 \, \text{N/m}^2$. Equations (2.3) and (2.4) are the equations of state for the two phases.

Denoting the dynamic perturbation by n, ρ_w, p, etc.,* i.e.

$$\bar{n} = n^0 + n, \qquad \bar{\rho}_w = \rho_w^0 + \rho_w, \qquad \bar{p} = p^0 + p, \text{etc.,} \tag{2.6}$$

where $(\ )^0$ refers to the values at static equilibrium after settlement, we may linearize (2.1) and (2.2) to give

$$-\frac{\partial n}{\partial t} + (1 - n^0)\nabla \cdot \mathbf{v} = 0 \tag{2.7}$$

and

$$\rho_w^0 \frac{\partial n}{\partial t} + n^0 \frac{\partial \rho_w}{\partial t} + \rho_w^0 n^0 \nabla \cdot \mathbf{u} = 0.$$

Using (2.4) we get

$$\frac{\partial n}{\partial t} + \frac{n^0}{K_f}\frac{\partial p}{\partial t} + n^0 \nabla \cdot \mathbf{u} = 0. \tag{2.8}$$

Equations (2.7) and (2.8) may be combined to give

$$n^0 \nabla \cdot \mathbf{u} + (1 - n^0)\nabla \cdot \mathbf{v} = -\frac{n^0}{K_f}\frac{\partial p}{\partial t}, \tag{2.9}$$

---

* p is just the excess pore pressure in soil mechanics.

which is one form of the storage equation. Physically the change of fluid pressure in a unit cube is due to the dilation of the pore fluid and the dilation of the skeleton.

Now consider the ith component of the linear momentum for a unit cube. The solid skeleton has the mass $(1-\bar{n})\bar{\rho}_s$; hence the inertia is $(1-\bar{n})\bar{\rho}_s \partial \bar{v}_i/\partial t$. The gravity force is clearly $-g(1-\bar{n})\bar{\rho}_s\delta_{i3}$. The fluid drag is, according to Darcy's law, $(\bar{n}^2/k)(\bar{u}_i - \bar{v}_i)$, where k is the coefficient of permeability. The surface force on the solid consists of two parts: one is due directly to the effective stress* $\bar{\sigma}'_{ij}$ of Terzaghi, i.e., $\partial \bar{\sigma}'_{ij}/\partial x_j$, while the other is due to the surrounding fluid pressure $-(1-\bar{n})\bar{p}\delta_{ij}$, that is $-(1-\bar{n})\partial \bar{p}/\partial x_i$. Ignoring convective inertia, Newton's law for the solid therefore reads,

$$(1-\bar{n})\bar{\rho}_s\dot{\bar{v}}_i = -g(1-\bar{n})\bar{\rho}_s\delta_{i3} + \frac{\partial \bar{\sigma}'_{ij}}{\partial x_j} - (1-\bar{n})\frac{\partial \bar{p}}{\partial x_i}$$

$$+ \frac{\bar{n}^2}{k}(\bar{u}_i - \bar{v}_i), \qquad \frac{\partial \bar{v}_i}{\partial t} \equiv \dot{\bar{v}}_i. \tag{2.10}$$

Similar application of Newton's law for the fluid in the unit cube gives†

$$\bar{n}\bar{\rho}_w\dot{\bar{u}}_i = -g\bar{n}\bar{\rho}_w\delta_{i3} - \bar{n}\frac{\partial \bar{p}}{\partial x_i} - \frac{\bar{n}^2}{k}(\bar{u}_i - \bar{v}_i). \tag{2.11}$$

In the original work of Biot,[2] an apparent mass term was also included in (2.10) and (2.11) to account for the relative acceleration between solid and fluid. This apparent mass has not been deduced theoretically and must be regarded as an extra empirical constant. It will be evident later that in the case of small permeability this quantity can be omitted everywhere. Therefore the simpler forms of (2.10) and (2.11) are given here.

Thus far the constitutive relation between the effective stress and strain is not specified. For small deformation and for mathematical simplicity, isotropic Hooke's law is assumed:

$$\bar{\sigma}'_{ij} = G\left(\frac{\partial \bar{V}_i}{\partial x_j} + \frac{\partial \bar{V}_j}{\partial x_i} + \frac{2\nu}{1-2\nu}\delta_{ij}\frac{\partial \bar{V}_k}{\partial x_k}\right), \quad \text{where} \quad \frac{\partial \bar{V}_i}{\partial t} = \bar{v}_i \tag{2.12}$$

and V_i are components of solid displacement.

Let us first consider the static equilibrium. Assume that the mud line, i.e., the free surface of the porous medium, is horizontal, $\partial/\partial x = \partial/\partial y = 0$. Denoting the static variables by $(\ )^0$, and assuming that $n^0 = $ constant, we have

---

* We follow the usual convention in applied mechanics and regard tensile stress σ_{ii} and compressive pressure p as positive.

† The reader will find it instructive to compare the above equations with equations (1.10) and (1.11), which in modified variables convey the exact equivalence.

from (2.10) and (2.11)

$$0 = -g(1-n^0)\rho_s^0 + \frac{\partial \sigma_{33}^0}{\partial z} - (1-n^0)\frac{\partial p^0}{\partial z}, \qquad (2.13)$$

$$0 = -n^0 \frac{\partial p^0}{\partial z} - g n^0 \rho_w^0. \qquad (2.14)$$

Elimination of $\partial p^0/\partial z$ gives

$$\frac{\partial \sigma_{33}^0}{\partial z} = g(1-n^0)(\rho_s^0 - \rho_w^0). \qquad (2.15)$$

Consistent with the assumption of constant p in (2.5), ρ_w^0 in (2.15) may be taken as constant so that

$$\sigma_{33}^0 = g(1-n^0)(\rho_s^0 - \rho_w^0)z, \qquad (2.16)$$

which of course gives rise to a static displacement.

Subtracting the static part from (2.10) and (2.11) we obtain for the dynamic momentum balance (where σ_{ij}' is the excess stress over the static solution)

$$(1-n^0)\rho_s^0 \frac{\partial v_i}{\partial t} = -gn\rho_s^0 \delta_{i3} + \frac{\partial \sigma_{ij}'}{\partial x_j}$$
$$- (1-n^0)\frac{\partial p}{\partial x_i} + \frac{(n^0)^2}{k}(u_i - v_i) \qquad (2.17)$$

and

$$n^0 \rho_w^0 \frac{\partial u_i}{\partial t} = -n^0 \frac{\partial p}{\partial x_i} - \frac{(n^0)^2}{k}(u_i - v_i) - g\delta_{i3}(n^0 \rho_w + n\rho_w^0) \qquad (2.18)$$

where use has been made of (2.14). Let us now estimate the importance of the gravity terms. From (2.8) and Hook's law, we have, in order of magnitude,

$$\frac{gn\rho_s^0}{(1-n^0)\nabla p} \sim \frac{(g/\omega)(v/L)\rho_s^0}{p_0/L} \sim \frac{gP_0}{G} \rho_s^0/(P_0/L) \sim \rho_s^0 \frac{gL}{G},$$

where L is the water wave length, P_0 the typical dynamic stress and v the typical velocity. The last ratio is typically of the order $O(10^{-5})$. Similarly, by using (2.4)

$$\frac{gn^0\rho_w}{n^0 \nabla p} \sim \frac{\rho_w gL}{\beta},$$

which is likewise small. Thus gravity may be ignored from (2.17) and (2.18), yielding, in vector form

$$(1-n^0)\rho_s^0 \frac{\partial \mathbf{v}}{\partial t} = \nabla \cdot \overline{\overline{\boldsymbol{\sigma}}} - (1-n^0)\nabla p + \frac{(n^0)^2}{k}(\mathbf{u}-\mathbf{v}), \qquad (2.19)$$

$$n^0 \rho_w^0 \frac{\partial \mathbf{u}}{\partial t} = -n^0 \nabla p - \frac{(n^0)^2}{k}(\mathbf{u}-\mathbf{v}). \qquad (2.20)$$

Upon adding (2.19) and (2.20) and using Hooke's law, we may show that

$$G\left(\nabla^2 \mathbf{v} + \frac{1}{1-2\nu}\nabla\nabla \cdot \mathbf{v}\right) - \nabla \dot{p} = n^0 \rho_w^0 \ddot{\mathbf{u}} + (1-n^0)\rho_s^0 \ddot{\mathbf{v}} \qquad (2.21)$$

$$k\nabla^2 p = \nabla \cdot \mathbf{v} + \frac{n^0}{\beta}\dot{p} - k\rho_w^0 (\nabla \cdot \dot{\mathbf{u}}), \qquad \frac{\partial u}{\partial t} \equiv \dot{u}, \text{ etc.} \qquad (2.22)$$

These equations together with (2.11) are equivalent to a linearized form of equations (1.15).[13]

When there is only the solid phase $n^0 = 0$, $p = 0$, $\mathbf{u} = 0$; equation (2.21) is the well-known equation for dynamic elasticity. When there is pore fluid the equations for solid and fluid phases are coupled and are naturally more difficult to solve analytically. In the following section an approximate theory of Mei and Foda[7] is outlined.

2.3   THE BOUNDARY LAYER APPROXIMATION

Although varying over a broad range, the permeability k of many soils is rather small, suggesting intuitively that for sufficiently high frequency nearly all the fluid in the interior of the soil is highly resisted by viscosity and hence cannot have a significant velocity relative to the solid skeleton. Near the free surface, however, drainage is of course much easier and relative motion must be appreciable within a thin layer. Thus one can expect two different regions where different physical mechanisms dominate. In the outer region, the length scale is L, which may be the wavelength or the dimension of a large structure, while in the boundary layer near the mud line the transverse length scale must be much less than the tangential scale. Let us decompose all unknowns into two parts: the outer part $(\ )_{\text{out}}$ and the boundary layer correction $(\ )_{\text{b.l.}}$, the latter being appreciable only within the boundary layer.

2.3.1   The outer field

In most of the porous medium the length scale is either the wavelength L or the dimension of the structure, assumed to be comparable here. Let us introduce the following outer variables:

$$\mathbf{x} = L\boldsymbol{\xi} \quad t = \tau/\omega, \quad \left(\begin{matrix} \tilde{\mathbf{u}} \\ \tilde{\mathbf{v}} \end{matrix}\right)_{\text{out}} = \frac{P_0 \omega L}{G}\left(\begin{matrix} \mathbf{u} \\ \mathbf{v} \end{matrix}\right), \quad \left(\begin{matrix} p \\ \sigma_{ij} \end{matrix}\right)_{\text{out}} = P_0\left(\begin{matrix} \tilde{p} \\ \tilde{\sigma}_{ij} \end{matrix}\right), \qquad (2.23)$$

where ω is the frequency and P_0 the typical amplitude of the applied stress. The scales of $\mathbf{u}$ and $\mathbf{v}$ are inferred from Hooke's law. Substituting into equation (2.8) we get a set of dimensionless equations. Two parameters appear in these equations.

One parameter is:

$$\rho_w^0 \omega L^2 / G \quad \text{or} \quad \rho_s^0 \omega L^2 / G,$$

which is the ratio of inertia to pressure or stress forces. Since $\lambda = \sqrt{(G/\rho_s)}/\omega$ is essentially the length of the shear wave, the above ratio is of the same order as

$$O(L/\lambda)^2.$$

For typical sea waves, $\omega = O(1\,\text{s}^{-1})$. Take $\rho_s^0 = 2.4 \times 10^3\,\text{kg/m}^3$ and $G = 10^7\,\text{N/m}^2$ for dense sand, then $\lambda = O(10^3)$. Hence the above parameter is very small and inertia terms are negligible if L is no greater than 130 m, say.

Another parameter is

$$\frac{\omega L^2}{Gk} \quad \text{or} \quad \frac{\omega L^2}{Kk},$$

which is the ratio of Darcy's drag force to the pressure gradient or the stress gradient. For generality we assume that the solid and fluid have comparable rigidity, i.e., $G = O(K)$. Now for sand, a typical value of permeability is $k = 10^{-8}\,\text{m}^3\,\text{s/kg}$. For $L \geqslant 10$ m, the above parameter is very large, which implies that

$$\tilde{\mathbf{u}} = \tilde{\mathbf{v}} \tag{2.24}$$

to the leading order. Thus fluid and solid move together.

Correspondingly one may show that

$$\nabla \cdot \tilde{\mathbf{v}} = -\frac{nG}{K_f} \frac{\partial \tilde{p}}{\partial t}, \tag{2.25}$$

so that solid dilation is directly related to the change of pore pressure, and

$$\nabla^2 \tilde{p} = 0, \tag{2.26}$$

$$\frac{\partial \tilde{\sigma}'_{ij}}{\partial \xi_j} + \frac{\partial \tilde{p}}{\partial \xi_i} = 0. \tag{2.27}$$

It is now expedient to use the total dynamic stress

$$\tilde{\sigma}_{ij} = \tilde{\sigma}'_{ij} - \tilde{p}\delta_{ij}, \tag{2.28}$$

so that

$$\partial \sigma_{ij}/\partial \xi_j = 0. \tag{2.29}$$

Hooke's law may be expressed as

$$\frac{\partial \tilde{\sigma}_{ij}}{\partial \tau_j} = \frac{\partial \tilde{v}_i}{\partial \xi_j} + \frac{\partial \tilde{v}_j}{\partial \xi_i} + \frac{\lambda_e}{G} \frac{\partial \tilde{v}_k}{\partial \xi_k} \delta_{ij}, \tag{2.30}$$

where λ_e may be called the effective Lamé constant:

$$\lambda_e = G \frac{2\nu}{1-\nu} + \frac{K_f}{n^0}. \tag{2.31}$$

Using (2.25) it may be shown that

$$\sigma_{kk} = \tilde{\sigma}'_{kk} - 3p = \left[\frac{2n^0 G(1+\nu)}{1-\nu} - 3 \right] \tilde{p}. \tag{2.32}$$

These results are more or less expected on intuitive grounds and merely justify the one-phase picture in most of the porous medium.

2.3.2 The boundary layer correction near a mud line

Within a thin layer $z = 0(\delta) \ll 0(L)$ near the mud line, drainage of pore fluid is much easier and (2.24) can no longer be true. We must add a correction $(\ )_{b.l.}$; that is,

$$(\ ) = (\ )_{out} + (\ )_{b.l.} \tag{2.33}$$

where $(\ )_{b.l.}$ are functions of x, z, y/δ and t. First we expect $(p)_{b.l.}$ and $(\sigma_{ij})_{b.l.}$ to be comparable to $(p)_{out}$ and $(\sigma_{ij})_{out}$, i.e., $O(P_0)$. In order that the strains shall produce stresses of such magnitude we must have

$$G \frac{\partial (v_3)_{b.l.}}{\partial z} = O(P_0 \omega), \qquad (v_3)_{b.l.} = O\left(\frac{P_0 \omega \delta}{G} \right)$$

from Hooke's law. Now with inertia ignored, the curl of equation (2.21) gives $\nabla^2 \nabla \times (\tilde{v})_{b.l.} = 0$, which means $\partial^2 (\nabla \times (v)_{b.l.})/\partial z^2 = 0$. Since $\nabla \times (\tilde{v})_{b.l.}$ vanishes outside the boundary layer it must be identically zero throughout the boundary layer. Thus $(v_i/v_3)_{b.l.} = O(\delta/L)$. We may now incorporate these order estimates in the definition of the boundary layer variables

$$(x, y) = L(X, Y), \qquad z = \delta Z, \qquad t = T/\omega,$$

$$(p, \sigma_{ij})_{b.l.} = P_0 (\hat{p}, \hat{\sigma}_{ij}),$$

$$(u_i, v_i)_{b.l.} = \frac{P_0 \omega L}{G} \frac{\delta^2}{L^2} (\hat{u}_i, \hat{v}_i), \qquad i = 1, 2,$$

$$(u_3, v_3)_{b.l.} = \frac{P_0 \omega L}{G} \frac{\delta}{L} (\hat{u}_3, \hat{v}_3). \tag{2.34}$$

Let us denote the small length ratio by

$$\varepsilon = \delta/L \ll 1. \tag{2.35}$$

Normalizing the governing equations it may then be proven that, with a relative error of $O(\varepsilon^2)$, the leading-order approximation gives rise to

$$\frac{1}{\varepsilon^2}\frac{\partial^2 \hat{p}}{\partial Z^2} = \frac{\omega L^2}{k}\left(\frac{1}{G}\frac{1-2\nu}{2(1-\nu)}+\frac{n^0}{K_f}\right)\frac{\partial \hat{p}}{\partial T}, \tag{2.36}$$

which is just the Terzaghi equation for one-dimensional consolidation. For sinusoidal motion, the boundary layer thickness may be defined as

$$\delta = \left(\frac{k}{\omega}\right)^{1/2}\left(\frac{1}{G}\frac{1-2\nu}{2(1-\nu)}+\frac{n^0}{K_f}\right)^{-1/2}, \qquad \varepsilon = \delta/L, \tag{2.37}$$

so that

$$\frac{\partial^2 \hat{p}}{\partial Z^2} = \frac{\partial \hat{p}}{\partial T}. \tag{2.38}$$

Once $\hat{p}$ is known, the transverse component $\hat{v}_3$ and the tangential component $\hat{v}_i$ can be easily found. For simple harmonic motion the solution of (2.38) is similar to that of Stokes' boundary layer in a viscous fluid. Written in outer variables, the general solution for the excess pore pressure is

$$\hat{p} = A(\xi_i)Ee^{-i\tau} \quad \text{with} \quad E = \exp\left(\frac{1-i}{\sqrt{2}}\frac{\zeta}{\varepsilon}\right), \tag{2.39}$$

where $A(\xi_i)$ is yet unknown. The stress components are dominated by $\partial \hat{v}_3/\partial \zeta$ and given by

$$\hat{\sigma}'_{ii} = \left(\frac{\nu}{1-\nu}A + O(\varepsilon^2)\right)Ee^{-i\tau},$$

$$\hat{\sigma}'_{33} = (A + O(\varepsilon^2))Ee^{-i\tau},$$

$$\hat{\sigma}'_{13} = (O(\varepsilon))Ee^{-i\tau}. \tag{2.40}$$

The total stresses, defined by $\hat{\sigma}_{ij} = \hat{\sigma}'_{ij} - \hat{p}\delta_{ij}$, are

$$\hat{\sigma}_{ii} = -\frac{1-2\nu}{1-\nu}AEe^{-i\tau},$$

$$\hat{\sigma}_{33} = 0,$$

$$\hat{\sigma}_{i3} = 0, \qquad i = 1, 2. \tag{2.41}$$

2.3.3 Outline for solving poro-elastic boundary value problems

The results of equation (2.41b, c) are particularly welcome as will now be shown in two classes of problems of practical interest.

(1) *Traction prescribed on the free surface.* Let the boundary values of σ'_{33}, p and σ'_{3i} be prescribed on the mud line: $\zeta = 0$:

$$\tilde{p} + \hat{p} = P(\xi_j), \quad \tilde{\sigma}'_{33} + \hat{\sigma}'_{33} = N(\xi_j), \quad \tilde{\sigma}' + \hat{\sigma}'_{3i} = T_i(\xi_j). \tag{2.42}$$

Using (2.41) the following boundary values are known for the outer problem:

$$\tilde{\sigma}_{33}(\xi_j, 0) = N - P \quad \text{and} \quad \tilde{\sigma}_{3i}(\xi_j, 0) = T_i(\xi_j). \tag{2.43}$$

Together with equation (2.29) and the stress–strain conditions, the outer problem for the total stresses is completely formulated and may be solved with traction prescribed on the mud line. Afterwards the outer pore pressure follows from (2.32), and the outer effective stresses from (2.28). Now the pressure boundary condition on the mud line may be invoked to determine the amplitude $A(\xi_i)$. The remaining corrections for the stresses in the boundary layer may then be obtained from (2.39) and (2.40).

(2) *Mixed boundary value problems.* Let the traction be prescribed only on part of the mud line S_1, and the displacements be prescribed on the rest of the mud line S_2. We then have a mixed boundary value problem which is of obvious relevance to offshore structures. For example, we may assume S_2 to be in welded contact with a base of rigid structure so that

$$v_i, v_3 \text{ given}; \quad \frac{\partial p}{\partial z} = 0, \quad z = 0. \tag{2.44}$$

The last condition follows from the fact that under the rigid base the normal components of fluid and solid velocities are equal and the inertia is negligible. Because the boundary-layer corrections for the displacements are small to the leading order, the boundary values of displacement may be approximated by the outer displacement alone. In view of the preceding subsection the outer boundary value problem is completely specified and can be solved by any of the known methods for elasto-static problems of the mixed type. To satisfy the condition of $\partial p/\partial z = 0$ a boundary layer correction can be easily added. Nevertheless, the correction $\hat{p}$ is $O(\varepsilon)$ because of the small thickness of the boundary layer.

(3) *Remarks on plane-strain elastostatics.* If both the medium properties and the external loading are homogeneous in the y direction, the entire problem is one of plane strain. In particular, $\varepsilon_{33} = 0$, so that

$$\tilde{\sigma}'_{22} = \nu(\tilde{\sigma}'_{11} + \tilde{\sigma}'_{33}). \tag{2.45}$$

For plane strain we may employ Airy's stress function F such that

$$\tilde{\sigma}_{11} = \frac{\partial^2 F}{\partial \xi^2}, \quad \tilde{\sigma}_{33} = \frac{\partial^2 F}{\partial \zeta^2}, \quad \tilde{\sigma}_{13} = -2\frac{\partial^2 F}{\partial \xi \, \partial \zeta}, \tag{2.46}$$

it then follows from compatibility that

$$\nabla^2 \nabla^2 F = 0. \qquad (2.47)$$

Note that because of (2.45), we may show by using Hooke's law and (2.25) that

$$\tilde{\sigma}'_{11} + \tilde{\sigma}'_{33} = -\frac{2n^0 G}{\beta(1-2\nu)} \tilde{p} \qquad (2.48)$$

and

$$\tilde{\sigma}_{11} + \tilde{\sigma}_{33} = -2(\lambda + m)\tilde{p}, \qquad (2.49)$$

where

$$m = \frac{n^0 G}{K_f} \frac{1}{1-2\nu} \quad \text{in general;} \quad \left(\approx \frac{G}{K_f} \text{ for } n^0 = \nu = \tfrac{1}{3} \right) \qquad (2.50)$$

is essentially the ratio of the solid elasticity to the fluid elasticity.

Using this approximation the following examples have been treated explicitly and analytically by Mei and Foda:[7]

(1) Progressive waves over a semi-infinite sea bed. The results check asymptotically with Yamamoto et al.[12] for small k.
(2) Progressive waves over a sea bed of finite thickness. The analytical results confirm the numerical results of Yamamoto.[11]
(3) Progressive waves past a small pipeline resting on the sea bed.

In addition the same method has been applied to

(4) Seismic waves impinging on a small tunnel of small radius.

In all of these problems the boundary layer corrections are important.

We give below another example in which the boundary layer correction is unimportant so that the outer solution alone suffices.

2.4 FORCED HEAVE OF A RIGID FOOTING

In the framework of the linearized theory the motion of a gravity structure in sea waves and the induced stresses in the sea bed may be solved by decomposing the whole problem into seven component problems: a scattering problem in which the structure is supposed to be stationary, and six radiation problems, each of which corresponds to one mode of oscillation with unit amplitude. The amplitudes of these modes may then be found by applying Newton's law to the structure.

Let us consider one of the radiation problems, namely that in which a rigid horizontal base of width $2L$ oscillates vertically. The boundary conditions on the mud line include

$$p = \sigma_{13} = \sigma_{33} = 0, \qquad |x| > L, \quad z = 0 \qquad (2.51)$$

over the part not under the base, and below the base

$$v = v_0 e^{-i\omega t}, \tag{2.52}$$

$$\frac{\partial p}{\partial z} = 0, \qquad |x| < L, \quad z = 0. \tag{2.53}$$

In addition, below the base one must impose a horizontal constraint. As an idealization we choose

$$\sigma'_{13} = 0 \qquad \text{(frictionless contact)} \qquad |x| < L, \quad z = 0. \tag{2.54}$$

There are many other possibilities involving friction.

This special case is particularly simple to solve, and the corresponding mixed boundary value problem, called the stamp problem in elastostatics, has a well known solution due to Sadowsky.[8] Specifically, the boundary conditions (2.51), (2.52) and (2.54) may be replaced by the following stress condition:

$$\left.\begin{array}{ll} \sigma_{33} = \dfrac{-p_0}{\sqrt{(a^2 - x^2)}} & |x| < L \\[2mm] \quad\;\; = 0 & |x| > L \end{array}\right\} z = 0, \tag{2.55}$$

$$\sigma_{13} = 0 \qquad\qquad |x| < \infty \tag{2.56}$$

in which, as well as in what follows, the time factor $e^{-i\tau}$ is omitted for brevity. Using normalized outer variables, with L as the characteristic outer length, we rewrite the stress condition:

$$\left.\begin{array}{ll} \tilde{\sigma}_{33} = -\mathbb{P} = -\dfrac{1}{\pi}\dfrac{1}{\sqrt{(1-\xi^2)}} & |\xi| < 1 \\[2mm] \quad\;\;\; = -\mathbb{P} = 0 & |\xi| > 1 \end{array}\right\} \zeta = 0 \tag{2.57}$$

$$\tilde{\sigma}_{13} = 0 \qquad\qquad\qquad |\xi| < \infty. \tag{2.58}$$

The problem is easily solved by applying the exponential Fourier transform, (see Ref. 9, §45.3). We only quote the result. The transform of $\mathbb{P}$ is

$$\bar{\mathbb{P}}(\lambda) = \frac{1}{\pi} \int_{-1}^{1} \frac{e^{-i\lambda\xi}}{\sqrt{(1-\xi^2)}} \, d\xi = \tfrac{1}{2} J_0(\lambda). \tag{2.59}$$

Since $\mathbb{P}(\xi)$ is even in ξ, $\bar{\mathbb{P}}(\lambda)$ is even in λ. It can be shown that

$$\tilde{\sigma}_{11} + \tilde{\sigma}_{33} = -2(1+m)\bar{p}$$

$$= -\frac{2}{\pi} \int_0^\infty d\lambda \, \cos \lambda\xi e^{\lambda\zeta} \bar{\mathbb{P}}(\lambda)$$

$$= -\frac{2}{\pi} \mathrm{Re} \int_0^\infty d\lambda \, e^{-s\lambda} \bar{\mathbb{P}}(\lambda), \tag{2.60}$$

$$\tilde{\sigma}_{33} - \tilde{\sigma}_{11} + 2i\tilde{\sigma}_{13} = -\frac{2}{\pi} \zeta \int_0^\infty d\lambda \, e^{-s\lambda} \bar{\mathbb{P}}(\lambda)\lambda \tag{2.61}$$

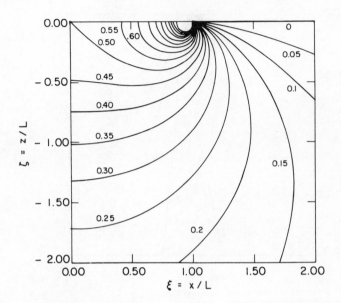

Figure 2.1 Contours of dynamic pore pressure amplitude: $\pi p/P_0$ for $m = 1$. For all other values of m, multiply $\pi p/P_0$ by $2(1+m)^{-1}$

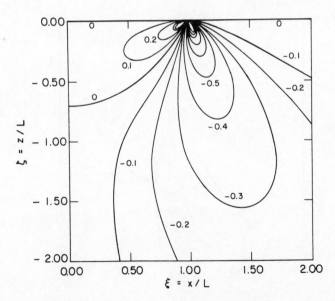

Figure 2.2 Contours of dynamic effective stress amplitude: $\pi\sigma'_{13}/p_0$

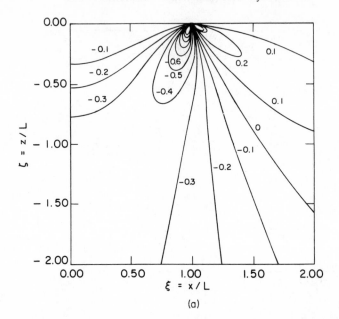

(a)

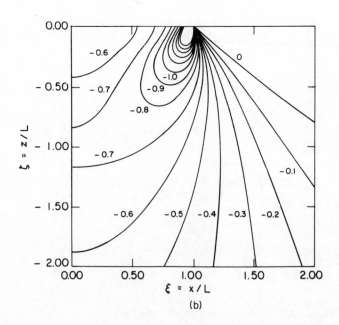

(b)

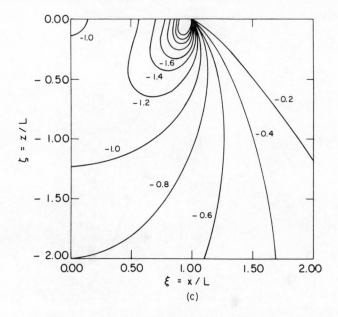

Figure 2.3 Contours of dynamic stress amplitude: $\pi\sigma'_{11}/p_0$ (a) $m = 0.0044$ (fully saturated pore water), (b) $m = 1$ (partially saturated pore water), (c) $m \to \infty$ (dry soil)

where

$$\tilde{p} = \frac{1}{1+m}\frac{1}{2\pi}\int_{-\infty}^{\infty} d\lambda \, e^{-s\lambda}\bar{\mathbb{P}}(\lambda), \qquad s \equiv i\tilde{z} \equiv i(\xi + i\zeta). \tag{2.62}$$

The following formula is an identity

$$\int_0^{\infty} d\lambda \, e^{-s\lambda}J_0(\lambda) = (1+s^2)^{-1/2}. \tag{2.63}$$

By differentiating with respect to s we also get

$$\int_0^{\infty} d\lambda \, e^{-s\lambda}\lambda J_0(\lambda) = s(1+s^2)^{-3/2}. \tag{2.64}$$

Let us define

$$\tilde{z} = re^{i\theta}, \qquad \tilde{z} - 1 = r_1 e^{i\theta_1}, \qquad \tilde{z} + 1 = r_2 e^{i\theta_2}, \tag{2.65}$$

where $-\pi < \theta, \theta_1, \theta_2 < 0$. When care is taken of the signs of the radicals we

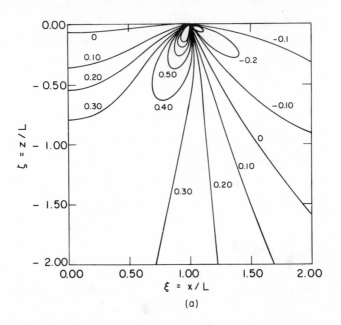

(a)

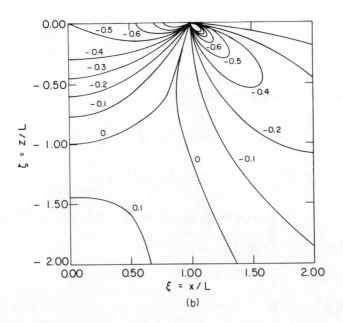

(b)

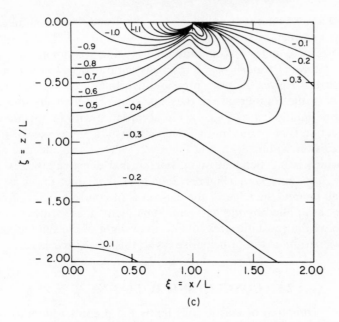

Figure 2.4 Contours of dynamic stress amplitude: $\pi\sigma'_{33}/p_0$ (a) $m = 0.0044$ (fully saturated pore water), (b) $m = 1$ (partially saturated pore water), (c) $m \to \infty$ (dry soil)

find

$$\tilde{\sigma}_{11} + \tilde{\sigma}_{33} = \frac{2}{\pi}(r_1 r_2)^{-1/2} \sin\tfrac{1}{2}(\theta_1 + \theta_2) = -2(1+m)\tilde{p}, \quad (2.66)$$

$$\tilde{\sigma}_{33} - \tilde{\sigma}_{11} + 2i\tilde{\sigma}_{13} = \frac{2}{\pi}\frac{\zeta r}{(r_1 r_2)^{1/2}} \exp\left[i(\theta - \tfrac{3}{2}(\theta_1 + \theta_2))\right]. \quad (2.67)$$

The outer effective stresses are easily found to be

$$\begin{pmatrix}\tilde{\sigma}'_{11} \\ \tilde{\sigma}'_{33}\end{pmatrix} = \frac{1}{\pi}(r_1 r_2)^{-1/2}\left\{\frac{m}{1+m}\sin\tfrac{1}{2}(\theta_1 + \theta_2)\right.$$

$$\left.\pm \frac{r^2}{(r_1 r_2)}\sin\theta\cos\left[\theta - \tfrac{3}{2}(\theta_1 + \theta_2)\right]\right\} \quad (2.68)$$

$$\tilde{\sigma}'_{13} = \frac{r^2}{\pi}(r_1 r_2)^{-3/2}\sin\theta\sin\left[\theta - \tfrac{3}{2}(\theta_1 + \theta_2)\right]. \quad (2.69)$$

The boundary layer correction near the mud line away from the base is given by (2.42). In order to satisfy (2.51a) we must have

$$A = -\tilde{p}(\xi, 0) = -\frac{1}{1+m}\frac{1}{2\pi}\int_{-\infty}^{\infty} d\lambda\, e^{i\lambda\xi}\bar{\mathbb{P}}(\lambda) = -\frac{1}{1+m}\mathbb{P}(\xi) \quad (2.70)$$

Since $\mathbb{P}(\xi) = 0$ away from the base, $A = 0$ for $|\xi| > 1$ and there is no need for boundary layer correction anywhere.

Contour lines of equal stress amplitudes have been computed for $m = 0.0044$ (fully saturated water), $m = 1$ (partially saturated water) and $m \approx \infty$ (dry soil). Figure 2.1 gives the pore pressure distribution πp for $m = 1$. For other values of m the contours are of the same shape but the values for each contour must be multiplied by $2/(1 + m)$ in accordance with (2.66). Because an $O(\varepsilon)$ correction is not made, not all the contours are perpendicular to the base of the caisson within $0 < \xi < 1$. Figure 2.2 shows the effective shear stress $\pi\sigma_{13}$ which is independent of m. For normal effective stresses $\pi\sigma_{11}$ (Figures 2.3a, b, c) and $\pi\sigma_{33}$ (Figures 2.4a, b, c) the effect of m is pronounced. In all figures, the edge of the caisson is of course a point of stress concentration. Multiplied by the harmonic time factor these stresses can be superimposed on the geostatic stress due to the weight of the soil and of the caisson in order to infer the principal stresses; this is omitted here.

2.5 CONCLUDING REMARKS

Although the assumption of elastic soils limits the present results to sandstones or infinitesimal strains under waves, two qualitative results of the boundary layer picture are probably valid even if non-linear constitutive equations are used. One is that away from a soil–water interface where free drainage is allowed, the solid skeleton and the pore fluid move together as a single phase, thus a single constitutive equation suffices. Near the interface the motion of the pore fluid differs from, but is coupled with, that of the solid skeleton, and a two-phase theory is necessary. While it is in principle possible to formulate a two-phase theory for the whole medium, nothing short of pure numerics will solve the problems of practical interest. The boundary layer theory outlined here will likely make it easier to achieve analytical solution, sharpen physical understanding and perhaps lead to more efficient numerical techniques.

ACKNOWLEDGEMENTS

The present work has been supported by the Earthquake Engineering Program and the Civil and Mechanical Engineering Program of the U.S. National Science Foundation. We also thank Arthur E. Mynett for computing and preparing the figures.

REFERENCES

1. Biot, M. A. General theory of three-dimensional consolidation, *J. Appl. Physics*, **12**, 155–64, 1941.

2. Biot, M. A. Theory of propagation of elastic waves in a fluid-saturated porous solid, Part I—Low frequency range, and Part II—Higher frequency range, *J. Acoust. Soc. Am.*, **28**, 168–191, 1956.
3. Deresiewicz, H. The effect of boundaries on wave propagation in a liquid filled porous solid, Part IV—Reflection of plane waves at free plane boundary (general case), *Bull. Seismol. Soc. Am.*, **52**, 595–626, 1962a.
4. Deresiewicz, H. The effect of boundaries on wave propagation in a liquid filled porous solid, Part IV—Surface waves in a half space, *Bull. Seismol. Soc. Am.*, **52**, 627–38, 1962b.
5. Geertsma, J., and Smith, D. C. Some aspects of elastic wave propagation in fluid-saturated porous solids, *Geophysics*, **26**, 169–81, 1961.
6. Madsen, O. S. Wave induced pore pressures and effective stresses in a porous bed, *Geotechnique*, **28**, 377–93, 1978.
7. Mei, C. C., and Foda, M. A. Wave-induced responses in a fluid-filled poro-elastic solid with a free surface—a boundary layer theory, *Geophy. J. Roy. Astr. Soc.* **66**, 597–637, 1981.
8. Sadowsky, M. *Z. angew. Math u. Mech.* **8**, 107, 1948.
9. Sneddon, I. N. *Fourier Transforms*, McGraw-Hill, New York, 1951.
10. Verruijt, A. Elastic storage of aquifers, Chapter 8 in de Wiest, R. J. M. (ed.) *Flow Through Porous Media*, Academic Press, New York; 1969. see also Chapter 3, this book.
11. Yamamoto, T. Wave induced instability in sea beds, *ASCE Symposium on Coastal Sediments '77*, Charleston, S.C., 1977.
12. Yamamoto, T., Koning, H. L., Sellmeigher, H., and Hijum, E. V., On the response of a poro-elastic bed to water waves, *J. Fluid Mech.* **78**, 193–206, 1978.
13. Zienkiewicz, O. C., and Bettess, P. Chapter 1, This book.

Soil Mechanics—Transient and Cyclic Loads
Edited by G. N. Pande and O. C. Zienkiewicz
© 1982 John Wiley & Sons Ltd

Chapter 3

Approximations of Cyclic Pore Pressures Caused by Sea Waves in a Poro-elastic Half-plane

A. Verruijt

SUMMARY

The analytical solution for the problem of linear elastic consolidation of a half-plane, carrying a load harmonic in both time and in space, reduces to rather simple forms if a characteristic wave parameter, defined in terms of the wave length, the wave period, and the coefficient of consolidation, is either very small or very large compared to unity. This enables one to identify 'long' and 'short' waves. In the case of a long wave the material behaves as if it were practically incompressible, time being too short for consolidation to take place, as discussed in Chapter 2. In the case of a short wave no excess pore pressures are generated by a tendency for volume changes, because the consolidation time corresponding to the characteristic length of the load is small compared to the time scale of the loading. Only if the order of magnitude of the wave parameter is close to unity is it worthwhile to go to the trouble of analysing the phenomenon of consolidation. Two special cases are investigated, namely an external force on a completely drained surface, and a standing water wave. In the first case the pore pressures are zero along the surface, but the total stresses vary harmonically with time. In the second case both the total stress and the pore pressure along the surface vary harmonically with time.

If the wave length is so large, or its frequency so high, that the material is practically incompressible, the approximation is not valid near a drained surface. Along such a surface a one-dimensional approximation may be used. As an example, the pore pressures in the foundation bed of the Oosterschelde storm surge barrier in the Netherlands are calculated.

3.1 INTRODUCTION

When a saturated, or almost saturated, soil is subjected to cyclic or transient loading, excess pore pressures may be generated, depending upon the deformation and drainage parameters. In a first approximation these pore pressures can be calculated using the linear theory of consolidation.[3] Only in a few cases this can be done analytically, and in many cases recourse is to be made to a numerical method, for instance based on a finite element approximation.[4-7,16,18] In particular, if one wishes to include some of the important non-linear effects of the mechanical behaviour of soils a numerical approach seems to be the only possible method of solution.

Computer programs for soil consolidation are not very easy to handle, however, and sometimes their operation may be rather expensive. Therefore it is worthwhile to pay some attention to the possibility of simple analytical approximations. It is of little use to apply an expensive and complicated computer program if the conditions of the problem are such that for instance no pore pressures are developed at all. In this paper some approximate analytical results are presented, which may make it possible to determine whether it is worthwhile to do a finite element analysis. The considerations are restricted to conditions of cyclic loading.

An analytical solution for the response of a porous elastic bed to water waves has been given by Yamamoto *et al.*[17] and by Madsen.[9] The limiting cases for an incompressible pore fluid and for a very compressible pore fluid (or an almost rigid matrix) were considered in some detail in these papers. It will be shown in the present paper that the solution of this type of problem also admits simple approximations if the wave length is either very long or very short. The paper deals with standing waves only, generated by cyclic surface loads or water pressures, but the conclusions are equally well applicable to the case of a running wave, as investigated by Yamamoto *et al.*[17] and Madsen.[9]

3.2 BASIC EQUATIONS

The basic equations of the theory of consolidation, which describe the deformation of a porous elastic material, the pores of which are filled with a viscous compressible fluid, are, in the case of plane-strain deformations in the x,y-plane[3,15]*

$$(1+m)G\frac{\partial e}{\partial t}+\theta(1+m)\frac{\partial p}{\partial t}-c\,\nabla^2 p=0, \tag{3.1}$$

$$mG\frac{\partial e}{\partial x}+G\,\nabla^2 u-\frac{\partial p}{\partial x}=0, \tag{3.2}$$

$$mG\frac{\partial e}{\partial y}+G\,\nabla^2 v-\frac{\partial p}{\partial y}=0. \tag{3.3}$$

In these equations u and v are the displacement components of the solid material, G is its shear modulus, and m is an auxiliary elastic coefficient, related to Poisson's ratio ν by

$$m=\frac{1}{1-2\nu}. \tag{3.4}$$

---

*These equations are a linearized form of consolidation equations of Chapter 1,[18] i.e. equation (1.17). As phenomena discussed in this chapter are slow all dynamic terms have been omitted.

Furthermore p denotes the stress (pressure) in the pore fluid (in excess of the hydrostatic stresses), and θ is a parameter indicating the relative compressibility of the pore fluid

$$\theta = n\beta G = nG/K_f \tag{3.5}$$

where n is the porosity and β is the compressibility of the pore fluid, that is, the inverse of the modulus of volumetric compression of the fluid (bulk modulus K_f). Finally, in equation (3.1), c is the coefficient of consolidation, defined by

$$c = \frac{k}{\mu} G(1+m), \tag{3.6}$$

where k is the permeability of the porous material, and μ is the viscosity of the fluid.

Equation (3.1) expresses conservation of the pore fluid, and is sometimes referred to as the storage equation. Equations (3.2) and (3.3) are the equations of equilibrium in the x and y directions, expressed in terms of the displacements u and v, and the volume strain e,

$$e = \frac{\partial u}{\partial x} + \frac{\partial v}{\partial y}. \tag{3.7}$$

The particular choice of the physical parameters in equations (3.1)–(3.3) has been made for future convenience.

The problems considered in this paper all refer to the half-plane $y > 0$. On the surface $y = 0$ a load is applied in the form of a standing wave in the normal stress and/or the pore stress. Thus the boundary conditions are assumed to be

$$y = 0: \qquad \sigma_{yx} = 0, \tag{3.8}$$

$$y = 0: \qquad \sigma_{yy} = \bar{q} \exp{(i\omega t)} \cos{(\lambda x)}, \tag{3.9}$$

$$y = 0: \qquad p = \bar{p} \exp{(i\omega t)} \cos{(\lambda x)}. \tag{3.10}$$

Because these boundary conditions involve the stresses, the stress–strain relationship is also needed in order to fully define the mathematical problem,

$$\sigma_{xx} = 2G\frac{\partial u}{\partial x} + (m-1)Ge - p, \tag{3.11}$$

$$\sigma_{yy} = 2G\frac{\partial v}{\partial y} + (m-1)Ge - p, \tag{3.12}$$

$$\sigma_{xy} = G\left(\frac{\partial u}{\partial y} + \frac{\partial v}{\partial x}\right). \tag{3.13}$$

In equations (3.11) and (3.12) the last term is a consequence of Terzaghi's principle of effective stress, which states that the difference of total stress and pore stress is a measure for the deformations. In this paper these deformations are assumed to be purely elastic, as a first approximation.

3.3 SOLUTION

The general solution of the problem described by the equations of the previous paragraph, vanishing for $y \to \infty$, and periodic in t and x by a factor $\exp(i\omega t)\cos(\lambda x)$ is

$$p/G = -\{2\lambda A_1 \exp(-\lambda y) + (1+m)(\alpha^2 - \lambda^2)A_3 \exp(-\alpha y)\} \exp(i\omega t)\cos(\lambda x),$$
(3.14)

$$u = \{(A_2 - A_1(1+m\theta)\lambda y)\exp(-\lambda y) + \lambda A_3 \exp(-\alpha y)\}\exp(i\omega t)\cos(\lambda x),$$
(3.15)

$$v = \{(A_2 - A_1(1+2\theta+m\theta) - A_1(1+m\theta)\lambda y)\exp(-\lambda y)$$
$$+ \alpha A_3 \exp(-\alpha y)\}\exp(i\omega t)\cos(\lambda x),$$
(3.16)

where α is defined by

$$\alpha^2 = \lambda^2 + i\frac{\omega}{c}(1+\theta+m\theta), \qquad \mathrm{Re}(\alpha) > 0.$$
(3.17)

The volume strain is, from (3.7),

$$e = \{2\lambda\theta A_1 \exp(-\lambda y) - (\alpha^2 - \lambda^2)A_3 \exp(-\alpha y)\}\exp(i\omega t)\cos(\lambda x).$$
(3.18)

And the stress components on horizontal planes are

$$\sigma_{yx}/2G = \lambda\{(A_1\theta - A_2 + A_1(1+m\theta)\lambda y)\exp(-\lambda y)$$
$$- \alpha A_3 \exp(-\alpha y)\}\exp(i\omega t)\cos(\lambda x),$$
(3.19)

$$\sigma_{yy}/2G = \lambda\{(A_1(1+\theta+m\theta) - A_2 + A_1(1+m\theta)\lambda y)\exp(-\lambda y)$$
$$- \lambda A_3 \exp(-\alpha y)\}\exp(i\omega t)\cos(\lambda x).$$
(3.20)

The three constants in the solution can be determined from the boundary conditions (3.8)–(3.10). The result is

$$A_1 = \frac{\bar{q}(1+m)(\alpha+\lambda) + 2\bar{p}\lambda}{2G\lambda\{(1+m)(1+m\theta)(\alpha+\lambda) - 2\lambda\}},$$
(3.21)

$$A_2 = \frac{\bar{q}\{(1+m)(\alpha^2 - \lambda^2)\theta + 2\alpha\lambda\} + 2\bar{p}\lambda\{\alpha(1+m\theta) + \theta(\alpha-\lambda)\}}{2G\lambda(\alpha-\lambda)\{(1+m)(1+m\theta)(\alpha+\lambda) - 2\lambda\}},$$
(3.22)

$$A_3 = \frac{-\bar{p}(1+m\theta) - \bar{q}}{G(\alpha-\lambda)\{(1+m)(1+m\theta)(\alpha+\lambda) - 2\lambda\}}.$$
(3.23)

Two special cases will be investigated in some more detail below.

3.4 CYCLIC LOADING ON A DRAINED SURFACE

A case of some engineering interest is that of a half-plane, fully drained at its surface, and loaded by a surface force which is both cyclic in time and in space (see Figure 3.1). The spatial variation of the load, through a factor $\cos(\lambda x)$, can be considered to represent the dominant term in the Fourier expansion of a more general type of loading, having a characteristic dimension $l = \pi/\lambda$. In this case the parameter $\bar{p}$ is equal to zero, because of the condition of full drainage at the surface $y = 0$ (see equation (3.10)).

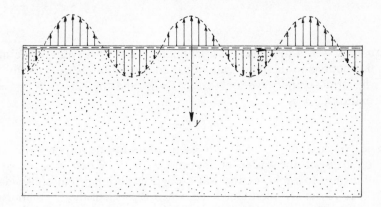

Figure 3.1 Cyclic load on a half-plane with a drained surface

In order to study the behaviour of the solution in this case, the vertical displacement of the surface $y = 0$ will be considered to be characteristic for the response. Using the values of the constants as given by (3.21)–(3.23) the expression for the vertical displacement v, equation (3.16), on the surface $y = 0$ can be written as

$$y = 0 \quad : \quad v = -f v_0 \exp(i\omega t) \cos(\lambda x), \tag{3.24}$$

where v_0 is the amplitude of the response in the purely elastic and incompressible case ($\nu = \frac{1}{2}$), that is

$$v_0 = \frac{\bar{q}}{2G\lambda}, \tag{3.25}$$

and where f is a dimensionless factor describing the deviation from the incompressible elastic response,

$$f = \frac{(1+m)(1+\theta+m\theta)(\alpha+\lambda)}{(1+m)(1+m\theta)(\alpha+\lambda)-2\lambda}. \tag{3.26}$$

In general this is a complex factor, because α is complex, see (3.17). The absolute value of f is a multiplication factor for the amplitude of the response, and the argument of f gives the phase shift of the periodic surface displacements, with respect to the loading.

In order to analyse the response it is now useful to introduce a characteristic wave parameter in the formula defining the value α. Therefore equation (3.17) is rewritten as

$$\alpha = \lambda\{1 + i\phi(1 + \theta + m\theta)\}^{1/2}, \qquad \text{Re}\,(\alpha) > 0, \tag{3.27}$$

where now

$$\phi = \omega/c\lambda^2. \tag{3.28}$$

The quantity ϕ is a dimensionless parameter, relating the time scale of the cyclic fluctuations to the time scale of the consolidation process, as given by $1/c\lambda^2$. In terms of the wave length $L(= 2\pi/\lambda)$ and the period $T\ (= 2\pi/\omega)$, the definition of ϕ is

$$\phi = L^2/2\pi cT. \tag{3.29}$$

It can be expected that consolidation is important for the response only if the value of ϕ is of the order of magnitude of 1. When ϕ is large the fluctuations are so rapid, or the wave length is so long, that consolidation cannot take place. On the other hand, when ϕ is small the fluctuations are so slow, or the wave length is so short, that consolidation occurs practically simultaneously with the loading. In that case no excess pore stresses are generated.

Indeed if $\phi \ll 1$ one obtains from (3.27)

$$\phi \ll 1: \quad \alpha = \lambda\{1 + \tfrac{1}{2}i\phi(1 + \theta + m\theta)\}, \tag{3.30}$$

which is very close to λ, because ϕ is small. In that case (3.26) reduces to

$$\phi \ll 1: \quad f = \frac{(1 + m)(1 + \theta + m\theta)}{(1 + m)(1 + m\theta) - 1}, \tag{3.31}$$

which is real, so that there is no phase shift. When it is furthermore assumed that the pore fluid is much stiffer than the soil, i.e. $\theta \ll 1$, equation (3.31) further reduces to

$$\phi \ll 1, \theta \ll 1: \quad f = \frac{1 + m}{m} = 2(1 - \nu). \tag{3.32}$$

It can easily be shown that this, combined with (3.24) and (3.25), just describes the fully elastic response, which applies in the absence of pore stresses.

When ϕ is large, one expects that there is no time for drainage, so that the material behaves as an elastic material as well, with a modified compression modulus (see equation (1.20d)):

$$K' = K + \frac{1}{n\beta} = K + K_f/n. \tag{3.33}$$

Indeed if $\phi \gg 1$ equation (3.17) reduces to

$$\phi \gg 1: \quad \alpha = (1+i)\lambda\{\tfrac{1}{2}\phi(1+\theta+m\theta)\}^{1/2}. \tag{3.34}$$

Both the real and imaginary parts of α now are large compared to λ, so that $\alpha + \lambda \approx \alpha$. Equation (26) now reduces to

$$\phi \gg 1: \quad f = \frac{1+\theta+m\theta}{1+m\theta}, \tag{3.35}$$

which is again real, showing that for rapid oscillations there is also no phase difference between the load and the surface displacement. When the pore fluid is practically incompressible equation (3.35) further reduces to

$$\phi \gg 1, \theta \ll 1: \quad f = 1. \tag{3.36}$$

This expresses the fact that for rapid waves on a medium filled with an incompressible fluid the response indeed corresponds to that of an incompressible elastic material.

In case the period of the fluctuations is of the same order of magnitude as the consolidation time, the value of ϕ is of order $O(1)$, and there will be both a phase shift and a multiplication factor for the amplitude. The basic parameters determining the value of f are Poisson's ratio ν, or the alternative parameter m, see equation (3.4), the relative compressibility of the pore fluid $\theta(=n\beta G)$, and the wave parameter ϕ introduced in equation (3.28).

As regards the influence of Poisson's ratio it is immediately evident from (3.17) and (3.26) that for an incompressible soil ($\nu = \tfrac{1}{2}$, or $m \to \infty$) the value of f again reduces to unity

$$\nu = \tfrac{1}{2}: \quad f = 1. \tag{3.37}$$

This confirms the statement made before that for an incompressible soil there is a unit response, determined by the factor $2G\lambda$ (see equation (3.25)). It may be noted that in this case the properties of the pore fluid are immaterial.

The influence of consolidation is greatest for small values of Poisson's ratio, i.e. for materials with a low resistance to compression, as compared to their shearing resistance. For the case $\nu = 0$, and $\theta = 0$ (indicating an incompressible pore fluid) the values of $|f|$ and $\arg(f)$, the multiplication factor and the phase shift, are shown in Figure 3.2, as a function of the parameter ϕ. It appears that the phase difference is never more than $12°$. For greater values of ν and θ the phase shift is even smaller.

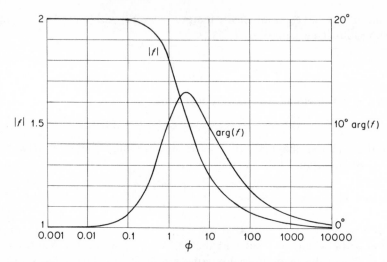

Figure 3.2   Variation of multiplication factor $|f|$ and phase shift arg (f) with
$$\phi = \omega/c\lambda^2$$

It can be concluded that the presence of the pore fluid in the materials makes itself felt in the response of the medium in two ways:

(1)   There will be a *reduction in the amplitude* of the displacements, i.e. an apparent increase of the rigidity of the material, but never more than that making the material behave as an incompressible solid. Thus for a dry soil one would have

$$p = 0: \quad f = 2(1 - \nu), \tag{3.38}$$

the latter varying between 1 and 2, if ν varies between 0 and $\frac{1}{2}$. The presence of a fluid in the pores will reduce the value of the factor f, but never below 1.

(2)   There may be a *phase lag* between the vertical displacements of the surface and the applied load, but this phase lag will not be greater than $12°$. Such a small phase lag is usually hardly noticeable.

It is noted that the solution given here strictly speaking applies only to an uncommon type of loading, namely that in which the total stress on the surface contains a factor cos (λx). This can be considered to be representative for a more general type of loading, however, with the amplitude being the coefficient of the first term in the Fourier series expansion of a periodic load, with period $L = 2\pi/\lambda$. For a block type of loading the coefficient of the second term in the Fourier series expansion is already $\frac{1}{3}$ of the first one. The solution given can in principle also be used to construct solutions for a periodic load in general, and even for an arbitrary load, by application of a

Fourier series expansion or the Fourier integral. A more elegant solution for problems of that type is to employ Fourier transforms from the outset, see McNamee and Gibson.[10] The evaluation of the final inverse transforms then remains as a mathematical problem, which has to be solved numerically, and which somewhat obscures the general character of the solution.

3.5 STANDING WATER WAVES

The solution given in equations (3.14)–(3.16) can also be applied to the case of a standing water wave on the surface of the half-plane. In this case the boundary conditions are such that in equation (3.9) and (3.10), $\bar{q} = -\bar{p}$. A quantity of special interest is the pore stress generated by the wave, for which one obtains from (3.14), (3.21) and (3.23):

$$\frac{p}{\bar{p}} = \frac{\{(1+m)(\alpha+\lambda)-2\lambda\}\exp{(-\lambda y)}+(1+m)(\alpha+\lambda)m\theta\exp{(-\alpha y)}}{(1+m)(1+m\theta)(\alpha+\lambda)-2\lambda}$$

$$\times \exp{(i\omega t)}\cos{(\lambda x)}. \tag{3.39}$$

Again the case of long and short waves (or rapid and slow fluctuations) will be investigated separately.

Short waves

If the wave length is small, or the fluctuation is very slow, the wave parameter ϕ, as defined by (3.28) or (3.29), is very small. The value of α is then very close to λ, see (3.27). Equation (3.39) then reduces to

$$\phi \ll 1: \quad p/\bar{p} = \exp{(-\lambda y)}\exp{(i\omega t)}\cos{(\lambda x)}. \tag{3.40}$$

This shows that in this case the variations in the pore pressure are just attenuated with depth, without any phase shift. The same approximation applies if the pore fluid is completely incompressible ($\theta = 0$) (see also Yamamoto *et al.*[17] and Madsen[9]). In fact, in all these cases (incompressible pore fluid or short waves) the pore pressure satisfies the Laplace equation. The soil deformations do not vanish completely, but the problem is uncoupled, i.e. the pore pressures are not affected by the deformations.

Long waves

If the wave length is large, or the fluctuation is very rapid, the wave parameter ϕ will be very large. In that case the value of α is very large compared to λ, see (3.34). The solution (3.39) then reduces to

$$\phi \gg 1: \quad \frac{p}{\bar{p}} = \frac{\exp{(-\lambda y)}+m\theta\exp{(-\alpha y)}}{1+m\theta}\exp{(i\omega t)}\cos{(\lambda x)}. \tag{3.41}$$

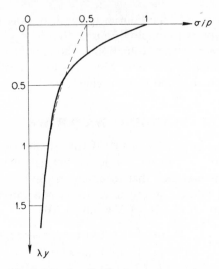

Figure 3.3   Amplitude of pore pressure response as a function of depth

Because α is complex (see (3.34), the second term in the numerator will exhibit a phase shift. This part of the solution is rapidly attenuated with depth, however, because the real part of α is large compared to λ. Near the surface the importance of the second term depends upon the value of the parameter $m\theta$, which in practice will usually be small. For a particular case, namely $m = 1$ (i.e. $\nu = 0$), $\theta = 1$ and $\phi = 10$, the variation of the pore pressure with depth, just under the crest of the wave ($x = 0$, $t = 0$), is given in Figure 3.3. Even in this extreme case, the oscillations in the second term do not manifest themselves in the final result.

3.6   APPROXIMATIONS OF THE DIFFERENTIAL EQUATIONS

The results obtained above can also be obtained, at least qualitatively, by introducing the dimensionless parameter ϕ into the differential equations. If the volume strain e and the pore stress p are written as

$$e = \bar{e} \exp{(i\omega t)} \cos{(\lambda x)}, \tag{3.42}$$

$$p = \bar{p} \exp{(i\omega t)} \cos{(\lambda x)}, \tag{3.43}$$

where $\bar{e}$ and $\bar{p}$ are (possibly complex) functions of the vertical coordinate y, then equation (3.1) reduces to the form

$$i\phi(1+m)G\{\bar{e}+n\beta\bar{p}\}+\left\{\bar{p}-\frac{1}{\lambda^2}\frac{\mathrm{d}^2\bar{p}}{\mathrm{d}y^2}\right\}=0. \tag{3.44}$$

It follows from this equation that if ϕ is very small ($\phi \ll 1$), then the first terms may be disregarded and the last two remain, which indicates that the pore stress p should satisfy the Laplace equation. This confirms the statements made above after deriving equation (3.40). The problem appears to be uncoupled, and if there is no generation of pore pressures by a time-dependent water level along the boundary, the pore pressures vanish throughout the body.

It also follows from equation (3.44) that if ϕ is very large ($\phi \gg 1$), then the first two terms dominate the last two, and one may write

$$\bar{p} = -\frac{\bar{e}}{n\beta}.$$

Substitution of this result into the equations of equilibrium (3.2) and (3.3) then confirms the conclusion drawn above that the response of the medium in this case (rapid waves, or waves with a relatively large wave length) is undrained. If the grain skeleton is elastic, then the response can be calculated using uncoupled elastic theory, with a modified compression modulus (see equation (3.33)).

3.7 ONE-DIMENSIONAL APPROXIMATIONS

It has been shown above that under certain conditions a poroelastic medium reacts undrained (for rapid waves) and under other conditions (very slow waves) it reacts fully drained. In the first case ($\phi \gg 1$) this result may be of limited value, because it may not apply in the immediate vicinity of the surface if this is drained. Indeed a practical estimate of the depth h over which a surface disturbance of time scale T influences the consolidation behaviour follows from the equality $cT/h^2 = O(1)$. For a periodic fluctuation, with frequency ω, this means that $h^2 = O(c/\omega)$, and because in the case considered $\omega/c\lambda^2 \gg 1$, it now follows that $h \ll 1/\lambda$. Hence the depth of the zone of consolidation near the surface is small compared to the wave length. If the surface is drained this means that the pore pressures near that surface can be considered to be independent of the wave length, and thus they can be calculated using a one-dimensional approach. A similar conclusion has been obtained, in a somewhat different way, by Mei and Foda.[11]

The equations describing the one-dimensional approximation can be obtained from the differential equations (3.1)–(3.3) by disregarding the horizontal displacement u and the derivatives of the volume strain e and the pore stress p in the x direction. The storage equation (3.1) now reduces to

$$(1+m)G\frac{\partial e}{\partial t} + \theta(1+m)\frac{\partial p}{\partial t} - c\frac{\partial^2 p}{\partial y^2} = 0. \tag{3.45}$$

Furthermore it follows from (3.12) that, if the volume strain e is equated to the vertical strain $\partial v/\partial y$,

$$(1+m)Ge = \sigma_{yy} + p. \qquad (3.46)$$

Hence one obtains

$$\{1+\theta(1+m)\}\frac{\partial p}{\partial t} = c\frac{\partial^2 p}{\partial y^2} - \frac{\partial \sigma_{yy}}{\partial t}. \qquad (3.47)$$

In case of a cyclic variation of the total stress on the surface, the boundary condition may be written as

$$y = 0: \quad \sigma_{yy} = q \sin(\omega t). \qquad (3.48)$$

Because of vertical equilibrium it now follows that this expression holds for all values of y, and thus equation (3.47) reduces to

$$\{1+\theta(1+m)\}\frac{\partial p}{\partial t} = c\frac{\partial^2 p}{\partial y^2} - q\omega \cos(\omega t), \qquad (3.49)$$

which is in fact the well-known Terzhagi consolidation equation.

A problem of practical interest is the development of pore pressures under the foundation structures of the storm-surge barrier in the Oosterschelde in the south-west of the Netherlands.[2,12] A sketch of a typical problem is represented in Figure 3.4. The problem is essentially one of cyclic loading of a heavy structure built upon a layer of coarse material (gravel) overlying a densified sand. Because of the cyclic wave the pore pressures at the top of the soil also vary sinusoidally, but they are not equal to the total stresses, because of the tilting of the pier structure. In fact the total stresses may be in counterphase with the pore stresses at the soil surface. If the compressibility of the gravel is disregarded, the boundary condition for

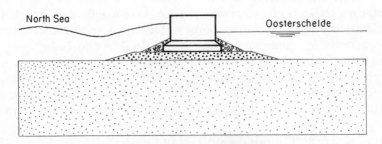

Figure 3.4 Schematic cross-section of Oosterschelde storm surge barrier

the pore pressure at the top of the sand layer can be written as

$$y = 0: \quad H\frac{\partial p}{\partial y} = p - \bar{p}\sin(\omega t) \qquad (3.50)$$

with

$$H = kd_g/k_g, \qquad (3.51)$$

where k is the permeability of the sand, k_g is the permeability of the gravel and d_g is the thickness of the gravel. In reality the filter consists of a system of graded layers. Then equation (3.51) is to be replaced by

$$H = k\sum (d_i/k_i), \qquad (3.52)$$

but the boundary condition (3.50) remains the same.

The solution of the problem is

$$p = p_\infty \sin(\omega t) - (p_\infty - \bar{p})\mu \exp(-y/L)\sin(\omega t - y/L - \psi), \qquad (3.53)$$

where

$$p_\infty = -\frac{1}{1 + \theta(1 + m)}q,$$

$$L = (2c/\omega)^{1/2}$$

$$\psi = \arctan\left(\frac{H}{L + H}\right),$$

$$\mu = \left(1 + 2\frac{H}{L} + 2\frac{H}{L}\right)^{-1/2}.$$

The physical meaning of H is that of a thickness of a layer of sand having the same resistance to flow as the actual filter. The other length parameter in the problem, L, represents a measure for the depth of influence of the surface fluctuations.

A quantity of special interest is the maximum gradient at the top surface of the sand, as this may control the internal stability of the material. For this quantity one obtains

$$i_{max} = \frac{1}{\gamma}\frac{\partial p}{\partial y} = \mu\sqrt{2}\frac{p_\infty - \bar{p}}{\gamma L}, \qquad (3.54)$$

where γ is the specific weight of water. For the local conditions at the barrier the following numerical data apply:

$$k = 10^{-4}\,\text{m/s}$$

$$d_g/k_g = 2500\,\text{s},$$

$$c = 3.4\,\text{m}^2/\text{s},$$

$$\omega = 2\pi/(10\,\text{s}).$$

From these data one obtains

$$H = 0.25 \text{ m,}$$
$$L = 3.29 \text{ m,}$$
$$\mu = 0.93,$$

and it then follows that

$$i_{max} = \frac{(p_\infty - \bar{p})/\gamma}{2.50 \text{ m}}. \tag{3.55}$$

This means that a water-level fluctuation of the order of magnitude of several meters may result in hydraulic gradients of about 2 or, depending upon the amplitude of the total stresses, even more. Gradients of that magnitude may well be critical for the stability of the filter, and therefore a thorough analysis of this problem is well justified. Such an analysis, involving two-dimensional numerical and analytical calculations, has indeed been executed for the Oosterschelde storm surge barrier.[2,8] The results of this analysis have had great influence upon the design.[1] One-dimensional calculations, as presented here, can be considered to be a valuable first indication of the relevance of more refined calculations. They are also useful, of course, to verify two-dimensional calculations.

3.8 CONCLUSION

It has been shown that the phenomenon of consolidation due to a surface wave on a porous bed is determined mainly by a dimensionless parameter $\phi = \omega/c\lambda^2$. If this parameter is either small or large compared to 1, the theoretical solutions can be approximated by simple analytical expressions.

ACKNOWLEDGEMENT

In analysing periodic solutions of the consolidation equations the author was assisted by some of his students. The approximation in terms of a dimensionless wave parameter was developed in 1971, in a project with C. Verspuy and P. Werksma.

REFERENCES

1. d'Angremond, K., and van der Does de Bye, M. R. Sill design for the Oosterschelde storm surge barrier, *Proc. Symp. on Foundation Aspects of Coastal Structures*, paper VI. 4, Delft, 1978.
2. Barends, F. B. J., and Thabet, R. A. H. Groundwater flow and dynamic gradients, *Proc. Symp. on Foundation Aspects of Coastal Structures*, paper VI. 1, Delft, 1978.

3. Biot, M. A. General theory of three-dimensional consolidation, *J. Appl. Phys.*, **12**, 155–64, 1941.

4. Booker, J. R. A numerical method for the solution of Biot's consolidation theory, *Quart. J. Mech. Appl. Math.* **26**, 457–70, 1973.

5. van Bijsterveld, J. J. Application of numerical analysis for two-dimensional consolidation, *LGM-mededelingen*, **17**, 97–116, 1976.

6. Christian, J. J., and Boehmer, J. W. Plane strain consolidation by finite elements, *J. Soil Mech. Found. Div. ASCE*, **96**, 1435–57, 1970.

7. Hwang, C. T., Morgenstern, N. R., and Murray, D. T. On solutions of plane strain consolidation problems by finite element methods, *Can. Geotechnical J.*, **8**, 109–18, 1971.

8. Koenders, M. A., and Saathof, L. E. B. Numerical and analytical computations of excess pore pressures, *Proc. Int. Symp. on Soils under Cyclic and Transient Loading*, edited by G. N. Pande and O. C. Zienkiewicz, 619–25, Balkema, Rotterdam, 1980.

9. Madsen, O. S. Wave-induced pore pressures and effective stresses in a porous bed, *Géotechnique*, **28**, 377–93, 1978.

10. McNamee J., and Gibson, R. E. Plane strain and axially symmetric problems of the consolidation of a semi-infinite clay stratum, *Quart. J. Mech. Appl. Math.*, **13**, 210–27, 1960.

11. Mei, C. C., and Foda, M. A. Boundary layer theory of waves in a poro-elastic sea bed, *Proc. Int. Symp. on Soils under Cyclic and Transient Loadings*, edited by G. N. Pande and O. C. Zienkiewicz, 609–18, Balkema, Rotterdam, 1980; see also Chapter 2, This book.

12. de Quelerij, L. Nieuwenhuis, J. K., and Koenders, M. A. Large model tests for the Oosterschelde storm surge barrier, *Proc. BOSS'79*, **2**, 257–78, BHRA Fluid Engineering, Cranfield, 1979.

13. Sandhu, R. S., and Wilson, E. L. Finite element analysis of seepage in elastic media, *J. Eng. Mech. Div. ASCE*, **95**, 641–52, 1969.

14. Smith, I. M., and Hobbs, R. Biot analysis of consolidation beneath embankments, *Géotechnique*, **26**, 149–71, 1976.

15. Verruijt, A., Elastic storage in aquifers, in de Wiest, R. J. M. (ed.), *Flow through Porous Media*, pp. 331–76, Academic Press, New York, 1969.

16. Verruijt, A., Generation and dissipation of pore water pressures, in Gudehus, G. (ed.), *Finite elements in Geomechanics*, pp. 293–317, Wiley, London, 1977.

17. Yamamoto, T., Koning, H. L., Sellmeyer, H., and van Hijum, E. On the response of a poro-elastic bed to water waves, *J. Fluid Mech.*, **87**, 193–206, 1978.

18. Zienkiewicz, O. C., and Bettess, P., Chapter 1, This book.

Soil Mechanics—Transient and Cyclic Loads
Edited by G. N. Pande and O. C. Zienkiewicz
© 1982 John Wiley & Sons Ltd

Chapter 4

On the Importance of Dissipation Effects in Evaluating Pore Pressure Changes due to Cyclic Loading

H. B. Seed and I. M. Idriss

4.1 INTRODUCTION

Much interest has developed during the past decade in the behaviour of soils, especially sands, under cyclic loading conditions, a situation which may be attributed to the increasing activity of geotechnical engineers in the fields of earthquake engineering and ocean engineering. In both of these areas of activity, the stresses developed on elements of soil in the ground, whether they be due to wave forces on an ocean bottom gravity structure as shown in Figure 4.1(a) or to wave motions resulting from ground shaking in an underlying rock formation as illustrated in Figures 4.1(b) and 4.1(c), are cyclic in nature and it becomes necessary to determine the manner in which the soil will respond to this particular type of stress condition.

It is clear that for the conditions depicted in Figures 4.1(a)–(c) the shear stresses induced either by wave loadings or earthquake loadings can be closely approximated by cyclic horizontal shear stresses applied on horizontal planes, as shown in Figures 4.1(d) and (e). However, depending on the location of the element in a soil deposit, initial shear stresses may or may not exist on horizontal planes before the cyclic loading effects are developed. Thus, for example, with no storm or earthquake loading, elements designated A in Figures 4.1(a) and Figure 4.1(b) have no initial shear stresses on horizontal planes. However, this is not so for elements designated B in Figures 4.1(a) and 4.1(c). The presence of these initial shear stresses can have a major effect on the response of the soil to a superimposed cyclic stress condition and, in general, the presence of initial shear stresses tends to reduce the effects of cyclic loading with regard to both pore pressure build-up and strain generation. Since the most critical conditions are likely to be those associated with no initial shear stress on horizontal planes, the following discussion will concentrate on these conditions, although it should be noted that the concepts presented are not limited to these conditions.

53

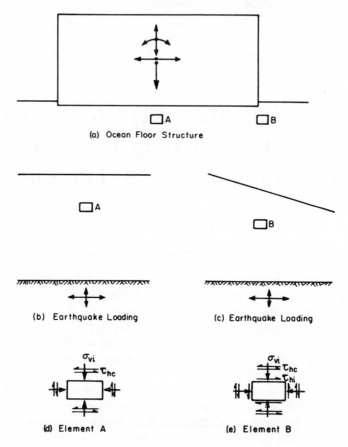

Figure 4.1 Stress conditions for soil elements in cyclic loading environments

One of the aspects of cyclic loading of primary importance in engineering design is the progressive build-up of pore water pressures which accompanies the sequence of shear stress reversals in the loading process. If enough cyclic stress reversals are applied, this process may culminate in the development of a pore pressure ratio ($r_p = \Delta p_g / \sigma'_{v0}$) of 100%—a condition representative of liquefaction or cyclic mobility depending on the density of the sand.

In recent years a number of procedures, both empirical and analytical, have been proposed for predicting the magnitude of the pore pressures that will develop under cyclic loading conditions (e.g. Martin, Finn, and Seed;[9] Seed, Martin, and Lysmer;[14] Liou, Streeter, and Richart;[8] Ghaboussi and Dikmen;[6] Zienkiewicz, Chang, and Hinton;[21] Sherif, Ishibashi, and Tsuchiya[17]). Of equal importance in many cases, however, is the fact that during the period of pore pressure generation there is, for many types of soil, a redistribution or some dissipation of the developed pore pressures which may have a very significant influence on the final magnitude of pore pressures developed. It is the purpose of this paper to emphasize and illustrate the importance of the latter aspect of many problems involving the response of soils to cyclic stress applications. The significance of pore pressure dissipation during and following a period of cyclic loading will be illustrated by typical examples including the generation of pore pressures in a sea-bed deposit of sand subjected to storm-wave loading, the generation of pore pressures under a gravity structure founded on the sea bottom, and the generation and dissipation of pore pressures in a deposit of sand during and following an earthquake.

The approaches used will be practical since they are intended to illustrate an important, though often neglected, aspect of cyclic loading, rather than to propose more sophisticated methods of analysing the response of soils subjected to this type of loading.

4.2 PORE PRESSURE GENERATION UNDER CYCLIC LOADING CONDITIONS

While there are many analytical approaches for predicting the rate of generation of pore pressures for any given soil, the simplest approach seems to be to measure the rate of generation of pore pressure directly in a cyclic simple shear test and use these results directly as the pore pressure generating function in a combined generation and dissipation analysis. Thus, for example, it has been found that the rate of pore pressure development in undrained cyclic simple shear tests on any given soil, and even on different soils in many cases, falls within a fairly narrow range when plotted in the normalized form shown in Figure 4.2. The average curve for the data shown

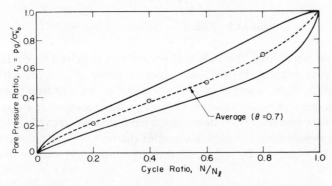

Figure 4.2 Rate of pore water pressure build-up in cyclic simple shear tests (after de Alba *et al.*, 1976)

in Figure 4.2 may be expressed by the relation

$$r_p = \frac{p_g}{\sigma'_{v0}} = \frac{2}{\pi} \arcsin \left(\frac{N}{N_l} \right)^{1/2\theta} \tag{4.1}$$

where N is the number of stress cycles applied to the sample

N_l is the number of stress cycles required to produce a pore pressure ratio of 100%

σ'_{v0} is the initial vertical effective stress

and

$\theta = 0.7$ for the data shown in Figure 4.2.

By varying the value of θ in this expression, the same equation can be made to fit any undrained pore pressure generation curve, as evidenced by the family of curves in Figure 4.3, drawn for different values of θ.[13]

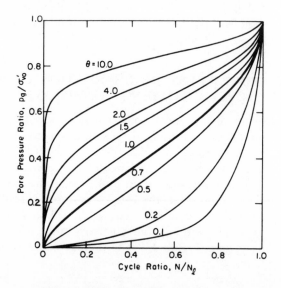

Figure 4.3 Rate of pore pressure geneneration

It may be noted that the only soil parameters required to evaluate pore pressure development in any soil element using this approach are the determination of the number of uniform stress cycles required to produce a peak cyclic pore pressure ratio of 100% and the appropriate value of θ for the soil. These can readily be determined from undrained cyclic simple shear tests on representative samples. Thereafter the rate of pore pressure generation may be obtained by differentiating equation (4.1) with respect to N, giving

$$\frac{\partial p_g}{\partial N} = \frac{\sigma'_{v0}}{\theta \pi N_l} \cdot \frac{1}{\sin^{2\theta-1}(\pi r_p/2) \cos(\pi r_p/2)}. \tag{4.2}$$

For practical purposes the irregular cyclic loading to which the soil is subjected at any stage of its loading history may be converted to an equivalent number of uniform stress cycles, N_{eq}, occurring in some duration of time, T_D (Seed et al.[12]). Thus

$$\frac{\partial N}{\partial t} = \frac{N_{eq}}{T_D} \qquad (4.3)$$

and combining equations (4.2) and (4.3) leads to

$$\frac{\partial p_g}{\partial t} = \frac{\partial p_g}{\partial N} \frac{\partial N}{\partial t} = \frac{\sigma'_{v0}}{\theta \pi T_D} \left(\frac{N_{eq}}{N_l} \right) \frac{1}{\sin^{2\theta-1}(\pi r_p/2) \cos(\pi r_p/2)}. \qquad (4.4)$$

In using this result it should be noted that the rate of pore pressure generation at any time t_1 depends on the previous cyclic history of the soil, and this may be represented approximately by the accumulated pore pressure p. Thus for any given time t_1, the appropriate rate of pore pressure generation must be determined from equation (4.4) corresponding to the value of p existing in the soil at that time. By this means the past history of strain cycles may be taken into account with a reasonable degree of accuracy.

4.3 COMBINED PORE PRESSURE GENERATION AND DISSIPATION

The combined effects of pore pressure generation and dissipation may be determined from the usual conditions of continuity of flow in soil elements. These conditions lead to the following equations:

For vertical flow only:

$$\frac{\partial p}{\partial t} = c_v \frac{\partial^2 p}{\partial z^2} + \frac{\partial p_g}{\partial t}. \qquad (4.5)$$

For radial flow only:

$$\frac{\partial p}{\partial t} = \frac{c_v}{r} \frac{\partial}{\partial r} \left[r \frac{\partial u}{\partial r} \right] + \frac{\partial p_g}{\partial t}. \qquad (4.6)$$

For combined vertical and radial flow:

$$\frac{\partial p}{\partial t} = \frac{c_v}{r} \frac{\partial}{\partial r} \left[r \frac{\partial p}{\partial r} \right] + c_v \frac{\partial^2 p}{\partial z^2} + \frac{\partial p_g}{\partial t}. \qquad (4.7)$$

These equations may readily be solved with the aid of appropriate computer programs. A general program suitable for solution of problems involving pore pressure generation and dissipation in three directions has been presented by Booker, Rahman, and Seed.[2]

[Editors' note: These equations are a simplified form of equation (1.16e), in which dynamic terms are suppressed and the total mean stress is assumed to remain constant during the motion. This form of the equations corresponds to the Terzhagi consolidation theory and is approximately valid if vertical acceleration effects are unimportant. In the above equations c_v is known as the coefficient of consolidation and is simply related to the bulk modulus and permeability of the drained material.]

4.4 WAVE-INDUCED PORE PRESSURES AND THEIR EFFECTS ON OCEAN FLOOR STABILITY

One of the important geotechnical considerations in connection with many engineering installations, such as pipelines and anchors, in an oceanic environment involving sand deposits, is that of potential ocean floor instability resulting from the development of high pore pressures caused by the direct action of waves. In the nearshore region, off the surf zone, where the waves are still stable and the water depth is not too great, the ocean floor will be subjected to travelling pressure waves (Figure 4.1) of significant magnitude. This will induce cyclic stresses within the profile which may cause a progressive build up of pore pressure. The pore pressure within the profile may build up to a stage where it becomes equal to the vertical effective stress. Structures founded in such soil deposits may then become unstable, depending on the location and extent of pore-pressure development in the supporting soil deposits. The rate and amount of pore pressure build-up will depend on three factors:

(1) the storm characteristics, i.e., the height, period, and lengths of different wave components;
(2) the cyclic loading characteristics of the soil deposits; and
(3) the drainage and compressibility characteristics of the different soil strata comprising the soil profile.

Evidence of such failures of a single 3.05 m (10 ft) diameter steel pipeline in Lake Ontario during storms is available, and it is believed that the backfill temporarily liquefied, causing the pipe to float to the water surface (Christain *et al.*[3]

A simple method for analysing problems of this type has been developed by Seed and Rahman.[16] The pressures induced on the ocean floor by the waves created by the design storm are evaluated as shown in Figure 4.4 and the shear stress changes on representative elements are then computed by elastic theory. Simple charts have been developed for this purpose. The cyclic loading characteristics of the soils below the ocean floor are then determined and used in conjunction with the preceding concepts of

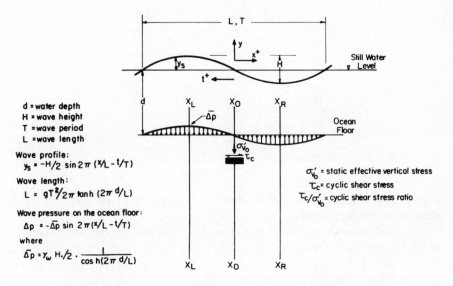

Figure 4.4   Wave characteristics and mechanism of pore pressure generation

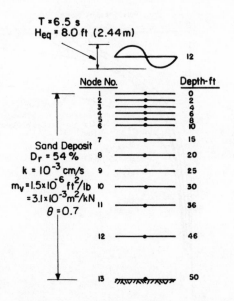

Figure 4.5   Soil profile and finite element
discretization

Table 4.1.   Wave composition of the design storm

Height m (ft) H_i	Period s T_i	Number of waves N_{w_i}	Duration (s) T_{D_i}
2.75 (9)	7	50	350
2.44 (8)	6.5	80	520
1.83 (6)	6	155	930
1.22 (4)	5	180	900
0.61 (2)	4	200	800

Total duration: $T_D = \sum T_{D_i} = 3500 \text{ s} \approx 58 \text{ min.}$

pore pressure generation and dissipation to determine the pore pressures induced in the soil profile.

A typical example is shown in Figure 4.5. This soil profile was subjected to the design storm shown in Table 4.1, which simple analyses showed could be represented by 232 uniform waves, each 8 ft high, occurring in a period of 58 minutes. The profile was represented by a discrete number of layers and analyses of response were made by the finite element method, using a computer program OCEAN1. The results of the analysis can be plotted to show the progressive build-up of pore pressures as the storm progresses (as shown in Figure 4.6 for sand having a coefficient of permeability of

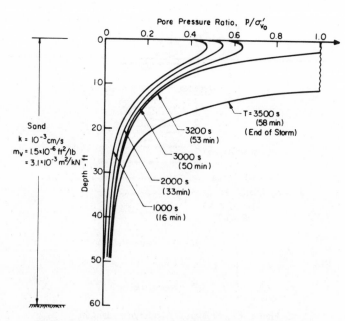

Figure 4.6   Pore pressure ratio at different stages of the storm

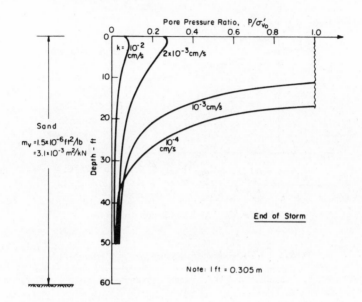

Figure 4.7 Influence of permeability on pore pressure response

10^{-3} cm/s) or alternatively the pore pressure distribution at the end of the storm for soils with different permeabilities (Figure 4.7). As the permeability decreases the results approach those for an undrained analysis.

It may readily be seen that the pore pressures developed are highly dependent on the permeability coefficient and thus on the rate of pore pressure dissipation. If the soil had a permeability of 2×10^{-3} cm/s but were analysed as an undrained system, it would be found to develop a pore pressure ratio of 100% to a depth of about 16 ft, whereas consideration of its pore pressure dissipation characteristics would show that the highest pore pressure ratio developed in the sand layer would be only about 25%.

4.5 PORE PRESSURE DEVELOPMENT UNDER OFF-SHORE OIL TANKS

A second example of the significance of dissipation effects is provided by computations of pore pressure development during major storms in sand deposits under off-shore oil tanks. A typical example is shown by the Ekofisk structure and the associated soil profile shown in Figure 4.8.

The characteristics of the waves in a design storm for this structure are shown in Table 4.2. The action of the waves on the sides of the structure causes a series of cyclically reversing stresses to be transmitted through the

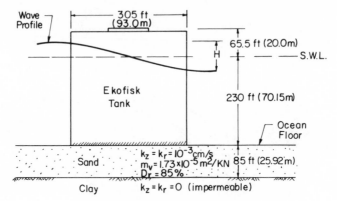

Figure 4.8   Ekofisk tank and soil profile

Table 4.2.   The characteristics of waves in a design storm

Wave Group N_{g_i}	Wave height H_i (ft)	Wave period T_i (s)	Number of waves N_{w_i}	Duration T_{D_i} (s)
1*	2	5.0	497	2485
2*	7	7.2	490	3528
3	20	10.0	485	4850
4	33	11.5	471	5417
5	46	12.5	282	3525
6	59	13.2	121	1597
7	72	13.4	32	429
8	82	13.5	3	40
				$T_D = 6$ h

* These wave groups of small waves are added to the design storm described by Bjerrum[1] to make the total duration of storm equal to 6 h.

base to the underlying sand deposit, with a resultant build-up of pore pressure in the sand.

At an early stage in the design, Bjerrum[1] analysed the average pore pressures which might be expected to develop under the base and concluded they would correspond to a pore pressure ratio of the order of 31%. This study was based on the assumption that there would be no dissipation of pore pressures from the sand during the storm.

In fact, the stress distribution and the resulting pore pressures which develop aross the base are reasonably uniform, but the maximum shear stresses develop just beyond the edge of the foundation and the pore pressures at this location are much higher than those at the base.

An analysis of the stresses and pore water pressures which might be expected to develop in the sand layer can readily be made from the results of cyclic simple shear tests on the sand and the stresses induced in the sand by the storm waves acting on the structure. Details of such an analysis are presented by Rahman *et al.*[11]

The results of the analyses for a point A under the centre of the tank and for a point B just off the edge of the tank are shown in Figure 4.9. Analytical results are shown for pore pressure development under conditions of no drainage and for conditions where dissipation is considered. Near the centre of the base the maximum computed pore pressure ratio allowing for dissipation is about 6%, while if no dissipation is allowed, it is about 33%. Just beyond the edge of the tank, the computed pore pressure ratio reaches a value of 100% after 2 h of the 6 h storm if dissipation is not considered, whereas the maximum pore pressure ratio at the same point attains a maximum value of only 28% if dissipation is taken into account.

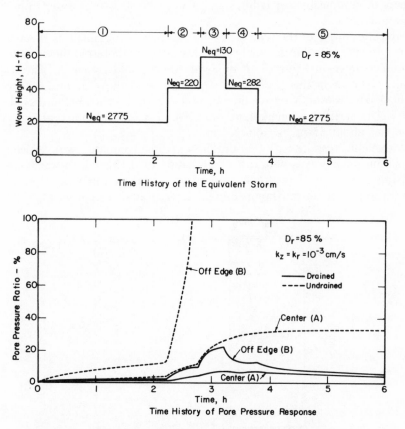

Figure 4.9 Analysis of pore pressure generation for Ekofisk tank on sand with relative density of 88%

4.6 PORE PRESSURE GENERATION AND DISSIPATION DURING EARTHQUAKES

Recent developments in studies of soil response to earthquake loadings have now made it possible to further extend these studies to include consideration of the pore water pressure redistribution that may take place during the period of earthquake shaking as well as the period following the earthquake. This may be extremely important in highly pervious materials such as gravels, and may, in fact, prevent significant pore water pressures from developing at all in such materials.[19] Furthermore, development of the ability to predict the generation of pore pressures as the earthquake progresses makes it possible to take into account the progressive changes in stress-deformation characteristics of the soils in computing the stresses induced by the earthquake motions. Finally, it permits the determination of the dissipation of pore pressures following the earthquakes—a phenomenon that may be responsible for failures and damage to structures founded near the ground surface.

A typical example of the type of results that may be obtained by this type of pore pressure distribution analysis is illustrated by those obtained for the soil profile shown in Figure 4.10 by Seed, Martin, and Lysmer.[14] The soil conditions shown in this profile are very similar to those in the areas of Niigata, Japan, where extensive liquefaction occurred during an earthquake in 1964. The earthquake magnitude was about 7.6 and the source of energy release was about 35 miles from Niigata.

The dynamic response of this profile was first computed for vertically propagating shear waves with a base excitation of 0.13g, which was considered representative of the rock motion developed by this type of earthquake. The computed maximum ground surface acceleration was 0.165 g,

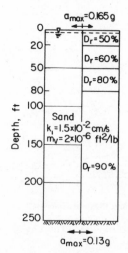

Figure 4.10 Soil profile and stress conditions used for analysis

which agrees well with the recorded maximum acceleration at the ground surface (0.16g).

For the purpose of the analysis, the coefficient of compressibility of the sands at low strains was estimated to be 2×10^{-6} ft^2/lb, the permeability of the sand below the water table corresponding to a grain diameter $D_{20} \simeq$ 0.15 mm to be 1.5×10^{-2} cm/s, and the permeability of the partially saturated sand above the water table to be 7.5×10^{-3} cm/s, which is in accord with measured values for such materials. Changes in coefficient of compressibility with increasing values of the pore pressure ratio, r_p, were taken into account in accordance with test data determined by Lee and Albaisa.[7]

The computed variations of pore water pressure with time at different depths in the deposit and over different time periods are shown in Figures 4.11 and 4.12. Figure 4.11 shows the build-up of pore pressures in soil layers at depths ranging from 3 ft to 50 ft due to both generation and dissipation of pore pressures during the period of earthquake shaking. It may be seen that the sand layer at a depth of 15 ft liquefies after about 21 s of shaking, followed by extension of the liquefied condition to depths of 20 ft, 30 ft, and 40 ft after about 23 s, 32 s, and 40 s of shaking, respectively. Although the layers above 15 ft depth continue to increase in pore pressure as the shaking progresses, the rate of increase is very slow after the 15 ft layer liquefies and is due only to upward flow of the pore water rather than shaking effects directly. At the end of the period of earthquake shaking, a pore pressure ratio of 100% would extend from a depth of 15 ft to 40 ft in the sand deposit but the excess pore water pressures at a depth of 3 ft would

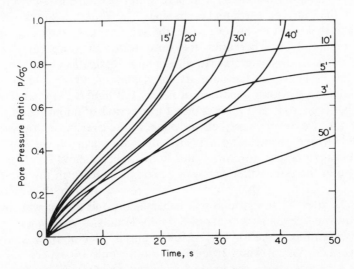

Figure 4.11 Computed development of pore water pressures during earthquake shaking for soil profile shown in Fig. 4.10

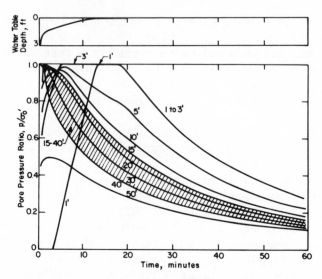

Figure 4.12 Computed variation of pore water pressures in 60 minute period following earthquake for soil profile shown in Fig. 4.10

at this point be only about 65% of the initial effective overburden pressure, while near the ground surface (depth of 1 ft) the pore water pressure would be unchanged.

Figure 4.12 shows the changes in pore water pressure in a period of 1 h following the start of earthquake shaking. The pore pressures when the earthquake shaking stops are of course, the same as those shown at a time of 50 s in Figure 4.11. Thereafter the pore water pressures at depths of 15–40 ft begin to dissipate slowly, with greater dissipation occurring in the deepest layers. However, the pore pressures continue to build up in the soil layers above 15 ft with a condition of nearly full liquefaction existing in all the soil between 10 ft and 30 ft depths for a period of about 5 min. At this time the pore water pressures become equal to the overburden pressure at a depth of 3 ft and it might be expected that the water would boil to the surface between cracks, forming water boils perhaps about 3 ft high. After about $3\frac{1}{2}$ min the pore water pressure begins to increase in the top foot of soil and it can be shown that when the pore pressure ratio in this soil reaches about 50% to 60%, the ground will become soft and a man will sink; this condition would develop after about 8.5 min. Finally after about 12 min the pore pressure ratio in the top foot of soil attains a value of 100% and at this stage a general flow of water from the ground would be expected.

Figure 4.12 also shows the gradual stabilization of conditions in a period of 1 h following the earthquake. The pore pressure ratio at the ground surface begins to decrease after about 20 min but it would not support a man until about 40–50 min after the earthquake shaking ended. Meanwhile from 5 min after the start of the shaking, pore pressures are progressively decreasing in the soil between 5 ft and 50 ft below the ground surface. However, it may be noted that significant excess pore pressures continue to exist in the soil deposit even as much as 1 h after the earthquake ceased.

It is interesting to compare the rate of pore pressure development and surface evidence of liquefaction effects in this example with those actually observed in Niigata.[18] Such a comparison clearly shows that the computed effects are in reasonable accord with those observed, at least with a degree of accuracy satisfactory for engineering purposes.

For purposes of comparison Figure 4.13 shows the development of pore water pressures in a similar deposit of very coarse sand, with a coefficient of permeability of 0.5 cm/s, subjected to the same shaking. Under undrained conditions, this soil would also develop pore pressure ratios of 100% between depths of about 15 ft and 40 ft during the earthquake. However, because of the high permeability and the corresponding rapid rate of dissipation of the pore water pressures induced by the shaking, there is no major build-up of water pressures and the soil would never reach a condition of 100% pore pressure ratio at any depth. A similar result is shown by Finn, Lee, and Martin.[5] The inability of coarse sands and gravels to develop high pore pressure induced by ground shaking due to the rapid rate of dissipation is probably a major factor leading to their superior performance under the effects of strong earthquake shaking. In fact, parametric studies[14]

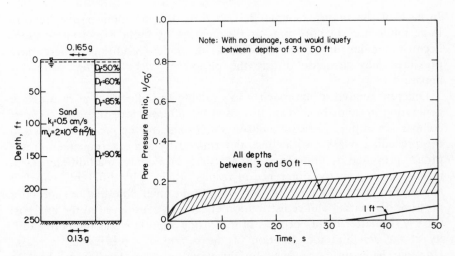

Figure 4.13 Development of pore water pressures in deposit of coarse sand subjected to earthquake shaking

indicate that for level ground deposits of cohesionless soils having $D_{20} >$ 0.7 mm, pore pressure ratios of 100% would be unable to develop in any earthquake since the rate of pore pressure dissipation would be able to keep pace with the rate of pore pressure generation due to earthquake shaking.

This assumes of course that the cohesionless soil deposit is uniform and thereby free-draining. However, a single layer of relatively impervious fine sand or silt in such a deposit would completely invalidate the results of pore pressure dissipation computations for vertical flow and for this reason analytical data of this type should be used with great care in engineering practice.

Much greater dependence can be placed on the results of dissipation analyses involving radial drainage,[13] since these effects are relatively independent of details of soil stratification. Thus, if pore pressures generated in a soil mass by cyclic loading can be dissipated almost as fast as they are created in pervious deposits, it is apparent that advantage may be taken of this effect to improve the permeability and pore pressure dissipation characteristics of potentially liquefiable deposits by the installation of a system of gravel or rock drains, so that pore pressures may be dissipated rapidly and liquefaction thereby prevented. This possibility appears to have been first proposed by Yoshimi and Kuwabara[20] and the theoretical developments required for its practical implementation have recently been presented by Seed and Booker.[13] In fact, such installations have already been used as a means of stabilizing potentially liquefiable deposits.

4.7 CONCLUSION

In each of the three examples described in the preceding pages the very large differences in pore pressures generated by cyclic stress applications occurring under undrained conditions, or under conditions where pore pressures may dissipate during the period of cyclic loading, is readily apparent.

Different analytical approaches to evaluate the effects of pore pressure generation in sands have been proposed by different investigators. A direct comparison of pore pressure build-up determined by the procedures used in the preceding pages[10] with those determined by a more fundamental approach proposed by Finn, Lee, and Martin.[5] shows only small differences in results and in fact the difference in results obtained by different analytical approaches may often be far less significant than whether or not pore pressure dissipation and redistribution effects are considered in the analysis. Clearly this latter point becomes of increasing importance as the permeability of the soil and the duration of the period of stress cycles increase. However in many cases, even for durations of cyclic loading as short as those produced by earthquakes, failure to take into account the dissipation part of a cyclic loading problem as well as the generation part of the problem can lead to serious errors in pore pressure evaluations.

REFERENCES

1. Bjerrum, L. Geotechnical problems involved in foundations of structures in the North Sea, *Géotechnique*, **23**(3), 319–58, 1973.
2. Booker, J. R., Rahman, M. S., and Seed, H. B., *GADFLEA A Computer Program for the Analysis of Pore Pressure Generation and Dissipation During Cyclic or Earthquake Loading*, Report No. EERC 76–24, Earthquake Engineering Research Center, University of California, Berkeley, October 1976.
3. Christian, J. T., Taylor, P. K., Yen, J. K. C., and Erali, D. R. Large diameter underwater pipeline for nuclear power plant designed against soil liquefaction, *Offshore Technology Conference Proceedings*, **2**, 597–602, 1974.
4. DeAlba, P. Seed, H. B., and Chan, C. K. Sand liquefaction in large-scale simple shear tests, *Journal, Geotechnical Engineering Division, ASCE*, **102** (GT9), 909–27, 1976.
5. Finn, W. D. L., Lee, K. W., and Martin, G. R., An effective stress model for liquefaction *Journal, Geotechnical Engineering Division, ASCE*, **103** (GT6), 517–33, 1977.
6. Ghaboussi, J., and Dikmen, S. U. Liquefaction analysis of horizontally layered sands, *Journal, Geotechnical Engineering Division, ASCE*, **104** (GT3), 341–56, 1978.
7. Lee, K. L., and Albaisa, A. Earthquake induced settlements in saturated sands, *Journal, Geotechnical Engineering Division, ASCE*, **100** (GT4), 387–406, 1974
8. Liou, C. P., Streeter, V. L., and Richart, F. E. A numerical model for liquefaction, *Journal, Geotechnical Engineering Division, ASCE*, **103** (GT6), 598–606, 1977
9. Martin, G. R., Finn, W. D. L., and Seed, H. B. Fundamentals of liquefaction under cyclic loading, *Journal, Geotechnical Engineering Division, ASCE*, **101** (GT5), 423–38, 1975.
10. Martin, P. P., and Seed, H. B. Simplified procedure for effective stress analysis of ground response, *Journal, Geotechnical Engineering Division, ASCE*, **103** (GT6), 739–58, 1979.
11. Rahman, M. S., Seed, H. B., and Booker, J. R. Pore pressure development under offshore gravity structures, *Journal, Geotechnical Engineering Division, ASCE*, **103** (GT12), 1419–36, 1977.
12. Seed, H. B., Idriss, I. M., Makdisi, F., and Banerjee, N. *Representation of irregular stress time histories by equivalent uniform stress series in liquefaction analyses*, Report No. EERC 75–29, Earthquake Engineering Research Center, University of California, Berkeley, October 1975.
13. Seed, H. B., and Booker, J. R. *Stabilization of Potentially Liquefiable Sand Deposits Using Gravel Drain Systems*, Report No. EERC 76–10, Earthquake Engineering Research Center, University of California, Berkeley, April 1976.
14. Seed, H. B., Martin, P. P., aand Lysmer, J. Pore water pressure changes during soil liquefaction, *Journal, Geotechnical Engineering Division, ASCE*, **102** (GT4), 323–46, 1976.
15. Seed, H. B., and Rahman, M. S. *Analysis for Wave-Induced Liquefaction in Relation to Ocean Floor Stability*, Report No. UCB/TE-77/02, Geotechnical Engineering, University of California, Berkeley, May 1977.

16. Seed, H. B. and Rahman, M. S. Wave-induced pore pressure in relation to ocean floor stability of cohesionless soils, *Marine Geotechnology*, **3**(2), February 1978.
17. Sherif, M. A., Ishibashi, I., and Tsuchiya, C. Pore pressure prediction during earthquake loadings, soils and foundations, *Japanese Society of Soil Mechanics and Foundation Engineering*, **18**(4), December 1978.
18. Soil and Foundation *Japanese Society of Soil Mechanics and Foundation Engineering*, **4**(1), January 1966.
19. Wong, R. T., Seed, H. B., and Chan, C. K. Cyclic loading liquefaction of gravelly soils, *Journal, Geotechnical Engineering Division, ASCE*, **101** (GT6), 571–83, 1975
20. Yoshimi, Y., and Kuwabara, F. Effect of subsurface liquefaction on the strength of surface soil, *Soils and Foundations*, **13**(2), 67–81, 1973
21. Zienkiewicz, O. C., Chang, C. T., and Hinton, E. Non-linear seismic response and liquefaction, *Int. J. for Numerical and Analytical Methods in Geomechanics*, **2** 381–404, 1978.

Soil Mechanics—Transient and Cyclic Loads
Edited by G. N. Pande and O. C. Zienkiewicz
© 1982 John Wiley & Sons Ltd

Chapter 5

Liquefaction and Permanent Deformation under Dynamic Conditions—Numerical Solution and Constitutive Relations

O. C. Zienkiewicz, K. H. Leung, E. Hinton and C. T. Chang

SUMMARY

(1) This chapter discusses the essential nature of soil behaviour under repeated loading, and briefly discusses appropriate constitutive relations;
(2) discusses the numerical formulation and solution techniques for dynamic soil problems which are governed by basic equations introduced in Chapter 1;
(3) presents some results of transient non-linear analysis for typical problems such as earth dams, foundation layers, etc., in which permanent deformation and liquefaction occurs under earthquake shocks.

5.1 INTRODUCTION

The description of deformation of soils under static conditions of stress is not easy and much research which has gone into this subject has not yet arrived at a universal, quantitative model which is generally acceptable. Nevertheless, reasonably accurate models are today available which can predict with sufficient accuracy collapse conditions and non-linear, permanent deformations occurring under load.[36]

The situation becomes more complex when fluctuating loads occur such as may be expected in offshore structures under wave action or in earthquake response of earth dams or structural foundations. Here two new phenomena are encountered. Referring to drained behaviour of the soil we note that

(1) permanent shear strains occur after each cycle of stress application;
(2) permanent volume contraction (densification) occurs after each cycle of loading (except for extremely dense materials).

Of the two effects, the second is the one of greatest importance in practice where saturated, undrained, or partially drained behaviour predominates.

71

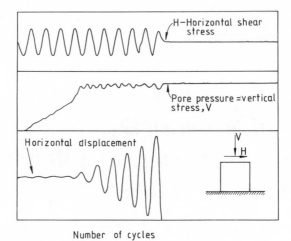

Figure 5.1 Typical soil behaviour under cyclic shear
stress (undrained)

It is this densification phenomenon that is responsible for the increase of
pore pressure during cyclic load applied to a sample in an undrained state,
and which accounts for large displacements occurring when the pore pressure builds up to a value of mean total, compressive, stress so that the
effective stress state reaches the failure (or critical state) of the material.
When this happens incipient liquefaction[36] is reached and material is at a
point of yielding. Some dilation which occurs in sands during plastic flow
counterbalances at this stage the pore pressure increase phenomena and
complete liquefaction follows only when such dilation is exhausted after a
considerable shearing strain. Figure 5.1 shows such behaviour in a typical
cyclic 'shear box' test carried out on a sand recently[15] and shows the relative
insignificance of cumulative shear strains before the onset of liquefaction
and the dramatic increase of strains following this.

Clearly it will not be an easy matter to derive a model capable of
reproducing fully all such phenomena. Many of the later chapters will in fact
be devoted to the description of rather complex constitutive relations which
attempt to reproduce this essential characteristic of the soil. In the first part
of this chapter we shall briefly review some of such models and introduce a
simple one which is capable of quantitative predictions concerning the pore
pressure increase evaluation.

In rapid loading phenomena such as those caused by earthquake (or other
shock loads such as for instance of explosions) it is impractical to use
techniques of limit load analysis, popular in estimating collapse under static
conditions. Now the duration of loading is so short that displacements may

well be small even if the strength capacity of the soil is exceeded. To answer questions concerning the safety of a structure the only practicable way involves the use of step by step numerical computation and we shall discuss in some detail such procedures in the second part of this chapter.

In Chapter 1 the basic, generally applicable, differential equations describing the coupling between soil deformations and the fluid motion have been introduced and we have shown under which circumstances undrained behaviour can be assumed. For typical structures in which sands or silts comprise an important part such assumptions are not generally adequate and we shall show how drainage can be readily included in the analysis.

The chapter will conclude with various examples of computation and a discussion of the failure of the Lower San Fernando Dam,[23,24] where liquefaction of fill and subsequent large deformation was nearly responsible for a major catastrophe.

Whilst the basic formulation is applicable to the slower cyclic loading phenomena caused by waves on offshore structures, for these a different computational treatment will in general have to be followed. This is due to:

(1)   the long period of cyclic loading and its slower rate;
(2)   the absence of dynamic effects resulting in quasi-static collapse possibilities;
(3)   excessive deformation being often a corollary of cumulative deformation (lack of shakedown or ratcheting) rather than of instantaneous failure.

We leave full discussions of such problems to Chapter 18.

PART 1   CONSTITUTIVE LAWS AND LIQUEFACTION

5.2   SOME CONSTITUTIVE LAWS FOR STATIC BEHAVIOUR

All constitutive relations for soil in which pore pressure changes play an important role are most conveniently defined in terms of effective stress variables. Undrained properties follow uniquely from such descriptions (see equation (1.6)).

In rate independent soils, i.e. when creep strains can be considered as negligible, the constitutive law for a non-linear material must be specified as an incremental relation between the changes of effective stress $d\sigma'$ and the changes of that part of strain which we consider to be directly related to stress, i.e. $d\varepsilon^\sigma$.*

---

* In this chapter we shall use vector–matrix notation for stresses and strains as defined in Appendix A of Chapter 1.

Thus we can write generally

$$d\boldsymbol{\sigma}' = \mathbf{D}\, d\boldsymbol{\varepsilon}^{\sigma}, \tag{5.1}$$

where $\mathbf{D}$ is the tangent modulus matrix. Such a tangent modulus matrix is in general a function of effective stress level, straining history, etc.

Many alternative models defining the constitutive law of the type given in equation (5.1) are available. Among these we have the classes of

(1)   non-linear elasticity (hyper-elasticity);
(2)   plasticity or viscoplasticity;
(3)   hypo-elasticity;

and finally

(4)   endochronic theory.

All have their proponents and antagonists, and each shows some philosophical or computational merits. If the application of loads on a structure is non-monotonic, non-linear elasticity can lead to entirely erroneous results and therefore such models are discarded *a priori*. As hypo-elasticity and endochronic theory are basically alternative descriptions of the plastic (or viscoplastic) phenomena, in what follows we shall entirely concentrate on the plasticity model.

This is defined by the following:

(a)   a yield surface

$$F(\boldsymbol{\sigma}', \boldsymbol{\kappa}) = 0 \tag{5.2}$$

which limits the field of elastic deformation. Here $\boldsymbol{\sigma}'$ is the effective stress tensor and $\boldsymbol{\kappa}$ represents a set of hardening parameters dependent on plastic strains $\boldsymbol{\varepsilon}_p$. Thus only elastic straining can occur if

$$F < 0, \tag{5.3a}$$

and when

$$F = 0 \tag{5.3b}$$

both plastic and elastic straining can be present.

(b) To complete plasticity definitions the directions of plastic straining have to be defined by a plastic potential

$$Q(\boldsymbol{\sigma}', \boldsymbol{\kappa}) = 0 \tag{5.4}$$

such that we can write the *flow rule* as

$$d\boldsymbol{\varepsilon}_p = \lambda \frac{\partial Q}{\partial \boldsymbol{\sigma}}, \tag{5.5}$$

where λ is a proportionality constant.

(c) Finally it is assumed that the total, stress dependent, strain increment $d\boldsymbol{\varepsilon}^\sigma$ can be divided into elastic and plastic parts, i.e. that

$$d\boldsymbol{\varepsilon}^\sigma = d\boldsymbol{\varepsilon}_e + d\boldsymbol{\varepsilon}_p \tag{5.6}$$

with

$$d\boldsymbol{\varepsilon}_e = \mathbf{D}_e^{-1} d\boldsymbol{\sigma}'$$

where $\mathbf{D}_e$ represents the matrix of elastic constants.

Plasticity rules are associative if

$$F = Q \tag{5.7a}$$

and non-associative if

$$F \neq Q. \tag{5.7b}$$

The definition of yield, potential and hardening rules suffices to determine the incremental $\mathbf{D}$ matrix of equation (5.1). Details of the necessary algebraic operations are presented elsewhere.[34]*

In attempting to model a soil by plasticity theory we note at the outset that the best-known characteristic of soils is that of their critical behaviour characterized by an envelope of Mohr's circles—and usually approximated to by values of drained cohesion c' and friction angle ϕ'. At such stress states, shown in the space of principal stress of Figure 5.2 as a pyramid,[36] deformation can continue without appreciable stress changes. Clearly it is therefore natural to associate this critical surface with the form of plasticity model used.

Three models of plasticity of a relatively simple kind have been used with success in static problems;

(A) In this model ideal (non-hardening) plasticity is assumed in which the critical Mohr–Coulomb surface plays the role of the yield surface $F(\boldsymbol{\sigma}') = 0$. Further fully associative behaviour is taken with $Q = F$.

(B) Similar to (A) but with non-associative flow rule assumed in which Q is taken in a form similar to that of F but with ψ now replacing the friction angle ϕ'. This rule for flow allows smaller dilatancy to be imposed on the material during plastic flow (the most frequent assumption being $\psi = 0$ at which no dilation occurs). This model is shown in Figure 5.3.

and finally

(C) The modified critical state model based on original work of Roscoe and his followers[20-22] amended by Zienkiewicz, Humpheson, and Lewis[31] to include a Mohr–Coulomb type critical surface in three-dimensional stress space.

* See also Chapter 8.

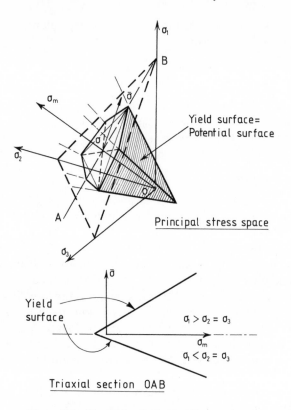

Figure 5.2   The Mohr–Coulomb failure surface in
principal stress space representing the yield surface
of an associated plastic model (Model A)

With this model strain hardening depends on the plastic volumetric
strain, i.e.

$$-(\varepsilon_{ii})_{\mathrm{p}} = -\mathbf{m}^{\mathrm{T}}\boldsymbol{\varepsilon}_{\mathrm{p}} \tag{5.8}$$

is assumed but a fully associative flow rule is taken. The yield surfaces
are given in this type of plasticity by elliptical (or similar) closed curves
in the sections of principal stress space containing the mean stress
direction and in such planes are defined in a manner giving zero
dilatancy for points lying on the Mohr–Coulomb 'critical' surface, as
shown in Figure 5.4. All loading paths starting in the elastic domain
for such a material will show either 'hardening' or 'softening' behaviour but will, at large strains, converge to the critical surface.

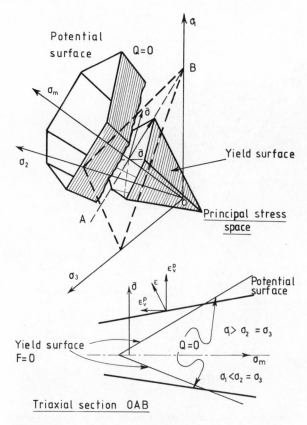

Figure 5.3 A non-associated ideal plastic model with potential and yield surfaces of similar, Mohr–Coulomb, form (Model B)

Numerical experiments reported elsewhere[31,32,36] lead one to the following conclusions applicable to static foundation or embankment analysis:

Static drained behaviour of normally consolidated materials

Models (A)–(C) give almost identical collapse loads and, with some adjustment of elastic constants, very similar displacement behaviour.

Static undrained behaviour of normally consolidated materials

Model (B) with $\psi = 0$ and model (C) give a very similar collapse and displacement performance, but model (A) shows no collapse due to negative

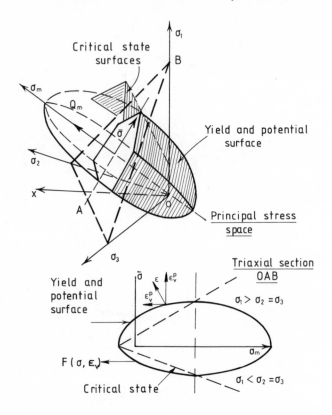

Figure 5.4 The critical state model—with a Mohr–Coulomb critical surface (Model C)

pressures developed on continuing dilation during yield, and finally

Static undrained behaviour—over-consolidated

Here only model (C) is capable of dealing with the over-consolidation phenomenon—and results recently obtained indicate excellent agreement with experiment.[36]*

From the results discussed above it would appear that adequate plasticity models exist for most static problems, with models (B) and (C) having a fairly universal applicability.

---

* Although the over-consolidated, drained behaviour has not been explicitly investigated, we note that on failure, strain softening will occur, with the residual strength values being the same for models (C) and (B). We would thus expect no advantage to arise for model (C) in estimating final strength.

We see that this is not the case for cyclic (or generally variable) load histories as *within an ideal (or isotropically hardening) yield surface no amount of stress cycling can produce any permanent strains.*

5.3 MODIFICATIONS OF STATIC PLASTICITY FOR CYCLIC (OR VARIABLE) LOADS

To reproduce the essential features of cyclic (or variable) load response the constitutive models discussed in previous sections need to be modified or replaced.

Two main lines of attack present themselves. In the first, the basic formulation used for static problems is replaced by a new one capable of accommodating the additional effects. Here, for instance, the critical-state model discussed in the previous section could be augmented by additional kinematically hardening yield surfaces as suggested by Mroz, Norris, and Zienkiewicz,[11-13] or other rules introduced for modifying the yield surface. The work of Ghaboussi *et al.*[6,7] Pender,[19] Nova,[16] and others[2-3] follows in this category and will be discussed later in this book. The addition of cumulative effects is natural in the context of endochronic models and here some success has been reported,[1,30] necessitating, however, the use of numerous measured parameters.

In the second approach a simpler philosophy is adopted. The static model with its well-tested structure is retained and the new effects are accounted for by a separate model.

Such an approach has the advantage of simplicity and of the most direct use of experimental evidence, and for this reason we have singled out its use in the present chapter.

In the introduction we have already stated that the most important feature of cyclic strain response is that of the *cumulative densification* which is responsible for such phenomena as liquefaction[4,8,29] and loss of strength. We shall therefore concentrate our efforts on including this aspect in our model.

In the previous section we have concluded that for static behaviour studies, the non-associated plasticity (Model B) or the 'critical' state models (Model C) were well applicable. Now we need simply to include an additional accumulation of strain ε^0 which is of volumetric nature, i.e.

$$\varepsilon^0 = \mathbf{m}\varepsilon_v^0/3 \tag{5.9}$$

and which is caused by *the history of elastic and plastic straining.*

Before considering detail formulae for this, it is essential to investigate the effect of such a volumetric strain on the changes of pore pressure under undrained conditions. Such conditions are characteristic of standard tests conducted under cyclic loads and allow the simplest correlation between ε_v^0 and the measured pore pressure changes.

In the next section we shall discuss this in detail, indicating the basic mechanism of liquefaction.

5.4 THE EFFECTS OF DENSIFICATION STRAIN AND SOME EMPIRICAL MODELS

Retaining the elastic plastic constitutive law of equation (5.1) and noting that the total strain $\mathbf{\varepsilon}$ is given on the sum of $\mathbf{\varepsilon}^\sigma$ and $\mathbf{\varepsilon}^0$, we can write

$$d\mathbf{\sigma}' = \mathbf{D}(d\mathbf{\varepsilon} - d\mathbf{\varepsilon}^0), \tag{5.10}$$

in which we assume that $d\mathbf{\varepsilon}^0$ has been independently determined.

Definition of effective and total stresses and their link with pore pressure gives

$$d\mathbf{\sigma} = d\mathbf{\sigma}' - \mathbf{m}\, dp, \tag{5.11}$$

and if undrained behaviour of a sample is considered,

$$dp = d\hat{p} = -\frac{K_f}{n}\mathbf{m}^\mathrm{T}\, d\mathbf{\varepsilon}. \tag{5.12}$$

Let us now investigate the behaviour of a sample of soil illustrated in Figure 5.5 on which the total stress does not change ($d\mathbf{\sigma} = 0$) and where an initial strain $d\mathbf{\varepsilon}^\circ = \mathbf{m}\, d\varepsilon_v^0/3$ occurs due to some extraneous causes. From

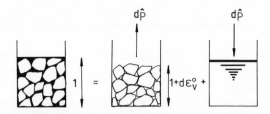

Figure 5.5 The basic mechanism of pore pressure changed due to densification with constant total stress

equations (5.10) and (5.11) we can write

$$\mathbf{m}\, d\hat{p} = \mathbf{D}\left(d\mathbf{\varepsilon} - \frac{\mathbf{m}\, d\varepsilon_v^0}{3}\right). \tag{5.13}$$

From equations (5.12) and (5.13) after eliminating $d\mathbf{\varepsilon}$ we can obtain a simple relation

$$d\hat{p} = -\beta\, d\varepsilon_v^0 \tag{5.14}$$

where

$$\beta = 1 \bigg/ \left(\frac{n}{K_f} + \frac{1}{K_T} \right) \simeq K_T \quad \text{if} \quad K_f \gg K_T \tag{5.15}$$

and

$$K_T = \text{tangent bulk modulus of skeleton}$$
$$= 1/\mathbf{m}^T \mathbf{D}^{-1} \mathbf{m}. \tag{5.16}$$

This mechanism of densification causing an increase of pore pressure is fundamental to our understanding of the liquefaction phenomena and we note that by virtue of equation (5.14) it is immaterial whether we specify with primary variables ε_v^0 or simply the undrained pressure increase $\hat{p}$ which can readily be measured in undrained, cyclic, tests.

Results of typical tests on sand under cyclic loading are reproduced in Figures 5.6(a) and (b),[35] in which

$$\theta = |\bar{\sigma}|/\sigma_{m0}, \tag{5.17}$$

where $|\bar{\sigma}|$ is the fluctuating cyclic shear (deviatoric invariant) and σ'_{m0} the mean effective stress at the start of loading.

As we have already mentioned, the development of the densification strain ε^0 must be related to the total strain history (and the level of the stresses). With this in mind a parameter ξ is defined such that

$$d\xi = \sqrt{(de_{ij} \, de_{ij})}, \tag{5.18}$$

where e_{ij} represents deviatoric strain components. In some earlier work[35] we show that a good correlation with experiment can be defined by taking an expression

$$d\varepsilon_v^0 = \frac{-A}{1 + B\kappa} \, d\kappa \tag{5.19}$$

and

$$d\kappa = e^{\gamma\theta} \, d\xi. \tag{5.20}$$

Here the constants A, B and γ define the characteristics of cyclic response.

In Figure 5.7 a correlation obtained using above expression is shown for the sand whose behaviour was given in Figure 5.6.

Other formulations have from time to time been proposed[14] and we must observe that development of ε_v^0 is very much dependent on the initial density of soil. It is clear that if the soil is packed to its maximum density no further densification is possible and thus $A \to 0$.

While the search for an optimal, all-embracing, expression continues, we note that by expression (5.14) we can interchangeably use ε_v^0 or $\hat{p}$ in all computations. The use of the latter is convenient as it represents a *most direct connection between experiment and subsequent calculations*. Seed *et al.*[10,25] show that for a wide variety of sands the relative pore pressure rise

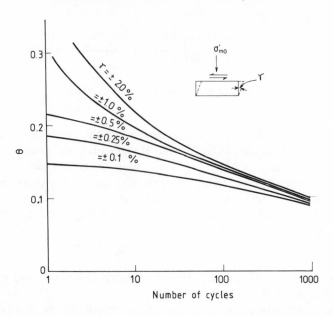

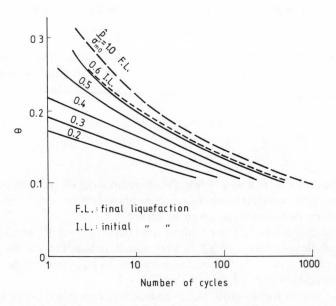

Figure 5.6 Results of a typical undrained cyclic shear test on sand (N.G.I. Sand[35]): (a) development of cyclic shear strains; (b) build up of excess pore pressure

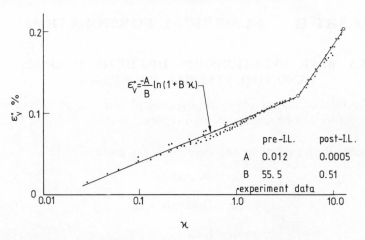

Figure 5.7 Correlation between volumetric densification ε_v^0 and the strain-loading parameter κ for tests of Fig. 5.6

$\hat{p}/\sigma_{m0}$ can be related to θ and to the number of stress cycles in a manner shown in Figure 5.8.

If we note that during a single cycle we can compute $\Delta\xi$ as twice the total absolute range of shear strain, it is easy to show that for N uniform cycles

$$\xi = N \, |\overline{\Delta\sigma}| \, 2/G = N\theta\sigma'_{m0} 2/G, \tag{5.21}$$

where N is the number of cycles and G the average shear modulus. The curves of Figure 5.8 can thus be adopted to compute $d\hat{p}$ directly for non-cyclic strains.

The foregoing discussion shows that with the information now available we have a reasonably good representation of the cyclic pore pressure (or ε_v^0) increase rate which can be used in non-linear analysis. General procedures of such an analysis can now be discussed.

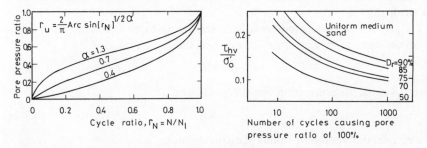

Figure 5.8 Pore pressure development for soils of various densities under cyclic load (after Martin and Seed[10])

PART II NUMERICAL FORMULATION

5.5 BASIC RELATIONSHIPS: DISCRETIZED FORMS AND TIME STEPPING PROCESSES

The differential equations governing dynamic phenomena in soils in which drainage occurs can be written as follows (see equations (1.16) or (A.2) of Chapter 1).

First; constitutive relations and definitions are given as

$$d\boldsymbol{\sigma} = d\boldsymbol{\sigma}' - \mathbf{m}\,dp, \qquad (5.22a)$$

$$d\boldsymbol{\sigma}' = \mathbf{D}(d\boldsymbol{\varepsilon} - d\boldsymbol{\varepsilon}^0), \qquad (5.22b)$$

$$d\boldsymbol{\varepsilon} = \mathbf{L}\,d\mathbf{u}, \qquad (5.22c)$$

where in general both $\mathbf{D}$ and $\boldsymbol{\varepsilon}^0$ depend on stresses, strains and their history in the manner already discussed, $\boldsymbol{\varepsilon}$ represents here the total strains, $\mathbf{u}$ displacements and $\mathbf{L}$ an appropriate differential operator defining strains in terms of displacements. Overall equilibrium can now be written as

$$\mathbf{L}^T\boldsymbol{\sigma} + \rho\mathbf{g} = \rho\ddot{\mathbf{u}}, \qquad (5.23)$$

and finally the porous fluid flow equations (defining the mass balance are introduced as

$$-\boldsymbol{\nabla}^T\mathbf{k}\boldsymbol{\nabla}p + \mathbf{m}^T\dot{\boldsymbol{\varepsilon}} + \boldsymbol{\nabla}^T(\mathbf{k}\rho_f\mathbf{g}) = \boldsymbol{\nabla}^T\mathbf{k}\rho_f\ddot{\mathbf{u}} - \frac{n}{K_f}\dot{\mathbf{p}} \qquad (5.24)$$

where $\mathbf{k}$ is the permeability, ρ_f and K_f respectively are the fluid density and bulk modulus of fluid, and n is the porosity. The above system with appropriate boundary conditions on displacements (or on total tractions) and pore pressures (or their normal gradients) can represent closely most dynamic phenomena encountered—and must be solved with the nonlinear characteristics already discussed (and possibly a permeability which depends on strains).

To do this a finite element discretization is here followed. In this the displacements $\mathbf{u}$ are described in terms of the nodal values $\bar{\mathbf{u}}$ as

$$\mathbf{u} = \mathbf{N}\bar{\mathbf{u}} \qquad (5.25)$$

with a similar discretization for the pressures,

$$= \bar{\mathbf{N}}\bar{\mathbf{h}} \qquad (5.26)$$

where $\bar{p}$ is the vector of nodal pressure values.

In the above $\mathbf{N}$ and $\bar{\mathbf{N}}$ are appropriate shape functions.

Following standard discretization procedures which the reader can find in appropriate texts, e.g. Ref. 34, we arrive simply at a semi-discrete system

$$\int_\Omega \mathbf{B}^T \boldsymbol{\sigma}' \, d\Omega + \mathbf{M}\ddot{\mathbf{u}} - \mathbf{Q}\bar{\mathbf{p}} = \mathbf{f} \tag{5.27}$$

with the stress changes related incrementally to the displacements by the discrete version of equation (5.22b).

$$d\boldsymbol{\sigma}' = \mathbf{D}(\mathbf{B}\, d\bar{\mathbf{u}} - d\boldsymbol{\varepsilon}^0). \tag{5.28}$$

In the above Ω represents the total problem domain resulting from the summation of element contributions. The flow equation (5.24) similarly yields a discrete form

$$\mathbf{H}\bar{\mathbf{p}} + \mathbf{S}\dot{\bar{\mathbf{p}}} + \mathbf{Q}^T\dot{\bar{\mathbf{u}}} - \hat{\mathbf{M}}\ddot{\mathbf{u}} = \bar{\mathbf{f}}. \tag{5.29}$$

In the above equations $\mathbf{B}$ is the standard strain matrix and $\mathbf{M}$ the mass matrix, both defined by appropriate components as

$$\mathbf{B}_i = \mathbf{L}\mathbf{N}_i, \tag{5.30a}$$

$$\mathbf{M}_{ij} = \int_\Omega \mathbf{N}_i^T \rho \mathbf{N}_j \, d\Omega. \tag{5.30b}$$

The other matrices are given by

$$\mathbf{Q}_{ij} = \int_\Omega \mathbf{B}_i^T \mathbf{m}\bar{\mathbf{N}}_j \, d\Omega, \tag{5.31a}$$

$$\mathbf{H}_{ij} = \int_\Omega (\boldsymbol{\nabla}\bar{\mathbf{N}}_i)\mathbf{k}\boldsymbol{\nabla}\bar{\mathbf{N}}_j \, d\Omega, \tag{5.31b}$$

$$\hat{\mathbf{M}}_{ij} = \int_\Omega \bar{\mathbf{N}}_i^T \boldsymbol{\nabla}^T \mathbf{k}\rho_f \mathbf{N}_j \, d\Omega, \tag{5.31c}$$

$$\mathbf{S}_{ij} = \int_\Omega \bar{\mathbf{N}}_i^T n/K_f \bar{\mathbf{N}}_j \, d\Omega. \tag{5.31d}$$

The system naturally has to be supplemented by appropriate elasto-plasticity relations for computing the $\mathbf{D}$ matrix and expressions such as those given by equations (5.9), (5.18)–(5.20) for computing the densification strains $\boldsymbol{\varepsilon}^0$.

Although certain 'symmetries' exist in the equation system, the primary variables $\bar{\mathbf{u}}$ and $\bar{\mathbf{p}}$ have a different structure (and indeed different physical units) and therefore standard time-step algorithms are not easy to apply. It is therefore best to adopt a 'staggered' solution process in which

(1) equation (5.27) is solved for the displacement changes $\Delta\bar{\mathbf{u}}$ using some 'extrapolated' value of $\bar{\mathbf{p}}$;
(2) equation (5.29) is solved for $\bar{\mathbf{p}}$ using the now available values of $\Delta\bar{\mathbf{u}}$.

Such staggered processes have been used successfully for other 'coupled' physical problems and a very full analysis of their characteristics with regard to stability and accuracy has recently been provided by Park and Felippa.[17,18]

Many alternative time-step procedures are obviously available and a number of alternative schemes for second-order equations (such as equation (5.27)) and first-order equations (such as equation (5.29)) are presented in Chapter 21 of Ref. 34 (see also Ref. 37).

For our purpose we find it convenient to use the explicit, central difference scheme for equation (5.27) which we shall write as

$$\mathbf{M}(\bar{\mathbf{u}}_{n+1} - 2\bar{\mathbf{u}}_n + \bar{\mathbf{u}}_{n+1})/\Delta t^2 + \left(\int_\Omega \mathbf{B}^\mathrm{T} \boldsymbol{\sigma} \, \mathrm{d}\Omega \right)_n - (\mathbf{Qp})_n = \mathbf{f}_n, \qquad (5.32)$$

where $n+1$, n, or $n-1$ denote a set of consecutive time points separated by increments Δt of time.

We note that in the above we have simply 'extrapolated' $\bar{\mathbf{p}}$ as the value of this vector at time n.

As initial values of $\bar{\mathbf{u}}$ and $\dot{\bar{\mathbf{u}}}$ as well as $\bar{\mathbf{p}}$ are given, i.e. values of $\bar{\mathbf{u}}_n$, $\bar{\mathbf{u}}_{n-1}$ and $\bar{\mathbf{p}}_n$ are available at the start of computation ($n = 0$), equation (5.32) can be solved for $\bar{\mathbf{u}}_{n+1}$.

This solution is almost trivial if $\mathbf{M}$ is a lumped, diagonal matrix. Various schemes for such diagonalization are discussed in Ref. 34 and throughout the numerical computation used here such schemes were employed allowing the programs to be implemented on small computers with minimal storage requirements.

It is well known that the central difference scheme is *conditionally* stable and requires

$$\Delta t \leqslant \Delta t_{\text{crit}}, \qquad (5.33)$$

where the value of Δt_{crit} is governed by the largest eigenvalue of the system in the linear cases.[34] Thus if equation (5.27) is considered to be linearized (with the elasticity matrix considered constant) and the term $\mathbf{Q\bar{p}}$ is neglected, we find that approximately

$$\begin{aligned} \Delta t_{\text{crit}} &\simeq h/c \\ c &= \sqrt{(K_\mathrm{T}/\rho)}, \end{aligned} \qquad (5.34)$$

in which h is the size of the smallest element and c is the compression-wave velocity determined by the tangent modulus K_T of the material. As in plastic processes a 'softening' of such a modulus tends to occur, it suffices to use

$$K_\mathrm{T} = K_\mathrm{e}, \qquad (5.35)$$

where K_e is the elastic bulk modulus of the (drained) material. As equation

(5.27) is coupled with equation (5.29) through the pore pressure $\bar{\mathbf{p}}$ where the value of $\bar{\mathbf{p}}$ depends on the permeability of the material and the free drainage boundary condition of the system respectively, it is apparent that the Δt_{crit} is also subjected to the influence of these factors. Therefore, the critical time step Δt_{crit} is governed by the stability of the system of coupled equations and has the value ranging from Δt_{crit} (undrained) to Δt_{crit} (drained) where the former is obtained by expression (5.34) using the undrained bulk modulus value.

Once $\bar{\mathbf{u}}_{n+1}$ is determined, equation (5.29) can be used for determination of $\bar{\mathbf{p}}_{n+1}$. Now we can compute, approximately,

$$\ddot{\bar{\mathbf{u}}}_n = (\bar{\mathbf{u}}_{n+1} - 2\bar{\mathbf{u}}_n + \bar{\mathbf{u}}_{n-1})/\Delta t^2,$$
$$\dot{\bar{\mathbf{u}}}_n = (\bar{\mathbf{u}}_{n+1} - \bar{\mathbf{u}}_n)/\Delta t \tag{5.36}$$

and introduce an appropriate time-step scheme for $\bar{\mathbf{p}}$. A wide family of such schemes can be written as

$$\mathbf{H}(\theta \bar{\mathbf{p}}_{n+1} + (1-\theta)\bar{\mathbf{p}}_n) + \mathbf{S}(\bar{\mathbf{p}}_{n+1} - \bar{\mathbf{p}}_n)/\Delta t = \bar{\mathbf{f}}_n - \mathbf{Q}^{\mathrm{T}}\dot{\bar{\mathbf{u}}}_n + \hat{\mathbf{M}}\ddot{\bar{\mathbf{u}}}_n, \tag{5.37}$$

with schemes for $\theta \geqslant \frac{1}{2}$ being unconditionally stable. The particular case of $\theta = 0$ leads, on suitable lumping of the matrix $\mathbf{S}$, to an explicit computation mode which is conditionally stable. As the compressibility matrix $\mathbf{S}$ can, on occasion, become very small in value, the critical time steps for the explicit scheme can become very low, so it is essential to adopt unconditionally stable schemes for this part of the operation, if the same time step as that used for solution of equation (5.32) is to be retained. We have thus adopted here a simple backward difference scheme with $\theta = 1$, which has the advantage of avoiding all oscillation (other schemes with $\theta = \frac{2}{3}$ for instance may prove to be even better and numerical experiments are proceeding).

5.6 SIMPLIFICATION OF UNDRAINED BEHAVIOUR

If undrained behaviour is assumed—and this is a tenable assumption when the permeability, $\mathbf{k}$, is low or if the frequencies to be considered are high—then a simplification occurs. We note from equation (5.24) that we now have simply

$$-n\dot{p}/K_f = \mathbf{m}^{\mathrm{T}}\dot{\boldsymbol{\varepsilon}}$$

or

$$-n\,\mathrm{d}p/K_f = \mathbf{m}^{\mathrm{T}}\,\mathrm{d}\boldsymbol{\varepsilon}. \tag{5.38}$$

Now, the constitutive relation can be written in total stress terms (using equations (5.22(a–b)) as

$$\mathrm{d}\boldsymbol{\sigma} = \bar{\mathbf{D}}\,\mathrm{d}\boldsymbol{\varepsilon} - \mathbf{D}\,\mathrm{d}\boldsymbol{\varepsilon}^0, \tag{5.39}$$

with

$$\bar{\mathbf{D}} = \mathbf{D} + \mathbf{m}\frac{K_f}{n}\mathbf{m}^T. \tag{5.40}$$

The discretized equation of equilibrium is now one involving only $\bar{\mathbf{u}}$ and can be written as

$$\int_\Omega \mathbf{B}^T\boldsymbol{\sigma}\,d\Omega + \mathbf{M}\ddot{\mathbf{u}} = \mathbf{f}, \tag{5.41}$$

with

$$d\boldsymbol{\sigma} = \bar{\mathbf{D}}\mathbf{B}\,d\mathbf{u} - \mathbf{D}\,d\boldsymbol{\varepsilon}^0. \tag{5.42}$$

This, once again, can be solved by time-step processes, and an obvious simplification occurs as $\bar{\mathbf{p}}$ solution is no longer needed. An explicit scheme for such a solution has been considered by Zienkiewicz, Chang, and Hinton[35] and used with considerable success since. However, we must note that now the critical time step is governed by the undrained compression modulus and the corresponding compression wave velocity $\bar{c}$; that is,

$$\Delta t \leqslant \Delta t_{crit} \approx h/\bar{c}, \qquad \bar{c} = \sqrt{\left(\frac{\bar{K}_T}{\rho}\right)}. \tag{5.43}$$

As the bulk modulus of the pore fluid, e.g. water, is of the order of $2500\ \mathrm{MN/m^2}$, an undrained analysis requires a very small value of Δt and is therefore expensive.

Although this critical time step appears to represent a lower bound for the coupled analysis of the previous section we have found that even with very small permeabilities ($k = 10^{-6}\ \mathrm{m/s}$) time steps approximately two times as large as in undrained analysis can be used, while with larger permeabilities a factor of ten is quite common. Thus from the computational viewpoint no economy is gained in the undrained formulation.

5.7 TEST EXAMPLE

As several exact solutions for full Biot's equations as well as for the approximate form given by equations (5.22)–(5.24) have been obtained analytically[38] (see also Chapter 1), it is convenient to test transient analysis computations on such examples.

The problem used concerns a layer of material of depth L placed on an impermeable base with free surface drainage. On the surface of this layer a vertical, total traction q is applied with a period T. Taking a porosity $n = \frac{1}{3}$ and $\rho_f/\rho = \frac{1}{3}$, the solution is found to depend on two non-dimensional parameters

$$\pi_1 = \frac{2}{\pi}\,k\rho\,\frac{T}{(\hat{T})^2},$$

$$\pi_2 = \pi^2\left(\frac{\hat{T}}{T}\right)^2, \qquad \hat{T} = \frac{2L}{\bar{c}}. \tag{5.44}$$

In Figure 5.9(a) and (b), we show the comparision between exact periodic solutions for pressure amplitudes and results of numerical computation which was carried out using 30 two-dimensional bilinear isoparametric quadrilateral elements in the layer and taking the time-step solution to the point where almost steady-state response was achieved. The excellent agreement for the type of approximation which we have here is worth noting.

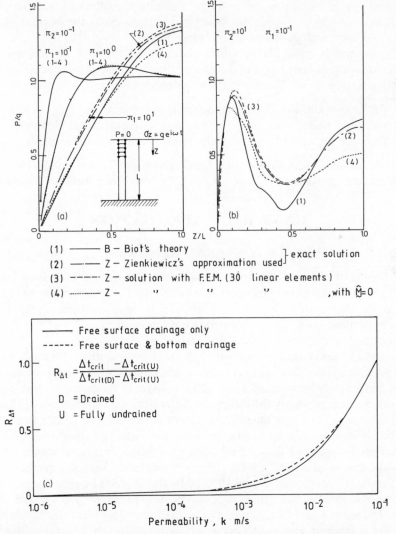

(1) ——— B — Biot's theory ⎤
(2) —·—· Z — Zienkiewicz's approximation used ⎦ exact solution
(3) ----- Z — solution with F.E.M. (30 linear elements)
(4) ········· Z — ,, ,, ,, ,with $\hat{M}=0$

Figure 5.9 A test problem for numerical computation—a soil layer subject to a periodic surface load: (a) $\pi_2 = \pi^2(\hat{T}/T)^2 = 10^{-1}$; (b) $\pi_2 = 10$; (c) critical time step for stability

In the same figure we also show results of a numerical computation carried out omitting the $\hat{\mathbf{M}}$ term in equation (5.29). We note that, at higher values of π_1, the error introduced is quite substantial and we do not recommend that this term should be generally suppressed, although at low permeabilities its importance is small.

In Figure 5.9(c) we show the relationship between the value of Δt_{crit} and the permeability. The increase of Δt_{crit} corresponding to the change of permeability is noticeable. In order to illustrate the effect of free drainage boundary conditions, a previous boundary is imposed at the bottom of the layer and the result is shown in the same figure.

The computer program in which the various features of analysis are incorporated is named LIQ2 and includes simple facilities for amendment of constitutive laws, etc. The basic elements used are

(1)   the four node, isoparametric, quadrilateral element;
(2)   the eight node, serendipity, isoparametric quadrilateral element;
(3)   the nine node, Lagrangian isoparametric quadrilateral element.

Mass lumping procedures described elsewhere[34], are used, and the performance of elements (2) or (3) using 'reduced' integration is optimal.

PART III   APPLICATION

5.8   GENERAL REMARKS

The formulation presented, together with the constitutive laws introduced, allows any problem of earthquake response to be studied quantitatively. Naturally, the results will only be as good as the physical data supplied and the idealization used for the constitutive model—irrespective of the degree of accuracy of the numerical representation.

In the examples that follow we shall deal with sand or silts subject to various earthquake shocks. In all, the simple densification model introduced in Section 5.3 (equations (5.18)–(5.20)) is used in conjunction with ideal, non-associative plasticity (Model B) or the critical state model (Model C). In real sands more complex models are generally needed as a certain degree of dilation will occur during plastic deformation (near the critical state). In a recent study, we find that by introducing a small degree of dilation in a non-associative plasticity model, a higher resistance against liquefaction is predicted. However, this will be affected by the degree of confinement of the system. Generally, the effect of dilation will be more pronounced in a two-dimensional problem than in a one dimensional problem. If the effect of dilation is ignored, the main effects will be more clearly illustrated and the solution will over-estimate the danger of liquefaction and thus be on the safe side.

Again we should remark that the strain hardening critical state model will be less prone to liquefaction (some of the plastic dilation counteracting the densification strains) and therefore in the study of the Lower San Fernando Dam such a model is not used.

5.9 A LAYER OF PERVIOUS SAND SUBJECTED TO A HORIZONTAL EARTHQUAKE SHOCK

A horizontal soil layer of depth 15.2 m is subjected to a base motion of the N–S component of the El Centro Earthquake (May, 1940) with the maximum acceleration scaled to 0.1g. The soil is modelled as a simple elastic-plastic material with zero dilation (Model B) and the values of the densification strain parameters A, B and γ based on Ref. 35 are adopted in this study. The relevant results of the solution are shown in Figures 5.10–13.

The same problem was first studied by Finn *et al.*[4,8] using a one-dimensional program called DESRA and more recently another approximate study was carried out by Martin and Seed[10] using programs MASH[9]

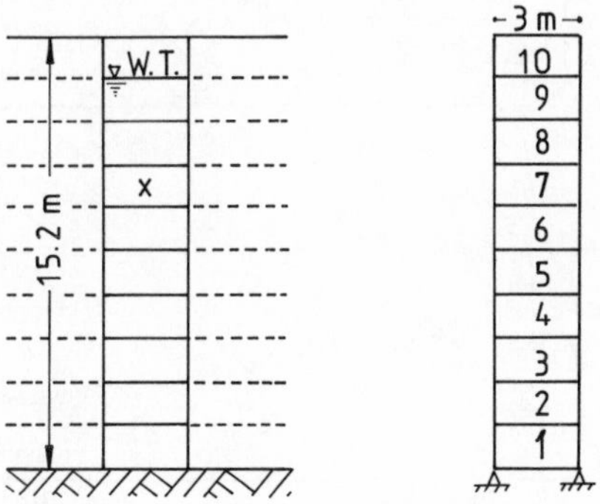

Material properties

Young's mod. , E	= 98 MN/m²	$\varepsilon^\circ = \dfrac{A}{B} \ln(1 + B\,e^{\gamma\theta})$	
Poisson's ratio, ν	= 0.15	A= 0.04	
Density $\rho_{sat.}$	= 1800 kg/m³	B= 55.5	
$\rho_{sub.}$	= 800 kg/m³	γ= 17.2	
ρ_b (above W.T.)	= 1600 kg/m³		

Bulk mod. of pore fluid, K_f = 2250 MN/m³

Cohesion , C' = 1.0 kN/m²

Friction angle, ϕ' = 40.0°

Void ratio, e = 0.6

Figure 5.10 A horizontal sand layer subject to an earthquake motion at base (El Centro, N–S component ×0.1g)—Finite element subdivision and material data used

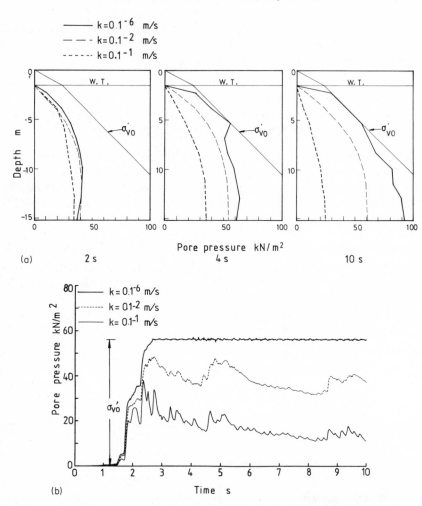

Figure 5.11 A horizontal sand layer subject to an earthquake motion at base (El Centro, N–S component ×0.1g): (a) pore pressure profiles; (b) pore pressure time history for element No. 7 of Figure 5.10

and APOLLO.[27] Unfortunately, there is no field pore pressure data available at present to provide a check on the numerical solution of any analysis of this type. However, the approach reproduces the basic features that one can expect during the generation and dissipation of pore pressure caused by an earthquake. For example, there is a 'cut-off' of surface motion after the liquefaction of any layer occurs. It is of interest to note that similar behaviour is observed in the results of the analysis by Finn, Lee, and Martin.[4]

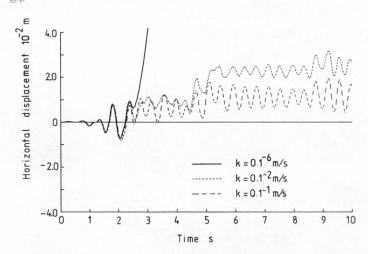

Figure 5.12 A horizontal sand layer subject to an earthquake motion at base (El Centro, N–S Component ×0.1g)—surface displacement)

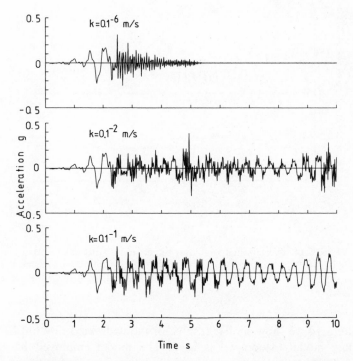

Figure 5.13 A horizontal sand layer subject to an earthquake motion at base (El Centro, N–S Component ×0.1g)—surface acceleration

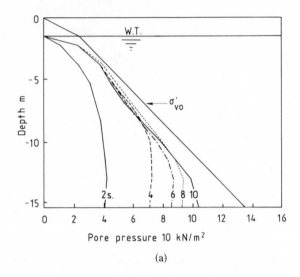

(a)

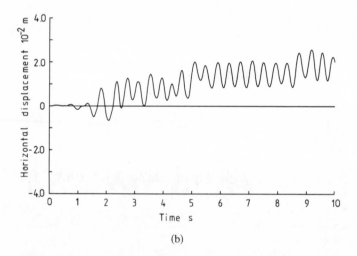

(b)

Figure 5.14 A horizontal sand layer subject to an earthquake motion at base (El Centro, N–S Component ×0.1g)—with a critical state model (Model C) (a) pore pressure profiles; (b) surface displacement history

In Figure 5.14 we show similar results for the undrained case but here the critical state model (Model C) is used. The model employed has a full elliptical yield surface and a normal consolidation condition is assumed. As anticipated earlier, complete liquefaction does not occur in this case.

5.10 A SIMULATION OF THE LOWER SAN FERNANDO DAM BEHAVIOUR

In this fully two-dimensional example all features of the analysis capability are demonstrated and we show that an approximate simulation of the behaviour of this dam together with an estimate of the permanent deformation occurring during the earthquake can be achieved.

The details of the dam and its failure in the San Fernando earthquake are described and analysed in reports prepared by Seed *et al.*[23,24] as well as in Seed's Rankine lecture of 1979.[28] As a result of the earthquake, the dam slid in the upstream shell, by good fortune retaining sufficient freeboard with a low reservoir level to prevent a disastrous spillage. The cross-section through the slide area and its reconstruction reproduced from Ref. 28 is shown in Figure 5.15.

The data for the present analysis were taken from Ref. 23 and re-interpreted in terms of the densification strain. Appendix A lists some of this information which is also summarized in Figure 5.16. This figure also shows the mesh of isoparametric 8-node elements used in the analysis. It should be noted that in the analysis, the zone above the original phreatic surface is assumed dry and the pore pressure there is simply set to zero.

Initial stresses and pore pressures are computed using an elasto-viscoplastic soil mechanics program described in Ref. 33. The analysis is carried out by a 'one-step gravity turn on' approach and the stages of construction of the dam are not taken into account. After determining the

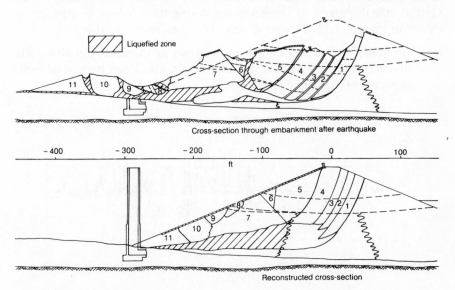

Figure 5.15 The failure of the Lower San Fernando dam (1971) (after Seed[28])

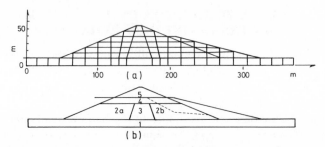

Material zone	Elastic modulus E MN/M²	V	φ' Degree	C' KN/M²	Unit weight T/M³	ρ̄=A/B(1+Be^γθ) A	B	γ
1	Alluvium	2·00	0,4	38	10	2·09			
2a	Hydraulic fill – sand	90	0,41	37	10	2·02	2914	1904	4,6
2b	" " "	1·10	0.41	37	10	2·02-sat 1·71-dry	2914	1904	46
3	Clay core	90	0,41	37	10	2·02			
4	Ground shale-hydraulic fill	90	0,41	37	10	2·02sat 1·71dry			
5	Rolled fill	60	0·3	25	126	2·0			

Figure 5.16 Data assumed for the analysis of the Lower San Fernando dam and the mesh of quadratic finite element

initial stresses, the dam is subjected to a base motion which was deduced from the abutment seismoscope record of the Lower San Fernando Dam by R. F. Scott. The same base motion was used by Seed *et al.*[23,24] and is shown in Fig. 5.17. The analysis is carried out for the first 10 s and the deformed mesh, contours of the build up of excess pore pressure, and contours of the vertical effective stress at various times during the earthquake passage are shown in Figures 5.18–20 respectively. The analysis shows that:

(1) substantial permanent deformation occurs at the upstream slope after 10 s from the start of motion (maximum vertical and horizontal measurements of 7.5 and 13.6 m respectively);

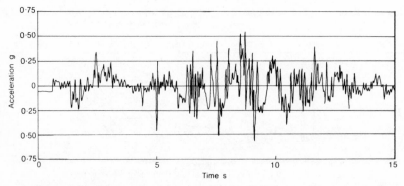

Figure 5.17 Base motion input for analysis of the Lower San Fernando dam

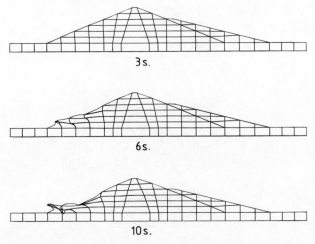

3 s.

6 s.

10 s.

Figure 5.18 The Lower San Fernando dam analysis—deformed
mesh plot (displacements ×2.0)

(2) a smaller amount of excess pore pressure is predicted in the inner
region of the upstream shell than in Seed's analysis.

There are two possible explanations for this behaviour. Firstly, permanent
deformations occur at the upstream toe of the dam and as a result decrease
the magnitude of horizontal vibration of the inner part of the upstream shell,

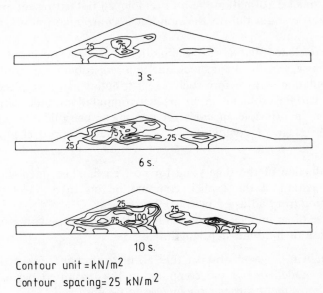

3 s.

6 s.

10 s.

Contour unit = kN/m^2
Contour spacing = 25 kN/m^2

Figure 5.19 The Lower San Fernando Dam analysis—contours
of excess pore pressure build-up

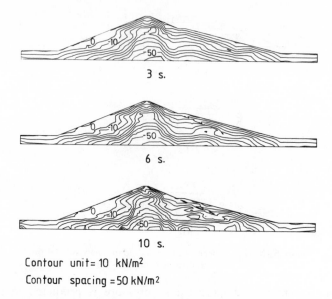

3 s.

6 s.

10 s.

Contour unit= 10 kN/m²
Contour spacing =50 kN/m²

Figure 5.20   The Lower San Fernando dam analysis—contours
of vertical effective stress

and therefore produce less densification strain and hence less pore pressure. Secondly, small deformation theory is employed in the present analysis and the volumetric strain due to the non-linear contribution is not taken into account.*

Seed[28] concluded that most of the motion of the slide occurred some time after the passage of the main shock due to a redistribution of pore pressure. Our computation stops before such a redistribution can take place. Indeed, this poses certain problems in an explicit computation mode, although the formulation in principle includes all non-linear consolidation effects. To include this stage of computation, two alternative approaches are being studied:

(1) A reduction of the time scale for post earthquake phenomena and a continuation of the explicit computation on an artificial time scale resulting from adjusted permeability values;
(2) the use of an implicit, unconditionally stable process with a subsequent increase of the time-step length.

The reader will however observe that all the essential forms of behaviour have been modelled and the computation has permitted permanent and significant movement due to liquefaction to be observed.

---

* Indeed this accounts for the somewhat unrealistic displacement plot near the base. In subsequent work large strain analysis will be introduced.

5.11 CONCLUDING REMARKS

This chapter shows that with the present stage of development of numerical methods and with the current understanding of liquefaction phenomena it is possible to carry out a full non-linear analysis to determine whether any permanent displacement will occur in a soil structure (such as a dam or foundation) after the passage of extreme earthquake motion.

Clearly when improved models of the material become available these can be readily inserted into the analysis—but at the present stage of development it is found that relatively simple elasto-plastic models including a separate densification computation are adequate.

Research continues on development of improved numerical algorithms for combining post-earthquake behaviour studies as well as on improvement of basic models to take account of effects of multi directional shaking such as that studied recently by Seed, Pyke, and Martin.[26]

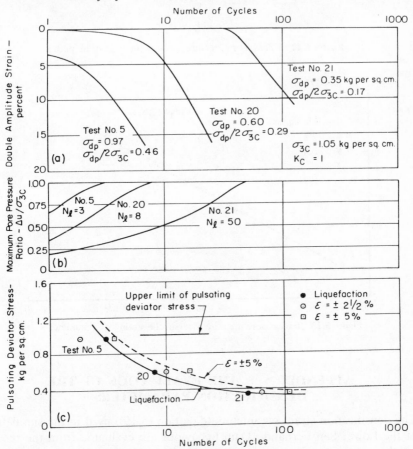

Figure 5.21 Results of cyclic load tests on isotropically consolidated samples of hydraulic sand fill—Upper San Fernando dam (after Seed *et al.*[23])

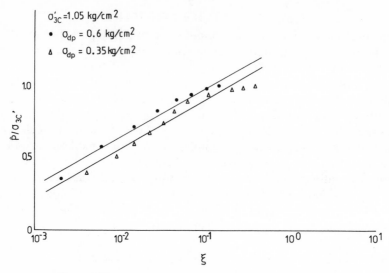

Figure 5.22   Excess pore pressure p versus the total strain path ξ

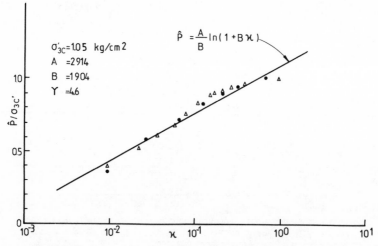

Figure 5.23   Excess pore pressure $\hat{p}$ versus the strain load parameter κ

APPENDIX A   THE EVALUATION OF THE DENSIFICATION PARAMETERS

The densification strain parameters (equation (5.19)) used in the simulation of the Lower San Fernando Dam behaviour are evaluated from the results of the cyclic triaxial test on isotropically consolidated samples of the hydraulic fill of the Upper San Fernando Dam.[23] The reason for adopting the Upper Dam data in this study is that the required information for the Lower Dam is not available from Ref. 23. Tests show the resistance to liquefaction defined as the maximum cyclic deviator stress required to cause a certain amount of strain in a number of cycles. For the Lower Dam this is found to be slightly higher than the value of the Upper Dam. Therefore the use of the Upper Dam data should provide conservative results.

Some results from Ref. 23 are shown in Figure 5.21. The strain amplitude and maximum deviator stress of each load cycle can be used to evaluate the value of $\theta, d\xi$ and ξ by the use of equation (5.17) and (5.18). The relationship between the excess pore pressure $\hat{p}$ and ξ is shown in Figure 5.22. Finally, the values in the $\hat{p},\xi$-plane can be transformed into the $\hat{p},\kappa$-plane by employing equation (5.20) and the result is shown in Figure 5.23. The parameters A, B and γ from Figure 5.23 are used in the present study.

REFERENCES

1. Bazant, Z. P., Ansal, A. M., and Krizek, J. Chapter 15, this book.
2. Carter, J. P., Booker, J. R., and Wroth, C. P. Chapter 9, this book.
3. Dafalias, Y. F., and Hermann, L. R. Chapter 10, this book.
4. Finn, W. D. L., Lee, K. W., and Martin, G. R. An effective stress model for liquefaction, *Proc. Am. Soc. Civ. Eng., J. Geotech. Eng. Div.*, **103,** 517–33, 1977
5. Finn, W. D. L., Martin, G. R., and Lee, M. R. W. *Comparison of Dynamic Analyses for Saturated Sands.* Presented at the 1978 Conference on Earthquake Engineering and Soil Dynamics, Pasadena, California.
6. Ghaboussi, J., and Dikmen, S. U. Liquefaction analysis of horizontally layered sands. *Proc. Amer. Soc. Civ. Engng., J. Geotech. Engng. Div.*, **104** (GT3), 341–56, 1978.
7. Ghaboussi, J., and Momen, H. Chapter 12, this book.
8. Martin, G. R., Finn, W. D. L., and Seed, H. B. Fundamentals of liquefaction under cyclic loading, *Proc. Am. Soc. Civ. Eng., J. Geotech. Eng. Div.*, **101,** 423–38, 1975.
9. Martin, P. P., and Seed, H. B. *MASH—A Computer Program for the Nonlinear Analysis of Vertically Propagating Shear Waves in Horizontally Layered Soil Deposits.* EERC Report No. UCB/EERC-78/23, Univ. of California, Berkeley, California, October 1978.
10. Martin, P. P., and Seed, H. B. Simplified procedure for effective stress analysis of ground response. *Proc. Am. Soc. Civ. Eng., J. Geotech. Engng. Div.*, **105** (GT6) 739–58, 1979.
11. Mroz, Z., Norris, V. A., and Zienkiewicz, O. C. An anisotropic hardening model for soils and its application in cyclic loading. *Int. J. Num. and Analytical Meth. Geomech.* **2,** 203–21, 1978.
12. Mroz, Z., Norris, V. A., and Zienkiewicz, O. C. Application of an anisotropic hardening model in the analysis of elastic-plastic deformation of soils. *Géotechnique,* **29,** 1–34, 1979.
13. Mroz, Z., Norris, V. A., and Zienkiewicz, O. C. Elastic plastic and viscoplastic constitutive models for soils with application to cyclic loading (to be published).

14. Nasser, S. Nemat, and Shokooh, A. A unified approach to densification and liquefaction of cohesionless sand in cyclic shearing. *Canadian Geotechnical Journal*, **16**, 659–78, November 1979.
15. Nasser, S. Nemat. Private communication, 1980.
16. Nova, R. Chapter 13, this book.
17. Park, K. C. Partitioned transient analysis procedures for coupled field problems—stability analysis *J. Appl. Mech.*, **47** 370–376, 1980.
18. Park, K. C., and Felippa, C. A. Partitioned transient analysis procedures for coupled field problems—accuracy analysis *J. Appl. Mech.*, **47** 919–926, 1980.
19. Pender, M. J. Cyclic mobility—a critical state model. *Proc. Int. Symposium on Soils under Cyclic and Transient Loading*, Swansea, 325–35, 7–11 January 1980.
20. Roscoe, K. H., Schofield, A. N., and Wroth, C. P. On the yielding of soils. *Géotechnique*, **8**, 22–53, 1958.
21. Roscoe, K. H., and Burland, J. B. On the generalised stress/strain behaviour of 'wet' clay. in Hayman, J., and Lockead, F. A. (eds.), *Engineering Plasticity*, Cambridge University Press, 535–609, 1968.
22. Schofield, A. N., and Wroth, C. P. *Critical State Soil Mechanics*, McGraw-Hill, 1968.
23. Seed, H. B., Lee, K. L., Idriss, I. M., and Makdisi, F. I. *Analysis of the Slides in the San Fernando Dam during the Earthquake of February 9, 1971*. EERC Report No. EERC73-2, University of California, Berkeley, California, 1973.
24. Seed, H. B., Idriss, I. M., Lee, K. L., and Makdisi, F. I. *Dynamic Analysis of the Slide in the Lower San Fernando Dam during the Earthquake of February 9, 1971*, Proc. Am. Soc. Civ. Eng., J. Geotech. Engng. Div., **101** (GT9) 889–911, 1975.
25. Seed, H. B., Martin, P. P., and Lysmer, J. Pore-water pressure changes during soil liquefaction Proc. Am. Soc. Civ. Eng., J. Geotech. Engng. Div., **192** (GT4) 323–46, 1976.
26. Seed, H. B., Pyke, R. A., and Martin, G. R. Effect of multi-directional shaking on pore pressure development in sands. Proc. Am. Soc. Civ. Eng. J. Geotech. Engng. Div., **104**, 27–44, 1978.
27. Seed, H. B., and Martin, P. P. *APOLLO—A Computer Program for the Analysis of Pressure Generation and Dissipation in Horizontal Sand Layers during Cyclic or Earthquake Loading*. EERC Report No. UCB/EERC-78/21, Univ. of California, Berkeley, California, October 1978.
28. Seed, H. B. Considerations in the earthquake-resistant design of earth and rockfill dams. 19th Rankine Lecture of the British Geotechnical Society, *Géotechnique*, **29**, 215–63, 1979.
29. Silver, M. L., and Seed, H. B. Volume changes in sands during cyclic loading. Proc. Am. Soc. Civ. Eng., J. Soil Mech. Found. Div., **97**, 1171–82, 1971.
30. Valanis, K. C., and Read, H. E. *Recent Development and Application of the Endochronic Theory to the Behaviour of Soils*. Presented in the Int. Symposium on Soils under Cyclic and Transient loading, Swansea, 7–11 January 1980.
31. Zienkiewicz, O. C., Humpheson, C., and Lewis, R. W. Associated and non-associated viscoplasticity and plasticity in soil mechanics. *Géotechnique*, **25**, 671–89, 1975.
32. Zienkiewicz, O. C., Humpheson, C., and Lewis, R. W. A unified approach to soil mechanics including plasticity and viscoplasticity, in Gudehus, G. (ed.), *Finite Elements in Geomechanics*, Wiley, 1977.
33. Zienkiewicz, O. C., and Humpheson, C. Viscoplasticity—A general model for description of soil behaviour. in 'Finite Elements in Geomechanics', Desai, C. S., and Christian, C., (eds) McGraw-Hill, 1977.

34. Zienkiewicz, O. C. *The Finite Element Method*, McGraw-Hill, London, 1977.
35. Zienkiewicz, O. C., Chang, C. T., and Hinton, E. Nonlinear seismic response and liquefaction. *Int. J. Num. and Analytical Meth. Geomechanics*, **2**(4), 381–404, 1978.
36. Zienkiewicz, O. C. Constitutive laws and numerical analysis for soil foundation under static, transient or cyclic loading. *Proc. Second International Conference on Behaviour of Offshore Structures* (BOSS), London, 1979.
37. Zienkiewicz, O. C. Finite elements in the time domain. State of art survey of finite element methods, *Proc. Am. Soc. Mech. Eng.*, to be published 1981.
38. Zienkiewicz, O. C., Chang, C. T., and Bettess, P. Drained, undrained, consolidating and dynamic behaviour assumptions in soils. Limits of validity. *Géotechnique*, **30**, 385–95, 1980.

Soil Mechanics—Transient and Cyclic Loads
Edited by G. N. Pande and O. C. Zienkiewicz
© 1982 John Wiley & Sons Ltd

Chapter 6

Dynamic Response Analyses of Saturated Sands

W. D. L. Finn

6.1 INTRODUCTION

A number of different methods are available for the analysis of the dynamic response of saturated cohesionless soils to earthquake loading. The methods differ in the simplifying assumptions that are made, in the representation of the stress–strain relations of soils, in the manner in which the development of pore water pressure is taken into account and in the methods used to integrate the equations of motion. They may be classified into two main categories; total stress methods and effective stress methods. In total stress methods no explicit account is taken of the effect of changes in pore water pressures. The first total stress analysis of saturated sands was reported by Seed and Idriss in 1967.[29]

Until recently, all dynamic analyses were carried out in terms of total stresses because of the lack of a model for predicting the pore water pressures developed by cyclic or seismic loading. The first model for predicting pore water pressures generated by cyclic loading was proposed in 1974 and 1975 by Martin, Finn, and Seed.[20] This model was coupled to a procedure for dynamic analysis by Finn, Lee, and Martin in 1976, 1977,[4] giving the first method for dynamic effective stress analysis. Many other methods for both total and effective stress analysis have been developed in the last few years. Most notable because of their potential for two- and three-dimensional analyses are those proposed by Prevost,[26,27] Zienkiewicz, Chang, and Hinton,[36] and Mroz, Norris, and Zienkiewicz.[23,24]

The development of existing methods of dynamic response analysis has been a gradual evolutionary process stimulated by the changing needs of practice and the increasing knowledge about the fundamental behaviour of soils under cyclic loading derived from field observation and laboratory testing. In this chapter, the evolution of the methods used in engineering practice today will be traced and their ability to meet the demands of practice will be assessed. Some of the more powerful of the newer methods proposed to overcome the limitations of current procedures will be discussed

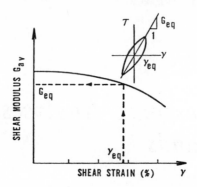

Figure 6.1 Secant modulus vs. shear
strains

also in the context of a comparative and critical assessment of the current
state-of-the-art of dynamic response analysis.

6.2 TOTAL STRESS EQUIVALENT LINEAR
METHODS OF ANALYSIS

Dynamic response analysis as practised today had its origins in the pioneering attempts of Seed and his co-workers at the University of California at
Berkeley to explain in a quantitative way the extensive liquefaction of
saturated sands that occured in 1964 during the earthquakes in Alaska and
Niigata, Japan. The liquefaction failures occurred primarily on level ground
and in formulating a procedure for analysis, Seed and Idriss[29] made two
basic assumptions:

(1) seismic excitation is primarily due to shear waves propagating vertically
     and
(2) level ground conditions may be approximated by horizontal layers with
     uniform properties.

Under these conditions the ground deforms in shear only and may be
analysed by treating a vertical column of soil as a shear beam.

Seed and Idriss[29] included the non-linear hysteretic stress–strain properties of the sand by using an equivalent linear elastic method of analysis. The
method was originally based on a lumped-mass mechanical model of the
sand deposit resting on a rigid base to which the seismic motions were
applied. Later the method was generalized to a wave propagation model
with an energy transmitting boundary. The seismic excitation could be
applied at any level in the new model.

The generalized method was incorporated in the computer program,

SHAKE.[28] This program is widely used to predict the distribution of ground motions due to seismic excitation for liquefaction studies and to provide input motions for soil–structure interaction analyses. The fundamental assumption of the equivalent linear method of analysis on which SHAKE is based is that non-linear response can be approximated satisfactorily by a damped linear elastic model if the properties of that model are chosen appropriately.

The stress–strain properties of the soil are defined by strain-dependent shear moduli and equivalent viscous damping factors. An equivalent modulus and damping ratio at any strain level are determined from the slope of the major axis of the hysteresis loop corresponding to that strain (Figure 6.1(a)) and the area of the loop respectively.[29,30] Since the vertical distribution of shear strain is unknown prior to analysis, initial values of moduli and damping are selected corresponding to small strain values or to strain levels judged appropriate for the anticipated earthquake loading, and an elastic analysis is carried out for the entire duration of the earthquake. The average strain (usually assumed to be about 65% of the maximum value) is computed at each level, and moduli and damping ratios are selected compatible with these average strains and used for the next iteration. The procedure is repeated until no significant changes in moduli or damping are necessary. The response determined during this last iteration is considered to be a reasonable approximation to the non-linear response. The integration of the equations of motion is based on classical wave-propagation theory using transfer functions.

For the case of saturated cohesionless soils, SHAKE is used in conjunction with laboratory data from undrained cyclic loading tests to assess liquefaction potential at any depth in a deposit. The time history of shear stresses at a given depth is converted to an equivalent number, N_{eq}, of uniform shear stress cycles τ_{av},[31] and τ_{av} is compared with the uniform cyclic shear stress, τ, to cause liquefaction of laboratory samples in N_{eq} cycles under the same initial effective stresses as in the field.

The decision in 1967 to model non-linear soil by an equivalent linear elastic model fixed the direction of development of dynamic response analysis for about the next 10 years. The success of the one-dimensional model in explaining the response of level ground, including liquefaction, during earthquakes led to generalizations of the basic model to 2 and 3 dimensions in the computer programs QUAD-4,[13] LUSH[18] and FLUSH.[19]

All of these programs are formulated in terms of total stresses and treat non-linear behaviour by iterative linear elastic procedures. They are the most frequently used methods for analysis of the seismic response of slopes and dams and the analysis of soil–structure interaction problems such as the response of deeply embedded nuclear reactor structures to earthquake loading.

6.3 ONE-DIMENSIONAL NON-LINEAR ANALYSIS

CHARSOIL

Streeter, Wylie, and Richart[34] produced the first true non-linear analysis of the seismic response of soils in 1974. They incorporated a Ramberg–Osgood representation of the stress–strain behaviour of soil into the equations of motion and solved the equations by the method of characteristics. The method of analysis is incorporated into the computer program, CHARSOIL.

DESRA

In 1976, 1977, Finn, Lee, and Martin[4] developed a non-linear method of analysis in which the equations of motion were integrated directly using the Newmark algorithm.[25] The stress–strain behaviour of sand was represented by the hyperbolic initial loading curve (skeleton curve) shown in Fig. 6.2(a). Stress–strain curves during loading and unloading (Fig. 6.2(b)) were defined by the Masing criterion;[22] i.e., if the loading stress–strain curve is given by

$$\tau = f(\gamma), \tag{6.1}$$

then the unloading–reloading curve is given by

$$\frac{\tau - \tau_r}{2} = f\left(\frac{\gamma - \gamma_r}{2}\right), \tag{6.2}$$

in which τ, γ = current shear stress and shear strain respectively and τ_r, γ_r defines the reversal point.

The pore water pressure generation model developed by Martin, Finn, and Seed[20] was coupled with the non-linear equations of motion to provide,

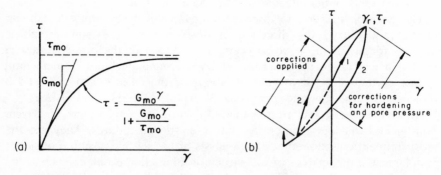

Figure 6.2   (a) Initial loading curve; (b) Masing stress–strain curves for unloading and reloading

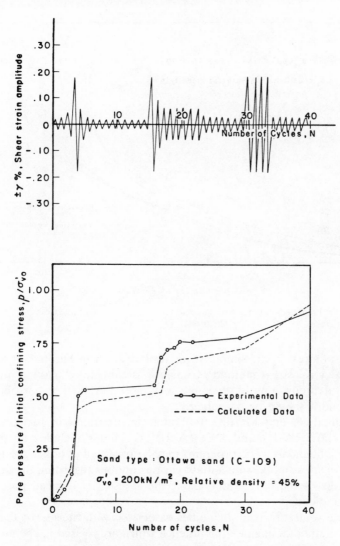

Figure 6.3 Predicted and measured pore water pressures under irregular cyclic loading

for the first time, the option of performing dynamic effective stress analyses if desired. An extensive verification of this method of analysis for saturated sands was conducted by Finn and Bhatia[9] in 1980 under cyclic simple shear conditions using a variety of cyclic loading patterns; constant stress, constant strain and irregular loading. A comparison of experimental data and response data calculated using the effective stress method of analysis for an irregular strain-history pattern is given in Fig. 6.3. A similar comparison for

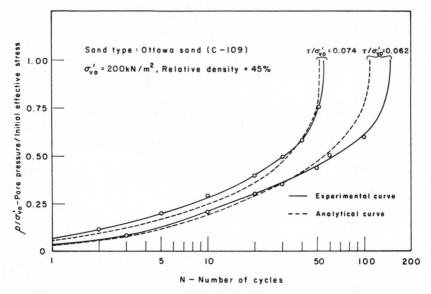

Figure 6.4 Predicted and measured pore water pressures in constant stress cyclic simple shear tests, $D_r = 45\%$

two constant stress cyclic simple shear tests is given in Fig. 6.4. It appears that the effective stress method of analysis is capable of making quantitatively useful predictions of dynamic response. For further corroboration Refs. 20 and 4 should be consulted.

The Finn, Lee, and Martin[4] method is incorporated in two computer programs, DESRA-1[16] and DESRA-2;[17] the latter includes an energy-transmitting boundary. The programs may be operated in either the total or the effective stress mode and contain various options regarding the diffusion and drainage of the pore water under the seismically induced pore water pressure gradients.

There are some practical difficulties associated with the use of the effective stress method in engineering practice which are related to the measurement of the parameters of the pore water pressure model. A brief description of the pore water pressure model is presented in order to clarify these difficulties and their resolution.

Pore pressure model[20]

The pore water pressures that develop in saturated undrained sands during seismic loading are caused by plastic volumetric strains generated by slips at grain contacts.[20] The plastic volumetric strains ε_v are prevented from actually occurring due to the presence of nearly incompressible water.

Thus, load is transferred to the water creating an increase in pore water pressure. This pressure reduces the effective stress regime in the sand and allows the sand skeleton to rebound elastically. The pore pressure increase is big enough to allow an elastic rebound equal to the total plastic volumetric strains if we assume that the bulk modulus of the water is an order of magnitude stiffer than that of the soil skeleton. Otherwise, it is sufficient to cause a rebound equal to the difference between the compressive volumetric strains in the pore fluid (water or water and gas) and the plastic volumetric strains.[20,21] This pore pressure model is described by the following incremental equations.

$$\Delta \varepsilon_v = C_1(\gamma - C_2 \varepsilon_v) + C_3 \varepsilon_v^2/(\gamma + C_4 \varepsilon_v), \tag{6.3}$$

in which $\Delta \varepsilon_v$ = increment in volumetric strain, γ = current shear strain amplitude, and C_1, C_2, C_3, and C_4 are constants for a given sand at a given density. $\Delta \varepsilon_v$, ε_v and γ are expressed in percentages. For undrained behaviour,

$$\Delta p = \Delta \varepsilon_v \bigg/ \left(\frac{1}{\bar{E}_r} + \frac{n_e}{K_w} \right), \tag{6.4}$$

in which Δp = increment in pore water pressure, $\Delta \varepsilon_v$ = increment in volumetric strain, $\bar{E}_r$ = one-dimensional rebound modulus of sand skeleton, n_e = porosity and K_w = bulk modulus of water.

$$\bar{E}_r = (\sigma_v')^{1-m}/mK_2(\sigma_{v0}')^{n-m}, \tag{6.5}$$

in which σ_{v0}', σ_v' = initial value and current value of effective stress, respectively, and n, m and K_2 are constants for a sand at a given density.

It will be noted that the pore water pressure model requires the determination of seven constants. The C-constants are measured in constant strain cyclic simple shear tests and most commercial laboratories still do not have the capability to conduct these tests. But the greatest difficulties are associated with the measurement of $\bar{E}_r$. Finn and Bhatia[9] have shown recently that the magnitude of the rebound modulus $\bar{E}_r$ is less under cyclic loading conditions than for unloading under static conditions. The measurement of $\bar{E}_r$ under cyclic loading conditions requires sophisticated equipment and is time-consuming.

In the application of the method to practical problems, the constants are always finally adjusted so that the model predicts the liquefaction resistance curve selected as representative of field conditions. This representative field curve is either determined on undisturbed samples or by adjusting laboratory curves on reconstituted samples as described by Seed.[32] This adjustment scales the statically determined rebound modulus towards the dynamic. Thus, except for very special problems, it is not necessary to measure the rebound modulus dynamically.

An alternative approach is to develop a direct link between pore water pressure and the dynamic response parameters of the sand–water system to obviate the need for determining the basic constants defining the volumetric strain and rebound characteristics of the sand skeleton. Endochronic theory provides the necessary link and will be discussed in a later section.

6.4 COMPARATIVE ANALYSES

Finn, Martin, and Lee,[5] in 1978 compared the response predictions for a given site using SHAKE, DESRA and CHARSOIL to compute the responses. The study illustrated that significant differences can occur between the results of total and effective stress analyses and also between the results of equivalent linear and non-linear analyses. This comparative study is important in elucidating the implications for engineering practice of the different assumptions underlying the methods of analyses.

SHAKE, CHARSOIL and DESRA were all used to estimate the dynamic response and liquefaction potential of a given idealized site profile under specified conditions. The soil profile used for the response analyses is shown in Figure 6.5. The soil is a uniform deposit of saturated sand with a relative density $D_r = 45\%$. The water table is 5 ft (1.5 m) below the surface. The distribution of the maximum shear modulus, corresponding to very small

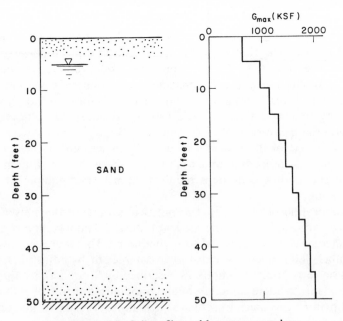

Figure 6.5 Soil profile used for response analyses

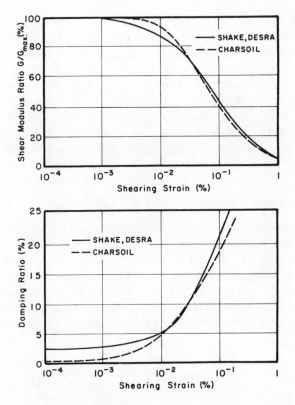

Figure 6.6 Modulus/damping relationships used for response analyses

shear strain, is also shown. The first 10 s of the N–S acceleration component of the 1940 El Centro earthquake, scaled to 0.1g, was used as input motion at the base of the deposit.

Each of the three methods uses a different representation of soil properties. If comparisons of dynamic responses by three methods are to be meaningful, the respective response analyses must be based on similar stiffness and damping characteristics. Therefore, values of the parameters in the stress–strain relations for CHARSOIL and DESRA were chosen so that, as closely as possible, they yielded similar curves of secant shear moduli and equivalent viscous damping ratios as functions of shear strain. The degree of similarity of moduli and damping curves achieved in the present study may be seen in Figure 6.6. It is clear that the curves of average shear modulus as functions of strain are very closely matched, those for DESRA and SHAKE having been chosen to be identical. It is more difficult to match the damping.

In DESRA, in addition to the inherent hysteretic damping, 2% viscous damping was included to make an allowance for any viscous damping due to the presence of water. The resulting total equivalent viscous damping curve was used in SHAKE. The 2% viscous damping could not be included in the current version of CHARSOIL. Figure 6.6 shows a very good match in damping for strains greater than $5 \times 10^{-3}\%$. Below that, SHAKE and DESRA are slightly over-damped with respect to CHARSOIL. Since liquefaction potential with the possibility of large strains is a matter of interest at this site, it is more important to have the better match in damping characteristics at the higher strains.

Acceleration and stress response

Using total stress analyses, the surface accelerations were determined by each method and the damped pseudo-acceleration response spectra are given in Figure 6.7. A stronger resonant response may be noted in the SHAKE spectrum which shows about a 50% increase in the maximum amplification of the pseudo-acceleration above the values computed by CHARSOIL and DESRA. This probably occurs because there is close coincidence between the predominant period of the El Centro motions and

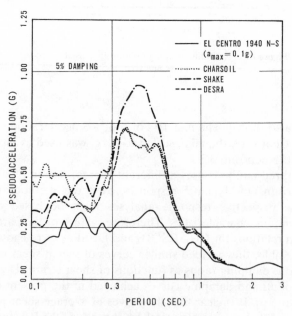

Figure 6.7 Acceleration response spectra—total stress analyses

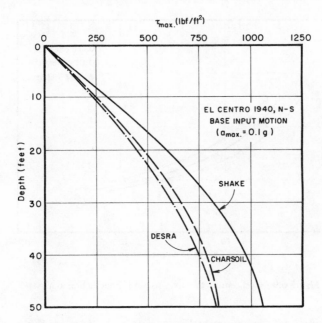

Figure 6.8 Maximum shear stress distributions—total stress analyses

the fundamental period of the site as determined from the strain-compatible moduli resulting from the elastic SHAKE analysis. The CHARSOIL and DESRA spectra are almost identical except for the low-period, high frequency range, in which CHARSOIL shows a high amplification. These high-frequency components are induced by the numerical procedures of integration, and when rounding the sharp corners of the Ramberg–Osgood curve. The components may be removed by digital filtering or the inclusion of a small amount of viscous damping.

The distribution of maximum shear stress with depth, a very important parameter in the estimation of liquefaction potential, is shown in Figure 6.8. Again, the results from CHARSOIL and DESRA are very similar. SHAKE predicts much larger shear stresses, probably due to the resonance effect.

Estimation of liquefaction potential

SHAKE uses a total stress analysis in conjunction with laboratory test data to assess liquefaction potential. To allow a comparison on common terms, therefore, both CHARSOIL and DESRA were also used in the total stress mode in conjunction with data on liquefaction resistance.

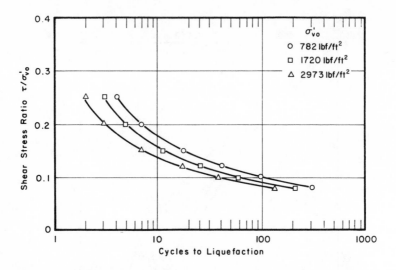

Figure 6.9 Cyclic strength curves used for liquefaction analyses

The resistance data is shown in Figure 6.9 in the form of the cyclic shear stress ratio, τ/σ'_{v0}, to cause liquefaction in a given number of cycles; τ is the uniform cyclic shear stress and σ'_{v0} is the vertical effective stress. The shear stress ratio is slightly dependent on the level of effective stress and three curves are used to cover the range of vertical effective stresses of interest.

The liquefaction potential was explored using the first 10 s of the N–S acceleration component of the 1940 El Centro earthquake scaled to maximum accelerations ranging from 0.05 g to 0.15 g. The shear stress histories at various depths were converted to an equivalent number of cycles, N_{eq}, of a uniform shear stress of magnitude $0.65\tau_{max}$.[15,31] The uniform cyclic shear stress to cause liquefaction in N_{eq} cycles is obtained from the appropriate liquefaction resistance curve in Figure 6.9. A comparison of this shear stress with the equivalent uniform cyclic shear stress, $0.65\tau_{max}$, is a measure of the liquefaction potential. A factor of safety against liquefaction FSL is now defined as:

$$\text{FSL} = \frac{\text{shear strength mobilized for } N_{eq} \text{ (Figure 6.9)}}{0.65 \times \text{maximum cyclic shear stress (Figure 6.8)}}. \tag{6.6}$$

The variation in FSL with the magnitude of base acceleration as computed by the various methods is shown in Figure 6.10. The results for CHARSOIL and DESRA are very similar, as might be expected since they give approximately the same values for maximum shear stress. Since SHAKE yielded higher values for the maximum dynamic shear stress it is not surprising that it indicates the lowest factors of safety against liquefaction.

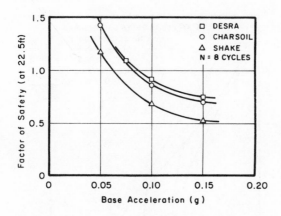

Figure 6.10 Factors of safety against liquefaction—
total stress analyses

Effective stress analyses were carried out using DESRA under three different drainage conditions:

(1) no internal redistribution of pore water pressure,
(2) internal redistribution but no dissipation across external drainage boundaries, and
(3) with dissipation.

The analyses were conducted assuming a uniform permeability $k = 0.003$ ft/s (0.1 m/s). This permeability is representative of medium sands. The vertical distributions of pore water pressures at a particular time during shaking (0.1g maximum acceleration) are shown in Figure 6.11 for each of the specified drainage conditions. Initial liquefaction is indicated when the curve of pore water pressure touches the line showing the distribution of initial vertical effective stress. The data show clearly the influence of drainage conditions on the development of pore water pressure. Near the surface, where the effective stresses are low, redistribution from deeper levels increases the pore water pressures and hence the liquefaction potential when no external drainage occurs. On the other hand, the diffusion of pore water pressure towards the surface tends to reduce the pore water pressure at depth. The marked reduction in pore water pressure due to redistribution is clearly evident at a depth of 32.5 ft (9.9 m). If redistribution were ignored, one would conclude that liquefaction was imminent at that depth. The influence of the drainage conditions, both internal and external, on development of pore water pressure with time of shaking at the depth of 32.5 ft (9.9 m) is evident in Figure 6.12. The influence of redistribution and drainage conditions on the development of pore water depends on the permeability. At extreme values of permeability representative of gravel,

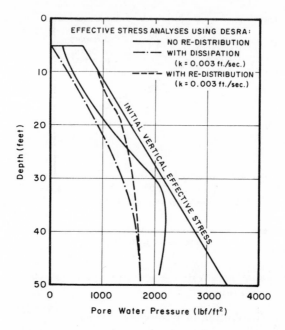

Figure 6.11   Distribution of pore water pressure at 7.0 seconds

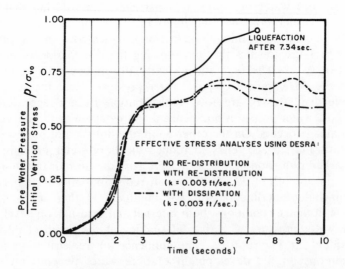

Figure 6.12   Pore water pressure response at 32.5 ft

very little pore water pressure will develop if the gravel is free-draining. However, if drainage is prevented, rapid redistribution of pore water pressure to areas of low effective stress near the surface can lead to liquefaction of the gravel deposits. The latter effect was noted in the Alaska earthquake of 1964 by Ferrians[2] in gravel deposits with a frozen surface layer, and the phenomenon has been analysed by Finn, Yong, and Lee.[6] Clearly, estimates of the liquefaction potential of saturated cohesionless deposits of relatively high permeability should take redistribution and drainage into account.

Effective stress estimates of liquefaction potential will now be compared with the previous estimates based on total stress analyses using the same input motions. The comparison is made using a similar concept of factor of safety as in the case of total stress analyses.

The factors of safety against liquefaction at a depth of 32.5 ft (9.9 m) computed by both total and effective stress methods for various levels of maximum acceleration are shown in Figure 6.13. This depth corresponds approximately to the zone of minimum factor of safety for the three cases considered in Figure 6.13. Factors of safety against liquefaction computed by DESRA operating in the effective stress mode were considerably higher than the factors obtained from total stress analyses, using CHARSOIL or DESRA in the total stress mode or using SHAKE. The limitations of SHAKE, when used in cases in which considerable pore water pressures develop and modify the dynamic response, are recognized by its authors.

The effect of drainage conditions is evident in the pseudo-acceleration spectra shown in Figure 6.14. The spectrum for total stress analysis indicates a maximum at a period determined only by the final strain-compatible shear moduli. The effect of increasing pore water pressure on the moduli is not

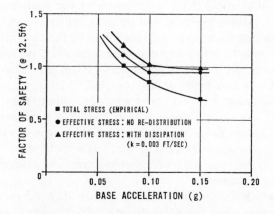

Figure 6.13 Factors of safety against liquefaction—comparison between total stress and effective stress analyses (DESRA only)

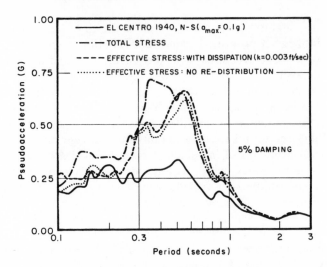

Figure 6.14 Acceleration response spectra—effective stress analyses vs. total stress using DESRA

included. The pore water pressure leads to a further softening or reduction in shear moduli and hence an increase in the fundamental period. This shift in period is indicated by the spectra determined by effective stress analyses. SHAKE will generally underestimate the factor of safety against liquefaction compared to DESRA and CHARSOIL because it tends to yield a higher level of dynamic response.

In most situations total and effective stress methods may be expected to yield similar results provided the magnitudes of the pore water pressures generated during shaking are such that they do not have a significant effect on the computed dynamic response.[3] This would be the case when the factor of safety against liquefaction computed on a total stress basis would be 1.5 or higher. However, an important exception is when saturated cohesionless material of relatively high permeability is sealed by an impermeable surface layer. In this case, redistribution of pore water pressure to the regions of low effective stresses can sharply increase the liquefaction potential. For this situation it would seem necessary to use an effective stress analysis. Assessments of the liquefaction potential by total stress methods are likely to be unconservative, in this case, unless computed results are modified considerably by professional judgment.

For cases in which the factors of safety are less than 1.5 on a total stress basis, it would also seem advisable to use effective stress methods. The inclusion of pore water pressure redistribution and drainage in these cases may move the computed factor of safety into a more acceptable range.

6.5 RESPONSE ANALYSES IN ENGINEERING PRACTICE

Engineering practice is dominated by total stress equivalent linear analyses. In the previous section it was noted that equivalent linear methods may overestimate the seismic response due to pseudo-resonance at periods corresponding to the strain-compatible stiffnesses used in the final elastic iteration analysis. Since the methods are elastic they cannot predict the permanent deformations that may occur. Furthermore, since they are formulated in terms of total stresses, they cannot predict the development of pore water pressure during seismic excitation.

In order to estimate liquefaction potential, the level of pore water pressure or the permanent deformations, the analyses must be used in conjunction with a bank of cyclic loading data on the soils in question. Since the data bank is based on constant stress cyclic loading tests, the stress-history outputs must be converted to an equivalent number of uniform stress cycles. The effects of these equivalent stresses are obtained from the data bank. This procedure decouples the stress analysis and the pore water pressure development and, as pointed out earlier, can lead to an overestimation of the pore water pressures. It also decouples the permanent deformations obtained from the data bank from the dynamic analysis and gives a distribution of non-compatible strains and displacements in the region of analysis. *Ad hoc* procedures are sometimes used to generate a compatible set of deformations from those obtained from the data bank[33] but, generally, judgment is applied to the raw non-compatible strains to assess their effects on a structure.[14]

The equivalent uniform stress approach to tapping a data bank has some limitations:

(1) the equivalent cycle concept has limited validity, and
(2) it is computationally inefficient to have continually to convert the computed irregular stress histories to equivalent uniform stress cycles in order to tap a data bank.

The weakness of the equivalent cycle approach is that equivalence can be defined only with respect to one point on the pore water pressure curve. This point is usually selected as initial liquefaction. The development of pore water pressure at levels below liquefaction may not be predicted with sufficient accuracy.

The development of pore water pressures under both a given cyclic loading and its mirror image is illustrated in Figure 6.15. The procedure for determining equivalent uniform cycles predicts that these loadings would have the same number of equivalent uniform cycles, yet it is clear that the pattern of pore water pressure is different in each.

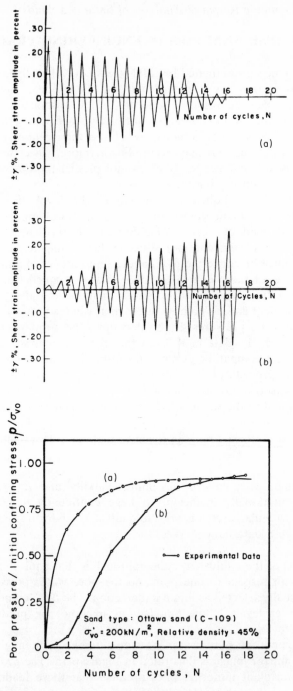

Figure 6.15 Measured pore water pressures under irregular cyclic loading

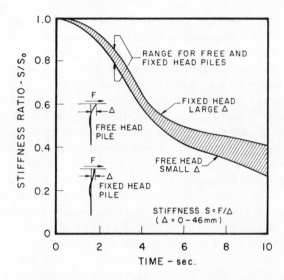

Figure 6.16 Lateral stiffness degradation characteristics with pore water pressure increases

Some of the limitations outlined above have lead to an interest in non-linear total stress and effective stress methods such as CHARSOIL and DESRA, and programs such as these are gradually coming into use. DESRA has been used in the evaluation of the liquefaction potential of offshore sites and has also been used in combination with soil–pile interaction programs to predict the performance of piled foundations for offshore platforms. Typical results for a single pile are shown in Figure 6.16, which shows how the initial global lateral stiffness, S_0, varies with the development of pore water pressure during an earthquake.[7,8]

However, as pointed out earlier, the determination of the material parameters for non-linear effective stress analysis is not an easy task, and the use of the DESRA program would be greatly simplified if this problem could be by-passed. This can be done successfully by the the use of certain endochronic variables.

6.6 ENDOCHRONIC THEORY OF SAND LIQUEFACTION

Endochronic theory is particularly useful in describing the volume changes and pore water pressures generated in saturated sands by earthquake motions. Endochronic theory was developed by Valanis[35] to describe non-linear material response. The non-linearity is described by means of parameters which describe the sequence of events leading to successive states of the material. Although not time variables, the parameters function as a kind of intrinsic time, hence the term 'endochronic'. The endochronic parameters are mathematical transformations of real physical variables of the problem but may themselves have no direct physical significance. For the liquefaction problem we will see that the relevant parameter is a transformation of deformation increments.

Bazant and Krizek[1] developed an endochronic constitutive law for the liquefaction of sand. They used an endochronic description of the stress–strain relations and represented the densification or volumetric strains caused by cyclic shearing in terms of endochronic variables. Zienkiewicz, Chang, and Hinton[36] used similar history-dependent damage parameter variables to describe the volumetric strains caused by cyclic loading. In both cases, the volumetric strains were then incorporated into a pore water pressure model which gave the increments of pore water pressure in terms of the volumetric strains and the bulk moduli of the soil skeleton and water. Both approaches were based on very limited test data. Finn and Bhatia, in 1980, applied endochronic methods to describe directly the pore water pressures generated by cyclic loading in two North American and four Japanese sands.[11]

It is clear from the experimental data in Figure 6.17 that the pore water pressures are a function of the strain amplitude γ and the number of shear strain cycles N. An alternative to N, which allows generalization to cycles of non-uniform loading, is the length of the strain path ξ, defined as $d\xi = (\frac{1}{2} d\varepsilon_{12} d\varepsilon_{12})^{1/2}$, in which ε_{12} are the deviatoric strains in the plane. Thus, for

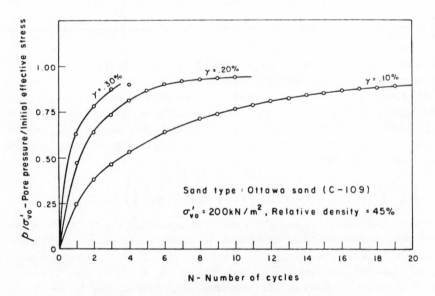

Figure 6.17 Experimental data on pore water pressures

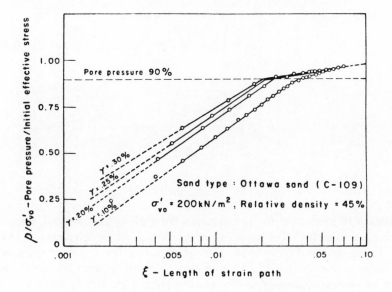

Figure 6.18 Pore water pressures vs. length of strain path

simple shear $d\xi = \frac{1}{2}|d\gamma|$. Therefore, we may write

$$p/\sigma'_{v0} = g(\gamma, \xi). \tag{6.7}$$

in which σ'_{v0} is the initial vertical effective stress (Figure 6.18). Alternatively, the length of the deviatoric stress path may be used. The pore water pressures, p, can be expressed as monotonically increasing functions of a single variable κ if a transformation T exists such that $\kappa = T\xi$. Then

$$p/\sigma'_{v0} = G(\kappa), \tag{6.8}$$

where the parameter κ is called the damage parameter. The transformation T and the function G will be obtained using the data in Figures 6.17 and 6.18.

The pore pressures generated in cyclic simple shear at various constant strain amplitudes ranging from $\gamma = 0.05\%$ to $\gamma = 0.3\%$ are shown in Figure 6.17 versus the number of load cycles N. The same data are shown in Figure 6.18 plotted against the logarithm of the length of the strain path, ξ. We now seek a transformation T which, when applied to ξ, will collapse the curves for different strain amplitudes into one curve. It is reasonable to assume that if such a transformation exists it will be a function of γ.

Consider a pore water pressure u_1 occurring at ξ_1 for a shear strain amplitude γ_1 and at ξ_2 for a shear strain γ_2. We ask whether p_1 can be associated with the value κ_1 of a variable κ such that

$$\kappa_1 = T\xi_1 = T\xi_2. \tag{6.9}$$

Consider

$$T = e^{\lambda\gamma},$$
$$p_1 = \xi_1 e^{\lambda\gamma_1} = \xi_2 e^{\lambda\gamma_2}, \tag{6.10}$$

$$e^{\lambda(\gamma_1-\gamma_2)} = \xi_2/\xi_1,$$

or

$$\lambda = \ln(\xi_2/\xi_1)/(\gamma_1 - \gamma_2). \tag{6.11}$$

For equation (6.8) to hold for the assumed form of T, the constant λ must have a unique value for all corresponding p, γ and ξ. When equation (6.11) is applied to the data in Figure 6.18, a range of values for λ was determined with a mean value of $\lambda = 3.56$. Using this value of λ the data points in Figure 6.18 were transformed to κ-space in Figure 6.19. It may be seen that the points do not define a unique curve but a narrow band. A non-linear least-squares curve-fitting method was used to define the curve in Figure 6.19, describing the relationship between Pore Pressure Ratio and the logarithm of κ. The relationship is given in Figure 6.20 to a natural scale and

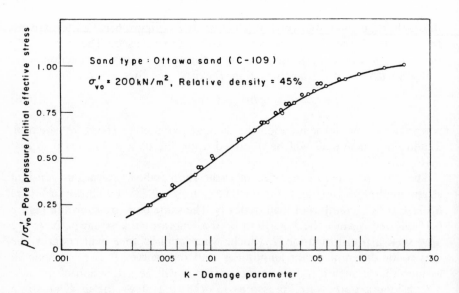

Figure 6.19 The pore water pressure function, $(u/\sigma'_{v0}) = G(\ln \kappa)$

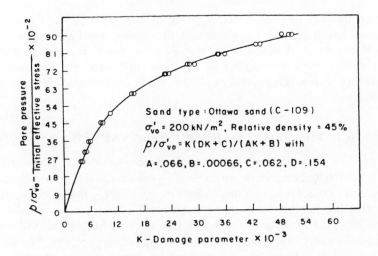

Figure 6.20 The pore water pressure function, $(u/\sigma'_{vo}) = G(\kappa)$

defines the function $G(\kappa)$ as

$$G(\kappa) = \kappa(D\kappa + C)/(A\kappa + B)$$

with

$$A = 0.066, \quad B = 0.00066, \quad C = 0.062, \quad D = 0.154.$$

A mean value of λ that will very closely transform the test data to a curve in κ-space without further manipulation is obtained by determining many values of λ over the whole range of likely interest before taking the mean. The broader the strain range the more difficult it is to select a single satisfactory value of λ by the above procedure. If adjustments in λ are required, a useful procedure is to perform the inverse transformation to ξ-space and note the regions of greatest divergence from experimental data. In these regions additional values of λ should be determined. These additional values will then weight the new mean value of λ in a way that will improve the representation in the areas of inferior definition.

The data discussed above is representative of data from six different sands. Therefore, it seems clear that pore water pressures generated in saturated sands due to seismic excitation can be expressed as continuous monotonically increasing functions of a damage parameter κ which is linked with the dynamic response of the ground through shear strain γ and the accumulated length of the shear strain path ξ. Thus, the increments in pore water pressure associated with the current value of κ can be readily included in a method of dynamic effective analysis such as that presented by Finn, Lee, and Martin.[4]

The determination of pore water pressure at any point in the dynamic analysis does not require a knowledge of the highly non-linear rebound modulus of the sand skeleton. This obviates the need to measure this parameter and reduces the number of constants that must be determined from laboratory testing. Discussion of similar models is included in Chapter 5 of this book.

6.7 MULTI-AXIAL RESPONSE ANALYSIS

At present only equivalent linear procedures are used for multi-axial analysis. Research is being done on extending one-dimensional non-linear methods such as DESRA to two dimensions by considering that response to all-round pressure is elastic and that shearing deformation is governed by the Masing model. But there are difficulties in such *ad hoc* extensions; for example, the difficulty in defining loading and unloading precisely. For this reason, there is increasing interest in methods based on the theory of plasticity and especially in the anisotropic kinematic theory of plasticity. In phenomenological terms this is the most comprehensive constitutive model for soils. Two particular formulations of the anisotropic theory of plasticity appear to have great potential for multi-axial analyses; the multi-yield surface model of Prevost[26,27] and the two-surface model of Mroz, Norris, and Zienkiewicz.[23,24] The models are very new and still in the process of development, but progress to date has been impressive. An extensive review of these methods was recently conducted by Finn and Martin[10] as part of a study for an offshore firm to select a constitutive relation for inclusion in a general program for the analysis of the behaviour of offshore foundations under wave and earthquake loading.

The study concluded that both models represent the response of soils to static loading very well. Both have demonstrated the potential to model the phenomenological aspects of pseudo-static cyclic loading, but verification has been limited. The two-surface model has not yet been verified for cyclic loading; the multi-surface model has been verified for the case of undrained cyclic loading of a lightly over-consolidated clay only—a fairly stable material. None of the models has yet been applied to the analysis of strongly degrading clays or readily liquefiable sands, or formulated for dynamic loading.

The search for a good constitutive model for soil is not yet over, but the anisotropic theory of plasticity is probably the best general model available at the moment for the analysis of the response of relatively stable soils to static and pseudo-static cyclic loading. The author has reservations about the capability of the method at its present stage of development to analyse the response of strongly degrading clays or readily liquefiable sands efficiently.

NOTATION

A, B, C, D	constants
C_1, C_2, C_3, C_4	constants
D_r	relative density
$\bar{E}_r$	tangent modulus of one-dimensional unloading curve
$G(\kappa)$	function of damage parameter
K	coefficient of permeability
K_2	constant
m, n	exponents
n_e	porosity
N_{eq}	equivalent number of cycles
S_0	initial global lateral stiffness
u	pore water pressure
Δu	increment of pore water pressure
u/σ'_{v0}	pore water pressure/initial confining stress
$\Delta \varepsilon_v$	increment of volumetric slip strain occurring during cyclic loading
ε_v	accumulated volumetric slip strain from cyclic loading
γ	cyclic shear strain
γ_r	shear strain value at which reversal occurs
σ'_v	vertical effective stress
σ'_{v0}	initial vertical effective stress
τ/σ'_{v0}	cyclic shear stress ratio
τ_{av}	average shear stress
τ_r	shear stress at reversal
τ_{max}	maximum shear stress
ξ	length of the strain path
λ	constant
κ	damage parameter
$d\varepsilon_{12}$	deviatoric strain

REFERENCES

1. Bazant, Z. P., and Krizek, R. J. Endochronic constitutive law for liquefaction of sand, *J. Engr. Mech. Div., ASCE,* **102** (EM2) 225–238, April 1976.
2. Ferrians, O. J., Jr. *Effects in the Copper River Basin Area,* U.S. Geological Survey Professional Paper 543-E, Superintendent of Documents, U.S. Government Printing Office, Washington, D.C., 1966.
3. Finn, W. D. L., Byrne, P. M., and Martin, G. R. Seismic response and liquefaction of sands, *J. Geotech. Engr. Div., ASCE,* **102** (GT8), 841–56, Proc. Paper 12323, August 1976.
4. Finn, W. D. L., Lee, K. W., and Martin, G. R. An effective stress model for liquefaction, *ASCE Annual Convention and Exposition,* Philadelphia, PA, September–October, 1976, Preprint 2752; also in *J. Geotech. Engr. Div., ASCE,* **103** (GT6) Proc. Paper 13008, 517–33 June 1977.
5. Finn, W. D. L., Martin, G. R., and Lee, M. K. W. Comparison of dynamic analyses for saturated sands, *Proceedings, ASCE Geotechnical Engineering Division Specialty Conference,* 472–91, Pasadena, California, June 1978.
6. Finn, W. D. L., Yong, R. N., and Lee, K. W. Liquefaction of thawed layers in frozen soil, *J. Engr. Div., ASCE,* **104** (GT10), 1243–55, Proc. Paper 14107, October 1978.

7. Finn, W. D. L., and Martin, G. R., Analysis of piled foundations for offshore structures under wave and earthquake loading, *Proceedings, BOSS '79 Conference*, London, England, **1**, 497–502, August 1979.
8. Finn, W. D. L., and Martin, G. R. Aspects of seismic design of pile-supported offshore platforms in sand, *Proceedings, Special Session, Soil Dynamics in the Marine Environment, ASCE National Convention and Exposition*, Boston, Preprint 3604, 1–36 April 1979.
9. Finn, W. D. L., and Bhatia, S. K. Verification of non-linear effective stress model in simple shear, Accepted for publication in *Proceedings, ASCE Fall Meeting*, Hollywood-by-the-sea, Florida, October 1980.
10. Finn, W. D. L., and Martin, G. R. Soil as an anisotropic kinematic hardening solid, accepted for publication in *Proceedings, ASCE Fall Meeting*, Hollywood-by-the-Sea, Florida, October 1980.
11. Finn, W. D. L., and Bhatia, S. K. Endochronic theory of sand liquefaction, accepted for publication in *Proceedings, 7th World Conf. on Earthquake Engineering*, Istanbul, Turkey, September 1980.
12. Finn, W. D. L., and Bhatia, S. K. Prediction of seismic pore-water pressures, accepted for publication in *Proceedings, 10th International Conference on Soil Mechanics and Foundation Engineering*, Stockholm, June 1981.
13. Idriss, I. M., Lysmer, J., Hwang, R., and Seed, H. B. *Quad-4: A Computer Program for Evaluating the Seismic Response of Soil-Structures by Variable Damping Finite Element Procedures*, Report No. EERC 73-16, Earthquake Engineering Research Center, University of California, Berkeley, California, July 1973.
14. Kramer, R. W., Macdonald, R. B., Tiedmann, D. A., and Viksne, A. "Dynamic Analysis of Tsengwen Dam", Taiwan, Republic of China, United States Department of the Interior, Bureau of Reclamation, 1975.
15. Lee, K. L., and Chan, K. (1972), Number of equivalent significant cycles in strong motion earthquake, *Proceedings, Int. Conf. on Microzonation*, Seattle, Washington, **II**, 609–27, October 1972.
16. Lee, M. K. W., and Finn, W. D. L. *DESRA-1, Program for the Dynamic Effective Stress Response Analysis of Soil Deposits Including Liquefaction Evaluation*, Soil Mechanics Series, No. 36, Dept. of Civil Engineering, University of British Columbia, Vancouver, B.C., 1975.
17. Lee, M. K. W., and Finn, W. D. L. *DESRA-2, Dynamic Effective Stress Response Analysis of Soil Deposits with Energy Transmitting Boundary Including Assessment of Liquefaction Potential*, Soil Mechanics Series, No. 38, Dept. of Civil Engineering, University of British Columbia, Vancouver, B.C., 1978.
18. Lysmer, J., Udaka, T., Seed, H. B., and Hwang, R. *LUSH: A Computer Program for Complex Response Analysis of Soil-Structure Systems*, Report No. EERC 74-4, Earthquake Engineering Research Center, University of California, Berkeley, California, 1974.
19. Lysmer, J., Udaka, T., Tsou, C. F., and Seed, H. B. *FLUSH: A Computer Program for Approximate 3-D Analysis of Soil-Structure Interaction Problems*, Report No. EERC 75-30, Earthquake Engineering Research Center, University of California, Berkeley, California, 1975.
20. Martin, G. R., Finn, W. D. L., and Seed, H. B. Fundamentals of liquefaction under cyclic loading, *Soil Mechanics Series*, No. 23, Dept. of Civil Engineering, University of British Columbia, 1974; also in *J. Geotech. Engr. Div.*, ASCE, **101** (GT5), Proc. Paper 11284, 423–38 May 1975.
21. Martin, G. R., Finn, W. D. L., and Seed, H. B. Effects of system compliance on

liquefaction tests, *J. Geotech. Engr. Div.*, *ASCE*, **104** (GT4) Proc. Paper 13667, 463–79, April 1978.

22. Masing, G. "Eigenspannungen und Verfestigung beim Messing", *Proceedings, 2nd International Congress of Applied Mechanics*, Zürich, Switzerland, 1926.
23. Mroz, Z. On the description of anisotropic hardening, *J. Mech. Phys. Solids*, **15**, 163–75, 1967.
24. Mroz, Z., Norris, V. A., and Zienkiewicz, O. C. An anisotropic hardening model for soils and its application to cyclic loading, *Int. J. Num. Anal. Meth. Geomechanics*, **2**, 203–21, 1978.
25. Newmark, N. M., and Rosenblueth, E. *Fundamentals of Earthquake Engineering*, Prentice-Hall, Englewood Cliffs, N.J., pp. 162–3, 1971.
26. Prevost, J. H. Anisotropic undrained stress–strain behaviour of clays, *J. Geotech. Engr. Div.*, *ASCE*, **104** (GT8), Proc. Paper 13942, 1075–90 August 1978.
27. Prevost, J. H. Mathematical modeling of soil stress–strain strength behaviour, *3rd Int. Conf. on Numerical Methods in Geomechanics*, Aachen, 347–61, 2–6 April 1979.
28. Schnabel, P. B., Lysmer, J., and Seed, H. B. *SHAKE: A Computer Program for Earthquake Response Analysis of Horizontally Layered Sites*, Report No. EERC 72–12, Earthquake Engineering Research Center, University of California, Berkeley, California, December 1972.
29. Seed, H. B., and Idriss, I. M. Analysis of soil liquefaction: Niigata earthquake, *J. Soil Mechanics and Foundations Div. ASCE*, **93** (SM3), 83–108, 1967.
30. Seed, H. B., and Idriss, I. M. *Soil Moduli and Damping Factors for Dynamic Response Analyses*, Report No. EERC 70–10, Earthquake Engineering Research Center, University of California, Berkeley, California, December 1970.
31. Seed, H. B., Idriss, I. M., Makdisi, F., and Banerjee, N., *Representation of Irregular Stress Time Histories by Equivalent Uniform Stress Series in Liquefaction Analyses*, Report No. EERC 75–29, Earthquake Engineering Research Center, University of California, Berkeley, California, October 1975.
32. Seed, H. B. *Evaluation of Soil Liquefaction Effects on Level Ground During Earthquakes*, ASCE Annual Meeting, October, 1976, Philadelphia, PA, Preprint 2752, 1976.
33. Serff, N., Seed, H. B., Makdisi, F. I., and Chang, C. K. *Earthquake Induced Deformation of Earthdam*, EERC 76-4, Earthquake Engineering Research Center, University of California, Berkeley, California, 1976.
34. Streeter, V. L., Wylie, E. B., and Richart, F. E., *Soil Motion Computations by Characteristics Method*, ASCE National Structural Engineering Meeting, San Francisco, California, Preprint 1952, April 1973.
35. Valanis, K. C. A theory of viscoplasticity without a yield surface, *Archivum Mechaniki Stosowanej*, **23** (4), 517–33, 1971.
36. Zienkiewicz, O. C., Chang, C. T., and Hinton, E. Non-linear seismic response and liquefaction, *Int. J. Num. Anal. Meth. Geomechanics*, **2**, 381–404, 1978.

Soil Mechanics—Transient and Cyclic Loads
Edited by G. N. Pande and O. C. Zienkiewicz
© 1982 John Wiley & Sons Ltd

Chapter 7

Dynamic Response Analysis of Level Ground based on the Effective Stress Method

K. Ishihara and I. Towhata

7.1 INTRODUCTION

In the discipline of the soil mechanics, two methods of approach have been widely used to evaluate the degree of potential stability of soil deposits and soil structures against failure. These are the effective stress method and the total stress method. The practice of soil dynamics has been handicapped by a lack of proper procedures enabling the effective stress analysis to be made in cyclic or seismic loading environments. Most of the dynamic analysis presently in use for the one-dimensional response of horizontal deposits of soils are based on the total stress principle. In recent years, with the growing recognition of the potential deficiency of the total stress analysis, several efforts have been made to explore the possibility of performing the dynamic analysis in terms of the effective stress principle. These approaches were developed by Martin, Finn, and Seed,[18] Finn, Lee, and Martin,[3] Liou, Streeter, and Richart,[17] Ghaboussi and Dikmen,[5] Zienkiewicz, Chang, and Hinton,[27] and Sato, Shihata, and Kosaka.[20] An excellent review of these approaches was presented by Finn.[4] In developing the effective stress methodology, of uppermost importance is the incorporation of an appropriate material model describing the real behaviour of the soils under consideration. Several attempts have also been made to establish the material model of soils under cyclic loading conditions based on comprehensive laboratory test data. This aspect of the effective stress approach was studied by Ishihara et al.,[8] Sherif, Ishibashi, and Tsuchida,[22] and Yoshimi and Tokimatsu.[25]

Whatever material model and analytical tool are used, a dynamic effective stress method should be able to offer a better understanding of the real phenomena, and more importantly help yield a plausible solution for the problems of engineering significance. The fundamental shortcomings of the dynamic total stress analysis hitherto used for the seismic response of the horizontally layered soil deposits appear to lie in the following points.

(1) Because the total stress analysis is unable to take into account the time-dependent stiffness-degrading effect caused by the pore water pressure build-up in the soil, it sometimes overestimates the intensity of the shaking of soft soil deposits, giving a justification to the notion that a higher seismic coefficient be incorporated in the design of structures to be built on the soft soil deposits.

(2) The sharp degradation in the stiffness of soft soils during and following a period of liquefaction produces motions with prolonged predominant periods on the ground surface. Large structures such as long bridges and oil storage tanks having long natural periods may thus experience a near-resonance condition during earthquakes.

In view of its engineering significance, the development of a useful effective stress method has been badly in need in recent years. The study described in the following pages is an attempt to shed some light on the questions as described above.

7.2 A STRESS PATH MODEL

7.2.1 Description of the model

The prediction of pore water pressure changes during seismic loading can be made by means of a modified version of the stress path model proposed previously by Ishihara, Tatsuoka, and Yasuda.[8] Consider an element of a saturated soil in a level ground consolidated to an effective overburden pressure, σ'_{v_0}, under the condition of lateral confinement shown in Figure 7.1(a). During earthquakes, the horizontal shear stress, τ, due to the upward propagation of shear waves will be superimposed on the soil element, causing it to deform horizontally in the simple shear manner illustrated in Figure 7.1(b). The procedure for predicting the increases in pore water

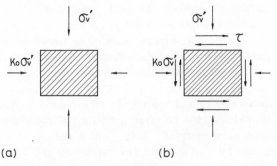

(a)   (b)

Figure 7.1 States of stress prior to and during seismic loading

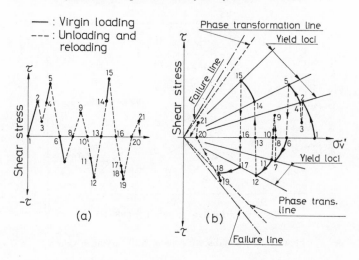

Figure 7.2 Illustration of the stress path model

pressure in the soil element subjected to a sequence of irregular loading under undrained conditions is illustrated in Figure 7.2. Figure 7.2(a) shows the irregular shear stress time history applied to a soil element. Figure 7.2(b) shows the effective stress path drawn in the $\sigma_v'-\tau$ stress space. The soil is first consolidated to the effective overburden pressure, σ_{v_0}', as indicated by point 1 in Figure 7.2(b) and then subjected to the shear stress to point 2. The loading from the initially consolidated state 1 to a new state 2 causes a plastic deformation and generates an increase in pore water pressure. During this virgin loading the yield surface is shifted upwards to a new yield surface, represented by a straight line connecting the origin and point 2 in Figure 7.2(b). The curve 1–2 is assumed to be a parabola given by

$$\sigma_v' = m - \frac{B_p'}{m}\tau^2, \qquad (7.1)$$

where B_p' is a soil constant representing the characteristics of a pore water pressure build-up, and m is a parameter for locating a current parabolic stress path at each time step of computation (see Appendix 1). Typical shapes of the parabolae for the case of $B_p' = 4.0$ are demonstrated in Figure 7.3, and the influence of B_p' on the parabolae for the case of $m = 80 \text{ kN/m}^2$ is shown in Figure 7.4. Unloading from point 2 to 3 and reloading back to point 4 takes place entirely inside the current yield surface. The deformation that occurs during this stress change is considered to be secondary plastic, and the pore water pressures are assumed to build up according to an

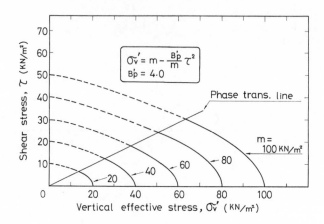

Figure 7.3 Parabolic stress paths with varying m values

empirical formula given by

$$\Delta\sigma_v' = -B_u'\left(\frac{\tau}{\sigma_{v_0}'} - \frac{\tau_m}{\sigma_{v_0}'}\right)\left(\frac{\sigma_v'}{\sigma_{v_0}'} - \kappa\right)\Delta\tau \qquad \text{for } \sigma_v' \geq \kappa\sigma_{v_0}',$$

$$\Delta\sigma_v' = 0 \qquad\qquad\qquad\qquad\qquad\qquad \text{for } \sigma_v' > \kappa\sigma_{v_0}', \tag{7.2}$$

where τ_m is the maximum shear stress that has been applied to the soil in the most recent cycle as illustrated in Figure 7.5 (see Appendix 2). The value, B_u', is a soil constant representing the pore water pressure build-up characteristics during the unloading and the reloading phases. The value, κ, was introduced to account for the fact that the pore water pressure ceases to

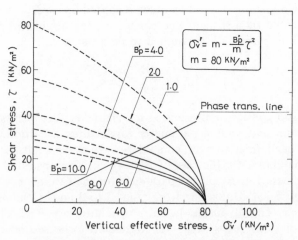

Figure 7.4 Parabolic stress path with varying B_p' values

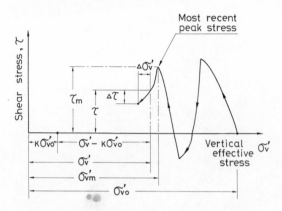

Figure 7.5 Change in the stress path during unloading

build up when the vertical effective stress decreases to a certain value. A continued loading from point 4 to 5 again takes a parabolic stress path (Appendix 1) and cuts across the yield surface, causing plastic deformation which in turn brings about an increase in the pore water pressure. Unloading from point 5 to 6 involves a secondary plastic deformation, resulting in an increase in pore water pressure (Appendix 2) which is calculated by equation (7.2). In the present model, it is assumed that yielding in the stress space above the σ_v' axis occurs independent of the yielding in the stress space below and vice versa. Therefore, loading from point 6 to 7 causes a virgin plastic deformation and increases pore water pressure, again following a parabolic curve. The stress path from point 7 to 18 follows the same pattern as above. Point 18 lies on what is termed the phase transformation line. After the stress path crosses the phase transformation line, the stress path during an increase in shear stress is assumed to trace a hyperbolic curve, regardless of whether the stress change is in the loading or reloading phase. The hyperbolic curve is given by

$$\left(\frac{\sigma_v'}{m}\right)^2 - \left(\frac{\tau}{m \tan \Phi_l'}\right)^2 = 1.0, \tag{7.3}$$

where ϕ_l' is the angle of internal friction of sand at small effective confining stress, and m is a parameter for locating a current hyperbolic stress path. It should be noted that the hyperbolic curve is constructed in such a way that it approaches the failure line asymptotically as illustrated in Figure 7.6. For the unloading phase, the stress path is assumed to follow a straight line which is tangent, at the point of stress reversal, to the hyperbolic curve along which the shear stress has been increased. The detailed illustration of how the stress paths are constructed is presented in Figure 7.6. When the pore

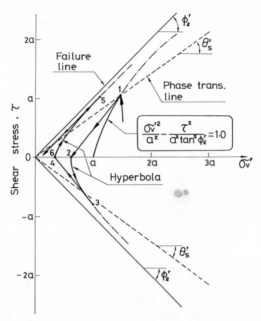

Figure 7.6 Stress paths after the phase transformation line has been cut across

water pressure has increased to 97% of the initial overburden pressure, it is kept from further increase and the stress path is made to follow the same hyperbolic curve back and forth. This last hyperbolic curve intersects the σ'_v axis at point, $\sigma'_v = 0.03\sigma'_{v_0}$. The pore water pressure build-up beyond this point is arrested because the actual shear stress application in the horizontal ground is not uni-directional but multi-directional. This means that even when the shear stress is reduced to zero in one direction near the incidence of initial liquefaction, there will always be some shear stress still being applied in the direction perpendicular to it. It may be mentioned, therefore, that at an instant close to the onset of the initial liquefaction, there is very little likelihood for the two mutually perpendicular components of shear stress in the horizontal plane to be simultaneously zero. With the shear stress being applied in one direction and the sand being kept deformed to the range of strain in which it tends to dilate, the cyclic stress application in the perpendicular direction cannot bring the sand to a state of complete liquefaction. This was clearly demonstrated in the rotational simple shear test performed by Ishihara and Yamazaki.[11] The simplest way to take this effect into account in one-dimensional response analysis would be to set a barrier in the stress path beyond which the pore water pressure is not allowed to build up. With reference to the result of the rotational shear tests, the barriers was set at point, $\sigma'_v = 0.03\sigma'_{v_0}$ in the stress space.

7.2.2 Determination of soil constant

(a) *Phase transformation angle, θ'_s, and angle of internal friction, ϕ'_l at very small effective confining stress*

Scrutiny of the triaxial shear and simple shear test results have shown that with a reasonable degree of accuracy the angle of phase transformation, θ'_s, is given as a function of the angle of internal friction at very small effective confining stress, ϕ'_l, as follows:

$$\tan \theta'_s = \tfrac{5}{8} \tan \phi'_l. \tag{7.4}$$

It is well known that the angle of internal friction of cohesionless soils mobilized at a very small confining pressure, ϕ'_l, is much larger than the angle of internal friction, ϕ', normally defined in the range of confining pressures of about $100\text{--}300\,\text{kN/m}^2$. Since the effective confining pressures existing in the sand at the stage near the incidence of initial liquefaction are very small, an increased value of the normally defined angle of internal friction must be used in determining the phase transformation angle, θ'_s, through equation (7.4). As a rough estimate, values of $\tan \phi'_1$ equal to 1.4 times $\tan \phi$ will be used in the following analyses.

(b) *Soil constant, κ*

The soil constant, κ, is determined as follows by considering the characteristic behaviour of the sand which is manifested in low-amplitude cyclic loading conditions. Figure 7.7 shows a typical result of cyclic triaxial shear or simple

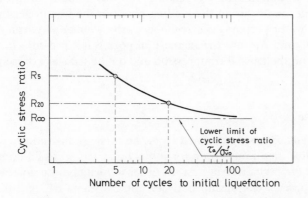

Figure 7.7 Illustration for reading off soil constants from an experimental curve of the cyclic stress ratio versus number of cycles

shear test in which the cyclic stress ratio causing initial liquefaction is plotted versus the number of cycles. A number of cyclic triaxial shear tests performed on undisturbed samples of sands has consistently indicated that in the above type of plot there always exists a minimum cyclic stress ratio, τ_a/σ'_{v0}, below which liquefaction does not occur even when an infinite number of cyclic loads are applied. Recalling the fact that the maximum shear stress that can be applied to the sand under a vertical effective stress σ'_{v0}, is given by $\sigma'_{v0} \tan \phi'$, the lower limiting shear stress, τ_a, may be

$$\tau_a = \kappa \sigma'_{v0} \tan \phi'. \tag{7.5}$$

This defines the constant, κ, as

$$\kappa = \frac{\tau_a}{\sigma'_{v0}} \frac{1}{\tan \phi'}. \tag{7.6}$$

The test results have shown that both the stress ratio, τ_a/σ'_{v0}, and $\tan \phi'$ increase approximately in the same proportion with the density of the sand, and it was found appropriate to give a constant value for κ. In the following analysis, the soil constant, κ, will be set

$$\kappa = 0.06. \tag{7.7}$$

(c)   Soil constant, B'_p

The soil constant, B'_p, is determined from the data of the cyclic triaxial tests. Figure 7.8 schematically shows the initial part of the recorded changes in the cyclic shear stress and pore water pressure obtained in the test. The constant, B'_p, is determined by reading off the amount of pore water pressure that develops during the loading phase in the first cycle. In Figure 7.8, this pore water pressure is calculated as the average between Δu_c and Δu_e, where Δu_c and Δu_e are the increase in pore water pressure in the first virgin loading on the triaxial compression and on the triaxial extension sides, respectively.

(d)   Soil constant, B'_u

The soil constant, B'_u, is determined from the cyclic stress ratio required to cause 5% double amplitude strain (or initial liquefaction) in 20 cycles, R'_{20}, obtained from the cyclic triaxial test. In order to facilitate the determination of B'_u, an empirical relationship between the constant, B'_u, and the cyclic stress ratio, R'_{20}, has been established for the predetermined values of B'_p and for the phase transformation angle of $\tan \theta'_s = \frac{5}{8} \tan \phi'_f = 0.5$. The constants, B'_p, B'_u and R'_{20} thus correlated with each other for the case of

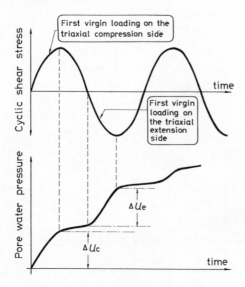

Figure 7.8 Procedures for data reading to determine the constant, B'_p

$\tan \theta'_s = 0.5$ are denoted by $\bar{B}_p$, $\bar{B}_u$ and $\bar{R}_{20}$, respectively, and will be used as a reference relationship.

In determining the reference relationship, a tentative set of values are assigned to the constants, $\bar{B}_p$ and $\bar{B}_u$. By incorporating these constants into the stress path model, together with the value of $\tan \theta'_s = 0.5$ and $\kappa = 0.06$, the number of cycles required to induce initial liquefaction (defined as the state of pore water pressure building up to 97% of the initial vertical effective stress) were computed for several given values of cyclic stress ratio. Thus, it becomes possible to obtain a curve of the cyclic stress ratio versus the number of cycles. Similar computations were repeated for the stress path model with different sets of preassigned values of $\bar{B}_p$ and $\bar{B}_u$. A typical result of the computation for various values of $\bar{B}_u$ is presented in Figure 7.9 with a value of $\bar{B}_p = 5.0$. From these curves, it is now possible to read off values of the cyclic stress ratio causing initial liquefaction in 20 cycles, $\bar{R}_{20}$. The relationship between $\bar{B}_p$, $\bar{B}_u$ and $\bar{R}_{20}$ thus established is summarized in a chart in Figure 7.10.

The following step must be followed to make use of the reference chart as above for the determination of the values of B'_p and B'_u for the general case of the phase transformation angle, θ'_s, different from $\tan^{-1} 0.5 = 26.6°$. The constant, B'_p, and the cyclic stress ratio, R'_{20}, determined from the cyclic triaxial test data, are once transformed into $\bar{B}_p$, and $\bar{R}_{20}$ through the

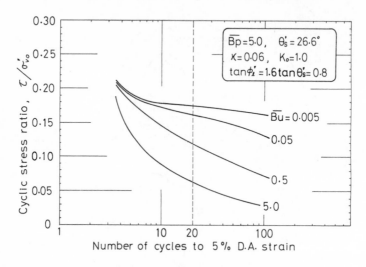

Figure 7.9 Computed relationship between cyclic stress ratio and number of cycles

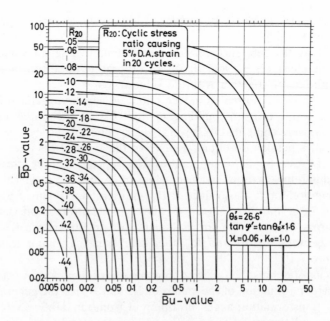

Figure 7.10 A chart to determine the soil constant $\bar{B}_u$ based on the result of the cyclic triaxial tests

formulae

$$\bar{B}_p = (2 \tan \theta_s')^2 B_p',$$

$$\bar{R}_{20} = \frac{1}{2 \tan \theta_s'} R_{20}', \tag{7.8}$$

where it is noted that $\bar{B}_p = B_p'$ and $\bar{R}_{20} = R_{20}'$ when $\theta_s' = 26.6°$. With the constant, $\bar{B}_p$, known on the ordinate in the reference chart, one can determine a point of intersection between the line of $\bar{B}_p = \text{constant}$ and a curve of $\bar{R}_{20} = \text{constant}$. By reading off the value of the abscissa corresponding to the point of intersection, the value of $\bar{B}_u$ is determined. This value must be transformed back to B_u' through the formula for the general case of the phase transformation angle different from $\theta_s' = 26.6°$, through the formula

$$B_u' = \frac{\bar{B}_u}{(2 \tan \theta_s')^2}. \tag{7.9}$$

7.2.3 Effects of K_0-conditions

The procedures for determining the constants in the stress path model as above were based on the cyclic triaxial test data obtained from samples consolidated isotropically. Therefore, the soil constants, θ_s', B_p' and B_u', for the conditions of K_0-values less than unity are estimated from the values obtained for $K_0 = 1.0$. In order to achieve the transformation, an assumption is made on the relationship between the cyclic stress ratios measured under isotropically consolidated state and under anisotropically consolidated state as follows (Ishihara *et al.*[9])

$$\frac{\tau}{\sigma_v'} = \frac{1 + 2K_0}{3} \frac{\tau'}{\sigma_v'} \tag{7.10}$$

where τ' and τ denote the horizontal shear stresses in a condition of $K_0 = 1.0$ and in a condition of $K_0 < 1.0$, respectively. Using equation (7.10) with reference to equations (7.1) and (7.2), the phase transformation angle, θ_s', and values of B_p' and B_u' for the general K_0-conditions can be obtained as follows:

$$\left.\begin{array}{l} \tan \theta_s = \dfrac{1 + 2K_0}{3} \tan \theta_s' \\[3mm] B_p = \left(\dfrac{3}{1 + 2K_0}\right)^2 B_p' = \left(\dfrac{3}{2(1 + 2K_0) \tan \theta_s'}\right)^2 \bar{B}_p \\[3mm] B_u = \left(\dfrac{3}{1 + 2K_0}\right)^2 B_u' = \left(\dfrac{3}{2(1 + 2K_0) \tan \theta_s'}\right)^2 \bar{B}_u \end{array}\right\} \tag{7.11}$$

in which θ_s, B_p, and B_u denote the values of θ'_s, B'_p and B'_u in the general case, in which the K_0-value and the phase transformation angle can take any value.

7.2.4 An example of stress path

To demonstrate the adequacy of the stress path model as described above, a picture of the stress path was depicted for a typical case of sand behaviour subjected to a constant-amplitude cyclic shear stress. This is shown in Figure 7.11, where it is noted that the computed stress path resembles the stress

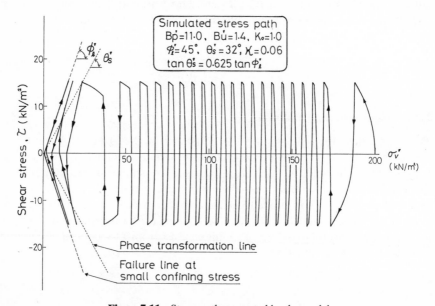

Figure 7.11 Stress path computed by the model

path observed in the laboratory cyclic triaxial or simple shear test with a relatively large amount of pore water pressure developing in the first cycle and in the last few cycles near the liquefaction.

7.3 STRESS VERSUS STRAIN MODEL

In order to compute the dynamic response of the horizontally layered deposit of soil subjected to the upward propagating shear waves, a non-linear relationship describing the hysteresis of stress and strain must be established. In this study the hyperbolic stress–strain relationship formulated by Konder and Zelasko[15] and also adopted by Hardin and Drnevich[6] will be used in conjunction with the Masing rule incorporated for modelling the hysteretic characteristics during unloading and reloading phases. The same hysteretic stress–strain relation was used by Finn, Lee, and Martin[3] in their dynamic effective stress analysis.

7.3.1 Initial loading

In the initial virgin-loading phase, it is assumed that the response of the soil follows the hyperbolic stress–strain relation,

$$\tau = \frac{G_t \gamma}{1 + |\gamma/\gamma_r|}, \qquad \gamma_r = \frac{\tau_f}{G_t}. \tag{7.12}$$

where τ is the shear stress at strain amplitude γ, G_t is the initial tangent shear modulus, τ_f is the strength of the soil, and γ_r is called the reference strain.[6] The curve represented by equation (7.12) is called the skeleton curve. Since the value of G_t can be chosen to be equal to the shear modulus of soil at infinitesimally small strains, the formula proposed by Iwasaki and Tatsuoka[13] will be used:

$$G_t = 14000 \frac{(2.17 - e)^2}{1 + e} \left(\frac{1 + 2K_0}{3} \sigma_v'\right)^{0.4} \quad (\text{kN/m}^2), \tag{7.13}$$

where e is the void ratio.

The strength of soil, τ_f, is assumed to be given by

$$\tau_f = (\sigma_v' \tan \phi' + c), \tag{7.14}$$

in which ϕ is the angle of internal friction and c is the cohesion of the soil.

7.3.2 Unloading and reloading

The stress–strain curve for initial loading given by equation (7.12) is called the skeleton curve and is schematically shown by a thick dotted line in Figure 7.12(a). For the unloading and the reloading phases, modified use is made of the skeleton curve according to the Masing rule. Suppose a loading reversal occurs at point A (γ_A, τ_A) in Figure 7.12(a), then the equation of the stress–strain curve during subsequent unloading is assumed to be given by

$$\frac{\tau - \tau_A}{2} = \frac{G_t(\gamma - \gamma_A)/2}{1 + (1/\gamma_r)|(\gamma - \gamma_A)/2|}. \tag{7.15}$$

This curve is drawn on the diagram by simply enlarging the portion of the skeleton curve from point 0 to point A by a factor of 2, and by attaching this curve, after reversing its direction, to the new starting point A. It will be

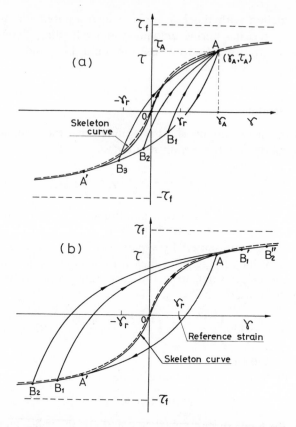

Figure 7.12 Stress–strain model based on the Masing rule

easily shown that the unloading curve merges on the skeleton curve at point A', which is located exactly at the opposite side of the point A. In case the unloading continues beyond the point A', the stress–strain follows the skeleton curve further down. It is to be noted that, where the unloading branch enters the skeleton curve at the point A', both curves meet smoothly together with an identical value of the tangent at the point A.

Let us consider next the reloading phase going back from point B (γ_B, τ_B) up toward point A, as shown in Figure 7.12(c). The equation of the stress–strain curve for this phase of reloading is given by

$$\frac{\tau - \tau_B}{2} = \frac{G_t(\gamma - \gamma_B)/2}{1 + (1/\gamma_r)\,|(\gamma - \gamma_B)/2|}. \tag{7.16}$$

This curve is drawn on the diagram in Figure 7.12(c) in a similar fashion to that used for equation (7.15): the unloading curve between points A and B

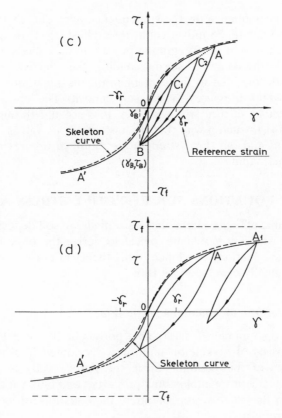

Figure 7.12 (c,d)

is reversed its direction and fitted so that the starting point is now from the point B. If the load is increased further, the stress–strain curve follows the original skeleton curve, indicating that the load application now regains the characteristics of the initial loading as described above. On the basis of the above principle, different branches can be drawn for the unloading and reloading curves. Figure 7.12(a) shows some examples of reloading curves coming back from different points such as points, B_1, B_2 and B_3. It is to be noted that all these reloading curves converge at the point A where the unloading has started from. Figure 7.12(b) also shows some examples of reloading curves returning from points on the skeleton curves beyond the point A'. Note that these reloading curves meet again with the skeleton curve at points B'_1 and B''_2, which are located beyond the point A. Figure 7.12(c) shows some curves of unloading from the reloading branch of the hysteretic curve. An example of an unload–reload cycle after the first cycle has been finished is indicated in Figure 7.12(d). It should be noted that the first cycle has no influence on the subsequent cycle. The characteristics as described above are taken into account in establishing a stress–strain relationship for each step of the integration procedure described later. It should be noted that in the course of seismic loading, the effective vertical stress, σ'_v, changes with time due to the contemporaneous generation and dissipation of pore water pressures in the soil deposit. The effect of the net decreasing effective vertical stress is taken into account through equations (7.13) and (7.14) in such a way that the instantaneous values of the initial tangent modulus and the shear strength are modified progressively at each time step of the loading.

7.4 BASIC EQUATIONS FOR EFFECTIVE STRESS ANALYSIS

Since the dynamic effective stress analysis is made on soil deposits which are saturated with water, it would be logical to derive the basic equations of motion from the equations developed in the theory of porous materials. The theory was originally developed by Biot.[1,2]

7.4.1 General form of equations of motion

In the macroscopic model of deformable porous bodies, it is assumed that the system consists of a skeleton uniformly distributed in space and fluid filling the pores of the skeleton. Since the porous body is a two-phase system, it is convenient to employ bulk properties and bulk variables of each phase. Then, in the plane strain condition, the equations of motion for the solid portion are written as

$$
\left.
\begin{aligned}
\rho_s \frac{\partial^2 u_x}{\partial t^2} &= -\frac{\partial \sigma_x}{\partial x} + \frac{\partial \tau}{\partial z} - b\left(\frac{\partial u_x}{\partial t} - \frac{\partial v_x}{\partial t}\right) \\
\rho_s \frac{\partial^2 u_z}{\partial t^2} &= -\frac{\partial \sigma_z}{\partial z} + \frac{\partial \tau}{\partial x} - b\left(\frac{\partial u_z}{\partial t} - \frac{\partial v_z}{\partial t}\right) + \rho_s g
\end{aligned}
\right\}
\tag{7.17}
$$

in which u_x, u_z, v_x, v_z denote the displacement components of the solid and fluid phases in the x and z directions, ρ_s is the bulk mass density of the solid phase, g is the gravitational acceleration, and b denotes a constant related to the permeability, k, porosity, n, and the unit weight of water, γ_w, in the following way,

$$
b = \frac{n^2 \gamma_w}{k}.
\tag{7.18}
$$

In (7.17) σ_x and σ_z denote the normal stresses acting on the planes perpendicular to the x and z directions. In what follows, the x direction will be chosen as horizontal and the z direction as vertical as shown in Figure 7.15. Therefore, σ_x and σ_z denote the normal stress acting on the solid portion in the vertical and horizontal planes, respectively. τ denotes the shear stress acting on the horizontal as well as vertical plane.

The equations of motion for the fluid phase are given by

$$\left.\begin{array}{l} \rho_w \dfrac{\partial^2 v_x}{\partial t^2} = -\dfrac{\partial p}{\partial x} + b\left(\dfrac{\partial u_x}{\partial t} - \dfrac{\partial v_x}{\partial t}\right) \\[4mm] \rho_w \dfrac{\partial^2 v_z}{\partial t^2} = -\dfrac{\partial p}{\partial z} + b\left(\dfrac{\partial u_z}{\partial t} - \dfrac{\partial v_z}{\partial t}\right) + \rho_w g \end{array}\right\} \tag{7.19}$$

in which ρ_w is the bulk unit mass of the fluid phase and p denotes the compressive stress acting on the fluid portion. In equations (7.17) and (7.19), σ_x and σ_z are taken positive when they are compressional stresses, and p is positive when it is compressive stress. It should be noted that σ_x and σ_z are equal to a sum of the intergranular stress (or effective stress) and the fluid pressure acting on the solid portion, and also that p is the fluid pressure acting only on the fluid part and is not equal to the pore water pressure in the usual sense (see Appendix 3). In equations (7.17) and (7.19), the terms with the coefficient b on the right-hand side indicate the drag force due to the seepage of the fluid. This force results inherently from the presence of viscosity in the pore fluid. Hence, it is introduced as being proportional to the difference in velocity between the two phases.

In the response analysis of a horizontally layered deposit of soil, it is generally assumed that any stress and displacement components change only in the vertical direction, and independently of the x coordinate in the horizontal plane. Therefore, the terms such as $\partial \sigma_x / \partial x$ and $\partial \tau / \partial x$ are disregarded in equations (7.17) and (7.19). Then we obtain

$$\left.\begin{array}{l} \rho_s \dfrac{\partial^2 u_x}{\partial t^2} = \dfrac{\partial \tau}{\partial z} - b\left(\dfrac{\partial u_x}{\partial t} - \dfrac{\partial v_x}{\partial t}\right) \\[4mm] \rho_s \dfrac{\partial^2 u_z}{\partial t^2} = -\dfrac{\partial \sigma_z}{\partial z} - b\left(\dfrac{\partial u_z}{\partial t} - \dfrac{\partial v_z}{\partial t}\right) + \rho_s g \end{array}\right\} \tag{7.20}$$

and

$$\left.\begin{array}{l} \rho_w \dfrac{\partial^2 v_x}{\partial t^2} = b\left(\dfrac{\partial u_x}{\partial t} - \dfrac{\partial v_x}{\partial t}\right) \\[4mm] \rho_w \dfrac{\partial^2 v_z}{\partial t^2} = -\dfrac{\partial p}{\partial z} + b\left(\dfrac{\partial u_z}{\partial t} - \dfrac{\partial v_z}{\partial t}\right) + \rho_w g \end{array}\right\} \tag{7.21}$$

7.4.2 Equation of motion in the horizontal direction

In the type of liquefaction problem being considered for the horizontal deposit of soil, the flow of water resulting from the build-up of pore water pressure can occur only in the vertical direction. Since no seepage flow occurs in the horizontal x direction, the horizontal velocity of the pore water must be equal to the horizontal velocity of the solid skeleton. It is to be noted here that in the theory of porous materials the seepage is defined to occur only when there is movement of the fluid phase relative to the solid phase. In other words, when the fluid phase is moving together with the solid phase at the same velocity, seepage is considered not to be occurring. Therefore, the fact that there is no seepage flow in the horizontal direction implies that

$$u_x = v_x. \tag{7.22}$$

Through the introduction of this identity relation in the first equations of equations (7.20) and (7.21), and adding both, we obtain

$$\rho \frac{\partial^2 u_x}{\partial t^2} = \frac{\partial \tau}{\partial z}, \tag{7.23}$$

$$\rho = \rho_s + \rho_w, \tag{7.24}$$

where ρ is the unit mass of the soil including the solid and fluid phase. Since there is no difference in the displacement between the solid and fluid phase, it is sufficient to solve the above equation embodying the motion of the both phases. It is to be noted that equation (7.23) is the well-known equation of shear wave propagation in the vertical z direction.

7.4.3 Equation of motion in the vertical direction

In the present study, the seepage flow of water is assumed to occur in the vertical direction as a result of the generation of hydraulic gradient in the saturated soil deposit. Therefore, the vertical velocity of the fluid phase is generally different from the velocity of the solid phase in the vertical direction. In this case, the two equations for the vertical motion must be treated independently. In order to simplify analytical procedures, it would be convenient to add the second equations of equations (7.20) and (7.21) as follows:

$$\rho_s \frac{\partial^2 u_z}{\partial t^2} + \rho_w \frac{\partial^2 v_z}{\partial t} = -\frac{\partial \sigma}{\partial z} + \rho g, \tag{7.25}$$

$$\sigma = \sigma_z + p.$$

Equation (7.25) is rewritten as

$$(\rho_s + \rho_w)\frac{\partial^2 u_z}{\partial t^2} + \rho_w\left(\frac{\partial^2 v_z}{\partial t} - \frac{\partial^2 u_z}{\partial t}\right) = -\frac{\partial\sigma}{\partial z} + \rho g. \tag{7.26}$$

The bulk unit mass, ρ_w, of the fluid phase is related to the unit weight of water, γ_w, as follows:

$$\rho_w = n\gamma_w/g. \tag{7.27}$$

Noting further from equation (7.24) that

$$\rho = \rho_s + \rho_w = \gamma_t/g, \tag{7.28}$$

where γ_t is the unit weight of soil, one obtains

$$\frac{\gamma_t}{g}\frac{\partial^2 u_z}{\partial t^2} + \frac{n\gamma_w}{g}\frac{\partial^2}{\partial t^2}(v_z - u_z) = -\frac{\partial\sigma}{\partial z} + \gamma_t. \tag{7.29}$$

Replacing the variables by

$$\frac{\partial u_z}{\partial t} = W, \qquad \frac{\partial}{\partial t}(v_z - u_z) = \frac{Q}{n}, \tag{7.30}$$

we eventually obtain

$$\frac{\gamma_t}{g}\frac{\partial W}{\partial t} + \frac{\gamma_w}{g}\frac{\partial Q}{\partial t} = -\frac{\partial\sigma}{\partial z} + \gamma_t. \tag{7.31}$$

In the notation introduced above, W represents the velocity of the solid skeleton and Q represents the velocity of water relative to the soil skeleton. In what follows, W will be called the velocity of skeleton and Q the seepage velocity. Equation (7.31) expresses the motion of the solid phase in the vertical direction. The presence of the second term on the left-hand side is due to the fact that the unbalanced force causes pore water to accelerate relative to the solid skeleton. Without this term, equation (7.31) is reduced to the classical equation of compressional wave propagation.

Another basic equation to be considered in the vertical motion is the second equation in equation (7.21). By substituting equation (7.30) into equation (7.21), transfomation of the variables u_z and v_z into W and Q can be achieved as follows:

$$\frac{n\gamma_w}{g}\frac{\partial W}{\partial t} + \frac{\gamma_w}{g}\frac{\partial Q}{\partial t} + b\left(\frac{Q}{n}\right) + \frac{\partial}{\partial z}(nu) - n\gamma_w = 0, \tag{7.32}$$

in which u denotes the pore water pressure which is related with the variable, p, by

$$p = nu. \tag{7.33}$$

7.4.4 Volumetric compatibility equation

In the type of problem being considered in this study, it is assumed that the lateral deformation is constrained and the volume changes of the skeleton and water can result only from deformation in the vertical direction. Therefore, the volumetric strains in the solid and the fluid phases viewed as bulk quantities are expressed by $\partial u_z/\partial z$ and $\partial v_z/\partial z$ respectively. In the development of the theory of porous media by Ishihara[7] it was shown that the bulk volumetric strains for the two phases are not allowed to occur independently without any kinematic interaction between them. The derivation of the kinematic relationship which the bulk volumetric strains in each of the solid and fluid phases must satisfy is explained in Appendix 4. Denoting the volumetric strains of the solid substance and the water by $\partial u^s/\partial z$ and $\partial v^w/\partial z$, respectively, this kinematic compatibility relationship given in equation (7A4.4) is rewritten as

$$(1-n)\frac{\partial u_z}{\partial z}+n\frac{\partial v_z}{\partial z}=(1-n)\frac{\partial u^s}{\partial z}+n\frac{\partial v^w}{\partial z}. \qquad (7.34)$$

Introducing the transformation of variables according to equation (7.30), and taking the derivative with respect to time, we obtain

$$\frac{\partial W}{\partial z}+\frac{\partial Q}{\partial z}=\frac{\partial}{\partial t}\left[(1-n)\frac{\partial u^s}{\partial z}+n\frac{\partial v^w}{\partial z}\right]. \qquad (7.35)$$

This is the most general form of equation expressing the volumetric compatibility law which must hold valid in the deformation of the two-phase body. It is to be noted that the left-hand side of equation (7.35) consists of the variables in terms of the bulk quantities used in the representation of the set of equations, whereas the right-hand side consists of the quantities determined from the intrinsic properties of water and solid. Depending upon the accuracy required in the analysis, one or two terms on the right-hand side of equation (7.35) may be disregarded. It has been shown[7] that the compressibility of solid material is smaller by an order of magnitude than that of the water and further by several orders of magnitude than the solid skeleton. Therefore, the term taking into account the effect of the volumetric strain of the solid material may be disregarded without significant loss of accuracy from equation (7.35). If the the term $\partial u^s/\partial z$ is neglected and the relationship between the volumetric strain of water and pore water pressure as given by

$$u=-\frac{1}{c_l}\frac{\partial v^w}{\partial z}. \qquad (7.36)$$

is used, equation (7.35) reduces to

$$\frac{\partial W}{\partial z}+\frac{\partial Q}{\partial z}=-nc_l\frac{\partial u}{\partial t}, \qquad (7.37)$$

where c_1 is the compressibility of water. This equation is the same as that used by Liou *et al.*[17] in their effective stress response analysis. The volumetric compatibility equation given by equation (7.37) may be further simplified by considering the fact that the compressibility of water itself is still small and the corresponding term can be deleted without significant loss of accuracy in the analysis. Then we obtain

$$\frac{\partial W}{\partial z} + \frac{\partial Q}{\partial z} = 0. \tag{7.38}$$

7.4.5 Volume change equation

The pore water pressure generated by the application of cyclic shear stress at any point in the sand deposit can be assessed by the use of the stress path model described in the previous section. The pore water pressure predicted in this way is generally different, however, from that which is actually occurring in the deposit, because of the dissipation of the pore water pressure taking place at the same time. Since the deformation of the deposit during shaking is not perfectly undrained, the rate of increase of pore water pressure will be less than that for completely undrained sand. Hence it would be necessary to have an equation for the volume change of the sand related with the pore water pressure dissipated as well as with the pore water pressure generated.[18]

In formulating the equation, the pore water pressure generated in a small interval of time is denoted by Δu_g. The value of Δu_g can be determined from the pore water pressure model. When the motion is assumed to take place in the drained condition, then the amount of volume change that could potentially occur in the soil skeleton would be $m_v \Delta u_g$, where m_v denotes the one-dimensional coefficient of volume compressibility. However, the current volume change, $\Delta(\partial u_z/\partial z)$, which is actually occurring in the soil deposit is generally smaller than $m_v \Delta u_g$, because the full drainage is prohibited from occurring. In other words, the soil skeleton is assumed to contract once by an amount, $m_v \Delta u_g$, under a constant effective vertical stress and expand back to the present level of volume contraction, $-\Delta(\partial u_z/\partial z)$. In order for the soil skeleton to expand by an amount, $m_v \Delta u_g + \Delta(\partial u_z/\partial z)$, there must be a reduction in the vertical effective stress. Denoting this decrease in the vertical effective stress by $-\Delta \sigma'_v$, the expansion is expressed by $-m_v \Delta \sigma'_v$. Hence, we have

$$m_v \Delta u_g + \Delta \frac{\partial u_z}{\partial z} = -m_v \Delta \sigma'_v. \tag{7.39}$$

By taking the time derivatives on the both sides and by using the expression of equation (7.38), we obtain

$$\frac{\partial W}{\partial z} = -m_v \left(\frac{\partial \sigma'_v}{\partial t} + \frac{\partial u_g}{\partial t} \right). \tag{7.40}$$

It should be noted that the value of the internally generated pore water pressure, u_g, is known from the effective stress path model by assuming an undrained condition during a small interval of time being considered. Equation (7.40) is further modified by introducing the relationship

$$\sigma'_v + u = \sigma, \tag{7.41}$$

which implies that the total vertical stress is equal to the existing vertical effective stress plus the pore water pressure (including the static hydraulic pressure). Inserting equation (7.41) into equation (7.40), we obtain

$$\frac{\partial W}{\partial z} = m_v \left(\frac{\partial u}{\partial t} - \frac{\partial u_g}{\partial t} - \frac{\partial \sigma}{\partial t} \right) \tag{7.42}$$

7.4.6 Soil properties for use in the basic equations

The major soil properties to be used for the basic equations introduced above are the permeability coefficient and the coefficient of volume compressibility. These constants are used in the set of equations governing the process of dissipation of the pore water pressure.

(a) *Permeability characteristics*

It is well known that the permeability of soils is closely related to the fraction of fines contained in the soils. The well-known empirical relationship proposed by Hazen expresses the coefficient of permeability, k, in terms of the 10% grain size, D_{10}. According to the grain size analysis performed on typical Niigata sands, the grain size, D_{10}, varies in a relatively narrow range of 0.1–0.3 mm. Therefore, Hazen's formula[23]

$$k = 100 D_{10}^2 \quad (D_{10} \text{ in centimetres}). \tag{7.43}$$

yields the coefficient of permeability ranging from $k = 10^{-2}$ to 10^{-1} cm/s.

(b) *Coefficient of volume compressibility*

The coefficient of volume compressibility, m_v, in equation (7.42) represents the soil characteristic that determines the amount of settlement or volume contraction due to the dissipation of pore water pressures. The previous study in the laboratory test[10] has shown that the magnitude of volumetric strain due to reconsolidation of triaxial test samples increases as the recon-solidation starts from smaller effective confining pressure. Yohimi, Kuwabara, and Tokimatsu[24] showed similar trends in the volume change characteristics for two kinds of sands and suggested the value of m_v to be in the range of 5×10^{-5} m²/kN to 20×10^{-5} m²/kN for a small confining pressure of 0.2–20 kN/m². Lee and Albeisa[16] showed that the grain size of the sand, the density of the sand, and the peak pore water pressure reached during the cyclic loading relative to the initial confining pressure were the important factors controlling the compressibility on reconsolidation. The results of this test are summarized in a paper by Seed, Martin, and Lysmer.[21] Martin, Finn, and Seed[18] also summarized the test data on the compressibility on reconsolidation and expressed the coefficient of volume compressibility as follows in terms of the initial confining pressure, σ'_{v0}, and the effective stress level, σ'_v, from which reconsolidation starts:

$$m_v = \frac{(\sigma'_{v0})^{1-b}}{aK_2},$$
(7.44)

in which K_2, a, and b are experimental constants for a given sand. In what follows, this relationship will be used for the pore water pressure analysis.

7.5 NUMERICAL INTEGRATION OF EQUATIONS

The dynamic response in the horizontal direction can be determined by solving equation (7.23) in combination with the hyperbolic stress–strain hysteresis model described in the preceding section. Since the stress–strain model expressed by equations (7.12), (7.15) and (7.16) depends upon the shear modulus at small strain, G_t, and the shear strength, τ_f, which in turn

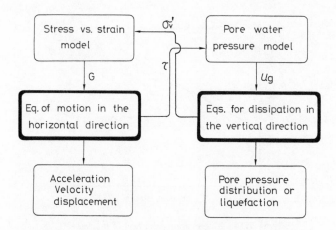

Figure 7.13 Flow chart of dynamic effective stress analysis

are functions of the current value of the effective vertical stress, σ_v', the stress–strain relationship must be modified successively in each interval of time of integration by considering the changing effective vertical stress which is determined by solving the set of equations (7.31), (7.32), (7.38) and (7.42). These four equations are solved for the four variables, W, Q, σ, and u, in combination with the stress path model described in the preceding section, in which the internally generated pore water pressure, u_g, is determined. To determine the value of u_g, the current change in shear stress must be in turn known. This information is supplied at each time interval through the solution of equation (7.23). A flow chart indicating the interplay between the models and the sets of equations is presented in Figure 7.13.

7.5.1 Numerical integration of the equation of motion in the horizontal direction

The motions of the horizontal stratum of sand deposits were determined by integrating the equation (7.23) with reference to the stress–strain relationships given by equation (7.12), (7.15) and (7.16). In carrying out the step-by-step integration with respect to time, it was postulated that in a small time interval, Δt, the stress–strain relation is approximated by a linear relationship as follows:

$$\tau = \tau_0 + G(\gamma - \gamma_0), \tag{7.45}$$

where τ_0 and γ_0 denote shear stress and shear strain at time, t, and G is the shear modulus. The value of the shear modulus, G, at time, t, is determined by drawing a straight line tangent to the stress–strain curve at point (γ_0, τ_0) as illustrated in Figure 7.14. Introducing equation (7.45) into (7.23) together

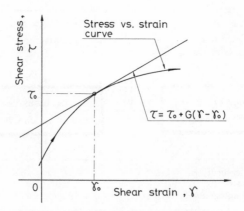

Figure 7.14   Linear approximation of a stress–strain curve

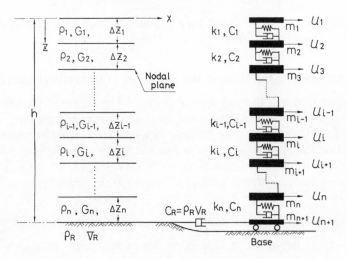

Figure 7.15 Lumped mass representation with energy-transmitting boundary

with the relation $\gamma = \partial u_x / \partial x$, we obtain

$$\rho \frac{\partial^2 u_x}{\partial t^2} = G \frac{\partial^2 u_x}{\partial x^2} + \frac{\partial}{\partial z}(\tau_0 - G\gamma_0). \qquad (7.46)$$

The above equation is a modified version of the original equation (7.23) to be adapted to the motion at time, t. In order to carry out integration with respect to the variable, z, the horizontal sand deposit was divided up into a number of layered elements, each with uniform properties as shown in Figure 7.15. This layered system is converted to a lumped mass system shown in Figure 7.15 by concentrating one-half of the mass of each layer at the layer boundaries. The one-half of the mass in the lowest layer is lumped at the level of the boundary between the soil stratum and the base rock. Input motions are applied to this mass through a viscous damping, $\rho_R V_R$, which is incorporated in order to take into account the effect of energy loss due to the dispersion of wave energy,[14] where ρ_R and V_R denote the unit mass and the velocity of shear wave propagation of the base material, respectively. The masses are connected by non-linear springs with stress–strain properties given by equation (7.12) for initial loading and by equations (7.15) and (7.16) for subsequent unloading and reloading. These equations reflect the non-linear, strain-dependent and hysteretic behaviour of sands. In addition to the hysteretic damping, viscous damping can be added independently if desired as illustrated in Figure 7.15. In the present study, the viscous damping was not employed.

The stepwise integration with time is made for equation (7.46) using the Newmark's β-method with $\beta = \frac{1}{4}$.[27]

7.5.2 Numerical integration of the equations in the vertical direction

The manner in which the soil deposit is divided into layered elements is the same as that adopted in the case of the horizontal response analysis illustrated in Figure 7.15. The seepage analysis was carried out only for the layers below the ground water table and not for the layer above. The set of basic equations given by equations (7.31), (7.32), (7.38) and (7.42) are integrated by means of the method of weighted residuals.[26]

7.5.3 Boundary conditions for the seepage equations

Since the soil deposit is assumed to be excited only in the horizontal direction at its base, the velocity of the solid skeleton, W, in the vertical direction is set equal to zero at the base. It is also assumed that the base underlying the soil deposit consists of impermeable materials, which means that the velocity of the water movement, Q, is zero at the base.

It is necessary to have two boundary conditions near the ground surface. They are specified at a fixed depth where the initial level of the water table has been located. As the dissipation of the internally generated pore water pressure progresses, the level of the water table migrates towards the ground surface. Then, time-dependent boundary conditions are established at the fixed elevation of the initial water table. The first boundary condition specifies, at the level of the initial water table, the time-dependent total stress which is computed as a sum of the static weight and the inertia forces of the soil and water lying above the level of the initial water table. The second boundary condition specifies the time-dependent velocity of water flow across the initial water table. This is calculated as the product of the permeability coefficient and the current value of the hydraulic gradient at the level of the initial water table. Since the pore water pressure difference and the distance between the initial level of the water table and the current level to which the water table has migrated is known, the hydraulic gradient is calculated as the ratio of the pore water pressure difference to the distance of the water table migration.

7.6 ANALYSIS OF A NIIGATA SITE DURING THE 1964 EARTHQUAKE

Detailed soil investigations were made recently, by means of a block sampling technique, at an open excavation site in Niigata where the construction of a sewage treatment facility was under way.[19] Signs of liquefac-

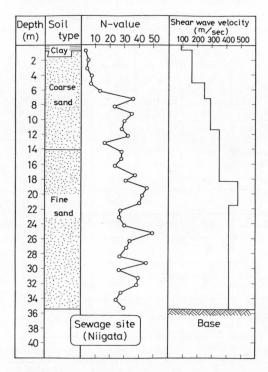

Figure 7.16   Soil profile at the sewage tratment site
in Niigata

tion such as surface crackings and sand volcanos were observed at this site
after the 1964 earthquake. Undisturbed sand samples were obtained to a
depth of 11.0 m and tested in the laboratory using a cyclic triaxial test
apparatus. The soil profile established in this investigation is shown in Figure
7.16, together with the velocity of shear wave propagation estimated from
the N-value of the standard penetration test. The layer division of this site
for the one-dimensional response analysis is shown in Table 7.1, along with
appropriate soil constants used in the analysis. The cyclic stress ratio causing
5% double amplitude strain in 20 cycles, $\bar{R}_{20}$, and the soil constant, B'_p,
were determined from the result of the laboratory cyclic triaxial test. The
unit weight, γ_t, porosity, n, and angle of internal friction, ϕ', were determined with reference to a number of test data obtained for the undisturbed
specimens of the sand deposits. The coefficient of permeability, k, was
roughly estimated on the basis of Hazen's empirical formula. The coefficient
of volume compressibility, m_v, was estimated using the formula of equation
(7.44). The shear wave velocity and unit weight of the base material were
estimated to be $V_R = 600$ m/s and $\gamma_R = 20$ kN/m^3, respectively.

Table 7.1   Soil properties used in the analysis

z (m)	Layer div.	Porosity n	Unit weight γ_t (kN/m³)	σ'_{v0} (kN/m²)	K_0 value	$G_t{}^*$ (kN/m²)	m_v† (m²/kN)	Permeability k (m/s)
2—	1	0.51	13	2.55	0.5	10000	—	—
	2	0.48	18	11.38	0.5	46×10^3	2.81×10^{-3}	10^{-4}
	3	0.48	18	23.54	0.5	46×10^3	2.13×10^{-3}	10^{-4}
4—	4	0.48	18	35.32	0.5	46×10^3	1.83×10^{-3}	10^{-4}
6—	5	0.44	19	50.03	0.6	120×10^3	1.60×10^{-3}	10^{-4}
8—	6	0.44	19	67.69	0.6	160×10^3	1.43×10^{-3}	10^{-4}
10—	7	0.44	19	85.35	0.6	160×10^3	1.31×10^{-3}	10^{-4}
12—	8	0.44	19	103.0	0.6	230×10^3	1.22×10^{-3}	10^{-4}
14—	9	0.42	20	121.6	0.7	250×10^3	1.14×10^{-3}	10^{-4}
16—	10	0.42	20	146.2	0.7	250×10^3	1.07×10^{-3}	10^{-4}
18— / 20—	11	0.42	20	175.6	0.7	460×10^3	0.994×10^{-3}	10^{-4}
22—	12	0.40	21	201.1	0.8	370×10^3	0.944×10^{-3}	10^{-4}
24— / 26—	13	0.40	21	228.1	0.8	370×10^3	0.900×10^{-3}	10^{-4}
28—	14	0.40	21	260.5	0.8	370×10^3	0.855×10^{-3}	10^{-4}
30— / 32—	15	0.40	21	292.8	0.8	370×10^3	0.818×10^{-3}	10^{-4}
34—	16	0.40	21	325.8	0.8	370×10^3	0.786×10^{-3}	10^{-4}

* Shear modulus at small strains
† Coefficient of volume compressibility
Sewage site (Niigata)

The base of the soil deposit to which input motion was applied was assumed to lie at a depth of 35 m.

At the time of the 1964 earthquake, acceleration records were obtained in the basement of the prefectural government office building standing on the hard deposit in the city of Akita located approximately 180 km from the epicentre. The recorded accelerations are presented in Figure 7.17. The N–S component of the accelerogram had a predominant period of 0.675 s and a

Table 7.1 (contd.)

Z (m)	Layer div.	ϕ'^* (deg.)	$\phi'_e{}^\dagger$ (deg.)	$\theta'_s{}^\ddagger$ (deg.)	Co-hesion (kN/m^2)	Cyclic str. ratio R'_{20}	Static strength τ_f(kN/m^2)	$\bar{B}_p$	$\bar{B}_u$
	1	20	27	17.6	10	—	7.29	—	—
2—	2	30	39	26.7	0	0.19	4.38	2.28	0.14
	3	30	39	26.7	0	0.19	9.06	2.28	0.14
4—	4	30	39	26.7	0	0.20	13.6	2.92	0.054
6—	5	35	44	31.3	0	0.23	25.7	4.15	0.009
8—	6	37	46	33.3	0	0.30	37.4	1.70	0.082
10—	7	37	46	33.3	0	0.33	47.2	0.92	0.09
12—	8	39	48	35.2	0	0.33	61.2	1.07	0.12
14—	9	40	49	36.1	0	0.35	81.6	1.02	0.10
16— 18—	10	40	49	36.1	0	0.35	98.1	1.02	0.10
20—	11	42	51	38.1	0	0.36	126.5	1.11	0.13
22—	12	40	49	36.1	0	0.38	146.2	0.87	0.08
24— 26—	13	40	49	36.1	0	0.38	165.9	0.87	0.08
28—	14	42	51	38.1	0	0.39	203.3	0.94	0.091
30— 32—	15	42	51	38.1	0	0.40	228.5	0.90	0.087
34—	16	42	51	38.1	0	0.40	253.8	0.90	0.087

* Angle of internal friction
† Angle of internal friction at very small confining stresses
‡ Angle of phase transformation tan $\theta'_s = 0.87$ tan $\phi' = 0.625$ tan ϕ'_2
Sewage site (Niigata)
Base rock: $\gamma_R = 20$ kN/m^3 $\bar{V}_R = 600$ m/s

maximum acceleration of 0.092g. Considering the sewage treatment site being located approximately 45 km from the epicentre, the original record of the N–S component was rescaled so as to have a maximum acceleration of 60 gal with a predominant period of 0.4 s. The motion thus modified was used as an incident input motion at the base of the deposit. The modified acceleration time history is shown in the lowest diagram in Figure 7.18.

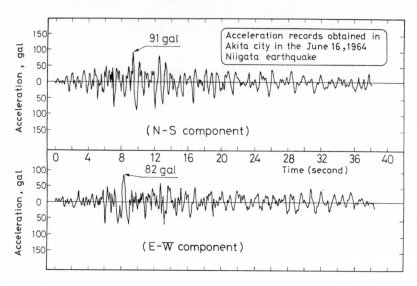

Figure 7.17 Acceleration records in the Akita city in the Niigata earthquake of 14 June 1964

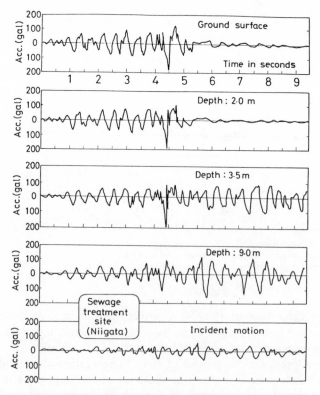

Figure 7.18 Computed accelerations at several depths of a deposit in Niigata

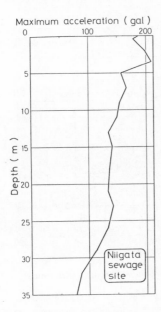

Figure 7.19 Distribution of the computed maximum acceleration

The computed acceleration time histories at several depths of the deposit are shown in Figure 7.18, where it is noted that around 4.5 s after the initiation of the shaking, liquefaction occurred in the soft layers near the surface. The distribution of the computed maximum acceleration is shown in Figure 7.19. The maximum acceleration on the ground is 0.187g and this is approximately in accord with the acceleration of 0.162g recorded during the 1964 earthquake at the basement of the Kawagishi-cho Apartment building

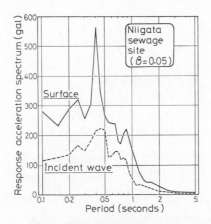

Figure   7.20 Response   acceleration spectrum of computed motions

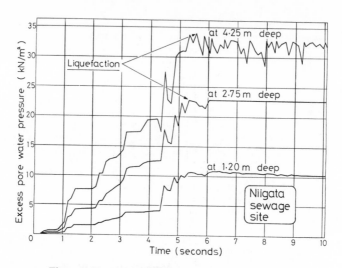

Figure 7.21   Variation of pore water pressures with time

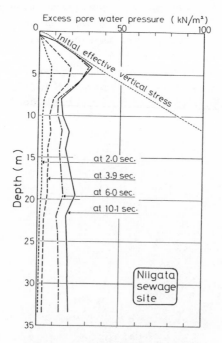

Figure 7.22   Pore water pressure distribution versus depth at several instances of time

which was located approximately 5 km from the sewage treatment site and where liquefaction was seen developing extensively during the 1964 event. For the accelerations on the surface and at the base, acceleration response spectra were computed as shown in Figure 7.20. The time changes of pore water pressures at three depths near the ground surface are shown in Figure 7.21, where it is noted that around 4.5 s the pore water pressure developed abruptly and at 6.0 s after the initiation of shaking, the surface deposit is put into a state of liquefaction. The distributions of the pore water pressure versus depth at several instances of time during seismic loading are shown in Figure 7.22. It is seen in these figures that the pore water pressure builds up gradually as the shaking proceeds and at 4.5 s after the initiation of the shaking, the loose layer near the surface is brought into the state of liquefaction, followed by an undulating wave with a prolonged period of approximately 1.0 s and with much smaller peak accelerations.

CONCLUSIONS

By integrating the stress path model and the continuum theory of two-phase media, a dynamic effective stress analysis scheme was organized for the one-dimensional level ground-response analysis in which the time-dependent stiffness degrading effect caused by the contemporaneous generation and dissipation of pore water pressures is taken into account. The proposed method was applied for a liquefaction analysis of a loose sand deposit at a sewage treatment site in Niigata, Japan, where signs of liquefaction were observed during the 1964 earthquake and also where detailed soil investigation were performed after the earthquake. The result of analysis showed that when the first peak acceleration in time history occurred the effective stress in the sand deposit became almost zero, inducing liquefaction throughout the loose deposit to a depth of 5.0 m. The computed maximum horizontal acceleration became 0.187 g on the ground surface which was approximately in agreement with the acceleration recorded at the time of the 1964 earthquake. The computed predominant period of motion on the ground following the liquefaction was approximately 1.0 s, indicating the effect of the stiffness degradation due to the pore water pressure build-up during the earthquake.

ACKNOWLEDGEMENTS

The kindness of Dr K. Mori in editing the original draft of this paper is sincerely acknowledged.

APPENDIX 1

In the actual time step of computation, the parameter m in equation (7.1) is determined in the following fashion. When loading starts from the initially consolidated state such as from point 1 in Figure 7.2(b), m is set equal to the consolidation pressure, σ'_{v0}. When the virgin loading is resumed following a reloading such as from stress point 4 in Figure 7.2(b), the current shear stress, τ_c, and the effective vertical stress, σ'_{vc}, as shown in Figure A7.1.1, are

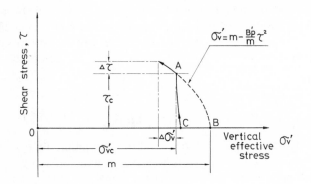

A7.1.1 Change in parabolic stress path during virgin loading

introduced in equation (7.1) as follows,

$$\sigma'_{vc} = m - \frac{B'_p}{m}\tau_c^2. \tag{A7.1.1}$$

Then the parameter m is determined by

$$\frac{m}{\sigma'_{vc}} = \tfrac{1}{2}\{1 + \sqrt{[1 + 4B'_p(\tau_c/\sigma'_{vc})^2]}\}. \tag{A7.1.2}$$

As illustrated in Figure A7.1.1, the parameter m specified a point of intercept, B, of the parabolic curve on the σ'_v axis. The next incremental step can be determined by a formula, as follows, obtained by differentiating equation (7.1),

$$\Delta\sigma'_v = \frac{2B'_p}{m}\tau_c\,\Delta\tau, \tag{A7.1.3}$$

where $\Delta\tau$ is an increment of shear stress.

APPENDIX 2

To visualize the pictures of the stress path during the unloading phase, equation (7.2) is integrated starting from the most recent peak state of stress (σ'_{vm}, τ_m). Integrating equation (7.2) with an initial condition, $\sigma'_v = \sigma'_{vm}$ and $\tau = \tau_m$, one obtains

$$\frac{\sigma'_v/\sigma'_{v0} - \kappa}{\sigma'_{vm}/\sigma'_{v0} - \kappa} = \exp\left[B'_u\left\{ -\frac{1}{2}\left(\frac{\tau}{\sigma'_{v0}}\right)^2 + \frac{\tau_m}{\sigma'_{v0}}\frac{\tau}{\sigma'_{v0}} - \frac{1}{2}\left(\frac{\tau_m}{\sigma'_{v0}}\right)^2 \right\} \right] \qquad (7.2.1)$$

Numerical examples for the case of $B'_u = 1.0$ and $\kappa = 0.06$ are demonstrated in Figure A7.2.1, in which the shear stress is reduced from the most recent peak value of 25 kN/m² to zero. It is noted in the figure that the rate of pore water pressure build-up during unloading decreases as the vertical effective stress, σ'_{vm}, at the most recent peak becomes smaller.

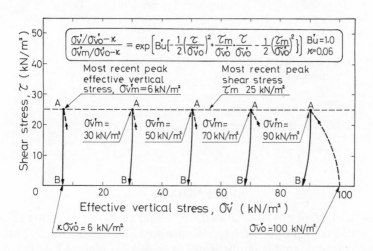

Figure A7.2.1 Examples of the stress paths during unloading

APPENDIX 3

In the theory of water-saturated porous media, equations of motions are first established by considering stresses and forces which act on the solid portion and fluid portion of the two-phase media. Since these stresses and forces are different in definition from the effective stresses, σ'_z and σ'_x, and the pore water pressure, u, commonly used in the discipline of the soil mechanics, it is desirable to clarify the relationships between them. It is to be noted that the pore water pressure acting on the solid portion is $(1-n)u$

and that on the fluid portion is nu, where n is the porosity of soil. Therefore, the stresses σ_z and σ_x in equation (7.25) consist of two parts:

$$\sigma_x = \sigma'_x + (1-n)u,$$
$$\sigma_z = \sigma'_z + (1-n)u. \qquad (A7.3.1)$$

Similarly, the stress p in equation (7.27) is given by

$$p = nu. \qquad (A7.3.2)$$

These relationships are summarized in Table A7.3.1 for the component in the z-direction.

Table A7.3–1. Relationships between two stress systems

	Effective stress	Pore water pressure	Total
Stresses acting on the solid phase	σ'_z	$(1-n)u$	σ_z
Stresses acting on the fluid phase	0	nu	p
Total	σ'_z	u	σ^*

$*\,\sigma$ is the total stress in equation (7.25)

APPENDIX 4

Consider an element of soil saturated with water as shown in Figure A7.4.1. The soil consists of porous skeleton having pores filled with water. Suppose the skeleton is separated from the water as shown in Figure A7.4.1(b). If the bulk volume of the skeleton is denoted by V_b, it is related to the net volume of the solid substance, V_s, and the porosity, n, as follows;

$$V_b = \frac{V_s}{1-n}. \qquad (A7.4.1)$$

Next, suppose the skeleton is taken away and the pores are left behind, giving the fluid phase the appearance of a bubble structure as shown in Figure A7.4.1(c). If the bulk volume of the bubble structure is denoted by V_a, it is related to the net volume of water, V_w, and the porosity, n, as follows;

$$V_a = \frac{V_w}{n}. \qquad (A7.4.2)$$

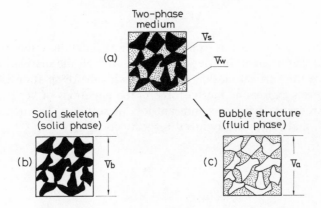

Figure A7.4.1

Suppose volume changes ΔV_b and ΔV_a occur in the bulk volumes of the solid skeleton and the pore bubble structures, respectively. Then, the volumetric strains, $\Delta V_b/V_b$ and $\Delta V_a/V_a$, for each phase can be easily derived by differentiating equations (A7.4.1) and (A7.4.2):

$$\frac{\Delta V_b}{V_b} = \frac{\Delta V_s}{V_s} + \frac{\Delta n}{1-n},$$

$$\frac{\Delta V_a}{V_a} = \frac{\Delta V_w}{V_w} - \frac{\Delta n}{n},$$

(A7.4.3)

in which $\Delta V_s/V_s$ and $\Delta V_w/V_w$ represent the volumetric strains of the solid substance and water itself, and Δn is the change in porosity. It has been assumed so far that the volume changes in the solid skeleton and the pore bubble structure can take place independently without any interaction between two phases. However, this assumption is not correct, because there exists a restrictive relationship as follows, derived by eliminating Δn from the two expressions in equation (A7.4.3):

$$(1-n)\frac{\Delta V_b}{V_b} + n\frac{\Delta V_a}{V_a} = (1-n)\frac{\Delta V_s}{V_s} + n\frac{\Delta V_w}{V_w}.$$

(A7.4.4)

The volumetric strains of the solid substance, $\Delta V_s/V_s$, and water, $\Delta V_w/V_w$, are known quantities if the stresses causing the volume changes are known. Therefore, bulk volumetric strains, $\Delta V_b/V_b$ and $\Delta V_a/V_a$, must occur in such a way that they satisfy the above relationship, in which the value on the right-hand side is given. This is what is meant by the volumetric compatibility equation in the theory of two-phase media.

APPENDIX 5

If a horizontal soil deposit is excited in the horizontal direction, the vertical motion that can occur in the deposit is only through the dissipation of pore water. Since the vertical motion associated with the dissipation is generally slow, the terms expressing inertia force in the equations of vertical motions may be disregarded without significant loss of accuracy. In this case, equations (7.31) and (7.32) are reduced respectively to

$$\frac{\partial \sigma}{\partial z} = \gamma_t \tag{A7.5.1}$$

and

$$bQ = -n^2 \frac{\partial u}{\partial z} + n^2 \gamma_w. \tag{A7.5.2}$$

Integrating equation (A7.5.1), with reference to a boundary condition $\sigma = 0$ on the ground surface, $z = 0$, one obtains

$$\sigma = \gamma_t z. \tag{A7.5.3}$$

Then, it follows that

$$\frac{\partial \sigma}{\partial t} = 0. \tag{A7.5.4}$$

This indicates the fact that the total vertical stress is simply determined by the weight of the soil existing above the elevation being considered, and does not change with time.

Taking the derivative of equation (A7.5.2) with respect to depth z,

$$\frac{\partial Q}{\partial z} = -\frac{n^2}{b} \frac{\partial^2 u}{\partial z^2}. \tag{A7.5.5}$$

Introducing this relation into equation (7.38), it follows that

$$\frac{\partial W}{\partial z} = \frac{n^2}{b} \frac{\partial^2 u}{\partial z^2}. \tag{A7.5.6}$$

By introducing equations (A7.5.4) and (A7.5.6) into equation (7.42), one obtains

$$\frac{\partial^2 u}{\partial z^2} = \frac{\gamma_w m_v}{k} \left(\frac{\partial u}{\partial t} - \frac{\partial u_g}{\partial t} \right), \tag{A7.5.7}$$

where equation (7.18) has been used. This equation is the same as that used by Finn, Lee, and Martin[3] and Seed, Martin, and Lysmer.[21]

REFERENCES

1. Biot, M. A. Theory of propagation of elastic waves in a fluid-saturated porous solid, *J. Acoustic Soc. America*, **28**(2), 168–78, 1956.
2. Biot, M. A. Mechanics of deformation and acoustic propagation in porous media, *J. App. Phys.* **33**(4), 1482–98, 1962.
3. Finn, W. D. L., Lee, K. W., and Martin, G. R. An effective stress model for liquefaction, *Proc. ASCE*, **103** (GT6) 517–33, 1977.
4. Finn, W. L. D. Critical review of dynamic effective stress analysis, *Proc. 2nd U.S. National Conference on Earthquake Engineering*, Stanford University, 853–67, 1979.
5. Ghaboussi, J., and Dikmen, S. U. Liquefaction analysis of horizontally layered sands, *Proc. ASCE*, **104** (GT3), 341–56, 1978.
6. Hardin, B. O., and Drnevich, V. P. Shear modulus and damping in soils; measurement and parameter effects, *Proc. ASCE*, **98** (SM6), 603–24, 1972.
7. Ishihara, K. Propagation of compressional waves in a saturated soil, *Proc. Int. Symp. Wave Propagation and Dynamic Properties of Earth Materials*, University of New Mexico Press, 195–206, 1968.
8. Ishihara, K., Tatsuoka, F., and Yasuda, S. Undrained deformation and liquefaction of sand under cyclic stress, *Soils and Foundations*, **15**(1), 29–44, 1975.
9. Ishihara, K., Iwamoto, S., Yasuda, S., and Takatsu, H. Liquefaction of anisotropically consolidated sand, *Proc. 9th Int. Conf. Soil Mechanics and Foundation Engineering*, Tokyo, **2**, 261–64, 1977.
10. Ishihara, K., and Okada, S. Effects of stress history on cyclic behavior of sand, *Soils and Foundations*, **14**(4), 31–45, 1978.
11. Ishihara, K., and Yamazaki, F. Cyclic simple shear tests on saturated sand in multi-directional loading, *Soils and Foundations*, **20**(1), 45–59, 1980.
12. Ishihara, K., and Towhata, I. Effective stress method in one-dimensional soil response analysis, *Proc. 7th Word Conf. Earthquake Engineeering*, Istanbul, Turkey, 1980.
13. Iwasaki, T., and Tatsuoka, F. Effects of grain size and grading on dynamic shear moduli of sands, *Soils and Foundations*, **17**(3), 19–35, 1977.
14. Joyner, W. B., and Chen, A. T. F. Calculation of nonlinear ground response in earthquakes, *Bull. Seismological Soc. America*, **65**(5) 1315–36, 1975.
15. Kondner, R. L., and Zelasko, J. S. A hyperbolic stress–strain formulation of sands, *Proc. 2nd Pan American Conf. Soil Mechanics and Foundations Engineering*, 289–324, 1963.
16. Lee, K. L., and Albeisa, A. Earthquake induced settlements in saturated sands, *Proc. ASCE*, **100**(GT4), 387–406, 1974.
17. Liou, C. P., Streeter, V. L., and Richart, F. E. Numerical model for liquefaction, *Proc. ASCE*, **103**(GT6), 589–606, 1977.
18. Martin, T. R., Finn, W. D. L., and Seed, H. B. Fundamentals of liquefaction under cyclic loading, *Proc. ASCE*, **101** (GT5), 423–38, 1975.
19. Mori, K., and Ishihara, K. Undisturbed block sampling of Niigata sand, *Proc. 6th Asian Regional Conf. Soil Mechanics and Foundation Engineering*, Singapore, **1**, 39–42, 1979.
20. Sato, T., Shibata, T., and Kosaka, M. Dynamic behavior and liquefaction of saturated soil, *Proc. Int. Symp. Soils under Cyclic and Transient Loading*, Swansea, **2**, 523–32, 1980.
21. Seed, H. B., Martin, P. P., and Lysmer, J. Pore water pressure changes during soil liquefaction, *Proc. ASCE*, **102**(GT4), 323–46, 1976.

22. Sherif, M. A., Ishibashi, I., and Tsuchida, C. Pore pressure prediction during earthquake loading, *Soils and Foundations*, **18**(4), 19–30, 1978.
23. Terzaghi, K., and Peck, R. B. *Soil Mechanics in Engineering Practice*, Wiley, p. 50, 1967.
24. Yoshimi, Y., Kuwabara, F., and Tokimatsu, K. One-dimensional volume change characteristics of sands under very low confining stresses, *Soils and Foundations*, **15**(3), 51–60, 1975.
25. Yoshimi, Y., and Tokimatsu, K. Two-dimensional pore pressure changes in sand deposits during earthquakes, *Proc. 2nd Int. Conf. Microzonation for Safer Construction, Research and Application*, San Francisco, **3**, 853–63, 1978.
26. Zienkiewicz, O. C. *The Finite Element Method*, 3rd edn, MacGraw-Hill, p. 49, 1977.
27. Zienkiewicz, O. C., Chang, C. T., and Hinton, E. Non-linear seismic response and liquefaction, *Int. J. Num. and Anal. Meth. in Geomechanics*, **2**, 381–404, 1978.

Soil Mechanics—Transient and Cyclic Loads
Edited by G. N. Pande and O. C. Zienkiewicz
© 1982 John Wiley & Sons Ltd

Chapter 8

Elastoplastic and Viscoplastic Constitutive Models for Soils with Application to Cyclic Loading

Z. Mroz and V. A. Norris

8.1 INTRODUCTION

The present chapter is aimed at discussing constitutive relations for soils and the hardening rules which follow from extensions of the classical plasticity theory. When viscous effects are neglected this theory results in constitutive equations between stress and strain increments (or rates) with instantaneous stiffness or compliance matrices being dependent on actual stress or strain states and the previous deformation history. It is also important that this theory allows for distinction between loading and unloading or reverse loading events for which different incremental relations are valid.

For soils under monotonic loading, the simple elastoplastic or non-linear elastic material models can be used to simulate the deformational response of a material with sufficient accuracy. For a perfectly plastic model both hardening and softening phenomena are neglected and the flow law is associated with the yield condition or the plastic potential by the gradient rule. For an isotropic hardening model, the irreversible void ratio or density is assumed to be a state parameter. Using this model, one is able to predict hardening, softening and critical state response. However, for more complex loading programmes involving one-way loading followed by unloading, or for repetitive action of loads, more complex hardening rules should be examined in order to describe the effects occurring for these loading conditions more realistically.

When using the concepts of the theory of plasticity we have to formulate

(1) the yield condition defining elastic and inelastic deformation domains;
(2) the flow rule relating the increments or rates of stress and irreversible strain;
(3) the hardening rule specifying the evolution of the yield surface in the course of plastic deformation and the evolution of hardening parameters defining the state of the material.

The formulation of proper evolution rules is a difficult part of constitutive modelling, since memory of particular loading events and progressive cyclic hardening or softening phenomena should be incorporated in the model.

The need for a more accurate description of inelastic behaviour follows for several reasons, namely

(1) a more complex model of a wide range of applicability, including monotonic and variable loading, allows for more rational experimental work in which identification and verification programmes can be selected and further improvements of the model can be introduced in subsequent steps,

(2) qualitative interpretation of material behaviour for various loading histories provided by a more complex model is useful in assessing and formulating design rules,

(3) a simpler version of the model for numerical implementation could be obtained by neglecting some interactions; the degree of simplification should be related to the type of problems investigated.

It is therefore important that the model should be formulated in a 'modular' way, that is, allowing for neglect or consideration of some effects without affecting other assumptions, or identified material parameters.

In Section 8.2 the general discussion of incremental constitutive relations will be presented, whereas in Section 8.3 simple and composite hardening rules will be discussed with emphasis on their applicability to variable loading conditions. In Sections 8.4 and 8.5, the static soil behaviour for loading–unloading and reverse loading histories will be considered in the light of model predictions, whereas in Section 8.7 the cyclic soil response will be discussed. In Section 8.6, a viscoplastic hardening rule will be formulated.

8.2 RATE (OR INCREMENTAL) CONSTITUTIVE RELATIONS

In the classical theory of plasticity it is usually assumed that a material exhibits elastic behaviour for stress states corresponding to the interior of the yield surface, whereas plastic deformation occurs for stress increments directed into the exterior of this surface. Thus each deformation process can be divided into an active or loading process involving variation of plastic strain and an unloading or elastic process corresponding to elastic behaviour. However, geological materials such as clay, sand or rock do not exhibit purely elastic behaviour during unloading and the yield surface, when defined by a small offset value, usually encloses an elastic domain lying in the vicinity of the loading point. Indeed, in some cases the yield surface may not exist at all. Therefore, a more extended formulation should allow for nullifying the elastic domain in the limiting case and for defining the subsequent reverse loading event.

Consider first the rate-independent behaviour of the material for which rates (or increments) of stress and strain are interrelated by the equations

$$\dot{\boldsymbol{\sigma}} = \mathbf{D}\dot{\boldsymbol{\varepsilon}} \quad \text{or} \quad \dot{\boldsymbol{\varepsilon}} = \mathbf{C}\dot{\boldsymbol{\sigma}}, \tag{8.1}$$

where the matrices $\mathbf{D}$ and $\mathbf{C}$ depend on stress, strain and hardening parameters but not on the stress or strain rates. Thus the relations (8.1) are *linear* and *homogeneous* in rates of stress and strain and the material behaviour does not depend on the natural time scale.

Our analysis will be limited to small strain theory, but all conclusions remain valid when account is taken of translation and rotation of the element in defining the strain rate.

Let there exist at each stage of the deformation process a yield (or loading) surface separating domains of active loading and elastic unloading (or reverse loading). In stress space, this surface is represented by the equation

$$f(\boldsymbol{\sigma}, \boldsymbol{\alpha}) = 0. \tag{8.2}$$

For strain-controlled processes, it is convenient to consider the loading surface in strain space, thus

$$\Phi(\boldsymbol{\varepsilon}, \boldsymbol{\varepsilon}^{\mathrm{p}}, \boldsymbol{\alpha}) = 0, \tag{8.3}$$

where $\boldsymbol{\sigma}$ and $\boldsymbol{\varepsilon}$ are stress and strain tensors and the usual decomposition

$$\boldsymbol{\varepsilon} = \boldsymbol{\varepsilon}^{\mathrm{e}} + \boldsymbol{\varepsilon}^{\mathrm{p}} \tag{8.4}$$

into elastic and plastic portions occurs in small strain theory. The symbol $\boldsymbol{\alpha}$ collectively denotes the hardening parameters.

Since different rate relations occur for active loading and unloading trajectories, we can write

$$\dot{\boldsymbol{\sigma}} = \mathbf{D}_2\dot{\boldsymbol{\varepsilon}} \quad \text{for } \Phi = 0, \quad \dot{\Phi}_\varepsilon \equiv \left(\frac{\partial \Phi}{\partial \boldsymbol{\varepsilon}}\right)^{\mathrm{T}} \cdot \dot{\boldsymbol{\varepsilon}} > 0, \quad \dot{\boldsymbol{\varepsilon}} \in V_{\varepsilon_2},$$

$$\dot{\boldsymbol{\sigma}} = \mathbf{D}_1\dot{\boldsymbol{\varepsilon}} \quad \text{for } \Phi < 0 \text{ or } \dot{\Phi}_\varepsilon < 0, \quad \Phi = 0, \quad \dot{\boldsymbol{\varepsilon}} \in V_{\varepsilon_1}, \tag{8.5}$$

and similar relations in stress rate space

$$\dot{\boldsymbol{\varepsilon}} = \mathbf{C}_2\dot{\boldsymbol{\sigma}} \quad \text{for } f = 0, \dot{f}_\sigma = \left(\frac{\partial f}{\partial \boldsymbol{\sigma}}\right)^{\mathrm{T}} \cdot \dot{\boldsymbol{\sigma}} > 0, \dot{\boldsymbol{\sigma}} \in V_{\sigma_2},$$

$$\dot{\boldsymbol{\varepsilon}} = \mathbf{C}_1\dot{\boldsymbol{\sigma}} \quad \text{for } f < 0 \text{ or } \dot{f}_\sigma < 0, f = 0, \quad \dot{\boldsymbol{\sigma}} \in V_{\sigma_1}. \tag{8.6}$$

In above we use a dot to denote scalar product, or a trace operator of a matrix product.

Thus there are two different matrices for stress or strain rates directed into two semispaces separated by the hyperplanes Π_Φ and Π_f, tangential to the yield surfaces, Figure 8.1(a, b). Assume now that the *continuity condition* is satisfied; that is, for stress or strain rates directed tangentially to yield surfaces

$$\mathbf{D}_2\dot{\boldsymbol{\varepsilon}} = \mathbf{D}_1\dot{\boldsymbol{\varepsilon}} \qquad \text{for } \Phi = \dot{\Phi}_\varepsilon = 0,$$
$$\mathbf{C}_2\dot{\boldsymbol{\sigma}} = \mathbf{C}_1\dot{\boldsymbol{\sigma}} \qquad \text{for } f = \dot{f}_\sigma = 0. \tag{8.7}$$

The continuity condition (8.7) imposes an essential constraint on the constitutive matrices which, must be interrelated: viz.[23,27]

$$\dot{\boldsymbol{\sigma}}_2 = \mathbf{D}_2\dot{\boldsymbol{\varepsilon}} = \mathbf{D}_1\dot{\boldsymbol{\varepsilon}} + \mathbf{h}(\mathbf{n}_\Phi^{\mathrm{T}} \cdot \dot{\boldsymbol{\varepsilon}}) = \dot{\boldsymbol{\sigma}}_1 - \dot{\boldsymbol{\sigma}}^{\mathrm{r}},$$
$$\dot{\boldsymbol{\varepsilon}}_2 = \mathbf{C}_2\dot{\boldsymbol{\sigma}} = \mathbf{C}_1\dot{\boldsymbol{\sigma}} + \mathbf{g}(\mathbf{n}_f^{\mathrm{T}} \cdot \dot{\boldsymbol{\sigma}}) = \dot{\boldsymbol{\varepsilon}}_1 + \dot{\boldsymbol{\varepsilon}}^{\mathrm{p}}, \tag{8.8}$$

where $\mathbf{h}$ and $\mathbf{g}$ are arbitrary tensors and $\mathbf{n}_\Phi$, $\mathbf{n}_f$ are the normalized gradients, i.e.

$$\mathbf{n}_\Phi = \frac{\partial\Phi/\partial\boldsymbol{\varepsilon}}{[(\partial\Phi/\partial\boldsymbol{\varepsilon})^{\mathrm{T}} \cdot (\partial\Phi/\partial\boldsymbol{\varepsilon})]^{1/2}}, \qquad \mathbf{n}_f = \frac{\partial f/\partial\boldsymbol{\sigma}}{[(\partial f/\partial\boldsymbol{\sigma})^{\mathrm{T}} \cdot \partial f/\partial\boldsymbol{\sigma}]^{1/2}}, \tag{8.9}$$

The relations (8.8) can be given an interpretation which is familiar in plasticity theory: identifying the matrices $\mathbf{D}_1$ and $\mathbf{C}_1$ with elastic stiffness and compliance matrices, we have $\dot{\boldsymbol{\varepsilon}}^{\mathrm{e}} = \dot{\boldsymbol{\varepsilon}}_1 = \mathbf{C}_1\dot{\boldsymbol{\sigma}}$, and $\dot{\boldsymbol{\sigma}}^{\mathrm{e}} = \dot{\boldsymbol{\sigma}}_1 = \mathbf{D}_1\dot{\boldsymbol{\varepsilon}}$; $\dot{\boldsymbol{\varepsilon}}^{\mathrm{p}}$ and $\dot{\boldsymbol{\sigma}}^{\mathrm{r}}$ now represent the plastic strain rate and the relaxation stress rate superimposed upon the elastic stress rate. Denoting

$$\mathbf{h} = -M\mathbf{n}_h, \qquad \mathbf{g} = \frac{1}{K}\mathbf{n}_g, \tag{8.10}$$

where $\mathbf{n}_h$ and $\mathbf{n}_g$ are tensors normalized according to (8.9) (or unit vectors in vector space), and M, K are scalar functions, we can rewrite (8.8) as follows:

$$\dot{\boldsymbol{\varepsilon}} = \dot{\boldsymbol{\varepsilon}}^{\mathrm{e}} + \dot{\boldsymbol{\varepsilon}}^{\mathrm{p}} = \mathbf{C}_1\dot{\boldsymbol{\sigma}} + \frac{1}{K}\mathbf{n}_g(\mathbf{n}_f^{\mathrm{T}} \cdot \dot{\boldsymbol{\sigma}}), \qquad f = 0, \dot{f}_\sigma > 0, \tag{8.11}$$

$$\dot{\boldsymbol{\sigma}} = \dot{\boldsymbol{\sigma}}^{\mathrm{e}} - \dot{\boldsymbol{\sigma}}^{\mathrm{r}} = \mathbf{D}_1\dot{\boldsymbol{\varepsilon}} - M\mathbf{n}_h(\mathbf{n}_\Phi^{\mathrm{T}} \cdot \dot{\boldsymbol{\varepsilon}}), \qquad \Phi = 0, \dot{\Phi}_\varepsilon > 0. \tag{8.12}$$

The scalar functions K and M will be called respectively the *hardening and relaxation moduli*. Similarly, the relations

$$\dot{\boldsymbol{\varepsilon}}^{\mathrm{p}} = \frac{1}{K}\mathbf{n}_g(\mathbf{n}_f^{\mathrm{T}} \cdot \dot{\boldsymbol{\sigma}}), \qquad \dot{\boldsymbol{\sigma}}^{\mathrm{r}} = M\mathbf{n}_h(\mathbf{n}_\Phi^{\mathrm{T}} \cdot \dot{\boldsymbol{\varepsilon}}) \tag{8.13}$$

are respectively the non-associated *flow rule* and the non-associated *relaxation rule*. From (8.13) it follows that

$$K = \frac{(\mathbf{n}_f^{\mathrm{T}} \cdot \dot{\boldsymbol{\sigma}})}{(\dot{\boldsymbol{\varepsilon}}^{\mathrm{p}} \cdot \dot{\boldsymbol{\varepsilon}}^{\mathrm{p}})^{1/2}}, \qquad M = \frac{(\dot{\boldsymbol{\sigma}}^{\mathrm{r}} \cdot \dot{\boldsymbol{\sigma}}^{\mathrm{r}})^{1/2}}{(\mathbf{n}_\Phi^{\mathrm{T}} \cdot \dot{\boldsymbol{\varepsilon}})}. \tag{8.14}$$

Since, in general, $\mathbf{n}_g \neq \mathbf{n}_f$ and $\mathbf{n}_h \neq \mathbf{n}_\Phi$ in (8.13), the plastic strain rate vector $\dot{\boldsymbol{\varepsilon}}^{\mathrm{p}}$ departs from the direction of the exterior normal vector $\mathbf{n}_f$ to the yield surface, and similarly the direction of $\dot{\boldsymbol{\sigma}}^r$ does not coincide with that of $\mathbf{n}_\Phi$ in the strain space. The case of *associated flow* and *relaxation rules* is obtained by postulating that $\mathbf{n}_g = \mathbf{n}_f$ and $\mathbf{n}_h = \mathbf{n}_\Phi$, which gives

$$\text{(a)} \quad \dot{\boldsymbol{\varepsilon}}^p = \frac{1}{K}\mathbf{n}_f(\mathbf{n}_f^T \cdot \dot{\boldsymbol{\sigma}}),$$

$$\text{(b)} \quad \dot{\boldsymbol{\sigma}}^r = M\mathbf{n}_\Phi(\mathbf{n}_\Phi^T \cdot \dot{\boldsymbol{\varepsilon}}). \tag{8.15}$$

The relations (8.15) are dual representations of the plastic deformation rule: whereas the flow rule (8.15(a)) specifies the plastic strain rate corresponding to a given stress rate, the relaxation rule (8.15(b)) specifies that portion of the stress rate which should be subtracted from the elastic stress rate associated with a given rate of strain. The hardening and relaxation moduli can be interrelated by inverting the rate equation (8.11) and identifying it with (8.12).

We have

$$\dot{\boldsymbol{\sigma}} = \mathbf{D}_1(\dot{\boldsymbol{\varepsilon}} - \dot{\boldsymbol{\varepsilon}}^p) = \dot{\boldsymbol{\sigma}}^e - \dot{\boldsymbol{\sigma}}^r = \left[\mathbf{D}_1 - \frac{\mathbf{D}_1\mathbf{n}_f\mathbf{n}_f^T \cdot \mathbf{D}_1}{K + \mathbf{n}_f^T \cdot \mathbf{D}_1\mathbf{n}_f}\right]\dot{\boldsymbol{\varepsilon}}, \tag{8.16}$$

limiting our discussion to the associated flow rule.

Since the yield condition can be expressed in terms of total strain as

$$\Phi(\boldsymbol{\varepsilon}, \boldsymbol{\varepsilon}^p, \boldsymbol{\alpha}) = f[\mathbf{D}_1(\boldsymbol{\varepsilon} - \boldsymbol{\varepsilon}^p), \boldsymbol{\alpha}] = 0, \tag{8.17}$$

we have

$$\mathbf{n}_\Phi = \frac{\mathbf{D}_1\mathbf{n}_f}{[(\mathbf{D}_1\mathbf{n}_f)^T \cdot (\mathbf{D}_1\mathbf{n}_f)]^{1/2}} \tag{8.18}$$

and the relation (8.16) can be rewritten as follows:

$$\dot{\boldsymbol{\sigma}} = \dot{\boldsymbol{\sigma}}^e - \dot{\boldsymbol{\sigma}}^r, \quad \dot{\boldsymbol{\varepsilon}}^e = \mathbf{D}_1\dot{\boldsymbol{\varepsilon}}, \quad \dot{\boldsymbol{\sigma}}^r = M\mathbf{n}_\Phi(\mathbf{n}_\Phi^T \cdot \dot{\boldsymbol{\varepsilon}}), \tag{8.19}$$

where

$$M = \frac{1}{K' + \mathbf{n}_\Phi^T \cdot \mathbf{C}_1\mathbf{n}_\Phi}, \quad \mathbf{C}_1 = \mathbf{D}_1^{-1}, \quad K' = \frac{K}{(\mathbf{D}_1\mathbf{n}_f)^T \cdot (\mathbf{D}_1\mathbf{n}_f)}. \tag{8.20}$$

Equations (8.20) provide the relationship between the hardening and relaxation moduli. Figure 8.1 presents the yield surfaces in stress and strain spaces. The hardening modulus K is obtained by projecting the stress rate onto the normal vector $\mathbf{n}_f$ and dividing by the modulus of $\dot{\boldsymbol{\varepsilon}}^p$. Similarly, the relaxation modulus M is defined as the ratio of the modulus of $\dot{\boldsymbol{\sigma}}^r$ and the projection of $\dot{\boldsymbol{\varepsilon}}$ onto the normal vector $\mathbf{n}_\Phi$. Figure 8.2 shows the uniaxial stress–strain curve and the variation of the tangent, plastic and relaxation moduli E_t, E_p

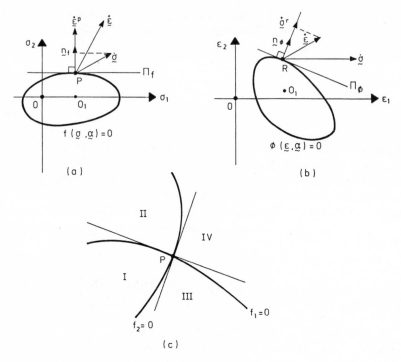

Figure 8.1   (a,b) Yield surfaces in stress and strain space; (c) intersection of two analytical surfaces

and E_r. Since

$$E_t = \frac{d\sigma}{d\varepsilon}, \quad E_p = \frac{d\sigma}{d\varepsilon^p}, \quad E_r = \frac{d\sigma^r}{d\varepsilon},$$

we have

$$E = E_r + E_t, \qquad \frac{E_r}{E} = \frac{1}{1 + E_p/E} \tag{8.21}$$

and the hardening modulus K corresponds to E_p in the uniaxial case.

Whereas the representation of the yield surface in stress space is common, the use of total strains rather than stresses does have certain advantages. The relaxation rule (8.15) is then characterized by a normality property in strain space, and the relaxation modulus specifies the rate of 'relaxation' of stress with respect to an elastic solution. This stress relaxation process is simulated numerically by means of the iterative 'initial stress' or 'initial strain' techniques applied to a boundary value problem. For softening material response, representation in strain space allows us to avoid the

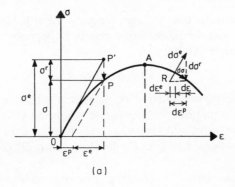

(a)

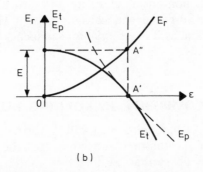

(b)

Figure 8.2 (a) Stress–strain curve; (b) variation of tangent, plastic and relaxation moduli

non-uniqueness which occurs for stress-controlled loading programmes. A more detailed discussion of constitutive relations expressed in terms of total strain and strain rate can be found in Ref. 28.

The rate equations (8.11) and (8.12) are the most general linear relations occurring for a regular loading surface when there is a unique normal vector at the loading point. A singular regime occurs when two or more analytical surfaces intersect at the loading point. Figure 8.1(c) presents the case when there are four regions I, II, III, and IV, where different constitutive matrices may occur in stress space. Following (8.8), we may write

$$\dot{\boldsymbol{\varepsilon}}_1 = \mathbf{C}_1 \dot{\boldsymbol{\sigma}}, \qquad\qquad\qquad \dot{\boldsymbol{\sigma}} \in V_{\mathrm{I}},$$

$$\dot{\boldsymbol{\varepsilon}}_2 = \mathbf{C}_2 \dot{\boldsymbol{\sigma}} = \mathbf{C}_1 \dot{\boldsymbol{\sigma}} + \mathbf{g}_1 \mathbf{n}_{f_1}^{\mathrm{T}} \cdot \dot{\boldsymbol{\sigma}}, \qquad\qquad \dot{\boldsymbol{\sigma}} \in V_{\mathrm{II}},$$

$$\dot{\boldsymbol{\varepsilon}}_3 = \mathbf{C}_3 \dot{\boldsymbol{\sigma}} = \mathbf{C}_1 \dot{\boldsymbol{\sigma}} + \mathbf{g}_2 \mathbf{n}_{f_2}^{\mathrm{T}} \cdot \dot{\boldsymbol{\sigma}}, \qquad\qquad \dot{\boldsymbol{\sigma}} \in V_{\mathrm{III}},$$

$$\dot{\boldsymbol{\varepsilon}}_4 = \mathbf{C}_4 \dot{\boldsymbol{\sigma}} = \mathbf{C}_1 \dot{\boldsymbol{\sigma}} + \mathbf{g}_1 \mathbf{n}_{f_1}^{\mathrm{T}} \cdot \dot{\boldsymbol{\sigma}} + \mathbf{g}_2 n_{f_2}^{\mathrm{T}} \cdot \dot{\boldsymbol{\sigma}}, \qquad \dot{\boldsymbol{\sigma}} \in V_{\mathrm{IV}}, \qquad (8.22)$$

and the continuity conditions between particular subdomains are satisfied.

As the number of intersecting surfaces tends to infinity, the rate equations become *non-linear* and the constitutive matrix $\mathbf{C}$ depends on the direction of the stress rate vector. Generally,

$$\dot{\boldsymbol{\varepsilon}} = \mathbf{C}(\boldsymbol{\sigma}, \boldsymbol{\alpha}, \dot{\boldsymbol{\sigma}})\dot{\boldsymbol{\sigma}}, \tag{8.23}$$

where $\mathbf{C}$ is a homogeneous function of the stress rate of order zero. Such non-linearity in rate equations will occur also when the yield surface shrinks to a point and an infinitesimal stress reversal produces an inelastic strain. Such non-linear equations arising in the limiting case will be discussed in the following section.

The continuity condition (8.7) following from the convexity of transformation[23] has frequently been violated in various formulations of rate equations using the concept of non-linear elasticity in particular sub-domains,[4] hypoelasticity,[36] or the so-called endochronic theory.[5] Here, however, we shall satisfy this condition, as it is essential for uniqueness of the solution of boundary-value problems.

8.3 SIMPLE AND COMPOSITE HARDENING RULES FOR SOILS

The present discussion of hardening rules will be carried out in the context of soil mechanics. Let the tensor of effective stress be σ'_{ij} and the pore pressure be denoted by p_p, so that

$$\sigma_{ij} = \sigma'_{ij} + p_p\delta_{ij}, \tag{8.24}$$

where δ_{ij} denotes the Kronecker delta. The compressive stresses and contractive strains will be assumed positive. The yield condition and constitutive relations can now be expressed in terms of the effective stress and hardening parameters or in terms of the total stress; thus

$$f(\boldsymbol{\sigma}', \boldsymbol{\alpha}) = 0 \tag{8.25}$$

or

$$f(\boldsymbol{\sigma}, p_p, \boldsymbol{\alpha}) = 0 \tag{8.26}$$

where p_p now becomes a new hardening function. In fact, when the pore pressure varies, the yield surface in total stress space translates along the hydrostatic axis and the material hardens or softens.

The total compressibility of soil depends on both the bulk modulus of the solid material and of the fluid, but a simple description is obtained when both the solid material and the fluid are regarded as incompressible and the macroscopic volume variation is due to closing or opening of voids and flow of fluid. Denoting the volumes of voids, material and the total volume of a representative element by V_v, V_m and V_t, the void ratio and the relative density are expressed as follows:

$$e = \frac{V_v}{V_m}, \qquad \eta = \frac{\rho}{\rho_m} = \frac{V_m}{V_t} = \frac{1}{1+e} \tag{8.27}$$

where ρ is the bulk density and ρ_m denotes the material density. Since $\dot{V}_m = 0$, $\dot{V}_v = \dot{V}_t$, the rate of variation of e and η is

$$\dot{e}^p = -(1+e)\,\mathrm{tr}\,\dot{\boldsymbol{\varepsilon}}^p, \quad \dot{e}^e = -(1+e)\,\mathrm{tr}\,\dot{\boldsymbol{\varepsilon}}^e, \quad \dot{e} = \dot{e}^p + \dot{e}^e,$$

$$\dot{\eta}^p = \eta\,\mathrm{tr}\,\dot{\boldsymbol{\varepsilon}}^p, \qquad\qquad\qquad \dot{\eta}^e = \eta\,\mathrm{tr}\,\dot{\boldsymbol{\varepsilon}}^e,$$

Thus the macroscopic volume variation is related to the variation of void ratio or relative density and the usual decomposition into elastic and plastic components applies.

8.3.1 Isotropic hardening rules

(a) *Critical state model (density-hardening)*

A simple hardening rule occurs when the yield condition (8.25) depends on one or several hardening functions whose evolution is directly related to the variation of plastic strain and no distinction is made between particular loading or reverse loading events. Let us discuss the simple assumption applicable to clays that the maximal consolidation pressure p'_c or irreversible void ratio variation is the only hardening parameter; thus

$$f(\boldsymbol{\sigma}', e^p) = 0 \qquad \text{or} \qquad f(\boldsymbol{\sigma}', p'_e) = 0 \tag{8.29}$$

where $p'_e = \frac{1}{3}\sigma'_{kk}$. The two forms (8.29) are equivalent, since for an isotropic consolidation process and subsequent unloading the irreversible void ratio change can be related to the maximal consolidation pressure.

Assuming the associated flow rule

$$\dot{\boldsymbol{\varepsilon}}^p = \frac{1}{K}\,\mathbf{n}(\dot{\boldsymbol{\sigma}}^T \cdot \mathbf{n}), \quad f = 0, \quad \dot{\boldsymbol{\sigma}}^T \cdot \mathbf{n} > 0, \tag{8.30}$$

and using the consistency condition

$$\left(\frac{\partial f}{\partial \boldsymbol{\sigma}'}\right)^T \cdot \dot{\boldsymbol{\sigma}}' + \frac{\partial f}{\partial e^p}\,\dot{e}^p = 0, \tag{8.31}$$

we obtain the expression for the hardening modulus in the form

$$K = \frac{\partial f}{\partial e^p}(1+e)\,\mathrm{tr}\,\mathbf{n}\left[\left(\frac{\partial f}{\partial \boldsymbol{\sigma}'}\right)^T \cdot \left(\frac{\partial f}{\partial \boldsymbol{\sigma}'}\right)\right]^{-1/2}, \tag{8.32}$$

where $\mathbf{n} = \mathbf{n}_f$ is the normalized gradient tensor $\partial f/\partial \boldsymbol{\sigma}'$. From (8.32) it follows that K varies along the yield surface, depending on the value of $\mathrm{tr}\,\mathbf{n} = \mathbf{n}_{kk}$.

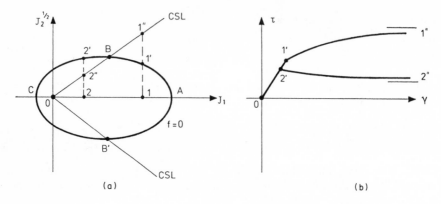

Figure 8.3 Isotropic hardening model: (a) yield surface; (b) material response under shear

Assuming that

$$\frac{\partial f}{\partial e^{p}} > 0, \tag{8.33}$$

it is seen that K has the same sign as tr **n**. Figure 8.3(a) presents a typical yield condition in the plane $(J_1, J_2^{1/2})$ for a particular value of e^p. Here J_1 and J_2' are respectively the first invariant of the effective stress tensor and the second invariant of the stress deviator. For a closed yield surface there exists a consolidation region for which tr **n** > 0 (domain $OBAB'$ in Figure 8.3(a)), a dilatancy or softening region for which tr **n** < 0 (domain $OBCB'$ in Figure 8.3(a)) and the critical state line for which tr **n** = 0 (lines OB and OB'). The hardening modulus K is positive in the consolidation domain $OBAB'$, negative in the softening domain $OBCB'$ and zero for the critical state lines OB or OB'. Figure 8.3(b) shows the stress–strain response for paths 1–1′–1″ and 2–2′–2″ when a shear stress is imposed upon an initial hydrostatic stress $p' = \frac{1}{3}J_1$.

The rate equations (8.30) should be supplemented by adding constitutive relations between the elastic strain rate $\dot{\boldsymbol{\varepsilon}}^e$ and the effective stress rate $\dot{\boldsymbol{\sigma}}'$. For instance, as is usually assumed for clays,[3,39] the rate equations for deviatoric and volumetric components take the form

$$\dot{e}^e_{ij} = \frac{\dot{s}_{ij}}{2G}, \qquad \dot{\varepsilon}^e_v = k\frac{\dot{p}'}{p'}\frac{1}{1+e}, \tag{8.34}$$

which is valid for $p' > 0$. Here **e** and **s** denote deviatoric strain and stress components. The volumetric rate relation follows the isotropic, elastic compression curve $e^e = e_0 - k \ln(p'/p'_0)$, where k is a constant parameter and e_0, p'_0 are the reference void ratio and pressure. In writing (8.34) it is assumed that the shear modulus G is constant; hence the rate equations

(8.34) can be regarded as being derived from the elastic potential,

$$W = \frac{s_{ij}s_{ij}}{4G} + \frac{k}{1+e} p' \left(\ln \frac{p'}{p_0'} - 1 \right).$$ (8.35)

However, it turns out that a more accurate description of elastic response is obtained by postulating that the shear modulus depends on effective pressure, $G = G(p')$. Substituting this function into (8.35) and applying the potential rule, we obtain

$$e_{ij}^{e} = \frac{s_{ij}}{2G(p')}, \qquad \varepsilon_v^{e} = \frac{-s_{ij}s_{ij}}{4G^2} \frac{\mathrm{d}G}{\mathrm{d}p'} + \frac{k}{1+e} \ln \frac{p'}{p_0'}.$$ (8.36)

Thus, an additional dilatancy term occurs due to shear modulus variation as usually $\mathrm{d}G/\mathrm{d}p' > 0$. Differentiating (8.36), we obtain the elastic strain rates expressed in a more complex way, namely

$$\begin{bmatrix} \dot{e}_{ij}^{e} \\ \dot{\varepsilon}_v^{e} \end{bmatrix} = \begin{bmatrix} \alpha_{ss} & \alpha_{sp} \\ \alpha_{sp} & \alpha_{pp} \end{bmatrix} \begin{bmatrix} \dot{s}_{ij} \\ \dot{p}' \end{bmatrix}$$ (8.37)

where the terms α_{ss}, α_{sp}, α_{pp} are obtained by differentiating (8.36). The use of more general relations of this type for a granular material has been discussed recently in Ref. 8, where the existence of a coupling effect due to α_{sp} is demonstrated. On the other hand, when using the simpler relations (8.34) with $G = G(p')$, it must be remembered that they describe recoverable but in general non-elastic (or hypo-elastic) strain rates and that energy dissipation would result for closed circuit stress paths.

The class of hardening models described by (8.29), (8.30) and (8.34) is applicable to clays under monotonic loading and can generally be described as *critical state (or density-hardening) models*. The particular formulations, such as the Cam-clay model[39] or cap model[12] differ only in the form of the yield condition, which is most important in the quantitative description of a stress–strain response. The disadvantage of this model lies in its inability to simulate properly the softening response and dilatancy as well as pore pressure variation in undrained tests on overconsolidated clays.

(b)   *Combined (deviatoric and density) hardening model*

As overconsolidated clays and dense sands exhibit stable behaviour despite dilatancy until the maximum stress is reached, one may expect that a better description would result from assuming that additional hardening occurs due to shear action. Let us introduce the combined hardening parameter κ, whose rate is expressed as follows:

$$\dot{\kappa} = \beta (\dot{\mathbf{e}}^{\mathrm{pT}} \cdot \dot{\mathbf{e}}^{\mathrm{p}})^{1/2} + \dot{\eta}^{\mathrm{p}} = \beta \dot{\lambda} + \dot{\eta}^{\mathrm{p}},$$ (8.38)

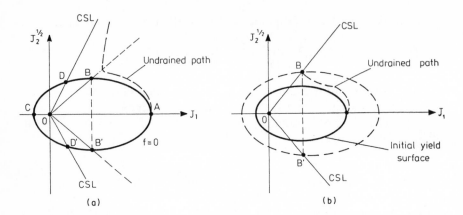

Figure 8.4 Isotropic hardening model with deviatoric and volumetric hardening; critical state line: (a) differs from zero dilatancy line; (b) coincides with zero dilatancy line

where the first term represents the deviatoric strain and the second corresponds to the irreversible density variation. Using the consistency condition (8.31), we obtain for the yield condition $f(\boldsymbol{\sigma}', \kappa) = 0$

$$K = \left(-\frac{\partial f}{\partial \kappa}\right)[\beta(\mathrm{dev}^{\mathrm{T}}\,\mathbf{n}\cdot\mathrm{dev}\,\mathbf{n})^{1/2} + \eta\,\mathrm{tr}\,\mathbf{n}]\left[\left(\frac{\partial f}{\partial \boldsymbol{\sigma}'}\right)^{\mathrm{T}}\cdot\left(\frac{\partial f}{\partial \boldsymbol{\sigma}'}\right)\right]^{-1/2}, \quad (8.39)$$

where dev denotes the deviatoric portion of $\mathbf{n}$, and β is a constant parameter. As $\partial f/\partial \kappa < 0$, the hardening modulus is positive when the second bracketed term is positive. Thus, in the consolidation domain, $K \geqslant 0$ for

$$\mathrm{tr}\,\mathbf{n} \geqslant -\frac{\beta}{\eta}(\mathrm{dev}^{\mathrm{T}}\,\mathbf{n}\cdot\mathrm{dev}\,\mathbf{n})^{1/2} \qquad (8.40)$$

and the critical state line, defined as a locus of points of vanishing hardening modulus, satisfies (8.39) as equality. In Figure 8.4(a) the consolidation domain $ODD'A$ is bounded by the critical state lines OD and OD' which do not coincide with the zero-dilatancy lines. The shear curves in consolidation and softening domains would be the same as in Figure 8.3(b). The undrained stress path would then tend asymptotically to the critical state line from below, which is a generally observed fact for dense sands. A combined hardening parameter of this type was applied by Nova and Wood[31] and Wilde[43] to describe the inelastic response of sands.

If, instead of the parameter κ, we used the plastic work as a hardening parameter, that is

$$\dot{W} = \mathbf{s}^{\mathrm{T}}\cdot\dot{\mathbf{e}}^{\mathrm{p}} + p'\dot{\varepsilon}_{\mathrm{v}}^{\mathrm{p}}, \qquad (8.41)$$

the hardening modulus would be expressed as follows:

$$K = \left(-\frac{\partial f}{\partial w_p}\right)[\mathbf{s}^{\mathrm{T}} \cdot \mathrm{dev}\,\mathbf{n} + p'\,\mathrm{tr}\,\mathbf{n}]\left[\left(\frac{\partial f}{\partial \boldsymbol{\sigma}'}\right)^{\mathrm{T}} \cdot \frac{\partial f}{\partial \boldsymbol{\sigma}'}\right]^{-1/2}, \tag{8.42}$$

and the consolidation domain would be described by the inequality

$$K \geqslant 0 \qquad \text{for } \mathrm{tr}\,\mathbf{n} \geqslant \frac{-\mathbf{s}^{\mathrm{T}} \cdot \mathrm{dev}\,\mathbf{n}}{p'}. \tag{8.43}$$

Thus, the critical state line would be positioned in the same way as in Figure 8.4(a), that is, departing from the zero-dilatancy line. The two hardening parameters κ and w_p would, therefore, yield similar results when applied in simulating an inelastic sand response. The plastic work w_p was used extensively as a hardening parameter by Lade[20] and Lade and Duncan[19] in modelling deformation of sands. The use of κ, however, is more convenient as there is more flexibility associated with the value of β; for $\beta = 0$ we obtain a critical state model, and the deviation between the critical state line and the zero-dilatancy line can be directly associated with the value of β.

Assume finally that λ and η^p act independently in the yield condition

$$f(\boldsymbol{\sigma}', \lambda, \eta^p) = 0 \tag{8.44}$$

and the shear stress–strain curve tends asymptotically to a steady value for any hydrostatic pressure. The initial yield locus and the critical state line for this type of hardening model are shown schematically in Figure 8.4(b). A two-parameter hardening rule was considered by Prevost and Hoeg.[34]

As follows from our brief review, the combined parameter or two-parameter description may improve the accuracy for monotonic loading paths in sands which cannot be fitted into the density-hardening model. However, the real improvement is achieved both for monotonic and cyclic loading histories when an anisotropic hardening model is used and the memory of particular loading events is properly incorporated.

8.3.2 Anisotropic hardening rules. Discrete material memory

For monotonic loading processes, the isotropic hardening model can be successfully applied in solving boundary value problems. However, for time-varying loads, and, in particular, for cyclic loading processes, when hysteretic phenomena are of essential importance, we must look for a more accurate description of material behaviour. In fact, the isotropic hardening surface is usually defined by a large offset value and its interior domain is regarded as elastic. However, when after initial pre-consolidation the stress is slightly decreased, reverse plastic flow is usually observed and the unloading stress–strain curve departs significantly from the elastic curve. Therefore, the elastic domain enclosed by the yield surface $f_0 = 0$, defined by a small offset value (say $\varepsilon = 10^{-4}$), is much smaller than that corresponding to the isotropic hardening surface, or may not exist at all. Moreover, this yield surface exhibits the material anisotropy (analogous to the Bauschinger effect in metals) which is induced by initial consolidation due to a stress state other than pure hydrostatic pressure.

Another important aspect of a more general material model is its memory of particular loading events, with the possibility of remembering 'important' events and erasing 'unimportant' events. In the case of isotropic hardening these events are of two types only, that is, active loading and elastic unloading. However, in a more general case, we have to make a distinction between events of larger 'intensity' and those of lesser 'intensity'. The following fundamental property of material memory will be assumed: *loading events of given intensity can only be erased from material memory by events of larger intensity.*

Figure 8.5 illustrates this memory property in the uniaxial case. Let the absolute value of a function $P = P(t)$ be assumed as its intensity measure. The first loading event 0–1 terminates at 1 with its maximal intensity P_1; the second loading event commences at 1 with decreasing intensity and terminates at 2 with $P_2 > P_1$, thus erasing the memory of the event 0–1. Similarly, the event 2–3 with $P_3 > P_2$ erases the memory of events 0–1 and 1–2. However, the subsequent events of decreasing or constant intensities $P_4, P_5, P_6, \ldots$ remain incorporated in the material memory until the event 8–9, with $P_9 > P_i$ ($i = 1, 2, \ldots, 8$), erases the effect of previous load changes. This memory property is an important component of the constitutive model and it is based on the physical assumption that large prestress determines a material structure (i.e. topology of distribution of contact forces) that can only be partially modified by stress variations of small amplitude. We shall use the term *discrete memory*, as only a discrete set of load reversal points needs to be remembered due to continuing erasure of less important events from the material memory.

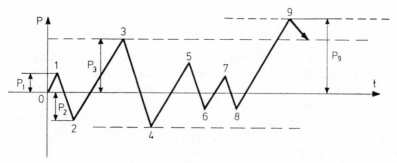

Figure 8.5 Memory rule for soil: large intensity loading event (P_3) cannot be erased by subsequent smaller loading events (P_4, P_5, P_6, P_7, P_8)

In Figure 8.5 the load level P_3 is kept fixed for all subsequent events of small intensity, thus becoming a *memory level* that can only be changed by a load of larger intensity. However, it may be assumed that this level is gradually modified by load cycles of smaller amplitude:

$$\bar{P}_3(\tau_f) = f(\mu), \qquad \text{where } \mu = \pi \int_0^{\tau_f} |\dot{P}|\, d\tau \quad \text{for } |P| \le P_3, \qquad (8.45)$$

where $\bar{P}_3(0) = P_3$ and $\tau = t - t_3$; that is, integration applies to all loading events of smaller intensity following the maximal load $P = P_3$. In this way, all cycles of smaller intensity, though they are later erased by larger loading events, can affect the memory level by means of 8.45).

Let us now discuss three versions of the anisotropic hardening model that possesses this multi-level memory rule.

(a) A multi-surface hardening model

Assume that the yield surface, if it exists (as defined by a small offset value), may translate and expand or contract in stress space; thus

$$f_0(\boldsymbol{\sigma}' - \boldsymbol{\alpha}^{(0)}, e^{\mathrm{p}}) = 0 \qquad (8.46)$$

and the associated flow rule (8.30) applies. In order to construct the complete set of model equations the translation rule of the yield surface and its dependence on the irreversible void ratio should be specified. It turns out, however, that this translation rule is a multi-valued function of plastic strain, and in order to incorporate sufficient memory into the model a set of nesting surfaces in stress space should be introduced to specify particular loading events and memory levels.

Assume that besides the yield surface $f_0 = 0$ enclosing the elastic domain there exists a consolidation or boundary surface defined by the degree of material consolidation. For clays the consolidation surface is developed during the initial consolidation process, whereas for sands it depends on material densification in the formation of the specimen. The domain enclosed by the consolidation surface is not elastic and for stress trajectories within this surface plastic flow occurs once the yield condition $f_0 = 0$ is satisfied. Assuming that the consolidation surface expands or contracts isotropically, we can identify it with the isotropic hardening surface discussed in the previous section; thus

$$F_{\mathrm{c}} = f(\boldsymbol{\sigma}', e^{\mathrm{p}}) = 0. \qquad (8.47)$$

It is further assumed that the hardening modulus on the consolidation surface varies according to (8.32), whereas the hardening modulus on the yield surface depends on the relative configuration of these two surfaces. In order to provide a more precise description of the variation of this modulus within the domain contained between the two surfaces $F = 0$ and $f = 0$, let us introduce a set of nesting surfaces that can translate and expand or contract due to density variation; that is

$$f_1(\boldsymbol{\sigma}' - \boldsymbol{\alpha}^{(1)}, e^{\mathrm{p}}) = 0, \qquad f_2(\boldsymbol{\sigma}' - \boldsymbol{\alpha}^{(2)}, e^{\mathrm{p}}) = 0, \ldots, f_k(\boldsymbol{\sigma}' - \boldsymbol{\alpha}^{(k)}, e^{\mathrm{p}}) = 0.$$
$$(8.48)$$

Assuming that all surfaces are similar, we can write

$$f_i(\boldsymbol{\sigma}' - \boldsymbol{\alpha}^{(i)}) - [a^{(i)}(e^{\mathrm{p}})]^2 = 0, \qquad i = 0, 1, 2, \ldots, n \qquad (8.49)$$

where f_i is a homogeneous function of order two. For any point P_0 on $f_0 = 0$, there exists, therefore, a set of conjugate points $P_1, P_2, \ldots, P_n = R$ on $f_1 = f_2 = f_n = 0$, for which the normal vector has the same direction. It is assumed that the particular surfaces do not intersect but may contact each other at the conjugate points.

The deformation process from any elastic state can thus be imagined as follows. The stress point first reaches the yield surface at P_0 and this surface translates toward a conjugate point P_1 on the nesting surface $f_1 = 0$. Before their contact, the hardening modulus K_0 applies in the flow rule (8.30); however, when $f_0 = 0$ engages $f_1 = 0$, the first nesting surface becomes the active loading surface and the hardening modulus K_1 occurs in the flow rule. For subsequent contacts of consecutive nesting surfaces, new corresponding values of hardening moduli apply until the boundary surface is reached. Since, for any point P_0 on $f_0 = 0$, there exists a set of conjugate points $P_1, P_2, \ldots, P_n = R$ on nesting surfaces and the moduli on $F_c = 0$ and $f_0 = 0$ are known, an interpolation rule is postulated that will specify the variation of K at conjugate points. For instance, it can be assumed that

$$K_K = K_R + \left(\frac{n - k}{n}\right)^{\alpha} (K_0 - K_R), \qquad k = 0, 1, 2, \ldots n, \qquad (8.50)$$

where K_K is the hardening modulus on the surface $f_k = 0$, K_R is the modulus at R on the consolidation surface and α denotes a material constant. Since K_R varies on the consolidation surface according to (8.32), the values K_K also depend on the position of P_K on the nesting surface. The configuration of nesting surfaces and the interpolation rule (8.50) define the *field of hardening moduli*, and the material response for any loading history may be studied by following the evolution of this field.

To formulate the translation rule for the yield surface or the active loading surface, assume that the surfaces $f_0 = 0$ to $f_l = 0$ are in contact at the point P_l and the surface $f_l = 0$ translates towards the conjugate point P_{l+1} (Figure 8.6). The position of the conjugate point is defined from the

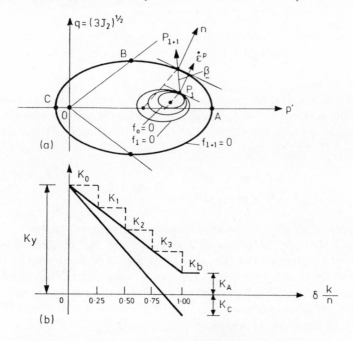

Figure 8.6 Anisotropic multi-surface hardening model: (a) yield surfaces and nesting surfaces; (b) variation of hardening moduli at conjugate points on nesting surfaces

proportionality condition

$$\sigma_P'^{(l+1)} - \alpha^{(l+1)} = \frac{a^{(l+1)}}{a^{(l)}} (\sigma_P'^{(l)} - \alpha^{(l)}) \tag{8.51}$$

and the vector connecting the points P_{l+1} and P_l is expressed as follows:

$$\beta = \frac{1}{a^{(l)}} [(a^{(l+1)} - a^{(l)})\sigma_P'^{(l)} - (a^{(l+1)}\alpha^{(l)} - a^{(l)}\alpha^{(l+1)})] \tag{8.52}$$

The relative motion of P_l with respect to P_{l+1} is assumed to occur along P_l–P_{l+1}; that is,

$$\sigma_P'^{(l)} - \sigma_P'^{(l+1)} = \beta \dot{\mu},$$

where $\dot{\mu}$ is a scalar factor. Since

$$\dot{\sigma}_P'^{(l)} = \dot{\alpha}^{(l)} + (\sigma_P'^{(l)} - \alpha^{(l)}) \frac{\dot{a}^{(l)}}{a^{(l)}},$$

$$\dot{\sigma}_P'^{(l+1)} = \dot{\alpha}^{(l+1)} + (\sigma_P'^{(l+1)} - \alpha^{(l+1)}) \frac{\dot{a}^{(l+1)}}{a^{(l+1)}}, \tag{8.53}$$

we obtain

$$\dot{\boldsymbol{\alpha}}^{(l)} = \dot{\boldsymbol{\alpha}}^{(l+1)} + \boldsymbol{\beta}\dot{\mu} + \frac{\dot{a}^{(l+1)} - \dot{a}^{(l)}}{a^{(l)}}(\boldsymbol{\sigma}_P'^{(l)} - \boldsymbol{\alpha}^{(l)}), \qquad (8.54)$$

and the scalar μ can be determined from the consistency condition (8.31).

If all the surfaces $f_0 = 0, f_1 = 0, \ldots, f_{l-1} = 0$ are in contact with the surface $f_l = 0$ and move with the stress point P_l their translation is governed by the motion of P_l, and we have

$$\frac{\boldsymbol{\sigma}_P'^{(l)} - \boldsymbol{\alpha}^{(k)}}{\boldsymbol{\sigma}_P'^{(l)} - \boldsymbol{\alpha}^{(l)}} = \frac{a^{(k)}}{a^{(l)}}, \qquad k = 0, 1, 2, \ldots, l. \qquad (8.55)$$

The values of all $a^{(k)}$ can be determined from equation (8.49).

A more detailed discussion of this model can be found in Ref. 25. It constitutes an extension of the concept of a field of hardening moduli first developed for metals by Mroz.[24] The case of undrained deformation of sills by using a similar model was treated by Prevost.[35] Petersson and Popov[32] applied a modified version of a multi-surface model to study the cyclic response of structural elements.

(b) A two-surface hardening model

The model presented possesses a multi-level memory structure, since for cyclically varying stress only a certain number of surfaces undergo translation; the other surfaces may change only due to density variation. For practical purposes it is possible to simplify this model, and we shall discuss two such simplified versions. The first possibility is to consider only the yield surface $f_0 = 0$ and the consolidation surface $F_c = 0$ to which the translation rule (8.54) is applied. Instead of the interpolation rule (8.50) specifying the hardening modulus on consecutive nesting surfaces, the field of hardening moduli is now described by prescribing the variation of K with the distance $\delta = PR$ between the stress point P on $f_0 = 0$ and the conjugate point R on $F_c = 0$. Such a modified description of the field of hardening moduli for metals was elaborated independently by Krieg[16] and Dafalias and Popov,[11] and here we shall extend this idea to soils.

Consider the initial consolidation process OA, after which the consolidation and yield surfaces are tangential to each other at A. The semi-diameters of these surfaces are respectively a_c and a_0, so that the maximal distance between the surfaces is $\delta_0 = P_0R = 2(a_c - a_0)$, Figure 8.7(a). If now the stress point moves along APD, the yield surface translates toward the conjugate point R on $F_c = 0$ and the hardening modulus is a function of δ and δ_0, so

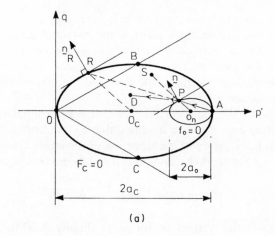

(a)

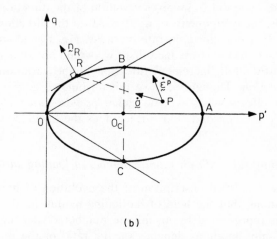

(b)

Figure 8.7 Two surface model: (a) yield and consolidation surfaces; (b) yield surface reduced to point P

that

$$K = K(\delta, \delta_0),$$
$$K = K_R \quad \text{for} \quad \delta = 0, \tag{8.56}$$
$$K = K_y \quad \text{for} \quad \delta = \delta_0,$$

where $\delta = f(\sigma'_R - \sigma'_P)^{1/2}$ is a scaled distance between P and R. For instance, it
can be assumed that

$$K = K_R + (K_y - K_R)\left(\frac{\delta}{\delta_0}\right)^\gamma, \tag{8.57}$$

so that the conditions (8.56) are satisfied; here γ is a constant parameter. During the plastic deformation process the maximal distance δ_0 changes only slightly due to density changes, whereas δ depends on the instantaneous position of the yield and consolidation surfaces. When the stress point reaches the consolidation surface, the flow and hardening rules corresponding to this surface are used.

The existence of the yield surface $f_0 = 0$ is not essential in our analysis, and in the limiting case it can be assumed that $a_0 = 0$. The conjugate point R then lies on the intersection of the direction of the stress rate vector with the consolidation surface, and the flow and translation rules become

$$\dot{\boldsymbol{\varepsilon}}^{\mathrm{p}} = \frac{1}{K} \mathbf{n}_R (\dot{\boldsymbol{\sigma}}'^{\mathrm{T}} \cdot \mathbf{n}_R), \qquad \dot{\boldsymbol{\alpha}} = \dot{\boldsymbol{\sigma}}', \tag{8.58}$$

where $\mathbf{n}_R$ denotes the normal vector at R (Figure 8.7(b)). It is seen that changing the direction of $\dot{\boldsymbol{\sigma}}'$ involves variation of the direction of $\mathbf{n}_R$ and $\dot{\boldsymbol{\varepsilon}}^{\mathrm{p}}$. In other words, the components of $\mathbf{n}_R$ depend on the direction of $\dot{\boldsymbol{\sigma}}'$ and the flow rule (8.58) is *non-linear* in rate of stress. For a yield surface which is small with respect to the stress increment we would obtain a strong sensitivity of the plastic strain increment direction to the variation of the stress increment direction. Though we have not formulated corner flow rules we obtain similar effects which will be called *pseudo-corner effects*. Only for stress states on the consolidation surface does the linear flow rule occur.

(c) A hardening rule with an infinite number of loading surfaces

An alternative simplified description of the evolution of hardening is provided by assuming that the field of hardening moduli inside the consolidation surface is represented by an infinite number of nesting surfaces, and that the hardening modulus depends on the ratio of the diameters of the instantaneous loading surface and the consolidation surface. Consider the case shown in Figure 8.8(a), where, after initial consolidation along OP_c, the stress path is next directed into the interior of the consolidation surface $F_c(\boldsymbol{\sigma}', e^{\mathrm{p}}) = 0$. We assume that the elastic domain is reduced to a point and that reverse plastic flow occurs immediately for any stress increment directed from P_c into the interior of the domain enclosed by the surface $F_c = 0$. A set of nesting surfaces created after reaching P_c is transformed by the reverse loading $P_c P$. At P the stress point touches the surface $F_{l1} = 0$, which now becomes the *active* reverse loading surface and all surfaces within $F_{l1} = 0$ are translated with the stress point. The plastic response at P is described by the flow rule (8.30) applied to $f_{l1} = 0$, with the plastic modulus governed by a relation similar to (8.57), namely

$$K = K_{\mathrm{R}} + (K_y - K_{\mathrm{R}}) R_1^{\gamma} \tag{8.59}$$

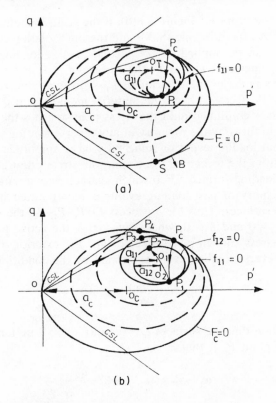

Figure 8.8 Model with an infinite number of surfaces: (a) first reverse loading; (b) second reverse loading

where

$$R_1 = \frac{a_c - a_{l1}}{a_c} = 1 - \frac{a_{l1}}{a_c} \tag{8.60}$$

and K_y is the initial modulus at P_c, that is, for $a_{l1} = 0$. The semidiameters of the active loading and consolidation surfaces are a_{l1} and a_c, respectively, and γ is a material parameter which will be assumed to vary during cyclic loading.

The *first reverse loading* programme occurs provided a_{l1} continues to increase. Thus, when P reaches the consolidation surface at S, the surfaces $f_{l1} = 0$ and $F_c = 0$ coincide and $K = K_s$. If, on the other hand, at P_1 the stress path reverses and is directed into the interior of the reverse loading surface $f_{l1} = 0$, then the *second reverse loading* programme commences. The active loading surface $f_{l2} = 0$ for this programme is a similar ellipse tangential at P_1 to the surface $f_{l1} = 0$, which now becomes the stress-reversal surface, passing through the stress point P_2, Figure 8.8(b). If the semi-diameter of this ellipse in denoted by a_{l2}, the flow rule (8.30) and the interpolation rule (8.59) with $R_2 = 1 - a_{l2}/a_c$ can be applied. If the second reverse loading process continues, then a_{l2} increases and at P_3 the surfaces $f_{l2} = 0$ and $f_{l1} = 0$ coincide. Thus, for stress paths in the exterior of $f_{l1} = 0$, the second reverse loading process P_1-P_2-P_3 is erased from the material memory and the first reverse loading process is continued until the stress point reaches the consolidation surface at P_4. In this way the memory of any reverse loading event is erased by a subsequent loading event of sufficiently large amplitude.

As is seen from the foregoing discussion, the stress point always remains on the active loading surface $f_{li} = 0$ which characterizes a particular loading event. The memory of past loading events is incorporated into a stack of stress reversal surfaces. Thus for a process O-P_c-P_1-P_2, the stress reversal surfaces are $F_c = 0$ and $f_{l1} = 0$, whereas $f_{l2} = 0$ is the active loading surface provided the stress trajectory is directed into the exterior of this surface. Thus the ith reverse loading event is defined by the conditions

$$f_{li} = 0, \quad \left(\frac{\partial f_{li}}{\partial \boldsymbol{\sigma}'}\right)^{\mathrm{T}} \cdot \dot{\boldsymbol{\sigma}}' \geq 0 \tag{8.61}$$

and the condition that the surfaces $f_{li} = 0$ and $f_{l(i-1)} = 0$ be tangential at the stress reversal point P_{l-1}; that is

$$\boldsymbol{\alpha}_i - \boldsymbol{\sigma}'_P = (\boldsymbol{\alpha}_{i-1} - \boldsymbol{\sigma}'_P) \frac{a_{li}}{a_{l(i-1)}} \tag{8.62}$$

where $\boldsymbol{\sigma}'_P$ is the position of the stress reversal point P_{l-1} and $\boldsymbol{\alpha}_i, \boldsymbol{\alpha}_{i-1}$ are the centres of the surfaces $f_{li} = 0$ and $f_{l(i-1)} = 0$. When the consolidation surface $F_c = 0$ is expanding or contracting, the positions of the stress reversal points $P_c, P_1, P_2, \ldots$ are translated appropriately.

8.4 PREDICTED AND OBSERVED STATIC RESPONSE OF CLAYS FOR AXISYMMETRIC 'TRIAXIAL' STRESS STATES

In this section we shall restrict our analysis to the case of a 'triaxial' test in which two effective principal stresses are equal, $\sigma'_2 = \sigma'_3$, and the principal stress directions are fixed with respect to the material element. The effective stress state can be described by two parameters

$$p' = \tfrac{1}{3}(\sigma'_1 + 2\sigma'_2), \qquad q' = \sigma'_1 - \sigma'_2 \tag{8.63}$$

and the associated strains are

$$\varepsilon_v = \varepsilon_1 + 2\varepsilon_2, \qquad \varepsilon_q = \tfrac{2}{3}(\varepsilon_1 - \varepsilon_2), \tag{8.64}$$

so that

$$p'\varepsilon_v + q'\varepsilon_q = \sigma_1'\varepsilon_1 + 2\sigma_2'\varepsilon_2. \tag{8.65}$$

The total stresses are $p = p' + p_p$, $q = q'$.

The constitutive relations can be presented in a fairly simple form if we use the elliptical yield and consolidation surfaces, that is

$$F_c(p', q, e^p) = (p' - c)^2 + \frac{q^2}{n^2} - a_c^2(e^p) = 0 \tag{8.66}$$

and

$$f_0(p', q, \alpha_p, \alpha_q, e^p) = (p' - \alpha_p)^2 + \frac{(q - \alpha_q)^2}{n^2} - a_0^2(e^p) = 0, \tag{8.67}$$

where

$$c = a_c \frac{n}{m}, \quad m = \tan \omega, \qquad n = \tan \xi \tag{8.68}$$

and the angles ξ and ω are shown in Figure 8.9. Equation (8.66) presents a one-parameter family of ellipses whose centres and semi-axes change with plastic density, but the ratio of semi-axes remains constant. The critical state lines OB and OD are inclined at an angle ω to the p' axis.

As was discussed in the previous section the initial consolidation of a clay from a slurry may be represented by a trajectory in the p', q plane, the end point of which lies on both the consolidation and yield surfaces. The reverse loading within the consolidation surface involves plastic deformation when the stress point moves with the yield surface $f_0 = 0$. The elastic domain enclosed by this surface may be very small, as compared to that enclosed by the surface $F_c = 0$, or vanish; in the latter case, plastic flow occurs for any stress increment within the surface $F_c = 0$.

For sands the consolidation process is not associated with the applied pressure, but with the process of forming a specimen at low stress, for instance, by pouring the sand from a silo. The consolidation surface is, therefore, related to the relative density or specific volume of the sand and the 'consolidation' pressure at A in Figure 8.9 may for dense sands be several orders of magnitude greater than the pressure used in testing or which occurs beneath a foundation. The stress state is therefore represented by a point within the consolidation surface, and even loose sands can be regarded as overconsolidated. Although the numerical testing of our model will be carried out for clays, the constitutive relations presented can be applied to sands with appropriate changes of material parameters.

We shall confine our discussion to the two-surface model (Ts model) and the model with an infinite number of loading surfaces (Ins model), which have been discussed in the previous section.

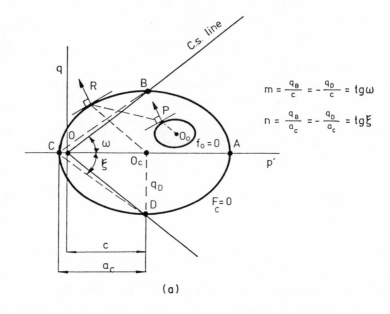

$$m = \frac{q_B}{c} = -\frac{q_D}{c} = tg\omega$$

$$n = \frac{q_B}{a_c} = -\frac{q_D}{a_c} = tg\xi$$

(a)

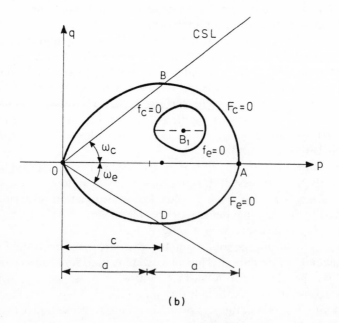

(b)

Figure 8.9 Anisotropic hardening model; surfaces formed by: (a) ellipses with axes parallel to p'-axis; (b) two partial ellipses with axes inclined to p'-axis

8.4.1   The two-surface model

For the yield surface (8.67), the plastic flow rule has the form

$$\dot{\varepsilon}_V^P = \frac{1}{K}\, n_p \dot{\sigma}_n, \qquad \dot{\varepsilon}_q^P = \frac{1}{K}\, n_q \dot{\sigma}_n \tag{8.69}$$

where

$$n_p = \frac{p' - \alpha_p}{G_f}, \qquad n_q = \frac{q - \alpha_q}{n^2 G_f} \tag{8.70}$$

are the components of the unit vector normal to the yield surface, and

$$\dot{\sigma}_n = \frac{1}{G_f}\left[(p' - \alpha_p)\dot{p}' + \frac{q - \alpha_q}{n^2}\,\dot{q}\right], \qquad G_f = \left[(p' - \alpha_p)^2 + \frac{(q - \alpha_q)^2}{n^4}\right]^{1/2}. \tag{8.71}$$

The translation rule (8.54) is now expressed as follows:

$$\dot{\alpha}_q = \dot{\mu}\left[-q_P + \frac{a_c}{a_0}(q_P - \alpha_q)\right] + \frac{\dot{a}_c - \dot{a}_0}{a_0}(q_P - \alpha_q),$$

$$\dot{\alpha}_p = \dot{\mu}\left[c - p'_P + \frac{a_c}{a_0}(p'_P - \alpha_p)\right] + \frac{\dot{a}_c - \dot{a}_0}{a_0}(p'_P - \alpha_p) + \dot{a}\,\frac{n}{m}, \tag{8.72}$$

where q_P and p'_P are the co-ordinates of the stress point P on the yield surface; the multiplier $\dot{\mu}$ is determined by satisfying the consistency condition $\dot{f}_0 = 0$. The expansion or contraction of surfaces $f_0 = 0$ and $F_c = 0$ is specified by

$$\dot{a}_c = a'_c\, d\dot{e}^P = -a'_c (1 + e)\dot{\varepsilon}_V^P = -\frac{1}{K}\, a'_c (1 + e) n_p \dot{\sigma}_n,$$

$$\dot{a}_0 = \dot{a}\,\frac{a_0}{a_c} = -\frac{a_0}{a_c}\frac{1}{K}\, a'_c (1 + e) n_p \dot{\sigma}_n, \tag{8.73}$$

where a'_c denotes the derivative of a_c with respect to e^P.

The rule for variation of the plastic modulus K is specified by assuming that K varies continuously from its initial value K_y on the yield surface to its value K_R on the consolidation surface, specified at the 'conjugate' point R. Using (8.32) we obtain

$$K_R = \frac{\partial F}{\partial e^P}(1 + e) n_p \frac{1}{2 G_F} = -\frac{(1 + e)a'_c}{G_F^2}\left[\left(p' - a_c \frac{n}{m}\right)\frac{n}{m} + a_c\right]\left[p' - a_c \frac{n}{m}\right], \tag{8.74}$$

where

$$G_F = \left[(p' - c)^2 + \frac{q^2}{n^4}\right]^{1/2}. \tag{8.75}$$

After isotropic consolidation OA, the maximal distance between the surfaces $f_0 = 0$ and $F_c = 0$, equals $\delta_0 = P_1 R_1 = 2(a_c - a_0)$. For any stress path the transformed distance between the stress point and the conjugate point R is expressed as follows:

$$\delta = \left[\left(\frac{q_R - q_P}{n} \right)^2 + (p'_R - p'_P)^2 \right]^{1/2} \tag{8.76}$$

and the interpolation rule (8.7) can be assumed. Elastic strains are given by (8.34); thus

$$\dot{\varepsilon}^e_q = \frac{\dot{q}}{3G_s}, \qquad \dot{\varepsilon}^e_V = \frac{\dot{p}'}{K_s} \tag{8.77}$$

where $K_s = (1 + e)p'/k$. Combining (8.69) with (8.77), the constitutive relations can be presented in the matrix form for stress-controlled programmes:

$$\begin{bmatrix} \dot{\varepsilon}_q \\ \dot{\varepsilon}_V \end{bmatrix} = \begin{bmatrix} A & B \\ B & D \end{bmatrix} \begin{bmatrix} \dot{q} \\ \dot{p}' \end{bmatrix}, \tag{8.78}$$

where

$$A = \frac{1}{3G_s} + \frac{1}{K} \frac{(q - \alpha_q)^2}{n^4 G_f^2}, \qquad B = \frac{1}{K} \frac{(q - \alpha_q)(p' - \alpha_p)}{b^2 G_f^2}$$

$$D = \frac{1}{K_s} + \frac{1}{K} \frac{(p' - \alpha_p)^2}{G_f^2} \tag{8.79}$$

and the relations (8.78) apply when $f_0 = 0$, $F_c < 0$ and $\dot{\sigma}_n > 0$. If $f_0 = F_c = 0$ and $\dot{\sigma}_n > 0$, the yield and consolidation surfaces are in contact; the relations (8.78) still apply provided we set $\alpha_q = 0$, $\alpha_p = c$ and replace G_f by G_F. Finally, when $f_0 < 0$ or $f_0 = 0$, $\dot{\sigma}_n < 0$, elastic deformations occur and $A = 1/3G_s$, $B = 0$, $D = 1/K_s$. The inverse relations for strain-controlled computation can be expressed as follows

$$\begin{bmatrix} \dot{q} \\ \dot{p}' \end{bmatrix} = \begin{bmatrix} M & N \\ N & P \end{bmatrix} \begin{bmatrix} \dot{\varepsilon}_q \\ \dot{\varepsilon}_V \end{bmatrix} \tag{8.80}$$

where

$$M = 3G_s \left[1 - \frac{3G_s}{nK} \frac{(q - \alpha_q)^2}{C} \right], \qquad N = -3G_s \frac{K_s}{K} \frac{(q - \alpha_q)(p' - \alpha_p)}{C}$$

$$P = K_s \left[1 - \frac{K_s}{K} \frac{(p' - \alpha_p)^2}{C} \right], \qquad C = (p' - \alpha_p)^2 \left(1 + \frac{K_s}{K} \right) + \frac{(q - \alpha_q)^2}{n^4} \left(1 + \frac{3G_s}{K} \right), \tag{8.81}$$

and the active loading conditions are

$$f_0 = 0, \quad F < 0, \quad 3G_s n_q \dot{\varepsilon}_q + K_s n_p \dot{\varepsilon}_V > 0. \tag{8.82}$$

Neglecting compressibility of soil grains and of pore water, it can be assumed that the undrained deformation corresponds to isochoric deformation, that is $\dot{\varepsilon}_V = 0$. Imposing this condition on (8.80) we obtain the differential equation for the undrained paths in the effective stress plane

$$\frac{dq}{dp'} = \frac{M}{N} = -\frac{1}{a_w} \tag{8.83}$$

and $dp_p = dp + a_w dq$. Thus, a_w can be identified as a pore pressure coefficient associated with shear stress variation.

When $\dot{\varepsilon}_2 = \dot{\varepsilon}_3 = 0$, $\dot{\varepsilon}_1 \neq 0$; that is, when $\dot{\varepsilon}_q - \frac{2}{3}\dot{\varepsilon}_V = 0$ we have the case of uniaxial K_0 consolidation. From (8.80) it follows that the corresponding stress path is defined by the equation

$$\frac{dq}{dp'} = \frac{2M + 3N}{2N + 3P} \tag{8.84}$$

and for the rigid-plastic material ($K_s \to \infty$, $G_s \to \infty$)

$$S_0 = \frac{q}{p} = \frac{2mn}{(9 + 4n^2)^{1/2} + 3m/n}. \tag{8.85}$$

The K_0 path is now a straight line inclined at an angle θ_0 to the p'-axis.

Figure 8.10(a) presents the undrained stress paths in the p', q plane for Weald clay, as predicted by the present model and measured by Henkel.[14] It is seen that the undrained paths possess a failure envelope which, for overconsolidated clays, lies above the critical state line. Figure 8.10(b) shows the variation of the pore pressure coefficient, $A_f = \Delta p_p / \Delta q_f + 1/3$, where q_f denotes the value of q on the critical state line. The predicted curve agrees very well with the experimental data of Bishop and Henkel[6] and is not sensitive to selection of material parameters.

The numerical tests carried out by Mroz, Norris, and Zienkiewicz[26] indicate that the elliptical yield condition does not allow for accurate approximation of the K_0 consolidation and rebound curves. A similar conclusion was also reached earlier by Roscoe and Burland.[37] Therefore, a significant improvement in the accuracy of the description is achieved by modifying the equations of the consolidation and yield surfaces. Such a modification was introduced by Pietruszczak and Mroz,[33] requiring that each surface $F_c = 0$ and $f_0 = 0$ be composed of two portions of rotated ellipses. The critical state lines are assumed to form angles ω_c and ω_e with the p' axis in the compression and extension domains and these angles follow from the Coulomb yield condition (Figure 8.9(b)),

$$m_c = \frac{q_B}{c} = \tan \omega_c = \frac{6 \sin \phi}{3 - \sin \phi},$$

$$m_e = -\frac{q_D}{c} = \tan \omega_e = \frac{6 \sin \phi}{3 + \sin \phi}, \tag{8.86}$$

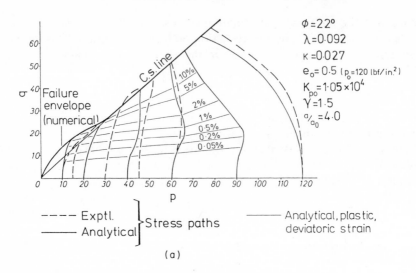

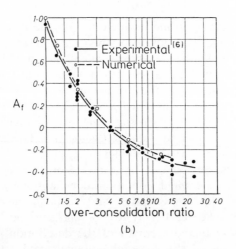

Figure 8.10 (a) Undrained stress paths in p', q plane; (b) variation of pore
pressure coefficient, A_f

where ϕ is the angle of internal friction at failure. The upper portion $(q>0)$
of the consolidation surface has the form

$$F_c = p'^2 + \frac{2(\zeta - 1)}{n_c} p'q + \frac{(1 - 2\zeta)^2 q^2}{n_c^2} - 2ap' - \frac{4(\zeta - 1)}{n_c} aq = 0, \quad (8.87)$$

where $\zeta = a/c$, $n_c = \tan \omega_c$, and $p' = 0$ for $q = 0$, $p = 2a$ when $\partial F_c/\partial q = 0$, $p = c$
when $\partial F/\partial p' = 0$. For $q < 0$ the lower portion is described by a similar

equation satisfying the condition $p = c$, $q = -n_e c$. We have $(q < 0)$:

$$F_e = p'^2 + \frac{2(1-\zeta)}{n_e} p'q + \frac{(1-2\zeta)^2}{n_e^2} q^2 - 2ap' - \frac{4(1-\zeta)}{n_e} aq = 0. \quad (8.88)$$

Let us note that for $\zeta = 1$, that is, for $a = c$, the principal axis of each ellipse coincides with the p' axis, and

$$F_c = (p-a)^2 + \frac{q^2}{n^2} - a^2 = 0; \quad (8.89)$$

that is, (8.89) is equivalent to (8.66) with $n = m$ $(c = a)$.

It is assumed that the yield surface is similar to the consolidation surface and is composed of portions of two ellipses:

$$f_c = (p'-\alpha_p)^2 + \frac{2(\zeta-1)}{n_c}(q-\alpha_q)(p'-\alpha_p) + \frac{(1-2\zeta)^2}{n_c^2}(q-\alpha_q)^2 + 2a_0(p'-\alpha_p)\frac{1-\zeta}{\zeta}$$

$$+ \frac{2(\zeta-1)(1-2\zeta)}{n_c\zeta} a_0(q-\alpha_q) + a_0^2 \frac{1-2\zeta}{\zeta^2} = 0, \quad q - \alpha_q > 0 \quad (8.90a)$$

$$f_e = (p'-\alpha_p)^2 + \frac{2(1-\zeta)}{n_e}(q-\alpha_q)(p'-\alpha_p) + \frac{(1-2\zeta)^2}{n_e^2}(q-\alpha_q)^2 + 2a_0(p'-\alpha_p)\frac{1-\zeta}{\zeta}$$

$$+ \frac{2(1-\zeta)(1-2\zeta)}{n_e\zeta} a_0(q-\alpha_q) + a_0^2 \frac{1-2\zeta}{\zeta^2} = 0, \quad q - \alpha_q < 0. \quad (8.90b)$$

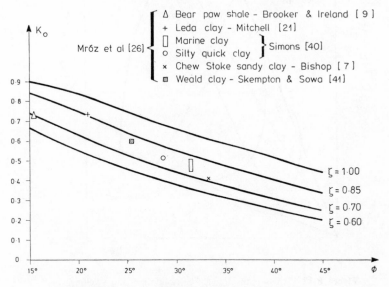

Figure 8.11 K_0 consolidation: variation of K_0 with ζ

For a rigid-plastic material the equation of the K_0 consolidation line now takes the form

$$S_0 = \frac{q}{p} = \frac{2n^2}{\{[4(\zeta-1)(3\zeta-n)+3]^2+4n(1-2\zeta)^2(n+3\zeta-3)}$$
$$\overline{-4n(\zeta-1)^2\}^{1/2}+4(\zeta-1)(3\zeta-n)+3} \qquad (8.91)$$

Figure 8.11 shows the predicted dependence of $K_0 = \sigma_2'/\sigma_1'$ on the angle of internal friction for four values of ζ. It is seen that collected experimental data are between the lines corresponding to $\zeta = 0.85$ and $\zeta = 0.70$. Figure 8.12 presents the K_0 consolidation and rebound curves of undrained stress paths in the plane of effective principal stresses σ_1', σ_3'. It is seen that the failure points 1, 2, 3, 4 for different overconsolidation ratios are accurately

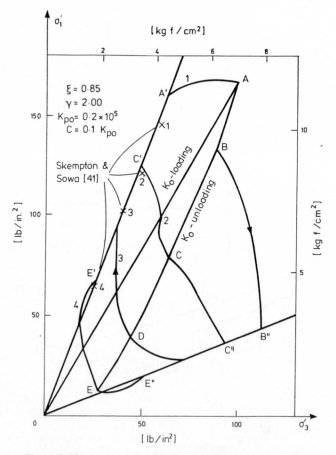

Figure 8.12 Stress paths for K_0 consolidation and swelling followed by undrained, triaxial compression and extension

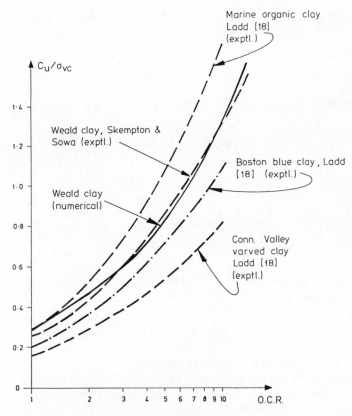

Figure 8.13   Variation of undrained strength with OCR

predicted when compared with Skempton and Sowa tests for Weald clay.[41] The material parameters for this clay were assumed as in the previous paper to be $\phi = 26^0$, $\lambda = 0.091$, $k = 0.030$, $e_0 = 0.57$, $\gamma = 2.1$, $\zeta = 0.85$, $K_y = 0.2 \cdot 10^5$ lbf/in², $a/a_0 = 5$, $G/K_s = 0.55$. The elastic shear modulus was, therefore, assumed to depend on hydrostatic pressure. Figure 8.13 shows the functions c_u/σ_{vc} versus OCR predicted and obtained experimentally for several clays. It is seen that the agreement with the data of Ref. 18 is very good, and the corresponding curves for other clays have a similar shape.

8.4.2   A model with an infinite number of loading surfaces (Ins model)

Using the elliptical consolidation surface (8.66), we obtain for the equation of the reverse active loading surface

$$f_{l1} = (p' - \alpha_p)^2 + \frac{(q - \alpha_q)^2}{n^2} - a_{l1}^2 = 0. \tag{8.92}$$

If the active loading surface is tangential at P_c to the consolidation surface and the stress point P is given, we have

$$\frac{p_c' - c}{p_c' - \alpha_p} = \frac{q_c}{q_c - \alpha_q} = \frac{a_c}{a_{l1}} \tag{8.93}$$

and satisfying equation (8.92), for $p' = p_P'$, $q = q_P$, we obtain

$$\frac{a_{l1}}{a_c} = \frac{|\boldsymbol{\delta}|^2}{2\boldsymbol{\delta} \cdot \mathbf{P_c O_c}}, \qquad \alpha_p = p_c' - \frac{a_{l1}}{a_c}(p_c' - c),$$

$$\alpha_q = q_c\left(1 - \frac{a_{l1}}{a_c}\right), \tag{8.94}$$

where

$$\boldsymbol{\delta} = \left(p_c' - p_P', \frac{q_c - q_P}{n}\right), \qquad \mathbf{P_c O_c} = \left(p_c' - c, \frac{q_c}{n}\right) \tag{8.95}$$

and p_c', q_c denote the stress values at the reversal point P_c. The co-ordinates of the conjugate point R on $F_c = 0$ are

$$p_R = c + \frac{a_c}{a_{l1}}(p_P' - \alpha_p), \qquad q_R = \frac{a_c}{a_{l1}}(q_P - \alpha_q) \tag{8.96}$$

and the interpolation rule (8.59) is applied.

The numerical results presented in Figures 8.10–13 for the Ts model would be the same for an Ins model using the same shape surfaces, as these models provide equivalent descriptions for the first reverse active loading process. However, when cyclic loading is involved, the prediction of steady cyclic states differs remarkably for Ts and Ins models.

8.5 YIELD AND CONSOLIDATION CONDITIONS FOR A GENERAL STRESS STATE

In Section 8.4 we have considered only 'triaxial' stress states for which the two principal stresses are equal. However, the applicability of the model can be extended considerably by formulating the yield condition and constitutive equations for a general stress state.

$$J_m' = \tfrac{1}{3}(\sigma_{ii}' - \alpha_{ii}),$$

$$\bar{\sigma} = [\tfrac{1}{2}(s_{ij} - \bar{\alpha}_{ij})(s_{ij} - \bar{\alpha}_{ij})]^{1/2}, \tag{8.97}$$

$$J_3 = \tfrac{1}{3}(s_{ij} - \bar{\alpha}_{ij})(s_{ki} - \bar{\alpha}_{ki})(s_{kj} - \bar{\alpha}_{kj})$$

of the translated stress $\sigma_{ij}' - \alpha_{ij}$. Here $s_{ij} = \sigma_{ij}' - \tfrac{1}{3}\sigma_{kk}'\delta_{ij}$ is the stress deviator and α_{ii}, $\bar{\alpha}_{ij}$ denote the spherical and deviatoric components of the translation tensor α_{ij}. Further, let us introduce the angular measure of the third invariant

$$\theta = \tfrac{1}{3}\arcsin\left(-\frac{3\sqrt{3}}{2}\frac{J_3}{\bar{\sigma}^3}\right), \qquad \frac{-\pi}{6} \leq \theta \leq \frac{\pi}{6}, \tag{8.98}$$

where the angle θ can be identified in the octahedral π-plane where $\sigma'_1 + \sigma'_2 + \sigma'_3 = \text{constant}$. For the 'triaxial' stress state, $\sigma_2 = \sigma_3$,

$$p' - \alpha_P = J'_m = \sigma'_m - \tfrac{1}{3}\alpha$$

$$q - \alpha_q = \sqrt{3}\,\bar{\sigma}_+ \tag{8.99}$$

where $\sigma'_m = \tfrac{1}{3}\sigma'_{ii}$, $\alpha = \alpha_{ii}$ and $\bar{\sigma}_+$ denotes the value of $\bar{\sigma}$ for $\sigma_2 = \sigma_3$, $\theta = \pi/6$. Generally, the principal stresses can be expressed in terms of J_m, $\bar{\sigma}$ and θ as follows:

$$\begin{bmatrix} \sigma'_1 \\ \sigma'_2 \\ \sigma'_3 \end{bmatrix} = \frac{2}{\sqrt{3}}\bar{\sigma}\begin{bmatrix} \sin(\theta + \tfrac{2}{3}\pi) \\ \sin\theta \\ \sin(\theta + \tfrac{4}{3}\pi) \end{bmatrix} + \sigma'_m. \tag{8.100}$$

Assume the form of the yield or active loading curve in the π-plane to be expressed by

$$\bar{\sigma} = \bar{\sigma}_+ g(\theta), \tag{8.101}$$

where

$$g(\theta) = \frac{2k}{(1+k) - (1-k)\sin 3\theta}, \qquad k = \frac{3 - \sin\phi}{3 + \sin\phi} \tag{8.102}$$

so that $g(\pi/6) = 1$. This relation follows from earlier developments by Zienkiewicz and Pande[44] and Gudehus.[13]

The yield condition (8.90) in the p', q plane can be rewritten as follows

$$f_0 = (\sigma'_m - \tfrac{1}{3}\alpha)^2 - \frac{2\sqrt{3}(\zeta - 1)}{n_c}(\sigma'_m - \tfrac{1}{3}\alpha)\bar{\sigma}_+$$

$$+ \frac{3(1 - 2\zeta)^2}{n_c^2}\bar{\sigma}_+^2 - 2a_0\frac{1 - \zeta}{\zeta}(\sigma'_m - \tfrac{1}{3}\alpha)$$

$$+ \frac{2\sqrt{3}(\zeta - 1)(1 - 2\zeta)}{n_c\zeta}a_0\bar{\sigma}_+ + a_0^2\frac{1 - 2\zeta}{\zeta^2} = 0, \tag{8.103}$$

and since $n_c = 6\sin\phi/(3 - \sin\phi)$ for $\theta = \pi/6$, we obtain

$$f_0 = 3\sin^2\phi(\sigma'_m - \tfrac{1}{3}\alpha)^2 - 6\sin^2\phi\frac{1 - \zeta}{\zeta}a_0(\sigma'_m - \tfrac{1}{3}\alpha)$$

$$- \sqrt{3}(\zeta - 1)\sin\phi(3 - \sin\phi\sin 3\theta) - \left[(\sigma'_m - \tfrac{1}{3}\alpha) - \frac{1 - 2\zeta}{\zeta}a_0\right]\bar{\sigma}$$

$$+ [\tfrac{1}{2}(1 - 2\zeta)(3 - \sin\phi\sin 3\theta)]^2\bar{\sigma}^2 + 3a_0^2\sin^2\phi\frac{1 - 2\zeta}{\zeta^2} = 0, \tag{8.104}$$

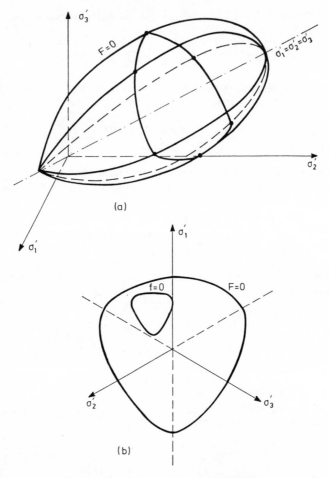

Figure 8.14 Anisotropic model for a general stress state: (a) boundary surface; (b) π-plane section of yield and boundary surfaces

and the equation of the consolidation surface is expressed as follows:

$$F_c = 3 \sin^2 \phi \, (\sigma'_m - a_c)^2 - \sqrt{3}(\zeta - 1) \sin \phi (3 - \sin \phi \sin 3\theta)(\sigma'_m - 2a_c)\bar{\sigma}$$
$$+ [\tfrac{1}{2}(1 - 2\zeta)(3 - \sin \phi \sin 3\theta)]^2 \bar{\sigma}^2 - 3a_c^2 \sin^2 \phi = 0, \quad (8.105)$$

where $\sigma'_m = \tfrac{1}{3}\sigma'_{ii}$ and $\bar{\sigma} = (\tfrac{1}{2}s_{ij}s_{ij})^{1/2}$. Equations (8.104) and (8.105) are represented by surfaces with similar cross-sectional shapes in all π-planes, Figure 8.14. The explicit expressions for gradient tensors to the surfaces $f_0 = 0$ and $F_c = 0$ were derived in Ref. 33. The flow and extrapolation rules can thus be formulated explicitly for any stress state.

8.6 A VISCOPLASTIC FLOW RULE

So far, we have assumed the mechanical response of soils under effective stresses to be time-independent. However, the creep and relaxation phenomena observed in clays are not only due to a time-dependent consolidation process but also to the viscous character of the irreversible deformation that accumulates in time. The viscous effects are of essential importance in cyclic loading conditions, when the effect of frequency of cycles on the shape and width of hysteresis loops is usually observed. In particular, the clays tested under 'rapid' cyclic loading with a frequency of several Hz may sustain many more cycles before failure than clays tested at 'slow' cycles with a frequency of the order of 0.01 Hz.

Now we shall briefly discuss the assumptions of a viscoplastic model based on the hardening rules developed in the previous sections. The total strain rate $\dot{\varepsilon}$ is decomposed into elastic and viscous portions $\dot{\varepsilon} = \dot{\varepsilon}^e + \dot{\varepsilon}^v$, where $\dot{\varepsilon}^v$ is not now the instantaneous strain rate but develops in time according to the distance of the instantaneous stress point from the static yield or loading surface.

Referring to Figure 8.7, we assume that the yield surface evolves according to rules described in Section 8.3, but the stress point S may remain outside the yield surface. For any position of S we may find the position of P on the radial line from the centre of the yield surface. The viscous flow rule now takes the form

$$\dot{\varepsilon}_V^v = \frac{1}{\nu} n_p \Phi(SP), \qquad \dot{\varepsilon}_q^v = \frac{1}{\nu} n_q \Phi(SP) \tag{8.106}$$

where ν denotes the material viscosity parameter and $\Phi(SP)$ is a function of the distance SP between the stress point and the yield surface. As the yield condition is a homogeneous function of order two of $p' - \alpha_p$ and $q - \alpha_q$, the relations (8.106) can be rewritten in the form

$$\dot{\varepsilon}_V^v = \frac{1}{\nu} n_p \Phi\left(\frac{\sigma_e}{a_0} - 1\right), \qquad \dot{\varepsilon}_q^v = \frac{1}{\nu} n_q \Phi\left(\frac{\sigma_e}{a_0} - 1\right) \qquad \text{for } \sigma_e > a_0,$$

$$\dot{\varepsilon}_V^v = 0, \qquad\qquad\qquad \dot{\varepsilon}_q^v = 0 \qquad\qquad \text{for } \sigma_e \leqslant a_0, \tag{8.107}$$

where

$$\sigma_e = \left[(p' - \alpha_p)^2 + \frac{(q - \alpha_q)^2}{n^2}\right]^{1/2}. \tag{8.108}$$

The evolution rules for a_c, a_0, α_p and α_q are identical to those stated by the relations (8.72) and (8.73). The factor $\dot{\mu}$ occurring in (8.72) should be related to viscous strain rates. This relation is determined by considering an infinitely slow deformation process for which the irreversible strain can be regarded as plastic and is governed by the flow rule (8.69). In this case, from

(8.69), it follows that

$$\dot{\sigma}_n = [(\dot{\varepsilon}_V^p)^2 + (\dot{\varepsilon}_q^p)^2]^{1/2} K = K[(\dot{\varepsilon}_V^v)^2 + (\dot{\varepsilon}_q^v)^2]^{1/2} = K\dot{\varepsilon}_e^v, \qquad (8.109)$$

where $\dot{\sigma}_n$ denotes the normal component of the stress increment vector. In the viscoplastic case, the stress point S may be outside the yield surface, but we may associate with S the point P lying on the same radial line from the centre of the yield surface. Since P always remains on the yield surface, we may write

$$\dot{\sigma}_{nP} = \dot{p}_P' n_P + \dot{q}_P n_q = K\dot{\varepsilon}_e^v = K[(\dot{\varepsilon}_V^v)^2 + (\dot{\varepsilon}_q^v)^2]^{1/2}. \qquad (8.110)$$

Satisfying the consistency condition $\dot{f}_0 = 0$ and using (8.72), (8.73) we obtain

$$\dot{\mu} = \frac{G_f K\dot{\varepsilon}_e^v - 2a_0(1+e)(a_c' - a_0')\varepsilon_V^v}{\dfrac{\partial f_0}{\partial p}\left[c - p_P' + \dfrac{a}{a_0}(p_P' - \alpha_p)\right] - \dfrac{\partial f_0}{\partial q}\left[-q_P + \dfrac{a}{a_0}(q_P - \alpha_q)\right]}. \qquad (8.111)$$

Numerical tests of this viscoplastic model were carried out in a report by Mroz and Sharma[29] where both monotonic and cyclic loading cases were discussed and the effect of the frequency of the cycles was studied.

8.7 MODELLING OF SOIL BEHAVIOUR FOR CYCLIC LOADING

In this final section, we shall discuss the application of the hardening model to the case of cyclically varying stress under undrained conditions. Our major interest will be concentrated on such features of cyclic deformation as (1) variation of pore water pressure in the course of cyclic loading, (2) evolution of hysteresis loops with associated growth of strain amplitude and accumulated strain, and (3) variation of material strength in static tests after a prescribed number of cycles. All these aspects are interrelated and it would be extremely convenient to simulate variation of all material properties within one model.

An extensive discussion of Ts and Ins models and their application to modelling of cyclic loading response of soils has been given in Ref. 30. Here we present only a few results by using the Ins model with the added sophistication of translation of nesting surfaces towards the origin as a function of accumulated deviatoric plastic strain (ε_A). This modification causes the effective stress path under undrained conditions to move further towards the origin and, given appropriate values of material parameters, reach the CSL. Such translation, superimposed upon universal contraction or expansion as specified by equation (8.73) for the Ts model, has been applied by means of the formula

$$\left(\dot{\boldsymbol{\alpha}}_i - \begin{bmatrix}\dot{a}_i \\ 0\end{bmatrix}\right) = \left(\boldsymbol{\alpha}_i - \begin{bmatrix}a_i \\ 0\end{bmatrix}\right)\left[-\frac{a_c'}{a_c}(1+e)\dot{\varepsilon}_V^p - h\right], \qquad (8.112)$$

where h is a function of ε_A.

In using this model for *monotonic* loading we have found that a useful expression for γ in Equation (8.59) takes the form

$$\gamma = R_1 \Lambda + 1 \tag{8.113}$$

when Λ is a constant. For qualitative demonstration of the model's potential we now allow Λ to vary as a function of ε_A in the course of *cyclic* loading, using the formula

$$\Lambda = \left[\frac{\Lambda_0 + \Lambda_u \chi \varepsilon_A}{1 + \chi \varepsilon_A} \right], \tag{8.114}$$

where Λ_0 is the initial value of Λ (when $\varepsilon_A = 0$)
$\quad \chi$ is a constant parameter
and Λ_u is a function of the stress history given by

$$\Lambda_u = \Lambda_0 \left(1 + \tau \frac{a_{l1}}{a} \right), \qquad d\Lambda_u \geqslant 0 \tag{8.115}$$

or else

$$\Lambda_u = \text{constant},$$

where τ is a constant parameter.

When $\varepsilon_A = 0$, $\Lambda = \Lambda_0$, whereas for $\varepsilon_A \to \infty$, $\Lambda \to \Lambda_u$; i.e. Λ increases from its initial value Λ_0 to its asymptotic value Λ_u. The dependence of Λ_u on a_{l1}/a has been found useful in the simulation of stress cycles of constant amplitude, since for such cases the stiffness degradation increases with the amplitude of the cycle.

Again purely for demonstration purposes we assume the following formula for h in Equation (8.112):

$$h = \rho \Lambda, \tag{8.116}$$

where ρ is a constant.

In order to use this model economically we have introduced an approximation by forgetting all stress-reversal surfaces except the most recent. All surfaces outside it are assumed centred on the line joining the centres of this surface and the consolidation surface, their locations varying linearly with their sizes.

The use of such a model will now be illustrated by its application to undrained, triaxial loading.

8.7.1 One-way, cyclic loading of normally consolidated and overconsolidated soil

Figures 8.15(a, b) show the effective stress paths associated with pore pressure variation for normally consolidated and overconsolidated soils subjected to cyclic loading of constant amplitude. In all cases the amplitude

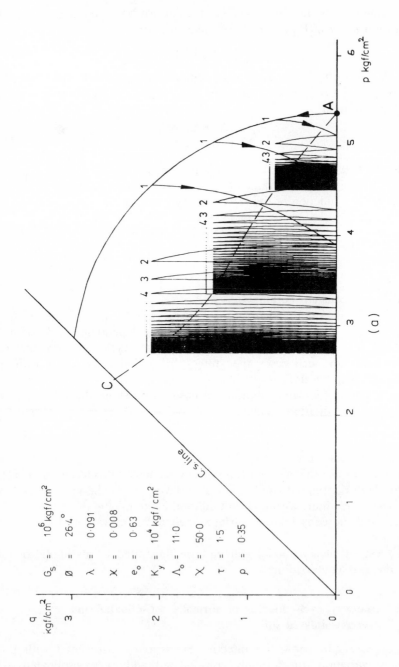

(a)

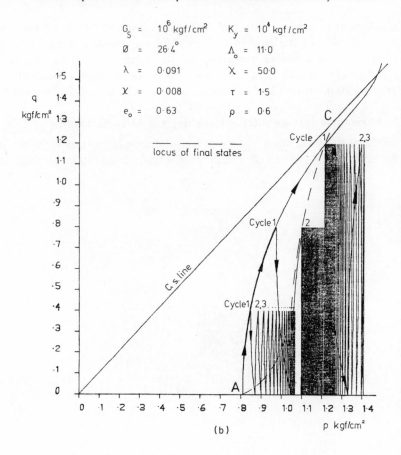

$G_S = 10^6 \, kgf/cm^2$ $K_y = 10^4 \, kgf/cm^2$

$\emptyset = 26.4°$ $\Lambda_o = 11.0$

$\lambda = 0.091$ $\chi = 50.0$

$\chi = 0.008$ $\tau = 1.5$

$e_o = 0.63$ $\rho = 0.6$

(b)

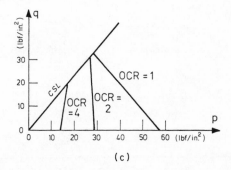

(c)

Figure 8.15 Stress paths for undrained, cyclic, triaxial loading: (a) normally consolidated; (b) overconsolidated (OCR = 6.5); (c) experimental final state bounding lines (Sangrey, Henkel and Esrig[38])

is not sufficient to cause failure and the resulting stable cycles are bounded by the lines *AC*. (In the case of overconsolidated soil the stress path first moves to the right before returning to the left). Comparison of these numerical results with the final state bounding lines obtained experimentally by Sangrey, Henkel, and Esrig,[38] Figure 8 15(c), shows good agreement.

8.7.2 Monotonic loading after various degrees of cyclic loading

Starting from the same state of normal consolidation these tests were carried out for different numbers of cycles of the same amplitude so that the

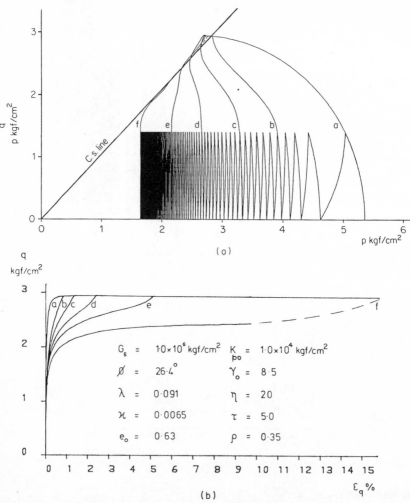

Figure 8.16 Undrained, cyclic, triaxial loading followed by monotonic loading: (a) stress paths; (b) deviatoric stress/strain plots

monotonic stages started at different values of mean effective stress. The resulting effective stress paths and stress/strain plots are shown in Figure 8.16.

The most obvious feature of the stress paths is that they meet almost at one point on the CSL. The tests involve no drainage, thus requiring that the plastic volumetric strains are of equal magnitude and opposite sign to the elastic volumetric strains. The volumetric plastic strains uniquely determine the expansion of the consolidation surface whose intersection with the CSL determines the end point of an undrained, triaxial stress path. Therefore, as all these tests start from the same mean stress, there is only one end point at which these conditions can be fulfilled.

The model thus predicts theoretically that the soil strength after cyclic loading will be the same as for a normally consolidated soil. Observation of the stress–strain plots for these tests shows, however, that for OCR approaching 4 the curve levels off, giving large strains well before the ultimate stress, thus predicting a much lower practical 'failure' stress. The magnitude of this stress depends of course on the material parameters used, and it is thus quite possible that the same model could predict both the strength reductions observed by Andersen *et al.*,[1] as much as 40%, and those recorded by Koutsoftas,[15] as little as 5–10%. Such softening has been observed experimentally by many authors, e.g. Andersen,[2] Thiers and Seed,[42] Castro and Christian[10] and Koutsoftas.[15]

8.7.3 Cyclic loading to failure

Figure 8.17 shows the effective stress paths and stress–strain plots for cyclic, triaxial loading predicted by both models. It may be seen that the INS model predicts almost the same deviatoric strain during unloading as during loading, in agreement with experimental findings, e.g. Kuntsche,[17] whereas the two-surface model predicts almost zero deviatoric strain during unloading. On the other hand the effective stress path predicted by the two-surface model shows a semi-stable stage before the final acceleration to failure, a feature of experimental observations.[22]

8.8. CONCLUDING REMARKS

In this paper we have attempted to show the potential of applying kinematic hardening to soils. The translation of the yield surface (of finite or zero size) within a consolidation surface is governed either by simple formulae or the use of a set of intermediate nesting surfaces, none of which is allowed to intersect the others or the yield and consolidation surfaces. Another simple formula is used for the variation of the plastic modulus between the consolidation surface and the point of initial yield. Whilst no attempt has yet been made to model any particular soil quantitatively for a wide range of

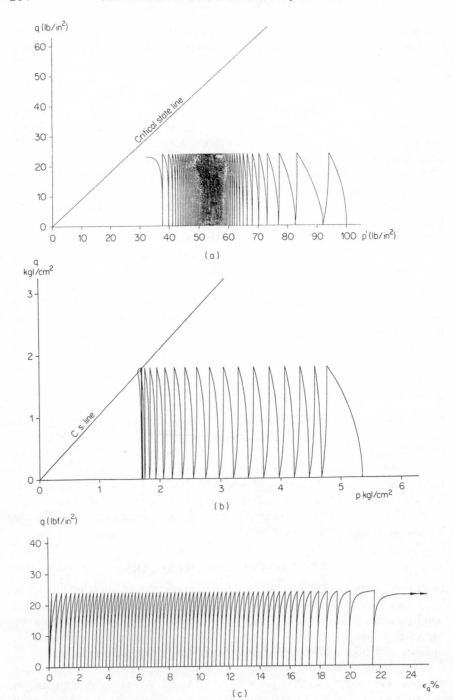

(a)

(b)

(c)

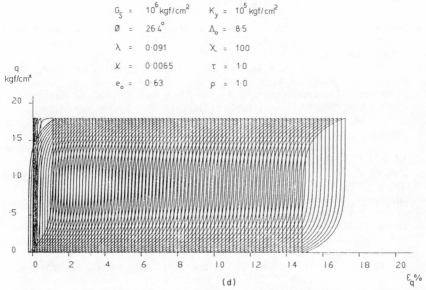

Figure 8.17 Undrained, cyclic, triaxial loading; stress paths predicted by: (a) Ts-model; (b) Ins-model; deviatoric stress–strain plots predicted by (c) Ts-model; (d) Ins-model

stress paths, considerable success has been obtained in predicting qualitative aspects of soil behaviour. This in itself could be useful in the analysis of real soil problems and at the same time provides a basis for the more time-consuming work involved in quantitative simulation. As in endochronic models the number of parameters required for the close simulation of real soils can be relatively large compared with those used in engineering practice, but this should be no numerical drawback at a time when computation is increasingly handled by computers. Once the nature of the required parameters has been satisfactorily determined, the main problem will remain that of developing practical techniques for determining these parameters. Whilst this is recognized to be a difficult field of research it is hoped that the development of numerical models will act as a guide in determining the nature of such parameters.

ACKNOWLEDGEMENTS

The major part of this chapter was written by the first author whilst he was a Visiting Fellow at the Engineering Department of the University of Cambridge. He wishes to express his thanks to Prof. A. N. Schofield and Dr. D. M. Wood for numerous discussions concerned with soil behaviour.

The development and application of the numerical models described was largely carried out at the Department of Civil Engineering, Univ. College of Swansea, under the direction of Professor O. C. Zienkiewicz and with the financial support of the Science Research Council.

REFERENCES

1. Andersen, K. H., Brown, S. F., Foss, I., Pool, J. H., and Rosenbrand, W. F. Effect of cyclic loading on clay behaviour, *Proc. conf. Design and Construction of Offshore Structures*, I.C.E., London, 1976.
2. Andersen, K. H. Behaviour of clay subjected to undrained cyclic loading, *BOSS Conf.*, Norwegian Inst. Tech., Trondheim, 1976.
3. Atkinson, J. H., and Bransby, P. L. *The Mechanics of Soils, An Introduction to Critical State Soil Mechanics*, McGraw-Hill, 1978.
4. Baron, M. I., and Nelson, I. Application of variable moduli models to soil behaviour, *Int. J. Solids, Struct.*, **7**, 399–417, 1971.
5. Bazant, Z. Endochronic in elasticity and incremental plasticity, *Int. J. Sol. Struct.* **14**, 691–714, 1978.
6. Bishop, A. W., and Henkel, D. J. *The Measurement of Soil Properties in the Triaxial Test*, William Arnold, London, 1957.
7. Bishop, A. W. Test requirements for measuring the coefficient of earth pressure at rest. *Proc. Conf. Earth Pressure Problems*, Brussels, **1**, 2–14, 1958.
8. Boyce, H. R. A non-linear model for the elastic behaviour of granular materials under repeated loading, *Proc. Int. Symp. Soils under Cyclic and Transient Loading*, Swansea, 1980.
9. Brooker, E. W., and Ireland, H. O. Earth pressures at rest related to stress history, *Can. Geotechn. J.*, **2**, 60–83, 1965.
10. Castro, G., and Christian, J. T. Shear strength of soils and cyclic loading, *J. Geotechn. Eng. Div.*, ASCE, **102** (GT9) 887–94, 1976.
11. Dafalias, Y. F., and Popov, E. P. A model of non-linearly hardening materials for complex loadings, *Acta Mechanica*, **21**, 173–92, 1975.
12. Di Maggio, F. L., and Sandler, S. Material model for granular soils, *J. Engn. Mech. Div., Proc. ASCE*, 935–40, June 1971.
13. Gudehus, G. Elastoplastische Stoffgleichungen für trockener Sand, *Ingenieur Archiv.*, **42**, 1973.
14. Henkel, D. J. The relationship between the effective stresses and water content in saturated clays, *Géotechnique*, **10**, 41–54, 1960.
15. Koutsoftas, D. C. Effect of cyclic loads on undrained strength of two marine clays, *J. Geotechn. Eng. Div.*, ASCE, **104** (GT5), 609–620, 1978.
16. Krieg, R. D. A practical two-surface plasticity theory, *J. Appl. Mech., Trans. ASME*, **E42**, 641–6, 1975.
17. Kuntsche, K. Response of kaolin to reversals of strain path in undrained triaxial tests, *Int. Symp. on Soils under Cyclic and Transient Loading*, Swansea, 179–86, 1980.
18. Ladd, C. C. Strength parameters and stress–strain behaviour of saturated clays, *Soils Publ. MIT*, 278–81, 1971.
19. Lade, P. V., and Duncan, J. M. Elastoplastic stress-strain theory for cohesionless soil, *J. Geotechn. Eng. Div., Proc. ASCE*, **101** (GT10) 1037–53, 1975.
20. Lade, P. V. Elastoplastic stress–strain theory for cohesionless soil with curved yield surfaces, *Int. J. Solids, Struct.*, **13**, 1019–35, 1975.
21. Mitchell, R. J. On the yielding and mechanical strength of Leda clays, *Can. Geotechn. J.*, **7** 297–312, 1970.
22. Mitchell, R. J., and King, R. D. Cyclic loading of an Ottawa area Champlain sea clay, *Can. Geotech. J.*, **14**, 52–63, 1977.
23. Mroz, Z. On forms of constitutive laws for elastic-plastic solids, *Arch. Mech. Stos.*, **18**, 3–35, 1966.

24. Mroz, Z. On the description of anisotropic work-hardening, *J. Mech. Phys. Solids*, **15**, 163–75, 1967.
25. Mroz, Z., Norris, V. A., and Zienkiewicz, O. C. An anisotropic hardening model for soils and its application to cyclic loading, *Int. J. Num. Anal. Meth. Geom.*, **2**, 203–21, 1978.
26. Mroz, Z., Norris, V. A., and Zienkiewicz, O. C. Application of an anistropic hardening model in the analysis of elasto-plastic deformation of soils, *Géotechnique*, **29**, 1–34, 1979.
27. Mroz, Z. On hypoelasticity and plasticity approaches to constitutive modelling of inelastic behaviour of soils, *Int. J. Num. Anal. Meth. Geom.* **4**, 45–55, 1980.
28. Mroz, Z. On hardening and relaxation moduli for elastoplastic and viscoplastic materials, *Arch. Mech. Appl.*, 1981 (in press).
29. Mroz, Z., and Sharma K. G. *A Viscoplastic Model for Soils with Anisotropic Hardening*, Technical Report, University of Swansea, 1980.
30. Mroz, Z., Norris, V. A., and Zienkiewicz, O. C. An anisotropic critical state model for soils subject to cyclic loading, *Géotechnique*, **31**, 451–69, 1981.
31. Nova, R., and Wood, D. M. A constitutive model for sand in triaxial compression, *Int. J. Num. Anal. Meth. Geom.*, **3**, 255–78, 1979.
32. Petersson, H., and Popov, E. P. Constitutive relations for generalised loadings, *Proc.*, *ASCE*, **103**, EM 611–27, 1977.
33. Pietruszczak, St., and Mroz, Z. Description of mechanical behaviour of anisotropically consolidated clays, *Proc. Euromech. Coll. Anisotropy in Mechanics*, Grenoble, Noordhoff Int. Publ, 1979.
34. Prevost, J. H., and Hoeg, K. Effective stress–strain strength model for soils, *Proc. ASCE*, **101** (GT3) 259–78, 1975.
35. Prevost, J. H. Mathematical modelling of monotonic and cyclic undrained clay behaviour, *Int. J. Num. Meth. Geom.*, **1**, 195–216, 1977.
36. Romano, M. A continuum theory for granular media with a critical state, *Arch. Mech. Stos.*, **26**, 1011–28, 1974.
37. Roscoe, K. H., and Burland, J. B. On the generalised stress–strain behaviour of 'wet' clay, in Heyman, J., and Leckie, F. A. (eds.), *Engineering Plasticity*, Cambridge University Press, 535–609, 1968.
38. Sangrey, D. A., Henkel, D. J., and Esrig, M. I. The effective stress response of a saturated clay soil to repeated loading, *Can. Geotechn. J.*, **6**, 241–52, 1969.
39. Schofield, A., and Wroth, P. *Critical State Soil Mechanics*, McGraw-Hill, London, 1968.
40. Simons, N. Discussion on coefficeint of earth pressure at rest, *Proc. Conf. Earth Pressure Problems*, Brussels, **3**, 50–3, 1958.
41. Skempton, A. W., and Sowa, V. A. The behaviour of saturated clays during sampling and testing, *Géotechnique*, **13**, 269–90, 1963.
42. Thiers, G. K., and Seed, H. B. Strength and stress–strain characteristics of clay subjected to seismic loading conditions, *ASTM STP 400, Amer. Soc. Testing, Mat.*, 3–56, 1968.
43. Wilde, P. Two invariant depending models of granular media, *Arch. Mech. Stos.* **29**, 799–809, 1977.
44. Zienkiewicz, O. C., and Pande, G. N. Some useful forms of isotropic yield surfaces for soil and rock mechanics, in Gudehus, G., *Int. Symp. Num. Meth. Soil, Rock Mech.* Wiley, New York, 1975.

Soil Mechanics—Transient and Cyclic Loads
Edited by G. N. Pande and O. C. Zienkiewicz
© 1982 John Wiley & Sons Ltd

Chapter 9

A Critical State Soil Model for Cyclic Loading

J. P. Carter, J. R. Booker, and C. P. Wroth

SUMMARY

Recently, several sophisticated constitutive models have been proposed for the prediction of the behaviour of soils under cyclic loading. In this paper the concepts of critical state soil mechanics have been used to develop a simple model which predicts many aspects of clays under repeated loading. The model employs the parameters that are usually associated with the Cam-clay family of models together with an additional parameter which characterizes the cyclic behaviour. This parameter can conveniently be determined by performing cyclic triaxial tests under undrained conditions.

The behaviour of soils which are either initially normally or initially overconsolidated is investigated for stress-controlled and strain-controlled loadings in the triaxial test. The results of this theoretical investigation show encouraging agreement with the results of laboratory tests on saturated clays.[1,2,30]

9.1 INTRODUCTION

A problem of considerable importance in geotechnical engineering is that of the prediction of the behaviour of soils under repeated loading. The necessity of understanding the response of soil under earthquake conditions has long been appreciated, but more recently the problems of offshore technology have accentuated the need for adequate descriptions of this aspect of soil behaviour. Highway engineers have also been interested in the response of soil and pavement materials to repeated loads of the type caused by rolling vehicles, and testing of these materials under simulated loading conditions has been carried out in the laboratory, e.g. Monismith, Ogawa, and Freeme.[11]

There exists a considerable body of data on the behaviour of sands under cyclic loading conditions[8,18,24,28] and engineering theories have been developed for particular classes of problems.[10,19,26,27]

Recently, data for the behaviour of clays under cyclic loading have been obtained.[1-5,9,21,23,29-33] Although the conclusions of these examinations are not identical, several facts emerge. The most important of these is that under undrained loading excess pore pressures are generated and if cyclic loading is continued for a sufficiently long time a failure or critical state condition may be reached.

A natural consequence of this interest in cyclic loading has been the attempt to develop constitutive models to predict this type of behaviour.[12,13,16,17] Generally, these models are complex, involving nested yield surfaces and both kinematic and isotropic hardening, and depend on the specification of a number of parameters. There seems to be no straightforward way of determining values for these parameters directly and this places a severe limitation on the use of these models in practical situations. A less complicated model, which is potentially applicable to cyclic loading, has been suggested by Pender.[14,15]

In this paper the concepts and models of critical state soil mechanics[22] have been extended to provide a description of the response of clay to cyclic loading. This new model predicts the generation of excess pore pressures and ultimate failure of the soil under repeated, undrained loading conditions. It requires the specification of only one additional soil parameter which can be conveniently determined from the number of cycles to failure in an undrained stress controlled triaxial test.

9.2 THEORETICAL DEVELOPMENT

9.2.1 Modified Cam clay

In order to clarify the presentation, some of the essential features of critical state soil mechanics will be summarized. In particular, the theory will be developed in terms of the modified Cam-clay soil model[20] and attention will largely be restricted to triaxial conditions. The extension to three dimensional conditions, using a von Mises failure condition is self-evident, and the extension to more general cases, e.g. the Mohr–Coulomb failure condition, is straightforward.[35]

The state of effective stress of a soil specimen will be expressed in terms of the stress invariants p' and q defined by

$$p' = \tfrac{1}{3}(\sigma_1' + \sigma_2' + \sigma_3'), \tag{9.1}$$

$$q = \sqrt{\{\tfrac{1}{2}[(\sigma_1' - \sigma_2')^2 + (\sigma_2' - \sigma_3')^2 + (\sigma_3' - \sigma_1')^2]\}}, \tag{9.2}$$

where $\sigma_1', \sigma_2', \sigma_3'$ are the principal effective stresses.

For the case of the triaxial test it is assumed that $\sigma_2' = \sigma_3'$ and these quantities reduce to

$$p' = \tfrac{1}{3}(\sigma_1' + 2\sigma_3'), \tag{9.3}$$

$$q = \sigma_1' - \sigma_3'. \tag{9.4a}$$

In the above description the subscripts 1 and 3 refer to the major and minor principal stresses respectively (compression positive). When presenting the results of triaxial tests it is often convenient to distinguish between so-called compression and extension. For this reason a stress difference q^* is defined as

$$q^* = \sigma'_z - \sigma'_r = \sigma_z - \sigma_r, \tag{9.4b}$$

where σ'_z and σ'_r are the axial and radial components of effective stress respectively. The total stress σ_r is equal to the cell pressure in a conventional triaxial test and σ_z is equal to the total axial stress. Under compression conditions $\sigma_1 = \sigma_z$, $\sigma_3 = \sigma_r$ and q^* is a positive quantity; while under extension conditions $\sigma_1 = \sigma_r$, $\sigma_3 = \sigma_z$ and q^* is a negative quantity.

The convenient measures of strain for triaxial conditions are v, the volume strain, and ε, a measure of octahedral shear strain, given by

$$v = \varepsilon_1 + 2\varepsilon_3, \tag{9.5}$$

$$\varepsilon = \tfrac{2}{3}(\varepsilon_1 - \varepsilon_3), \tag{9.6}$$

where ε_1 and ε_3 are the major and minor principal strains, respectively.

Only saturated clays will be dealt with here and the symbol e is used as usual to denote the voids ratio. With compressive strains taken as positive, the incremental volume strain dv is related to the change in voids ratio de by

$$dv = -\frac{de}{1+e}. \tag{9.7}$$

The modified Cam-clay model requires the specification of five parameters, values of which may be readily obtained from standard oedometer and triaxial compression tests. These parameters are:

λ the gradient of the normal consolidation line in $e - \ln p'$ space,

κ the gradient of the swelling and recompression line in $e - \ln p'$ space

e_{cs} a value of voids ratio which locates the consolidation lines in $e - \ln p'$ space, conveniently taken as the value of e at unit p' on the critical state line

M the value of the stress ratio q/p' at the critical state condition; M is related to ϕ', the angle of friction obtained in triaxial compression tests, by

$$M = \frac{6 \sin \phi'}{3 - \sin \phi'}, \tag{9.8}$$

G the elastic shear modulus.

For states of stress within the current yield surface the soil responds elastically and the incremental effective stress–strain law may be written as

$$\begin{bmatrix} dp' \\ dq \end{bmatrix} = \begin{bmatrix} K & 0 \\ 0 & 3G \end{bmatrix} \begin{bmatrix} dv \\ d\varepsilon \end{bmatrix}, \tag{9.9}$$

where the bulk modulus K is given by

$$K = \frac{(1+e)p'}{\kappa} \tag{9.10}$$

and the shear modulus G is constant.

Yielding of the material occurs whenever the stresses satisfy the following criterion

$$q^2 - M^2\{p'(p_c' - p')\} = 0, \tag{9.11}$$

where p_c' is a hardening parameter—analogous to a preconsolidation pressure—which defines the non-zero intersection of the current elliptical yield locus and the p' axis in effective stress space—see Figure 9.1. Plastic flow is determined by an associated flow rule and the permanent volume strain dv^p is related to the change in the hardening parameter p_c' as follows:

$$dv^p = \left(\frac{\lambda - \kappa}{1 + e}\right) \frac{dp_c'}{p_c'}. \tag{9.12}$$

Types of loading can be categorized in terms of a variable p_y', defined as

$$p_y' = p' + \left(\frac{q}{M}\right)^2 \frac{1}{p'}. \tag{9.13}$$

Equation (9.13) is also the locus of an ellipse in p', q space which passes through the current stress point and the origin, and is centred on the p' axis, i.e. it has the same shape as the yield locus—see Figure 9.1. This variable p_y' is the (non-zero) value of p' at which the ellipse cuts the p'-axis and is a convenient way of comparing the current stress state with the current yield locus represented by p_c'.

The material is elastic whenever $p_y' < p_c'$ and during any elastic deformation

$$\frac{dp_c'}{p_c'} = 0. \tag{9.14}$$

The material behaves plastically whenever $p_y' = p_c'$ and three conditions can be identified. These are

(1)   hardening whenever $dp_y' = dp_c' > 0$;
(2)   softening whenever $dp_y' = dp_c' < 0$;
(3)   'neutral loading', when the yield locus does not change while plastic behaviour occurs, $dp_y' = dp_c' = 0$.

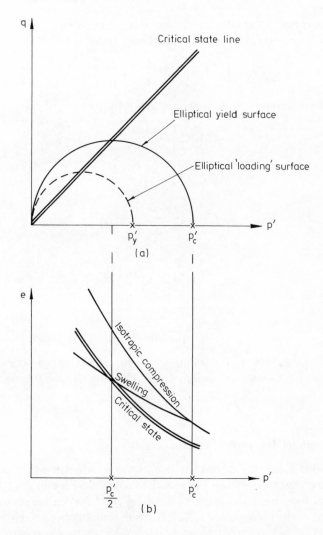

Figure 9.1 Some aspects of the modified Cam-clay model for triaxial conditions

Condition (1) requires $p' > p'_c/2$, and the material is said to be 'wet' of critical, and (2) requires $p' < p'_c/2$, and the material is said to be 'dry' of critical.

During plastic behaviour the yield locus changes according to the law

$$\frac{\mathrm{d}p'_c}{p'_c} = \frac{\mathrm{d}p'_y}{p'_y}. \tag{9.15}$$

The incremental stress–strain relation during yielding may be shown by conditions of associated plasticity (see Chapter 8) to be

$$\begin{bmatrix} dv \\ d\varepsilon \end{bmatrix} = \begin{bmatrix} C_{11} & C_{12} \\ C_{21} & C_{22} \end{bmatrix} \begin{bmatrix} dp' \\ dq \end{bmatrix}, \tag{9.16}$$

where the compliance coefficients are given by

$$C_{11} = \left(\frac{\lambda - \kappa}{1 + e} \right) \frac{a}{p'} + \left(\frac{\kappa}{1 + e} \right) \frac{1}{p'},$$

$$C_{12} = C_{21} = \left(\frac{\lambda - \kappa}{1 + e} \right) \left(\frac{1 - a}{p'} \right),$$

$$C_{22} = \left(\frac{\lambda - \kappa}{1 + e} \right) \frac{b}{p'} + \frac{1}{3G},$$

and

$$a = \frac{M^2 - \eta^2}{M^2 + \eta^2},$$

$$b = \frac{4\eta^2}{M^4 - \eta^4},$$

$$\eta = \text{the stress ratio } q/p'.$$

As would be expected, the relation (9.16) breaks down when the soil reaches the critical state condition $\eta = M$, showing there infinite increments of ε.

9.2.2 A model for cyclic loading

The modified Cam-clay model has been shown to match well the observed behaviour of insensitive clays subjected to monotonic loading for which the stress level increases, and in particular, was used for the successful prediction of the performance of the M.I.T. Trial Embankment.[34] However, the predictions are not as satisfactory when the soil undergoes repeated loading.

When saturated clay is unloaded and then reloaded it is found that permanent strains occur earlier than predicted by the Cam-clay model. One way of interpreting this real behaviour is to assume that the position and perhaps the shape of the yield surface have been affected in some way by the elastic unloading.

For the sake of simplicity in developing a new model it is assumed that the form of the yield surface remains unchanged but that its size has been reduced in an isotropic manner by the elastic unloading. This can then only mean that the hardening parameter p'_c has been reduced by this process. In order to specify how this reduction occurs a relation is proposed between the hardening parameter p_c' and the loading parameter p_y'. From consideration of equation (9.14) it seems reasonable to postulate that when the material is elastic ($p_y' < p_c'$) and when $dp_y' < 0$, the following relation holds:

$$\frac{dp_c'}{p_c'} = \theta \frac{dp_y'}{p_y'}. \tag{9.17a}$$

If θ takes a value of unity, then the yield surface would shrink back in such a way that the stress state always lay on it. It is to be expected that the yield surface will recede only a fraction of this amount and the values of θ will tend to be quite small.

It is postulated, however, that if the material is elastic, but $dp_y' \geqslant 0$, then the current yield surface is not changed, i.e.

$$\frac{dp_c'}{p_c'} = 0. \tag{9.17b}$$

The distinction between these two types of behaviour is shown schematically in Figure 9.2.

The mode of behaviour can be illustrated by considering a simple example of isotropic effective stress change so that p_y' is always equal to p'. Suppose that a clay specimen is isotropically normally consolidated to a mean

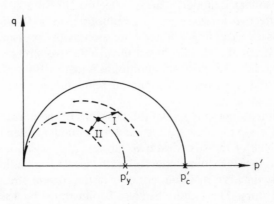

•   Current stress state

⌒⌐   Current yield surface

⌐⌐⌐. Current 'loading' surface

⌐⌐⌐⌐ New 'loading' surface

I   Elastic 'loading' — p_c' constant

II   Elastic 'unloading' — p_c' decreases

Figure 9.2 The yield surface and the 'loading' surface in p'–q space

effective stress of $p' = p'_c = \alpha_0$ and is subsequently allowed to swell elastically by reducing the mean effective pressure to a value $p' = \alpha_1$. During the swelling, equation (9.17a) predicts that the value of p'_c is reduced to the value $p'_c = \alpha_0(\alpha_1/\alpha_0)^\theta$. If the specimen is then reconsolidated, p'_c will remain unchanged and the material will behave elastically until $p' = p'_c = \alpha_0(\alpha_1/\alpha_0)^\theta$. If loading is continued indefinitely, the material will deform plastically and thereafter p'_c will be equal to p'. This means that in a laboratory test, yielding of the soil will be observed during reloading at a value of isotropic pressure which is smaller than the actual preconsolidation pressure. Hence for this isotropic test the *measured* overconsolidation ratio will be

$$\text{OCR} = \alpha_0 \left(\frac{\alpha_1}{\alpha_0}\right)^\theta \Big/ \alpha_1 = \left(\frac{\alpha_0}{\alpha_1}\right)^{1-\theta}. \qquad (9.18)$$

This is less than the value given by the conventional definition, i.e.

$$\text{OCR} = \frac{\alpha_0}{\alpha_1}. \qquad (9.19)$$

The magnitude of this effect (usually quite small for one cycle) depends on the value of θ, which can be thought of as an OCR degradation parameter. For example, consider a specimen having a value of $\theta = 0.5$, which is initially normally consolidated under a mean effective pressure of $\alpha_0 = 100$ units. This mean effective pressure is then reduced to $\alpha_1 = 50$ units and subsequently increased. Yielding will occur as soon as the pressure reaches 96.6 units again. When the sample has a mean effective stress of 50 units the OCR given by the conventional definition is equal to $100 \div 50 = 2$; the value of OCR inferred from the behaviour on reloading is equal to $96.6 \div 50 = 1.93$.

In the modified Cam-clay model the positions of the normal consolidation or 'λ lines' in $e - \ln p'$ space are assumed to be uniquely determined for any clay by the value of the stress ratio η. However, a consequence of the new model is that these λ lines 'migrate' with 'elastic unloading' and so the position at any time is a function, not only of the current stress ratio but also of the stress history. This feature can also be illustrated by a simple example.

The behaviour of both models under repeated consolidation and swelling at constant stress ratio η, between the limits $p' = p'_A$ and $p' = p'_B$, is shown schematically in Figures 9.3 and 9.4. In modified Cam-clay (Figure 9.3) the yield locus does not change during elastic swelling, and so after the initial normal consolidation to $p' = p'_B$ the material is always elastic in this test. As a result the path in $e - \ln p'$ space varies continuously along a 'κ line' between points A_i and B_i of Figure 9.3, i.e. the voids ratio oscillates about some mean. In the new model a shrinkage of the yield surface is predicted during each period of elastic swelling or 'unloading', as explained above.

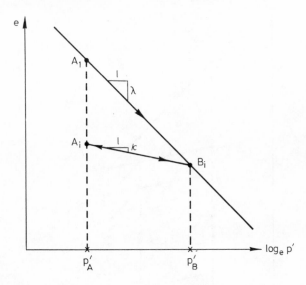

Figure 9.3   Repeated consolidation and swelling of modified
Cam-clay at constant stress ratio

Hence, on reconsolidation in each cycle the material will yield at some value
of $p' = p'_D$ less than p'_B, i.e. at points $D_2, D_3, \ldots, D_n$ in Figure 9.4. Movement from D_i to B_i down a 'λ line' implies some irreversible or plastic
volume change, so that the average voids ratio of the material is reduced
with each cycle of this type, i.e. the material becomes more dense. As a
result the normal consolidation or λ line is seen to migrate. In contrast to
this, the λ line of Figure 9.3 remains fixed. In both models the λ lines
corresponding to different stress ratios will always remain parallel to each
other in $e - \ln p'$ space.

The amount of densification per cycle predicted by the new model will
depend on the value of the degradation parameter θ. It is emphasized
again that for most soils θ will be small, so that the shift in any one cycle is
likely to be small, and thus laboratory specimens may require many such
cycles before they exhibit a measurable densification.

One of the most important features common to modified Cam-clay and
the new model is that concerned with the prediction of the undrained
strength c_u under increasing deviator stress in a triaxial test. For both
models this strength is uniquely related to the current mean effective stress
p' and the hardening parameter p'_c by

$$c_u = \frac{M}{4} p'_c \left(\frac{2p'}{p'_c}\right)^{\kappa/\lambda}, \qquad (9.20)$$

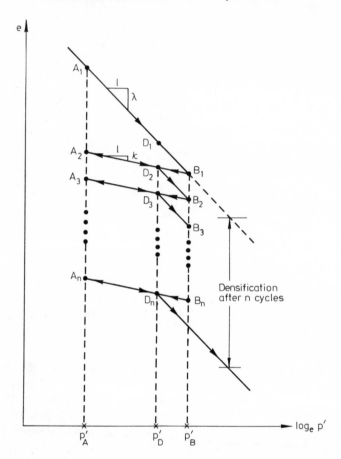

Figure 9.4 Repeated consolidation and swelling of cyclic Cam-clay
at constant stress ratio

where c_u is one half of the deviator stress at failure. It should be emphasised
that the values of p' and p'_c occurring in equation (9.20) are those that exist
in the sample before undrained testing proceeds. Given an initial stress state
modified Cam-clay predicts a unique undrained stress path in p', q space and
a unique failure point. However, for the new model this is true only as long
as the deviator stress q always increases. Thus cyclic loading will have an
influence not only on the effective stress path, but also on the generation of
excess pore pressure and the value of the undrained strength.

It has been shown in this section that the modified Cam-clay model and
the new soil model have many features in common. The criterion for
yielding is the same, the flow rule and the hardening law are the same, and
the incremental elastic and elastoplastic stress–strain relations are the same.

The only difference is the modification to the yield surface associated with 'elastic unloading' (p'_y decreasing). This slight modification has important consequences to the repeated loading problem; some relevant to drained conditions have already been discussed. Others relevant to undrained conditions are dealt with in the following sections.

9.3 PREDICTION OF THE BEHAVIOUR OF NORMALLY CONSOLIDATED CLAY

In order to illustrate the behaviour predicted by this model, one set of values for the conventional Cam-clay parameters has been selected. In this and subsequent sections the following values have been adopted:

$$\lambda = 0.25, \quad \kappa = 0.05, \quad M = 1.2(\phi' = 30°), \quad G = 200c_{u0},$$

where

$$c_{u0} = \frac{M}{4} p'_{c0} \left(\frac{2p'_0}{p'_{c0}}\right)^{\kappa/\lambda}$$

is the initial value of the undrained strength predicted by the modified Cam-clay model. The subscript zero indicates an initial value. For all calculations in which the soil is initially in a normally consolidated state, the initial voids ratio is taken as $e_0 = 0.6$.

The effects of cyclic loading are most dramatic when the soil is loaded in an undrained manner and so attention will be concentrated on undrained triaxial conditions for both total stress- and strain-controlled loadings. In all cases reported here the excess pore pressure u has been determined from the effective stress principle, i.e. it is the difference between the applied total stress and the effective stress. The latter has been calculated using equations (9.9) or (9.16), as appropriate, together with the constant volume condition.

9.3.1 Stress-controlled loading

The model presented in this paper involves the specification of the additional degradation parameter θ. In order to examine its effects, calculations have been performed for the case of cyclic axial load at constant cell pressure in the triaxial test. In each case loading is applied so that the deviator stress q^* is varied continuously between limits of 0 and q_c, i.e. one-way compression loading where $\sigma_z \geq \sigma_r$ with σ_r constant.

Typical results for calculations with $\theta = 0.1$ and $q_c/2c_{u0} = 0.75$ are shown in Figure 9.5. The effective stress path, plotted in p', q space, is shown in Figure 9.5(a). In the first half of the first cycle the yield surface expands, i.e. the material work hardens, and the stress path is identical to that predicted by modified Cam-clay. During the second half of the first cycle the soil is unloaded (q decreasing) and it responds elastically. As no drainage occurs

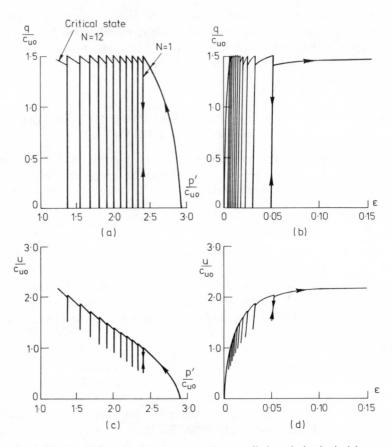

Figure 9.5 Predictions for a one-way, stress-controlled, undrained triaxial test: $OCR = 1$, $\theta = 0.1$

there is no change in p'; however, the value of p'_c has decreased according to equation (9.17a), i.e. the yield surface has contracted slightly. On reloading in the second cycle the material behaves elastically until the stress point reaches the yield locus again,* thereafter the material yields, the yield surface expands, further plastic deformations occur, the stress state migrates toward the critical state condition and additional excess pore pressure is generated. This sequence is repeated at each additional load cycle and ultimately, if this process is continued, a critical state condition is reached. In every cycle there is yielding and associated permanent strains and, in particular, during any cycle there is an increment of permanent volume

---

* In modified Cam-clay the yield surface remains fixed during the unloading and elastic behaviour is predicted for all subsequent cycles; there is no further increase in pore pressure.

strain. Because the deformation occurs at constant volume there must be a corresponding elastic volume increase and this implies a decrease in mean effective stress, i.e. an increase in pore pressure. The accumulation of excess pore pressure with each cycle is plotted against mean effective stress in Figure 9.5(c) and against shear strain in Figure 9.5(d). The relation between deviator stress and shear strain is also shown in Figure 9.5(b).

For this material, which has $\theta = 0.1$, failure occurs on the loading portion (q increasing) of the 12th cycle. In general the number of cycles to failure N_f will be dependent not only on the value of θ but also on the cyclic load level q_c. Results are presented in Figure 9.6 for a number of values of θ and a range of different load levels. It can be seen that for a given material, i.e. a particular value of θ, the number of cycles to failure increases as the amplitude of loading is decreased. For a given amplitude of loading the number of cycles to failure decreases as θ increases. This is as expected, since a larger value of θ implies a greater contraction of the yield surface with elastic 'unloading', i.e. a greater decrease in p'_c. Consequently there are greater permanent volume strains and greater excess pore pressures generated per cycle and thus the material reaches the critical state after fewer cycles.

The number of cycles to failure in this particular type of test is independent of the elastic shear modulus G. The value of this quantity only affects the magnitude of the shear strains.

Another important feature predicted by this model is indicated in Figure 9.7, where the ratio of the undrained shear strength c_u, measured immediately after the Nth cycle, to the original undrained strength c_{u0},

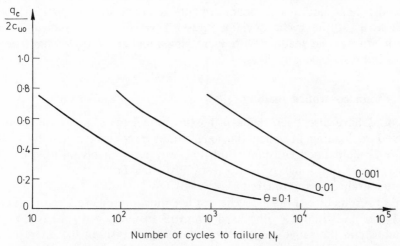

Figure 9.6 Variation of the number of cycles to failure with cyclic stress amplitude q_c, in a one-way, stress-controlled, undrained triaxial test: OCR = 1

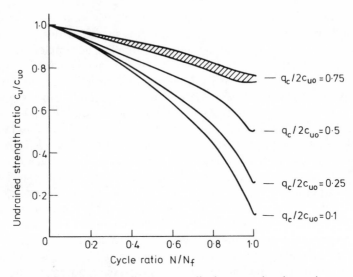

Figure 9.7 Effect of cyclic stress amplitude q_c on the change in undrained strength during a one-way, stress-controlled triaxial test

measured before cycling, is plotted against the ratio N/N_f. The results show a continual reduction in the undrained shear strength for soils subjected to repeated increments of deviator stress. Each of the curves of Figure 9.7 corresponds to a different amplitude of cyclic deviator stress and results for materials with θ in the range $0.001 \leqslant \theta \leqslant 0.1$ appear to lie on either a unique curve or in a narrow region as shown. When the soil reaches failure after N_f cycles, the final undrained shear strength is equal to one half of the amplitude q_c of the cyclic deviator stress. This effect of a reduction in strength after cyclic loading with increasing number of cycles has been observed in tests on many clays.[1,2,30]

9.3.2 Strain controlled loading

Predictions have also been made using this model for samples which are subjected to loading in which the axial strain is controlled and the cell pressure is maintained constant. Both one-way tests, involving total compressive strain only, and two-way tests, involving both compression and extension strains, are considered.

Typical results of a two-way cyclic test are shown in Figure 9.8 for which $\theta = 0.1$ and the strain ε ($=$ the axial strain in an undrained test) is varied continuously in the range $-\varepsilon_c \leqslant \varepsilon \leqslant \varepsilon_c$, where $\varepsilon_c = 0.001$. In the interest of clarity results for only 25 cycles have been plotted. Figure 9.8(a) shows the effective stress path during the test. When the stress path is parallel to the

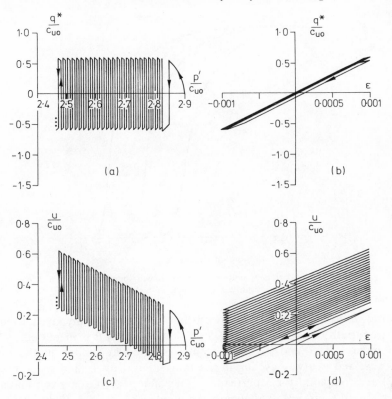

Figure 9.8 Predictions for a two-way, strain-controlled, undrained triaxial test— first 25 cycles only: OCR = 1, $\theta = 0.1$, $G = 200c_{u0}$

deviator stress axis, i.e. p' constant, the soil is responding elastically. It can be seen that yielding occurs in each cycle when q is largest in both compression and extension. In this type of strain-controlled test the stress path migrates toward the critical state condition, oscillating between compression and extension, with the mean effective stress gradually reducing to zero, i.e. the soil liquefies. This trend is also illustrated in Figure 9.10, where the line of peaks of the stress path is also plotted. As this reduction in p' occurs, the excess pore pressure is gradually increased as shown in Figures 9.8(c) and 9.8(d).

Cycling in this type of test also causes reduction in the undrained shear strength and the strength ratio c_u/c_{u0} is shown in Figure 9.9 against the number of cycles N, plotted as $\log_{10}(N+1)$. Results are given for three different values of θ (0.01, 0.03, 0.1) and it can be seen that all of those materials undergo a rapid reduction in shear strength, i.e. they tend to liquefy as the number of cycles N is increased. In Figure 9.10 all of the

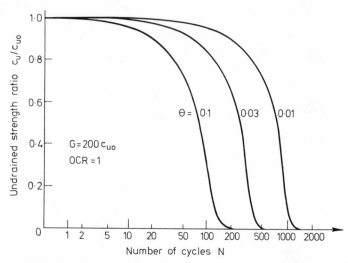

Figure 9.9 Variation of undrained strength with number of cycles in a
two-way, strain-controlled triaxial test: $\varepsilon_c = 0.001$

materials follow the same line of peaks but movement to any given point on
the stress path occurs in fewer cycles as the value of θ is increased.

In Figures 9.11 and 9.12 the results of one-way cycling in the range
$0 \leqslant \varepsilon \leqslant \varepsilon_c$ and two-way cycling in the ranges $-\varepsilon_c \leqslant \varepsilon \leqslant \varepsilon_c$ and $(-\varepsilon_c/2) \leqslant \varepsilon \leqslant$
$(\varepsilon_c/2)$ are compared for a particular value of $\theta = 0.1$, i.e. a given material.
The strength ratio is plotted against the number of cycles in Figure 9.11 and

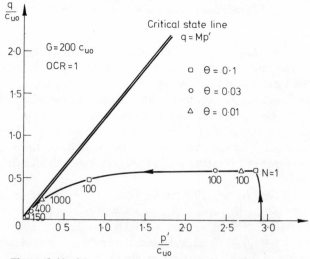

Figure 9.10 Line of peaks in the effective stress path in a
two-way, strain-controlled, undrained triaxial test: $\varepsilon_c = 0.001$

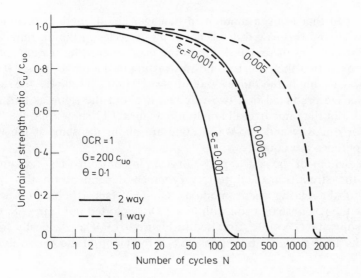

Figure 9.11 Effect of cyclic strain amplitude on the undrained strength in one- and two-way triaxial tests

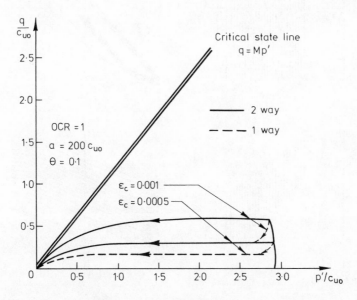

Figure 9.12 Effect of cyclic strain amplitude on the line of peaks in the effective stress path in one- and two-way, undrained triaxial tests

it can be seen that a given amount of damage, i.e. strength reduction, in both one-way and two-way tests occurs in fewer cycles as the magnitude of the cyclic strain ε_c is increased. Perhaps the most interesting feature shown by this figure is that the plots for one-way testing in the range $0 \leqslant \varepsilon \leqslant 0.001$ and two-way testing in the range $-0.0005 \leqslant \varepsilon \leqslant 0.0005$ are almost the same, i.e. the damage predicted in a two-way test is about the same as that in a one-way test of the same overall amplitude. Figure 9.12 shows that the lines of peak for one-way and two-way testing are about the same if the total strain amplitudes are equal.

The predictions of the soil response in a strain-controlled test are, unlike those in the stress-controlled test, very much dependent on the value assigned to the elastic shear modulus G. This fact is observed when comparing the response of a soil with $G = 200c_{u0}$, plotted in Figure 9.8, with that of a soil with $G = 400c_{u0}$, plotted in Figure 9.13. The elastically stiffer soil exhibits a softening in its response, i.e. a reduction in the peak value of

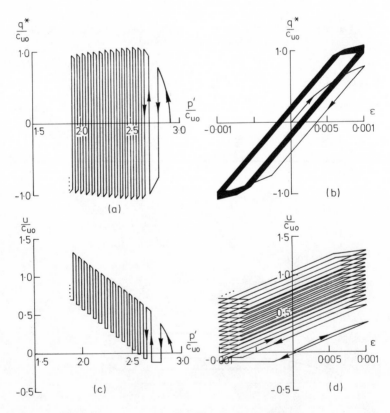

Figure 9.13 Predictions for a two-way, strain-controlled, undrained triaxial test—first 15 cycles only: OCR = 1, $\theta = 0.1$, $G = 400c_{u0}$

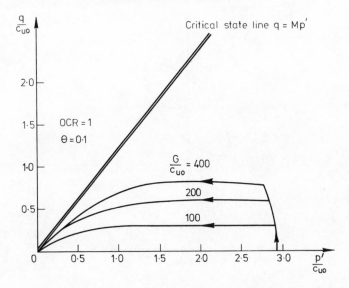

Figure 9.14 Effect of the elastic shear modulus on the line of peaks in the effective stress path in two-way, undrained triaxial tests, $\varepsilon_c = 0.001$

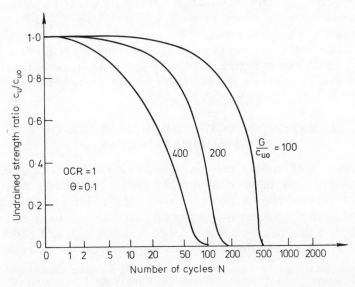

Figure 9.15 Effect of the elastic shear modulus on the undrained strength in a two-way, strain-controlled triaxial test: $\varepsilon_c = 0.001$

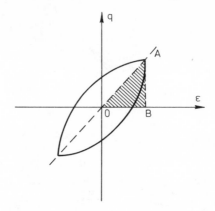

Apparent shear modulus = $\frac{1}{3}$ × slope of OA

Figure 9.16   Definition of the apparent shear modulus

q, in fewer cycles than does the soil with a lower value of G—see Figure 9.14, and the rate of increase of average excess pore pressure is also greater for the stiffer soil—compare Figure 9.13(c), (d) with Figure 9.8(c), (d). A given reduction in undrained shear strength also occurs in fewer cycles if the soil is elastically stiffer. This effect is illustrated in Figure 9.15.

The results presented in Figure 9.13 also exhibit other features observed in laboratory tests on many clays. These include hysteresis of the stress–strain behaviour and the effects of cyclic loading on the apparent shear modulus which is defined in Figure 9.16. It can be seen in Figure 9.13(b) that the apparent shear modulus G decreases as the loading is repeated, and it is noted that the rate of decrease is greater as the amplitude of the shear strain increases. This feature has been observed in experiments by many workers.[1,6,25,29–32]

9.4   PREDICTIONS OF THE BEHAVIOUR OF OVERCONSOLIDATED CLAY

The behaviour of an initially overconsolidated sample when subjected to repeated loading may be contrasted with that of an initially normally consolidated soil. In Figure 9.17 results are presented for a material with $\theta = 0.001$ which has been initially isotropically consolidated to an effective stress of $3.85c_{u0}$ and has then been allowed to swell to a mean effective stress equal to $0.961c_{u0}$, so that the conventional overconsolidation ratio* is

---

* The conventional definition of overconsolidation ratio is the maximum value of the major principal effective stress divided by the value at the beginning of the test. For this model, the preconsolidation pressure (and hence the OCR) detected during reloading is smaller than the actual value on account of the small migration of the normal consolidation line.

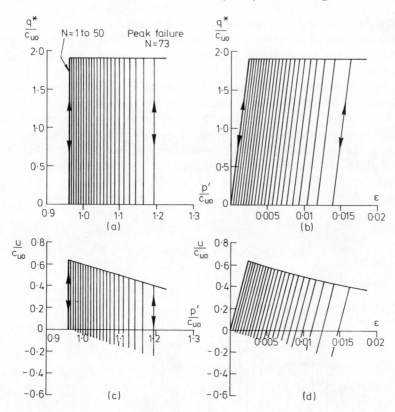

Figure 9.17 Predictions for a one-way, stress-controlled, undrained triaxial test: OCR = 4, $\theta = 0.001$, $G = 200c_{u0}$

equal to 4. The soil has then been subjected to a continuous variation of deviator stress between the limits $0 \leqslant q \leqslant q_c$, where $q_c = 1.9c_{u0}$ under undrained conditions. All stress levels quoted here have been expressed as multiples of the undrained strength c_{u0}, which is the value after swelling to an OCR of 4 but prior to cyclic loading.

The initial swelling and the period when q decreases in each cycle constitute elastic 'unloading' as defined above, i.e. p'_y decreasing. During each of these unloading events the yield surface contracts until eventually the stress point reaches the yield surface. Thereafter, there are periods of plastic loading in each cycle. In this particular example the first plastic strains were observed in the 51st cycle. Thus during the first 50 cycles the material responds entirely elastically; there are no permanent strains and the excess pore pressure oscillates between 0 and $\frac{1}{3}q_c$. After 51 cycles, permanent strains occur and in this particular case the material dilates and softens plastically because the stress state is on the 'dry' side of critical. Since the deformation is occurring at constant volume, the increase in (compressive) plastic volume strain must be compensated by a decrease in elastic volume strain, i.e. the stress state migrates towards critical state and the pore pressure decreases. In common with modified Cam-clay the cyclic model predicts a peak strength in a stress-controlled test under certain circumstances; hence failure may occur either when the stress state reaches the critical state or when it reaches this peak undrained strength, whichever occurs first. In samples which have an overconsolidation ratio greater than 2, such as the one considered here, peak failure is likely to occur. In contrast, soils which are slightly overconsolidated, i.e. on the 'wet' side of critical, behave after sufficient cycles, in the manner of initially normally consolidated soils.

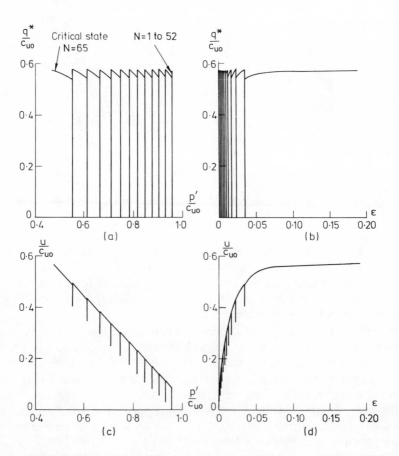

Figure 9.18 Prediction for a one-way, stress-controlled, undrained triaxial test: $OCR = 4$, $\theta = 0.1$, $G = 200c_{u0}$

It is also a feature of this model that all initially overconsolidated soils eventually respond to cyclic loading in the same manner as an initially normally consolidated soil, as long as the deviator stress q is never greater than M times p'. To illustrate this feature consider the predictions of Figure 9.18. The initial value of conventional OCR is 4, the initial value of p' is $p'_0 = 0.96c_{u0}$ and θ has a value of 0.1. The material is otherwise the same as that for the previous example given in Figure 9.17. The cyclic deviator stress level in the present case is given by $q_c = 0.58c_{u0}$, and hence prior to failure q is always less than p' times the friction constant M, i.e. below the critical state level. As shown in Figure 9.18, the first 52 cycles are elastic, while during cycles 53–64 elastic and plastic behaviour is predicted. Failure occurs during cycle 64 when the soil sample comes into a critical state condition with $q = Mp'$. In Figure 9.18(c) it can be seen that, even in this case of an initially overconsolidated soil, the excess pore pressure can gradually build up as cyclic loading is continued. Hence, in overconsolidated soils the cyclic stress level has a significant effect on the predicted response.

9.4.1 The effect of initial OCR on cyclic behaviour

Calculations have been performed for a number of ideal soils which all had different values of θ but all had the same conventional overconsolidation ratio of 4. Before swelling each soil was considered to be normally consolidated with a voids ratio $e = e_{nc} = 0.6$ at a mean pressure $p' = 2.902c_{u0}(\text{n.c.})$ where $c_{u0}(\text{n.c.})$ is the value of the undrained strength of the soil in the normally consolidated condition. All samples were subsequently allowed to swell so that they had a voids ratio $e_0 = 0.669$ and a mean pressure $p' = 0.7255c_u(\text{n.c.})$, i.e. OCR = 4. During the elastic swelling the value of p'_y will have decreased and thus so will p'_c. The new value of p'_c, and hence c_{u0}, the undrained strength in the overconsolidation condition, depends on the vaue of θ. This dependence is shown in Table 9.1 for the several values of θ considered.

Table 9.1 Variation of p'_c and strength
with θ at OCR = 4

θ	$p'_c/c_u(\text{n.c.})$	$c_{u0}/c_u(\text{n.c.})$
0	2.902	0.758
0.001	2.898	0.757
0.003	2.890	0.755
0.01	2.862	0.750
0.03	2.784	0.733
0.1	2.526	0.678

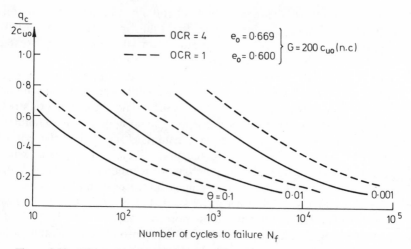

Figure 9.19 Effect of initial OCR on the number of cycles to failure in a one-way, stress-controlled, undrained triaxial test

Figure 9.19 shows the prediction of the number of cycles to failure N_f in a one-way stress-controlled test plotted against the magnitude of the applied deviator stress q_c. Curves have been plotted for three different materials corresponding to $\theta = 0.001$, 0.01 and 0.1. The trend is the same as that for normally consolidated soils, i.e. the number of cycles to failure increases as

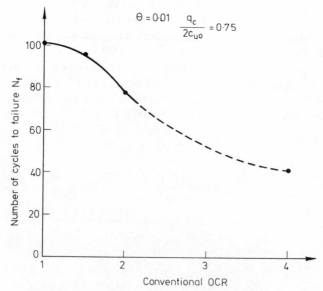

Figure 9.20 Effect of initial OCR on the number of cycles to failure in a one-way, stress-controlled, undrained triaxial test

q_c decreases and as θ decreases. Broken curves have also been plotted in Figure 9.19 for soils with OCR = 1 and the same values of θ. A comparision of the three pairs of curves shows that the number of cycles to failure is also a function of the initial overconsolidation ratio of any soil. The model predicts that overconsolidated soils fail sooner, i.e. in fewer cycles, in repeated load tests than do normally consolidated samples of the same soil; see, for example, Figure 9.20. This prediction is in agreement with the trends shown in laboratory tests on Drammen clay, e.g. Andersen.[1]

9.5 EXPERIMENTAL DETERMINATION OF THE MODEL PARAMETERS

For calculations under fully drained or undrained conditions the model requires the specification of the five basic soil parameters λ, κ, M, G and θ. In addition, the initial state of effective stress and the initial voids ratio e_0 are required. Values for λ and κ may be obtained directly from the results of oedometer or drained triaxial tests; both tests must include unloading–reloading cycles. The frictional parameter M is directly related to ϕ', see equation (9.8), and is conveniently determined from good quality drained triaxial tests. The elastic constant G can be determined as one third of the gradient of the deviator stress–axial strain curve on an unloading portion of an undrained triaxial test.

In principle, it is possible to determine a value for θ from the results of one unloading–reloading cycle in a consolidation test. However, in practice it seems more reasonable to base the estimate of θ on the results of a large number of cycles rather than on one single cycle. While it may be possible to interpret the results of many cycles of consolidation, it is probably more convenient to use the results of undrained cyclic tests. It is possible to replot the results of Figure 9.6, which are for a one-way, undrained, stress-controlled compression test in the triaxial apparatus, in the form shown in Figure 9.21. For example, in a test in which the stress level $q_c/2c_{u0} = 0.5$ has been repeated, failure is indicated for the particular sample after 458 cycles. Using Figure 9.21, as shown, a value of $\theta = 0.001$ could be inferred for this soil.

Although it is convenient, it is not necessary to use the number of cycles to failure as a means of determining θ. It is possible to calculate and to plot a family of curves similar to those in Figure 9.21 for the number of cycles required for the permanent axial strain to reach some chosen value, 3% for example. Alternatively, the number of cycles required to generate a given excess pore pressure in a strain controlled test could be used to deduce a value for θ. These tests, to determine the value of the additional degradation parameter, are straightforward and easily performed in the laboratory. The

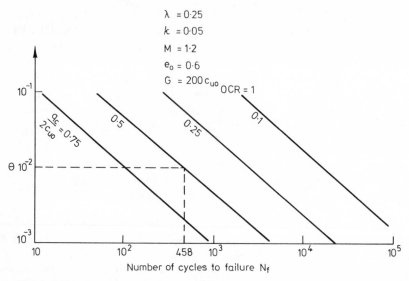

Figure 9.21 Determination of the degradation parameter θ from a one-way, stress-controlled, undrained triaxial test

physical significance which can be attached to this parameter is an advantage which is likely to be appreciated by engineers.

9.6 COMPARISON OF PREDICTIONS WITH EXPERIMENTAL RESULTS

9.6.1 Tests of Taylor and Bacchus

Taylor and Bacchus[30] reported the results of cyclic triaxial tests in which 100 sinusoidal strain-controlled cycles were applied to artificially prepared saturated clay samples. The significant effect on normally consolidated clay was to reduce the mean effective stress p' by an amount which depended on the applied strain amplitude. The results of one of these tests in which the initial OCR = 1, $e_0 = 0.962$ and $p'_0 = 64$ lbf/in^2, are plotted in Figure 9.22 for the case in which the strain was varied continuously in the range $-0.003 \leqslant \varepsilon \leqslant 0.003$. Also shown on this plot are some predictions made using the new model. Values for the model parameters and the source for each are given in Table 9.2. It can be seen that the predictions in these strain-controlled tests are very dependent on the value selected for the elastic shear modulus G. In both predictions the rate of decrease in p' is overpredicted in the latter stages of both tests and possible reasons for this behaviour are discussed

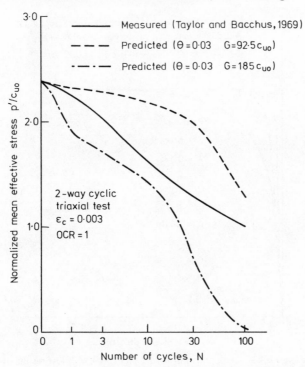

Figure 9.22 Comparison of model predictions with test results of Taylor and Bacchus (1969) for a two-way, strain-controlled, undrained triaxial test

below. Nevertheless, the model displays the general trend in this type of cyclic test.

Table 9.2   Parameters used in predictions of Taylor and Bacchus[30] test

Parameter	Value	Source
λ	0.132	Consolidation plot, Figure 2
κ	0.021	of Taylor and Bacchus
M	1.5	Effective stress state at failure, Figure 10 of Taylor and Bacchus
G	5000 lbf/in^2	Averages from unloading curves,
	2500 lbf/in^2	Figure 7 of Taylor and Bacchus
θ	0.03	Estimated

Test results and predictions are shown in Figure 9.23 for the case of a monotonic triaxial compression test under undrained conditions. It can be seen that, although the predictions for the ultimate deviator stress are very

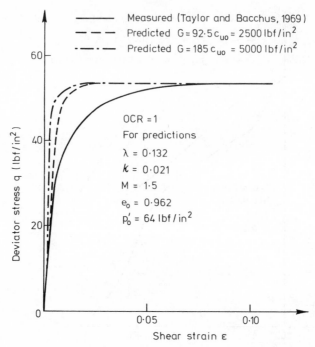

Figure 9.23 Comparison of model prediction with test results of Taylor and Bacchus (1969) for an undrained, monotonic triaxial compression test

accurate, the predicted shear stress–strain responses are both too stiff prior to failure. These predictions, which are the same as would be provided by the modified Cam-clay model, do not show enough plastic shear strain and in fact overpredict the plastic volume strain. As a result a given drop in p' (or increase in u) is predicted in a cyclic test in fewer cycles than is observed. Both static and cyclic tests suggest that the elliptical yield locus used in the model (which is identical to the plastic potential because of an associated flow rule) is not an accurate representation of the actual behaviour. Better predictions might be obtained, for this particular material if some other shape is used for a yield locus—one in which plastic shear strains are greater at lower values of deviator stress q than predicted by the ellipse. A shape like the original Cam-clay yield locus[22] might be better as long as the singularity formed by the vertex at the isotropic axis is removed.

9.6.2 Tests on Drammen clay

An extensive programme of cyclic testing has been carried out on Drammen clay and the results have been brought together in a report published by the Norwegian Geotechnical Institute.[1] Included in this programme were cyclic

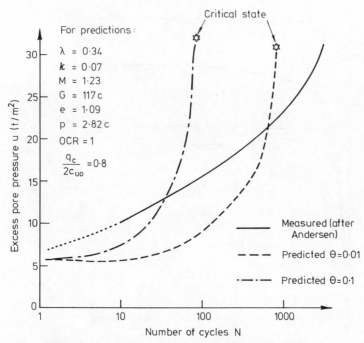

Figure 9.24 Comparison of model predictions with test results for Drammen clay (after Anderson, 1975)—one-way, stress-controlled, undrained triaxial test

triaxial tests during which pore pressure measurements were made. The results of one of these tests are plotted in Figure 9.24 for the case in which the initial OCR = 1, $e_0 = 1.092$ and $p_0' = 40$ t/m². In this test the deviator stress was varied continuously in compression in the range $0 \leq q \leq 0.8c_{uo}$. Also shown in this plot are some predictions made using the new model. Values for the model parameters and the source for each are given in Table 9.3. In both predictions the rate of increase in excess pore pressure is

Table 9.3 Parameters used in predictions of tests on Drammen clay

Parameter	Value	Source
λ	0.34	Interpretation of triaxial
κ	0.07	test data made by van Eekelen
M	1.23	and Potts[5]
	($\phi' = 30°$)	
G	1667 t/m²	From static test data
	($= 117c_{uo}$)	
θ	0.1	Estimated
	0.01	

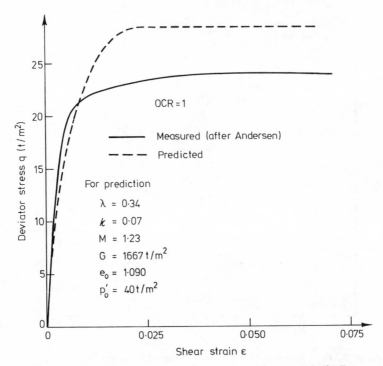

Figure 9.25 Comparison of model prediction with test results for Drammen clay (after Anderson, 1975)—undrained, monotonic triaxial compression test

overpredicted in the latter stages of the test, i.e. at larger cycle numbers. This means that the rate of decrease in p' will also be overpredicted at larger cycle numbers and this is the same trend as noted in the predictions of the tests of Taylor and Bacchus. Again a possible explanation for the discrepancy between predicted and observed results may be the choice of the yield surface used in the model. It is proposed to investigate this matter in future research work.

Figure 9.25 shows the comparison between observed and predicted deviator stress–strain curves for a normally consolidated sample of Drammen clay subjected to undrained monotonic triaxial compression. With the values of soil parameters given in Table 9.3 it can be seen that the prediction of the initial stiffness appears to be adequate but the prediction of the ultimate strength is too high.

9.7 SUGGESTIONS FOR FUTURE RESEARCH

There are several matters which require further attention. Some of the more important are:

(1) From the previous section it is obvious that more work must be done to determine accurately the shape of the yield surface. High-quality testing is required in addition to the theoretical interpretation. It is apparent from the calculations presented in this paper that yield surfaces which give reasonable predictions for monotonic tests may not be satisfactory for predictions of cyclic behaviour. Any small error in the static test predictions is likely to become a significant error after a large number of load repetitions. Although the elliptical form of the modified Cam-clay yield locus has been used here, it is also possible to adopt alternative shapes.

(2) Further computations should be performed to determine the significance of the type of 'loading surface' adopted in the model. Specially designed laboratory tests may provide information that will help in the choice of the best shape.

(3) The present model assumes that the elastic shear modulus G for the soil is constant. Houlsby[7] has suggested that more accurate predictions might be obtained if the value of G is taken as a function of p'_c, so that as the preconsolidation pressure is increased, then the soil becomes elastically stiffer in shear. This small adjustment in the details of the model will enable hysteresis to be more closely predicted.

(4) When the points (1)–(3) have been resolved it will then be possible to extend the model to more general stress conditions.

It is the intention of the authors to include an investigation of these above-mentioned points in future research.

9.8 CONCLUSIONS

A soil model, capable of predicting many of the observed features of the behaviour of clay when subjected to repeated loading, has been presented. The model possesses most of the characteristics of the former critical state models but with a simple, yet important modification. This involves a specified contraction of the yield surface as the soil sample is unloaded (with the definition of unloading as given above). With the introduction of this modification an additional parameter must also be defined. It has been shown that a value for this parameter may be determined, in a straightforward manner, from a laboratory triaxial test involving repeated, undrained loading.

Calculations have been made using this model and the results have been presented in parametric form. The predictions exhibit many of the same trends that have been observed in laboratory tests involving the repeated loading of saturated clays. In addition, predictions have been made of the behaviour of two particular clays and the results have been compared with the actual test results. A limited degree of agreement was found between the measured and predicted behaviour.

As a result of the parametric study and the predictions of laboratory behaviour, some suggestions for future research have been made. The most important of these is concerned with the need for an accurate determination of the yield surface and plastic potential, for any particular clay, under conditions of static loading. It is suggested that the shape of this surface must be known in some detail before good quality predictions can be expected for the behaviour of the same soil under repeated loading.

It should be emphasized that the model described in this paper cannot be expected to reproduce accurately all features of the behaviour of a real clay under monotonic and cyclic loading. Indeed, it is believed that no mathematical model that can be used sensibly and economically for design calculations is likely to achieve this aim. The philosophy behind this work has been the need to develop as simple a family of models as possible that reproduce qualitatively the salient features of cyclic behaviour of soils, *and* that are expressed in terms of soil parameters that have physical meaning and can be easily measured in conventional laboratory tests.

ACKNOWLEDGEMENTS

The authors wish to acknowledge that this work was produced on a Science Research Council contract with the University of Cambridge, which provided financial support for J. P. Carter as a research assistant and support for J. R. Booker as a visitor.

The authors are grateful to C. M. Szalwinski for the provision of computer graphics routines.

REFERENCES

1. Andersen, K. H. *Research Project, Repeated Loading on Clay—Summary and Interpretation of Test Results*, Norwegian Geotechnical Institute, Report No. 74037–9, Oslo, 1975.
2. Andersen, K. H. Behaviour of clay subjected to undrained cyclic loading, *Proc. BOSS Conf.*, NGI, 392–403, 1976.
3. Brown, S. F., Lashine, A. K. F., and Hyde, A. F. L. Repeated load triaxial testing of a silty clay, *Géotechnique*, **25**(1), 95–114, 1975.
4. Brown, S. F., Andersen, K. H. and McElvaney J. The effect of drainage on cyclic loading of clay, *Proc. 9th. Int. Conf. on Soil Mech. and Found. Eng.*, Tokyo **2**, 195–200, 1977.
5. Eekelen, H. A. M. van, and Potts, D. M. The behaviour of Drammen clay under cyclic loading, *Géotechnique*, **28**(2), 173–96, 1978.
6. Hardin, B. O. and Drnevich, V. P. Shear modulus and damping in soils: Measurement and parameter effects, *J. Soil Mech. and Found. Divn.*, ASCE, **98** (SM6), 603–24, 1972.

7. Houlsby, G. T. Private Communication, 1979.
8. Lee, K. H., and Seed, H. B., Cyclic stress conditions causing liquefaction of sand, *J. Soil Mechs. Found. Divn.*, ASCE, **93** (SM1), 47–70, 1967.
9. Lewin, P. I. *The Deformation of Soft Clay under Generalised Stress Conditions*, Ph.D. Thesis, King's College, University of London, 1978.
10. Martin G. R., Finn, W. D. L., and Seed, H. B. *Fundamentals of Liquefaction under Cyclic Loading*, University of British Columbia, Dept. of Civil Engineering, Soil. Mech. Series No. 23, 1974.
11. Monismith, C. L., Ogawa, N. and Freeme, C. R. *Permanent Deformation Characteristics of Subgrade Soils in Repeated Loading*, Transportation Research Record 537, 1–17, 1975.
12. Mroz, Z., Norris, V. A., and Zienkiewicz, O. C. An anisotropic hardening model for soils and its application to cyclic loading *Int. J. Num. and Analytical Methods in Geomechanics*, **2**, 203–21, 1978.
13. Mroz, Z., Norris, V. A., and Zienkiewicz, O. C. Application of an anistropic hardening model in the analysis of elasto-plastic deformation, *Géotechnique*, **29**(1), 1–34, 1979.
14. Pender, M. J. Modelling soil behaviour under cyclic loading, *Proc. 9th Int. Conf. Soil Mech. Found. Engng.*, Tokyo, **2**, 325–31, 1977.
15. Pender, M. J. A model for the behaviour of over-consolidated clay, *Géotechnique*, **28**(1), 1–25, 1978.
16. Prévost, J. H. Mathematical modelling of monotonic and cyclic undrained clay behaviour, *Int. J. Num. Anal. Meth. Geomech.*, **1**, 195–216.
17. Prevost, J. H. Plasticity theory for soil stress–strain behaviour, *J. Eng. Mech. Div.*, ASCE, **104** (EM3), 1117–94, 1978.
18. Pyke, R. M. *Settlement and Liquefaction of Sands under Multidirectional Loading*, Ph.D. Thesis, University of California, Berkeley, 1973.
19. Rahman, M. S., Seed, H. B., and Booker, J. R. Pore pressure development under offshore gravity structures, *J. Geotech. Eng. Div.*, ASCE, **103** (GT12), 1419–36, 1977.
20. Roscoe, K. H. and Burland, J. B. On the generalised stress–strain behaviour of 'wet' clay, in Heyman, J., and Leckie, F. A. (eds.) *Engineering Plasticity*, Cambridge University Press, pp. 535–609, 1968.
21. Sangrey, D. A., Henkel, D. J., and Esrig, M. I. The effective stress response of a saturated clay to repeated loading, *Can. Geotech. J.*, **6**(3), 241–57, 1969.
22. Schofield, A. N., and Wroth, C. P. *Critical State Soil Mechanics*, McGraw-Hill, London, 1968.
23. Seed, H. B., Chan, C. K., and Monismith, C. L. Effect of repeated loading on the strength and deformation of compacted clay, *Proc. Highway Research Board*, **34**, 1955.
24. Seed, H. B., and Lee, K. H. Liquefaction of saturated sands during cyclic loading, *J. Soil Mech. Found. Divn.*, ASCE, **92**, (SM6), 105–34, 1966.
25. Seed, H. B., and Idriss, I. M. *Soil Moduli and Damping Factors for Dynamic Response Analysis*, Earthquake Engineering Research Centre EERC, Report No. EERC 70–10, University of California, Berkeley, 1970.
26. Seed, H. B., Martin, P. P., and Lysmer, J. *The Generation and Dissipation of Pore Water Pressures during Soil Liquefaction*, Earthquake Engineering Research Centre, Report No. EERC 75–26, University of California, Berkeley, 1975.
27. Seed, H. B., and Booker, J. R. 'Stabilisation of potentially liquefiable sand deposits using gravel drains, *J. Geotech. Eng. Div.*, ASCE, **103** (GT7), 757–63, 1977.
28. Seed, H. B. Nineteenth Rankine Lecture, Considerations in the Earthquake

resistant design of earth and rock fill dams. *Géotechnique*, **29,** 215–263, 1979.
29. Taylor, P. W., and Hughes, J. M. O. Dynamic properties of foundation subsoils as determined from laboratory test, 3rd *World Conf. Earthquake Engng.*, **1,** 196–211, 1965.
30. Taylor, P. W., and Bacchus, D. R. Dynamic cyclic strain tests on a clay, *Proc. 7th Int. Conf. Soil Mechs. Found. Engng.*, Mexico, **1,** 401–409, 1969.
31. Thiers, G. R. *The Behaviour of Saturated Clay under Seismic Loading Conditions*, Ph.D. Thesis, University of California, Berkeley, 1966.
32. Thiers, G. R. and Seed, H. B. Cyclic stress–strain characteristics of clay, *J. Soil Mech. Found. Div.*, ASCE, **94** (SM2), 555–69, 1968.
33. Wilson, N. E., and Greenwood, J. R. Pore pressures and strains after repeated loading of saturated clay, *Can. Geotech. J.*, **11**(2), 269–77, 1974.
34. Wroth, C. P. The predicted performance of a soft clay under a trial embankment loading based on the Cam-clay model, in Gudehus, G. (ed.) *Finite Elements in Geomechanics*, Wiley, London, Ch. 6.
35. Zienkiewicz, O. C., Humpheson, C., and Lewis, R. W. Associated and Non-associated visco-plasticity and plasticity in soil mechanics, *Géotechnique*, **25,** 57–89, 1975.

Soil Mechanics—Transient and Cyclic Loads
Edited by G. N. Pande and O. C. Zienkiewicz
© 1982 John Wiley & Sons Ltd

Chapter 10

Bounding Surface Formulation of Soil Plasticity

Y. F. Dafalias and L. R. Herrmann

SUMMARY

The concept of the bounding surface in soil plasticity is presented and its advantages over the classical yield surface are discussed, particularly in connection with the soil response under cyclic loading. A general framework is provided within which a classical yield surface formulation can be easily transformed into a corresponding bounding surface formulation on the basis of the following idea: for any stress point inside the surface, a unique 'image' point is defined on the surface by means of a specific rule. The value of the plastic modulus depends on the distance between the stress point and its 'image', while the gradient of the bounding surface at the 'image' point defines the loading–unloading direction. The salient feature of the bounding surface formulation is the occurrence of plastic deformation for stress states inside the surface.

Within this general framework, a particularly simple model for isotropic clays is constructed in invariant stress space employing the concepts of critical state soil mechanics. Only three new material constants are introduced and a specific calibration procedure for their determination is suggested and applied. The general behaviour and predictive capabilities of the model for monotonic, cyclic, drained and undrained loading are presented in the triaxial space.

10.1 INTRODUCTION

A common shortcoming of many stress–strain laws for soils is that they are applicable only to loading conditions of a rather specific nature. This weakness becomes particularly strong when an artificial distinction is made between monotonic and cyclic loading for practical purposes. For example, studies of cyclic loading response deal primarily with gross overall soil behaviour under specifically chosen cyclic conditions, and the corresponding constitutive relations are useless for monotonic or interchange of monotonic and cyclic loading. But what is, after all, a cyclic loading but a sequence of monotonic ones? This shows the necessity to develop the constitutive laws within a more fundamental framework, such that they are applicable to monotonic or cyclic, drained or undrained, or any other form of loading conditions in order to be of value for the analysis of soil structures under complex loading.

The classical mathematical theory of plasticity provides such a framework and great advances have been made in the last 25 years, especially after the establishment of the critical state theory.[25] Still, however, some very important aspects of soil behaviour, mainly in relation to the cyclic response, cannot be adequately described. The principal reason is that the classical concept of a yield surface provides little flexibility in describing the change of the plastic modulus with loading directions and implies a purely elastic stress range contrary to the reality for many soils. The last feature is responsible for the inability of the classical theory to predict, even qualitatively, strain accumulation for drained, or pore water pressure build up for undrained cyclic deviatoric loading within a stress domain which has been defined as elastic.

The need for new concepts in plasticity became a necessity. Eisenberg and Phillips[14] introduced the concept of loading surfaces distinct from the yield surface and Mroz[16] introduced the field of work-hardening moduli which had similarities, but it is not equivalent to the sublayer models. Mroz's idea was originally applied to metal plasticity and subsequently to soils by Prevost[22] and Mroz, Norris, and Zienkiewicz.[17] The endochronic concept[26] presents a totally novel approach which is discussed elsewhere in this volume.[4,27]

Among the new concepts is that of the 'bounding surface' originally introduced by Dafalias and Popov[5,8] and Dafalias,[6] and independently by Krieg[15] in conjunction with an enclosed yield surface for metal plasticity. The concept and the name were motivated by the observation that the stress–strain curves converge with specific 'bounds' at a rate which depends on the distance of the stress point from the bounds. The original bounding surface model has been extended to include materials with a vanishing yield surface,[7,9] rate effects[10] and other aspects of material behaviour.[20] The bounding surface bears a similarity with the outer surface used in the field of work-hardening moduli formulation,[16] but it is not equivalent and in general provides a simpler constitutive model. In addition, the concept of the bounding surface occupies a more fundamental position in the development of plasticity theory due to its interpretation in terms of micromechanics.[21] The salient features of a bounding surface formulation are that plastic deformation may occur for stress states within the surface, and the possibility to have a very flexible variation of the plastic modulus. These features yield definite advantages over a classical yield surface formulation, particularly with regard to soil plasticity.

A bounding/yield surface plasticity formulation for soils was fully developed by Mroz, Norris, and Zienkiewicz[18] within the triaxial space of critical state soil mechanics. This is further elaborated in Chapter 8.[19] The reader is referred to the above references for a comprehensive discussion on the physical meaning of the bounding surface in relation to the soil structure. Two different direct bounding surface formulations within the framework of critical state soil plasticity were also presented qualitatively by Dafalias for the case of zero elastic range[11] and quasi-elastic range.[12] The latter was fully developed and applied to clays by Dafalias and Herrmann[13] in the triaxial space.

This chapter presents a generalization and further analysis of the previous work[13] in a general stress space by means of stress invariants. Once the basic equations are obtained, the new material parameters are identified and methods for their calibration are proposed. The qualitative behaviour of the model is presented in detail under monotonic, cyclic, drained, and undrained loading conditions for zero, light and heavy overconsolidation in the triaxial space. Using the proposed calibration methods the new parameters introduced are obtained from certain experimental data, and subsequently the predictions of the model are successfully compared with other experiments which are restricted here to undrained triaxial loading. Cyclic predictions are also presented.

It should be mentioned finally that in this presentation it is attempted not only to introduce a new soil plasticity model with a definite form and certain applications, but perhaps more important, it is attempted to present a new framework within which many different constitutive models for soils can be developed by employing the concept of the bounding surface.

10.2 GENERAL FORMULATION OF ELASTOPLASTICITY FOR SOILS

In this section a general formulation of rate independent elastoplasticity for soils will be summarized. The role of the bounding surface in this formulation will be presented in the following section. Since the constitutive relations refer to the deformation of the soil skeleton, the state of the material is defined in terms of the effective stress σ'_{ij} and plastic internal variables q_n accounting for the past loading history. The q_n are usually scalar or second-order tensor quantities such as the plastic work, the plastic strains, etc. With p denoting the pore water pressure, the total stress σ_{ij} is related to σ'_{ij} and p by

$$\sigma_{ij} = \sigma'_{ij} + \delta_{ij} p, \tag{10.1}$$

where δ_{ij} is the Kronecker delta. Observe that the deviatoric components s_{ij} and s'_{ij} are equal. If now ε_{ij}, ε'_{ij}, ε''_{ij} are the total, elastic and plastic strains respectively and a dot indicates the rate, the relation

$$\dot{\varepsilon}_{ij} = \dot{\varepsilon}'_{ij} + \dot{\varepsilon}''_{ij} \tag{10.2}$$

is assumed. Small strain formulation is presented henceforth, but the extension to large deformations is straightforward if a proper stress measure and a conjugate rate of deformation measure for which the decomposition (10.2) holds true are employed.

The elastic incremental constitutive relations are given by

$$\dot{\varepsilon}'_{ij} = C_{ijkl}\dot{\sigma}'_{kl}, \qquad \dot{\sigma}'_{ij} = E_{ijkl}\dot{\varepsilon}'_{kl}, \tag{10.3}$$

where C_{ijkl}, E_{ijkl} are the tensors of elastic compliance and moduli respectively, being in general functions of the state with $E_{ijkl}C_{klpq} = (\delta_{ip}\delta_{jq} + \delta_{iq}\delta_{jp})/2$. The summation convention over repeated indices applies.

The plastic constitutive relations require the definition of the direction (or vector) of plastic loading L_{ij} and the plastic modulus K_p, both functions of the state, which in turn determine the loading function L as

$$L = \frac{1}{K_p} L_{ij}\dot{\sigma}'_{ij}. \tag{10.4}$$

Plastic loading, unloading, and neutral loading occur when $L > 0$, $L < 0$, and $L = 0$, respectively. The inclusion of K_p in L allows for the description of unstable behaviour when both scalar quantities $L_{ij}\dot{\sigma}'_{ij}$ and K_p are negative but $L > 0$. Then, assuming linear dependence of $\dot{q}_n$ on $\dot{\sigma}'_{ij}$ for rate independence (homogeneity of order one would guarantee rate independence for a more general development), considering ε''_{ij} as one of the q_n but distinguishing it for emphasis, and imposing the requirement of continuous material response with respect to a changing direction of $\dot{\sigma}'_{ij}$ across neutral loading,[8] the constitutive relations are given by

$$\dot{\varepsilon}''_{ij} = \langle L \rangle R_{ij}, \tag{10.5a}$$

$$\dot{q}_n = \langle L \rangle r_n, \tag{10.5b}$$

where the brackets $\langle \rangle$ define the operation $\langle z \rangle = zh^*(z)$, h^* being the Heavyside step function, and R_{ij}, r_n are functions of the state. The R_{ij} is usually assumed to be the gradient of a plastic potential. In classical plasticity, the L_{ij} is defined as the gradient of the yield surface $f = 0$ and K_p is obtained by means of the consistency condition $\dot{f} = 0$.

The inversion of equations (10.2), (10.3), (10.4), and (10.5) yields

$$L = \frac{L_{kl}E_{klmn}}{K_p + E_{abcd}L_{ab}R_{cd}} \dot{\varepsilon}_{mn}, \tag{10.6}$$

$$\dot{\sigma}'_{kl} = E_{klij}[\dot{\varepsilon}_{ij} - \langle L \rangle R_{ij}]. \tag{10.7}$$

Equation (10.6) offers an easy computation of L when $K_p = 0$ at the initiation of unstable behaviour.

For undrained conditions and incompressible fluid and solid phases, the internal constraint $\dot{\varepsilon}_{kk} = 0$ must be satisfied, which by means of equations (10.2), (10.3a), and (10.5a) yields

$$\dot{\varepsilon}_{kk} = \dot{\varepsilon}'_{kk} + \dot{\varepsilon}''_{kk} = [C_{kkij} + (1/K_p)L_{ij}R_{kk}]\dot{\sigma}'_{ij} = 0, \tag{10.8}$$

where of course for $L \leq 0$, $\dot{\varepsilon}''_{kk} = 0$. Equation (10.8) is a differential equation for the undrained effective stress path which is defined by the tensor inner product of $\dot{\sigma}'_{ij}$ and the bracketed quantity. The effective stress is not independently controlled any more, and the constitutive relations should be expressed in terms of the total strain and stress rates. Using equations (10.1), (10.4), (10.6), and (10.8), one has

$$L = \frac{C_{aabb}L_{ij} - L_{qq}C_{kkij}}{K_p C_{rrss} + R_{mm}L_{nn}} \dot{\sigma}_{ij} = \frac{L_{kl}E_{klmn}}{K_p + E_{abcd}L_{ab}R_{cd}} \dot{e}_{mn}, \tag{10.9}$$

$$\dot{p} = \frac{1}{C_{aabb}} [R_{kk}\langle L \rangle + C_{kkij}\dot{s}_{ij}] + \tfrac{1}{3}\dot{\sigma}_{mm}$$
$$= \tfrac{1}{3}E_{kkij}[R_{ij}\langle L \rangle - \dot{e}_{ij}] + \tfrac{1}{3}\dot{\sigma}_{mm} \tag{10.10}$$

with s_{ij}, e_{ij} the deviatoric stress and strain components. Observe that $\dot{s}_{ij}$ can be used instead of $\dot{\sigma}_{ij}$ in (10.9). The constitutive relations (10.5a) and (10.7) hold true for undrained loading with L defined from (10.9) and $\dot{e}_{ij}$ substituting $\dot{\varepsilon}_{ij}$ since $\dot{\varepsilon}_{kk} = 0$. The total mean normal stress rate $\tfrac{1}{3}\dot{\sigma}_{mm}$ is indeterminate (incompressibility) and affects only the pore water pressure rate $\dot{p}$ through equation (10.10).

10.3 THE BOUNDING SURFACE

A novel approach of defining the key quantities L_{ij} and K_p is presented in the following, which is not subjected to the limitations of a yield surface plasticity formulation as explained in the introduction. The previous loading history, expressed quantitatively by means of the values of q_n, determines a 'bounding surface' in stress space, Figure 10.1, analytically described by

$$F(\bar{\sigma}'_{ij}, q_n) = 0, \tag{10.11}$$

where a bar over stress quantities indicates points on $F = 0$. The actual stress point σ'_{ij} lies always within or on the bounding surface. To each σ'_{ij}, a unique 'image' point $\bar{\sigma}'_{ij}$ on $F = 0$ is defined according to a specific rule which is part of the constitutive relations. It is possible to define different rules according to the material under consideration. One such rule which was found to provide the attractive features of simplicity and predictive capability for

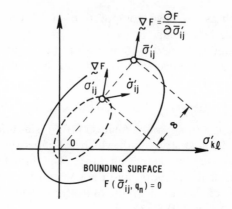

Figure 10.1 Schematic illustration of the bounding surface in a general stress space

applications to clays is the simple 'radial' rule: assuming that the origin O lies always within a convex bounding surface, $\bar{\sigma}'_{ij}$ is defined as the intersection of $F = 0$ and the straight line connecting the origin with σ'_{ij}, Figure 10.1. Analytically, this can be expressed by

$$\bar{\sigma}'_{ij} = \beta(\sigma'_{kl}, q_n)\sigma'_{ij} \tag{10.12}$$

with β obtained from $F(\beta\sigma'_{ij}, q_n) = 0$. Only for the origin $\sigma'_{ij} = 0$ and $\bar{\sigma}'_{ij}$ is undefined without any further consequence. It is important to emphasize that the following general formulation does not depend on the particular rule chosen to associate the 'image' with the actual stress point. The use of equation (10.12), whenever necessary, will be explicitly stated.

The direction of plastic loading L_{ij} at σ'_{ij} is defined as the gradient of F at the 'image' point $\bar{\sigma}'_{ij}$, i.e.

$$L_{ij} = \frac{\partial F}{\partial \bar{\sigma}'_{ij}}. \tag{10.13}$$

For any stress rate $\dot{\sigma}'_{ij}$ causing plastic loading, a corresponding 'image' stress rate $\dot{\bar{\sigma}}'_{ij}$ occurs through the hardening of $F = 0$ by means of the $\dot{q}_n$. Thus, the following three key equations complete the bounding surface formulation:

(1)   The loading function L, equation (10.4), is defined in terms of L_{ij}, equation (10.13), the two stress rates $\dot{\sigma}'_{ij}$, $\dot{\bar{\sigma}}'_{ij}$, and two plastic moduli: the actual one K_p associated with $\dot{\sigma}'_{ij}$ and a 'bounding' plastic modulus $\bar{K}_p$ associated with $\dot{\bar{\sigma}}'_{ij}$ as follows:

$$L = \frac{1}{K_p} \frac{\partial F}{\partial \bar{\sigma}'_{ij}} \dot{\sigma}'_{ij} = \frac{1}{\bar{K}_p} \frac{\partial F}{\partial \bar{\sigma}'_{ij}} \dot{\bar{\sigma}}'_{ij}. \tag{10.14}$$

(2) The 'bounding' plastic modulus $\bar{K}_p$ is obtained from the consistency condition $\dot{F} = 0$. Using equations (10.5b), (10.11), and the second part of (10.14), one has

$$\bar{K}_p = -\frac{\partial F}{\partial q_n} r_n. \tag{10.15}$$

(3) A state dependent relation $\hat{K}_p$ between K_p and $\bar{K}_p$ is established as a function of the distance $\delta = [(\bar{\sigma}'_{ij} - \sigma'_{ij})(\bar{\sigma}'_{ij} - \sigma'_{ij})]^{1/2}$ between the current stress state and its 'image', i.e.

$$K_p = \hat{K}_p(\bar{K}_p, \delta, \sigma'_{ij}, q_n) \tag{10.16}$$

such that $K_p > \bar{K}_p$ for $\delta > 0$ (the stress point inside the bounding surface) and $\bar{K}_p = K_p$ for $\delta = 0$ (the stress point on the bounding surface, identical with its 'image').

The last equation (10.16) embodies the meaning of the bounding surface concept. It allows for plastic deformation to occur for points within the surface at a progressive rate which depends on δ. The closer is the stress point to the surface, the smaller is the K_p approaching the corresponding $\bar{K}_p$, and the greater is the plastic strain rate for a given stress rate. The stress σ'_{ij} may eventually reach the bounding surface in the course of plastic loading as can be seen from equation (10.14) where the projection of $\dot{\sigma}'_{ij}$ on the gradient of $F = 0$ is greater than the corresponding projection of $\dot{\bar{\sigma}}'_{ij}$ since $K_p \geqslant \bar{K}_p$. The stress point remains on $F = 0$ if loading continues, and upon unloading it detaches from $F = 0$ moving inwards and so forth.

As a result of the definition of L_{ij} and L, it follows that at each point σ'_{ij}, a surface homeothetic to the bounding surface with respect to the origin is indirectly defined, shown by the dashed curve in Figure 10.1, which determines all the paths of neutral loading emanating from σ'_{ij}. This surface defines a quasi-elastic domain, but it is not a yield surface since the stress point may move first elastically inwards and then cause plastic loading before it reaches this surface again. It is not a loading surface[14] either, since no associated consistency condition is required. As a matter of fact it never enters the present formulation explicitly.

Finally, it is worth mentioning that a classical yield surface formulation can be obtained easily in the limit if the functional dependence of K_p, equation (10.16), on δ is such that $K_p = \hat{K}_p \simeq \infty$ for $\delta > 0$ and $K_p = \hat{K}_p \to \bar{K}_p$ as $\delta \to 0$. Then the bounding surface becomes a yield surface in the classical sense. More important, this leads to the inverse conclusion that any classical yield surface formulation can be transformed easily into a bounding surface formulation by identifying the yield surface as the corresponding bounding surface and using the set of three key equations (10.14), (10.15), (10.16) together with a proper association of an 'image' stress point to any actual stress point within or on the surface.

10.4   ISOTROPIC SOILS

The formulation so far has been very general. As a matter of fact it can be applied to any material and it is only the concept of the effective stress under undrained conditions which makes it appropriate for soils in particular, i.e. equations (10.8), (10.9), (10.10). Any type of material symmetries can be incorporated by properly defining the elastic moduli, equation (10.3), and rendering F a function of proper invariant quantities of $\bar{\sigma}'_{ij}$ and q_n, equation (10.11). Further consideration will be restricted to isotropic soils. Assuming that elastic isotropy is not altered by plastic deformation, the elastic moduli are given by

$$C_{ijkl} = \frac{2G-3K}{18KG}\delta_{ij}\delta_{kl} + \frac{1}{4G}(\delta_{ik}\delta_{jl} + \delta_{il}\delta_{jk}), \tag{10.17a}$$

$$E_{ijkl} = (K-\tfrac{2}{3}G)\delta_{ij}\delta_{kl} + G(\delta_{ik}\delta_{jl} + \delta_{il}\delta_{jk}), \tag{10.17b}$$

where the bulk modulus K and the shear modulus G are independent of q_n but can depend on the isotropic stress invariants. Equation (10.3) can now be written

$$\dot{\sigma}'_{kk} = 3K\dot{\varepsilon}'_{ii}, \qquad \dot{s}_{ij} = 2G\dot{e}'_{ij}. \tag{10.18}$$

Subsequently, compressive stress and strain are considered positive.

Isotropy requires that $F=0$ must be a function of the basic and mixed isotropic invariants of $\bar{\sigma}'_{ij}$ and q_n. This does not exclude, however, the possibility to describe developed plastic anisotropy with respect to subsequent reference states by a proper choice of q_n. For example, if the bounding surface undergoes general kinematic hardening, this is obtained by choosing some of the q_n to be the coordinates of its centre, thus providing a built-in feature which generates anisotropy in the course of plastic deformation.[6,8,18] Here a simpler model will be suggested.

It will be first assumed that $F=0$ depends only on the basic isotropic invariants of σ'_{ij} and q_n (not the mixed). Furthermore, the stress dependence will be restricted to the first effective stress invariant I' and the square root of the second deviatoric stress invariant $J^{1/2}$, defined as

$$I' = \operatorname{tr} \sigma'_{ij} = \sigma'_{ii} \qquad \sqrt{J} = [\tfrac{1}{2}s_{ij}s_{ij}]^{1/2}. \tag{10.19}$$

A more general model with an asymmetric bounding surface around the hydrostatic axis $J^{1/2}=0$ would require the introduction of the third stress invariant.

It will also be assumed that the bounding surface undergoes isotropic and kinematic hardening along the hydrostatic axis, described by one single scalar q_n which measures the plastic volumetric strain ε''_{kk}. If e is the total void ratio, it follows that

$$\dot{e} = -(1+e_0)\dot{\varepsilon}_{kk} = -(1+e_0)\dot{\varepsilon}'_{kk} - (1+e_0)\dot{\varepsilon}''_{kk} = \dot{e}' + \dot{e}'', \tag{10.20}$$

where e_0 is the initial void ratio corresponding to the reference configuration with respect to which strains are measured (for natural strains: $e_0 = e$). One can easily now identify the quantities $\dot{e}' = -(1+e_0)\dot{\varepsilon}'_{kk}$ and $\dot{e}'' = -(1+e_0)\dot{\varepsilon}''_{kk}$ as the elastic and plastic void ratio rates respectively, and choose e'' as the only q_n. Thus, equation (10.11) becomes

$$F(\bar{I}', \sqrt{\bar{J}}, e'') = 0, \tag{10.21}$$

where a bar indicates again stress invariants on $F = 0$.

Using equations (10.13), (10.14), (10.19), and (10.21) and assuming the associated flow rule $L_{ij} = R_{ij}$, a straightforward computation yields

$$L_{ij} = R_{ij} = \frac{\partial F}{\partial \bar{I}'} \delta_{ij} + \frac{1}{2\sqrt{\bar{J}}} \frac{\partial F}{\partial \sqrt{\bar{J}}} s_{ij}, \tag{10.22}$$

$$L = \frac{1}{K_p}\left[\frac{\partial F}{\partial \bar{I}'} \dot{I}' + \frac{\partial F}{\partial \sqrt{\bar{J}}} \sqrt{\dot{J}}\right] = \frac{1}{\bar{K}_p}\left[\frac{\partial F}{\partial \bar{I}'} \dot{\bar{I}}' + \frac{\partial F}{\partial \sqrt{\bar{J}}} \sqrt{\dot{\bar{J}}}\right], \tag{10.23}$$

where the relation $(\bar{s}_{ij}/2\bar{J}) = s_{ij}/2\sqrt{J}$ pertaining to the 'radial' rule of associating $\bar{\sigma}'_{ij}$ with σ'_{ij} was used. The form of L can easily be interpreted as indicating loading whenever the inner product of the rate of the stress invariants $\dot{I}'$, $\sqrt{\dot{J}}$ with the gradient of F in invariant space divided by K_p is positive. The plastic strain rate is given from equations (10.5a), (10.22), and (10.23). Non-associated flow rules can also be used by defining R_{ij} otherwise.

Writing now, according to equations (10.5b) and (10.20),

$$\dot{e}'' = r\langle L \rangle = -(1+e_0)\dot{\varepsilon}''_{kk} = -(1+e_0)R_{kk}\langle L \rangle,$$

r can be easily identified and with R_{kk} expressed from equation (10.22) the expression (10.15) yields

$$\bar{K}_p = 3(1+e_0)\frac{\partial F}{\partial e''}\frac{\partial F}{\partial \bar{I}'}. \tag{10.24}$$

Observe from (10.24) that with $\partial F/\partial e'' > 0$, the bounding plastic modulus $\bar{K}_p$ is positive (consolidation), negative (dilatation), or zero (unrestricted shear flow) according to the value of $\partial F/\partial \bar{I}'$. Correspondingly, the bounding surface expands, contracts or does not harden. This is a particularly interesting property which allows the easy incorporation of the present formulation into a critical state framework. It is a drawback of the above formulation that developed anisotropy cannot be accounted for. On the other hand, the simplicity involved and the fact that many soils can be adequately described by monitoring only e, plus all the advantages obtained by using a bounding surface, renders the present formulation a useful constitutive model. Developed anisotropy can still be incorporated in terms of a varying non-associated flow rule but this will be presented elsewhere.

The changes of $\bar{K}_p$ on $F = 0$ reflect into the values of the actual plastic modulus K_p by means of equation (10.16). Here the following form of this equation will be assumed

$$K_p = \bar{K}_p + H(I', \sqrt{J}, e'')\frac{\delta}{\delta_0 - \delta} \tag{10.25}$$

where H is a positive 'shape' hardening function of the state, δ is the distance between actual and 'image' stress points in either the stress or stress invariants space, and δ_0 is a properly chosen reference stress or distance in the corresponding space such that $\delta_0 - \delta \geq 0$ and δ_0/δ remains invariant from the space used to calibrate H. The exact definition of H will require the identification and experimental determination of certain material parameters. The H and the associated parameters constitute the 'new' elements of the present formulation with regard to classical yield surface formulations, and are intimately related to the soil response for states within $F = 0$ (overconsolidation). For $H \to \infty$ observe that $K_p \to \infty$, and $K_p = \bar{K}_p$ only for $\delta = 0$. In this case, the bounding surface behaves as a yield surface. It is possible to have $\bar{K}_p < 0$ and $K_p > 0$ if δ is large enough, which allows the description of an initially rising stress–strain curve as the stress point approaches the contracting bounding surface ($\bar{K}_p < 0$), and the subsequent unstable falling curve behaviour when eventually δ becomes small enough to have both $\bar{K}_p$, $K_p < 0$ as in the case of heavily overconsolidated clays.

Using now equations (10.17) and (10.22), the inverse relations (10.6), (10.7) become

$$L = \frac{3K(\partial F/\partial\bar{I})\dot{\varepsilon}_{kk} + (G/\sqrt{J})(\partial F/\partial\sqrt{J})s_{pq}\dot{\varepsilon}_{pq}}{K_p + 9K(\partial F/\partial\bar{I}')^2 + G(\partial F/\partial\sqrt{J})^2}, \tag{10.26}$$

$$\dot{\sigma}'_{ij} = 2G\dot{\varepsilon}_{ij} + (K - \tfrac{2}{3}G)\dot{\varepsilon}_{kk}\delta_{ij} - \langle L\rangle\left[3K\frac{\partial F}{\partial\bar{I}'}\delta_{ij} + \frac{G}{\sqrt{J}}\frac{\partial F}{\partial\sqrt{J}}s_{ij}\right]. \tag{10.27}$$

Similarly, the undrained relations (10.9) and (10.10) become

$$L = \frac{(\partial F/\partial\sqrt{J})\sqrt{J}}{K_p + 9K(\partial F/\partial\bar{I}')^2} = \frac{(G/\sqrt{J})(\partial F/\partial\sqrt{J})s_{rs}\dot{\varepsilon}_{rs}}{K_p + 9K(\partial F/\partial\bar{I}')^2 + G(\partial F/\partial\sqrt{J})^2}, \tag{10.28}$$

$$\dot{p} = 3K\frac{\partial F}{\partial\bar{I}'}\langle L\rangle + \tfrac{1}{3}\dot{I}, \tag{10.29}$$

where again $\dot{I}$ is indeterminate. With L defined from (10.28) and R_{ij} from (10.22), $\dot{\varepsilon}_{ij}$ is obtained from (10.5a). Also, with $\dot{\varepsilon}_{kk} = 0$ and L from (10.28), equation (10.27) yields $\dot{\sigma}'_{ij}$ in the undrained inverse formulation. Observe that the loading condition $L \geq 0$ depends on $\sqrt{J}$ only, since $\dot{I}'$ is not controlled independently. Finally, equation (10.8) yields for the undrained invariant stress path the differential equation

$$\frac{\sqrt{\dot{J}}}{\dot{I}'} = -\left[\frac{\partial F/\partial \bar{I}'}{\partial F/\partial \sqrt{\bar{J}}} + \frac{K_p}{9K(\partial F/\partial \bar{I}')(\partial F/\partial \sqrt{\bar{J}})}\right]. \tag{10.30}$$

For any given increment $\dot{\sigma}'_{ij} \, dt$ or $\dot{\varepsilon}_{ij} \, dt$, the soil response is fully determined from the above incremental relations. Observe that the material state expressed in terms of the seven quantities σ'_{ij}, e'' suffices to define completely these relations beginning with the determination of the 'image' stress point $\bar{\sigma}'_{ij}$, equation (10.12), and subsequently applying in a straightforward way the corresponding expressions.

10.5 RELATIONS BETWEEN QUANTITIES IN INVARIANT AND TRIAXIAL SPACES

The soil response under triaxial loading conditions can be obtained as a special case of the general development. Therefore, there is no need to recast the incremental relations in terms of the triaxial variables. However, since the material parameters associated with equation (10.25) will be determined from triaxial experiments within a 'unit normal' formulation to be defined subsequently, there is a need to relate the plastic moduli and the distances δ, δ_0 between invariant and triaxial spaces.

The usual stress and strain triaxial measures[25] p', q and ε_p, ε_q are given in terms of the principal stresses and strains by

$$p' = \tfrac{1}{3}(\sigma'_1 + 2\sigma'_3), \qquad q = \sigma_1 - \sigma_3 = \sigma'_1 - \sigma'_3, \tag{10.31a}$$

$$\varepsilon_p = \varepsilon_1 + 2\varepsilon_3, \qquad \varepsilon_q = \tfrac{2}{3}(\varepsilon_1 - \varepsilon_3). \tag{10.31b}$$

The sets of the triaxial and invariant stress measures are related by

$$I' = 3p', \qquad \sqrt{J} = (1/\sqrt{3})|q|, \tag{10.32}$$

and corresponding relations hold between the partial derivatives with respect to the two sets. The bounding surface in triaxial space is described by $F^* = 0$, where

$$F(\bar{I}', \sqrt{\bar{J}}, e'') = F(3\bar{p}', (1/\sqrt{3})|\bar{q}|, e'') = F^*(\bar{p}', \bar{q}, e'') = 0. \tag{10.33}$$

The components n_p, n_q of the unit normal to $F^* = 0$ are defined by

$$n_p = \frac{1}{g^*}\frac{\partial F^*}{\partial \bar{p}'}, \qquad n_q = \frac{1}{g^*}\frac{\partial F^*}{\partial \bar{q}}, \qquad n_p^2 + n_q^2 = 1, \tag{10.34}$$

where

$$g^{*2} = \left(\frac{\partial F^*}{\partial \bar{p}'}\right)^2 + \left(\frac{\partial F^*}{\partial \bar{q}}\right)^2 = 9\left(\frac{\partial F}{\partial \bar{I}'}\right)^2 + \frac{1}{3}\left(\frac{\partial F}{\partial \sqrt{\bar{J}}}\right)^2 \tag{10.35}$$

Using now equations (10.22), (10.23), (10.32), the basic constitutive law (10.4), (10.5a) in the 'unit normal' triaxial space formulation is given by

$$L^* = g^*L = \frac{1}{K_p^*}(n_p\dot{p}' + n_q\dot{q}) = \frac{1}{\bar{K}_p^*}(n_p\dot{\bar{p}}' + n_q\dot{\bar{q}}) \qquad (10.36a)$$

$$\dot{\varepsilon}_p'' = \langle L^* \rangle n_p, \qquad \dot{\varepsilon}_q'' = \langle L^* \rangle n_q. \qquad (10.36b)$$

where the triaxial moduli K_p^*, $\bar{K}_p^*$ are related to K_p, $\bar{K}_p$ by

$$K_p = g^{*2}K_p^*, \qquad \bar{K}_p = g^{*2}\bar{K}_p^*. \qquad (10.37)$$

The triaxial moduli K_p^*, $\bar{K}_p^*$ have the proper dimension of stress, as can be seen from equation (10.36), and this is a definite advantage when a relation similar to equation (10.25) is established between them from triaxial experiments.

Attention will be focused on the definition of δ_0, equation (10.25). For further use, we introduce the quantities

$$\theta = \frac{\sqrt{J}}{I'}, \qquad \eta = \frac{q}{p'}, \qquad \eta^2 = 27\theta^2. \qquad (10.38)$$

Using (10.12), (10.32), and (10.38), the distance between actual and 'image' stress points in the invariant and triaxial spaces, denoted by δ and δ^* respectively, are expressed by

$$\delta = (\beta - 1)(1 + \theta^2)^{1/2}I', \qquad (10.39a)$$

$$\delta^* = (\beta - 1)(1 + \eta^2)^{1/2}p' = \tfrac{1}{3}(\beta - 1)(1 + 27\theta^2)^{1/2}I'. \qquad (10.39b)$$

Obtaining the relation corresponding to (10.25) between K_p^* and $\bar{K}_p^*$ requires the definition of a reference distance $\delta_0^*(p', q, e'')$ in triaxial space. Generalizing this relation to the invariant space by means of equation (10.37), a corresponding reference distance δ_0 must be defined in such a way that the ratio $\delta/(\delta_0 - \delta)$ remains invariant under the transformation (10.32). In other words, one must have $\delta_0/\delta = \delta_0^*/\delta^*$, which by means of equations (10.32) and (10.39) yields:

$$\delta_0 = 3\left(\frac{1+\theta^2}{1+27\theta^2}\right)^{1/2} \delta_0^*(I'/3, \sqrt{3J}, e''). \qquad (10.40)$$

The conclusion of this section can be summarized as follows: establishing for the triaxial space a relation of the form

$$K_p^* = \bar{K}_p^* + H^*(p', q, e'')\frac{\delta^*}{\delta_0^* - \delta^*}, \qquad (10.41)$$

the corresponding relation for the invariant space is given by equation (10.25), where δ_0 is defined by equation (10.40), δ is defined by equation

(10.39a) and according to equation (10.37) H is given by

$$H(I', \sqrt{J}, e'') = g^{*2} H^*(I'/3, \sqrt{3J}, e''). \tag{10.42}$$

The important quantity therefore which is to be determined from experimental data is the material function H^* (or H). It must be emphasized on the basis of the above conclusion that in order to calibrate the different material constants of H^* from triaxial experiments, the general formulation of Section 10.4 can be used in combination with equations (10.25) and (10.42) without recasting the constitutive relations in terms of the triaxial variables.

10.6 SPECIFIC FORM OF THE BOUNDING SURFACE

Subsequently the development will focus on clay soil behaviour. Postponing the determination of H and δ_0 until the next section, specific analytical expressions for the bounding surface will be given here together with the corresponding expressions for the $\partial F/\partial \bar{I}'$, $\partial F/\partial \sqrt{\bar{J}}$ and the $\bar{K}_p$ from equation (10.24). Extending the ideas of the critical state soil mechanics from the triaxial to the invariant space, the bounding surface is shown in Figure 10.2 intersecting the projection of the critical state line, which has a constant slope N, at point C where $\partial F/\partial \bar{I}' = 0$ and the I' axis at points where $\partial F/\partial \sqrt{\bar{J}} = 0$. The intersection with the positive I' axis is denoted by I_0. It follows immediately from (10.32) that $N = (1/3\sqrt{3})M$ with M the slope of the critical state line in triaxial space.

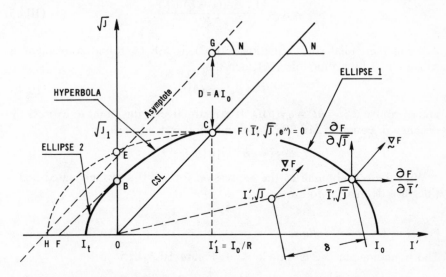

Figure 10.2 The bounding surface in the space of stress invariants

The dependence of $F = 0$ on e'' is introduced by means of the dependence of I_0 on e'', which accounts simultaneously for isotropic and kinematic hardening along the I' axis. The centre of the bounding surface, which is the projection I_1' of the point C on I', is related to I_0 by $I_1' = I_0/R$, where R is a material constant. The value of R has been taken in the past equal to 2.72 (natural logarithm)[25] or 2,[23] but any other value in the range $1 < R < \infty$ may be suitable for particular soils. Let κ and λ denote the slopes of the rebound and isotropic consolidation curves in the e–ln I' plot (same as in the e–ln p' plot). Assuming that there is a limit value $I_l > 0$ (usually taken equal to one atmosphere) such that for $I' \le I_l$ the relation between I' and the elastic part of the void ratio e' changes continuously from logarithmic to linear in order to prevent excessive softening of the elastic stiffness around $I' = 0$ for cohesive soils, one has

$$\frac{\dot{I}'}{\dot{e}'} = -\frac{\langle I' - I_l \rangle + I_l}{\kappa}, \qquad \frac{\dot{I}_0}{\dot{e}} = -\frac{I_0}{\lambda}. \tag{10.43}$$

The sought expression $\dot{I}_0/\dot{e}''$ follows immediately from $\dot{e}'' = \dot{e} - \dot{e}'$ and equation (10.43):

$$\frac{\dot{I}_0}{\dot{e}''} = -\frac{I_0}{\lambda - \omega\kappa} \qquad \text{with} \qquad \omega = \frac{I_0}{\langle I_0 - I_l \rangle + I_l}. \tag{10.44}$$

The first part of equation (10.43) in combination with equations (10.18), (10.20) yields an expression of the bulk modulus K in terms of κ as

$$K = \frac{(1 + e_0)(\langle I' - I_l \rangle + I_l)}{3\kappa}. \tag{10.45}$$

Using the radial rule, equation (10.12), it will be found convenient to substitute β by a function $\gamma(\theta)$ such that

$$\bar{I}' = \beta I' = \gamma(\theta) I_0 \tag{10.46}$$

and of course $\bar{J}^{1/2} = \theta \bar{I}' = \theta \gamma(\theta) I_0$. It is instructive at this point to express the distance δ, equation (10.39a), in terms of γ instead of β by

$$\delta = (1 + \theta^2)^{1/2} [\gamma(\theta) I_0 - I']. \tag{10.47}$$

With these preliminaries the expression for $F = 0$ and the related quantities are defined as follows:

(1) *For $0 \le \theta \le N$ (ellipse 1)*

The bounding surface is the ellipse 1, Figure 10.2, defined by

$$\frac{F}{I_0^2} = \left(\frac{\bar{I}'}{I_0}\right)^2 + \left(\frac{R-1}{N}\right)^2 \left(\frac{\sqrt{\bar{J}}}{I_0}\right)^2 - \frac{2}{R}\frac{\bar{I}'}{I_0} + \frac{2-R}{R} = 0 \tag{10.48}$$

and with $x = \theta/N$ it follows that

$$\gamma(\theta) = \frac{1 + (R-1)[1 + R(R-2)x^2]^{1/2}}{R[1 + x^2 + R(R-2)x^2]}, \qquad (10.49)$$

$$\frac{\partial F}{\partial \bar{I}'} = 2I_0\left(\gamma - \frac{1}{R}\right), \qquad (10.50a)$$

$$\frac{\partial F}{\partial \sqrt{\bar{J}}} = 2I_0\theta\gamma\left(\frac{R-1}{N}\right)^2, \qquad (10.50b)$$

$$\bar{K}_p = \frac{1+e_0}{\lambda - \omega\kappa}\frac{12I_0^3}{R}\left(\gamma - \frac{1}{R}\right)(\gamma + R - 2), \qquad (10.51)$$

where for the last equation use of equations (10.24), (10.44), (10.46), and (10.50a) was made. It can easily be shown that as θ varies from 0 to N, $\bar{K}_p$ remains non-negative, varying continously from $\bar{K}_p = (1+e_0)12I_0^3(R-1)^2/(\lambda - \omega\kappa)R^2$ to $\bar{K}_p = 0$.

(2)   For $N \le \theta < +\infty$ (hyperbola)

It is possible to extend ellipse 1 to this range as shown by the dashed curve *CE* in Figure 10.2, using all the previous equations with the restriction $R \ge 2$ since the origin must lie within or on the bounding surface which crosses the origin for $R = 2$. It was found, however, that the prediction of the model was unsatisfactory for deviatoric loading of heavily over-consolidated clays, suggesting a curve with a shape more parallel to the critical line *OC*, which is tangent to the ellipse at *C*. This would yield the advantage of having the possibility to choose $R < 2$ if necessary, at the expense of one additional parameter for the new curve. A hyperbola is therefore proposed whose apex *C* is at a distance D from its centre *G* and its asymptote is parallel to *OC*. Assuming $D = AI_0$, the equation of the hyperbola is defined in terms of one additional material parameter A by

$$\frac{F}{I_0^2} = \left(\frac{\bar{I}'}{I_0}\right)^2 - \frac{1}{N^2}\left(\frac{\sqrt{\bar{J}}}{I_0}\right)^2 - \frac{2}{R}\frac{\bar{I}'}{I_0} + \frac{2}{N}\left[\frac{1}{R} + \frac{A}{N}\right]\frac{\sqrt{\bar{J}}}{I_0} - \frac{2A}{RN} = 0, \qquad (10.52)$$

and with $y = RA/N$ one has (recall $x = \theta/N$):

$$\gamma(\theta) = \frac{x - 1 + xy - [(x-y-1)^2 + (x^2-1)y^2]^{1/2}}{R(x^2-1)}, \qquad (10.53)$$

$$\frac{\partial F}{\partial \bar{I}'} = 2I_0\left(\gamma - \frac{1}{R}\right), \qquad (10.54a)$$

$$\frac{\partial F}{\partial \sqrt{\bar{J}}} = \frac{2I_0}{N^2}[A + (N/R) - \theta\gamma], \qquad (10.54b)$$

$$\bar{K}_p = \frac{1+e_0}{\lambda - \omega\kappa}\frac{12I_0^3}{R}\left(\gamma - \frac{1}{R}\right)\left[(1 - x - xy)\gamma + \frac{2A}{N}\right]. \qquad (10.55)$$

It can be shown that as x varies from 1 to $+\infty$ (θ varies from N to $+\infty$), $\bar{K}_\mathrm{p}$ is negative, decreasing from 0 at point C to

$$\bar{K}_\mathrm{p} = -(1+e_0)12I_0^3\left[\frac{2A}{N}-(1+y)[1+y-(1+y^2)^{1/2}]\frac{1}{R}\right]\bigg/(\lambda-\omega\kappa)R^2$$

at point B.

(3) *For $-\infty < \theta \leq 0$ (ellipse 2)*

In order to describe the material behaviour in tension,[1] it is possible to extend the bounding surface into the $I' < 0$ range as a smooth curve tangent to the hyperbola at point B, Figure 10.2, and intersecting the I' axis at point I_t with a tangent parallel to $\sqrt{J}$ axis. I_t measures the tensile strength of the soil. Such an extension as a second ellipse, Figure 10.2, was proposed by Dafalias and Herrmann[13] but other shapes may also be suitable. In view of insufficient evidence for the tensile behaviour, further discussion is postponed with the observation that the material response will be described by equations similar to the ones developed in the previous cases, with attention given in securing the continuity of constitutive relations at point B.

10.7 IDENTIFICATION AND CALIBRATION OF THE MATERIAL CONSTANTS

It remains only to specify the shape hardening function H and the reference distance δ_0 of equation (10.25), in terms of the state and certain material constants. Recalling equation (10.41), the following form of H^* is proposed for the triaxial space:

$$H^* = hp_\mathrm{a}\left(1+\left|\frac{M}{\eta}\right|^m\right), \tag{10.56}$$

where p_a is the atmospheric pressure providing the proper stress units and h, m are dimensionless material constants. The introduction of the absolute value of the ratio M/η does not allow plastic deformation to occur within $F = 0$ for $\eta = 0$ (zero deviatoric stress) rendering H^*, and by consequence K_p^*, infinite except when $\delta^* = 0$, where of course one must define $\lim[|M/\eta|^m \delta^*] = 0$ as η, $\delta^* \to 0$. This can be easily achieved numerically by changing η to $\eta + \varepsilon$ with ε a very small positive number. A small value of m eliminates the influence of $|M/\eta|$ for $\eta > 0$, and it is the shape-hardening constant h which mainly bears the responsibility for the material response within the bounding surface.

If p_0 is the point of intersection of $F^* = 0$ with the p' axis measuring the amount of past preconsolidation, the choice $\delta_0^* = p_0 = I_0/3$ is plausible. One can easily establish the condition $M \leq [R(R-1)]^{1/2}$ in order to guarantee that $p_0 - \delta^* \geq 0$. Other choices of δ_0^* are possible but will not be presented here. Recalling the discussion of Section 10.5 and using equations (10.25), (10.32), (10.35), (10.38), (10.42), and (10.56), the following relation is established between K_p and $\bar{K}_p$ in invariant stress space:

$$K_p = \bar{K}_p + hp_a \left[1 + \left| \frac{N}{\theta} \right|^m \right] \left[9 \left(\frac{\partial F}{\partial \bar{I}'} \right)^2 + \frac{1}{3} \left(\frac{\partial F}{\partial \sqrt{\bar{J}}} \right)^2 \right] \left[\frac{\delta}{\delta_0 - \delta} \right], \quad (10.57)$$

where δ is given by equation (10.47) and δ_0 by equation (10.40), in which $\delta_0^* = p_0 = I_0/3$. Observe that all the quantities appearing in the above relation are functions of the state I', $J^{1/2}$, e'' and two material constants m and h.

It is now possible to identify and suggest methods for calibration of the material constants entering the formulation. The set of these constants will be divided into two groups.

(1) 'Old' material constants

This group includes the elastic constants K (or κ) and G (or Poisson's ratio ν), and the critical state soil mechanics material constants λ, R, and $M = 3\sqrt{3} N$. Their determination follows well-known methods although there is still considerable discussion for K and G. Here K is defined by means of equation (10.45) in terms of κ, and G is computed from K and a constant ν. The objections raised on the basis of energy dissipation[29] against such a dependence of G on I' do not apply straightforwardly here, since no purely elastic range exists in the usual sense for $\theta > 0$. However, there may still be some cyclic stress paths near $\theta = 0$ causing problems, and further investigation will be necessary.

(2) 'New' material constants

The new constants are I_l, A, m, and h. The constant I_l is not related to the bounding surface concept and refers to a better description of the elastic response near the origin. It can be usually taken equal to the atmospheric pressure p_a. The constant A defines the shape of the hyperbolic part of the bounding surface for heavily overconsolidated states, and it can also be considered as an appropriate parameter for a classical yield surface formulation improving the surface shape.

The role of m has been explained earlier in connection with the role of $|M/\eta|$ or $|N/\theta|$. It was found that $m = 0.2$ can be used for most clays. For this value, the first bracket of equation (10.57) varies between 2.58 and 1, as θ changes from $N/10$ to ∞, a variation which is negligible compared to the values of $\bar{K}_p$ and the other quantities appearing in equation (10.57). The constant h is the most important material constant defining the response for stresses within the surface which is the salient feature of the bounding surface concept.

The following concrete steps are now suggested for the calibration of h and A, once the 'old' material constants have been defined and $I_l = p_a$, $m = 0.2$.

(1) Obtain the experimental curves ($q-p'$, $q-\varepsilon_1$, $p-\varepsilon_1$, etc.) for deviatoric loading of an isotropically overconsolidated sample at an overconsolidation ratio between 1 and R, preferably at $OCR = 3R/(2R+1)$. Determine h by curve fitting the experimental data using the developed incremental relations. This can be done by a trial-and-error process, observing that increasing h implies stiffer response. For this range of OCR, A does not appear in the equations.

(2) Obtain the experimental curve for deviatoric loading after heavy overconsolidation, that is $OCR \geqslant 5$ at least. With h known from step 1, A is determined similarly by a trial-and-error numerical process. Increasing A implies a 'flatter' hyperbola with a stiffer response and reduced dilatation, while a very small A can make the hyperbola almost identical with the critical state line for material with small cohesion.

It is important to emphasize that the response to cyclic loading is obtained on the basis of the above state-dependent formulation as a sequence of monotonic loading/unloading events without introducing any additional cyclic empirical parameter.

The following two sets of material constants will be used subsequently:

(a) Set No. 1

$$M = 3\sqrt{3} \; N = 1.05 \qquad R = 2.72$$
$$\kappa = 0.05 \qquad\qquad m = 0.20$$
$$\lambda = 0.14 \qquad\qquad h = 44$$
$$\nu = 0.15 \qquad\qquad A = 0.06$$

(b) Set No. 2

$$M = 3\sqrt{3} \; N = 0.95 \qquad R = 2$$
$$\kappa = 0.05 \qquad\qquad m = 0.20$$
$$\lambda = 0.26 \qquad\qquad h = 58 \text{ (or 20)}$$
$$\nu = 0.15 \qquad\qquad A = 0.06$$

Some of the above constants were obtained by the calibration procedure outlined above in connection with experimental data, as will be shown subsequently.

10.8 MODEL BEHAVIOUR IN THE TRIAXIAL SPACE

The response of the model under monotonic and cyclic loading is subsequently presented in the triaxial space $p' - q$, by using the general formulation of Section 10.4.

10.8.1 Monotonic loading

Using the set No. 1 of material constants, the drained and undrained behaviour of the model with increasing q up to critical failure at OCR = 1, 1.2, 2, 5, 10, and 20 for initial void ratios $e_0 = 0.94$, 0.95, 0.97, 1.02, 1.06, and 1.08 correspondingly, is shown in Figures 10.3 and 10.4 respectively, where also the initial position of the bounding surface is marked. Stress or strain increments are imposed, and the model incremental response is obtained from the developed constitutive relations.

The smooth undrained stress paths, Figure 10.4(a), show a p' reduction indicating plastic consolidation within the bounding surface at the very early stages of loading. In a classical yield surface formulation, the stress paths would 'shoot' upwards until they reach the surface first and then plastic loading occurs. The same 'smoothness' reflects in all other curves, without sharp transitions from elastic to elasto-plastic behaviour. For OCR = 5, 10, 20 the response is of particular interest. For drained or undrained cases the stress path crosses the critical state line becoming associated with the contracting hyperbolic part of the bounding surface, where $\bar{K}_p \leqslant 0$ but K_p is still positive. Eventually δ becomes small enough and $\bar{K}_p$ negative enough to yield $K_p = 0$ according to equation (10.57). The locus of all these points where $K_p = 0$ marks the so-called 'failure envelope', although failure has not yet occurred, lying between the critical state line and the initial bounding surface, Figures 10.3(a) and 10.4(a). The behaviour subsequently changes.

For drained loading K_p becomes negative indicating a falling $q-\varepsilon_1$ curve (Figure 10.3(b)). Simultaneously dilatation, which has begun the moment the stress crossed the critical state line while q was still increasing, becomes predominant and this is shown in the $\varepsilon_p-\varepsilon_1$ curve especially for OCR = 10, 20. The stress eventually falls back on the critical state line together with the contracting bounding surface, and critical failure occurs.

It is interesting to study now the undrained stress path in connection with its differential equation (10.30), where I', $J^{1/2}$ correspond to p', q. When the stress is on the critical line, the 'image' point is point C, Figure 10.2, where $(\partial F/\partial \bar{I}') = 0$; thus $(\sqrt{\bar{J}})/\bar{I}' = \infty$ and the path crosses vertically the line. When eventually $K_p = 0$, observe from equations (10.23) and (10.30) that this implies motion along a near neutral loading direction while negative pore water pressure develops, Figure 10.4(c), and the bounding surface contracts.

The stress path moves towards increasing mean effective stress, Figure 10.4(a), until it reaches the critical state line almost horizontally and then critical failure occurs. This is the well-known stabilizing effect of negative increase in pore water pressure. Accordingly, instability is not pronounced as in the drained case and this is shown in Figure 10.4(b). These observations on equation (10.30) explain also the characteristic 'hook' of the stress path, Figure 10.4(a).

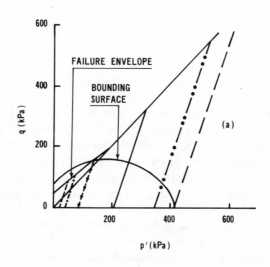

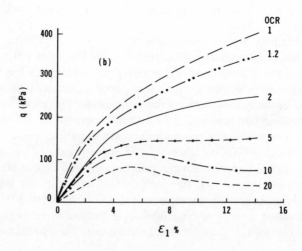

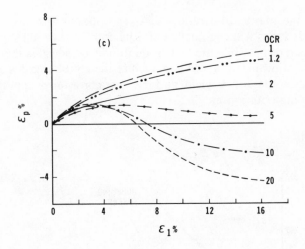

Figure 10.3 (a)(b)(c) Drained behaviour of the model at different OCR

10.8.2 Cyclic loading

Using the set No. 2 of soil parameters with $h = 20$ and $e_0 = 1$, the undrained cyclic deviatoric loading for 8 cycles and for amplitudes $q/p_0 = 0.25$ and $q/p_0 = 0.42$ yields the soil response as shown in Figure 10.5. Observe the progressive motion of the effective stress towards the critical state line and the simultaneous expansion of the bounding surface shown by discontinuous lines. At the end of the 8th cycle and for $q/p_0 = 0.25$, the stress state has not reached the critical state line. On the other hand for $q/p_0 = 0.42$, four cycles suffice to bring the soil to a state where the effective stress loop no longer progresses, while pore water pressure increases and decreases cyclically with zero mean net increase, and axial strain ε_1 accumulates continuously. For failure a final increase in q is necessary.

A problem here is that the stress state will be brought to the critical line (not necessarily with failure) for any amplitude of q if a sufficient number of cycles is applied. Exactly the opposite problem may arise for a bounding/yield surface formulation[18] with kinematic hardening, i.e. the stress loops stabilize when the stress alternates between the points of zero gradient component along p' of the yield surface, and this may be undesirable for the particular amplitude of q.[24] In the present formulation the corresponding problem can be easily remedied, introducing the concept of the elastic nucleus, Figure 10.5. The elastic nucleus defines a domain of purely elastic response (recall that irreversible plastic deformation is responsible for the phenomenon of cyclic mobility) within the bounding surface, in the sense that $K_p = \infty$ inside the nucleus. Its growth can depend on the magnitude of the plastic void ratio e''. As the nucleus size increases, the lower amplitude loops will eventually enter its domain with full stabilization (no accumulation of ε_1) while the higher amplitude loops will still be capable of reaching the critical state line. Observe that the elastic nucleus is not the same concept as that of a yield surface (no consistency condition, no loading–unloading criterion etc.).

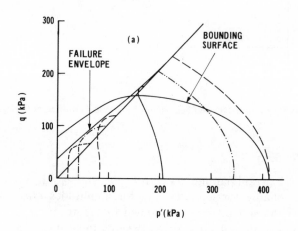

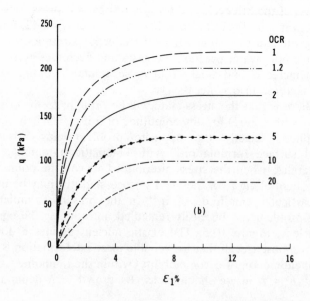

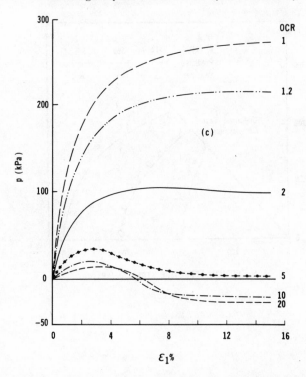

Figure 10.4 (a)(b)(c) Undrained behaviour of the model at different OCR

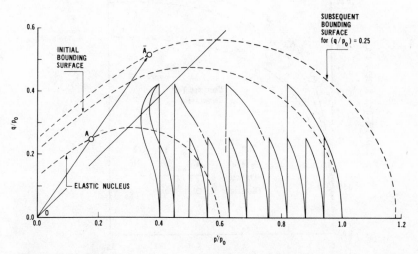

Figure 10.5 Undrained cyclic behaviour of the model for different cyclic stress amplitudes

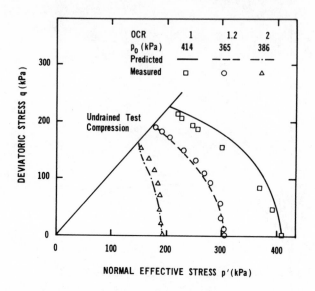

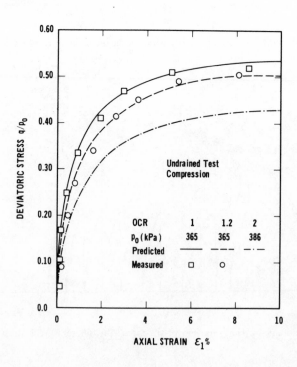

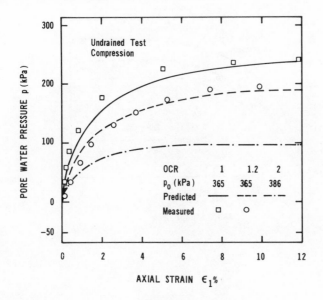

Figure 10.6 Predictions of the model for lightly overconsolidated clay (experimental data from Banerjee and Stipho[2])

As a matter of fact, it is not necessary to explicitly write an equation for the growth of the elastic nucleus. Its effect can be conveniently built into the form of equation (10.57) interpreted in the triaxial space as follows. With $\bar{r} = O\bar{A}$, Figure 10.5, a purely elastic response i.e. $K_p = \infty$ can be assumed whenever $\delta = A\bar{A} \geq \bar{r}/s$, where $s > 1$ is a stabilization factor, possibly a function of the state. This can be achieved by substituting for $\delta_0 - \delta$ in equation (10.57) the quantity $\langle p_0 - (sp_0/\bar{r})\delta \rangle$ if $\delta_0 = p_0$. Then, whenever $\delta \geq \bar{r}/s$, the brackets yield a zero value and $K_p = \infty$. Observe that $K_p \to \infty$ in a continuous way as $\delta \to \bar{r}/s$. Further investigation of this modification is necessary.

10.9 COMPARISON WITH EXPERIMENTS

The experimental results of soft clay response under undrained monotonic deviatoric loading in compression are shown by corresponding symbols in Figures 10.6 and 10.7 for OCR = 1, 1.2, 2, 5, 8, and 12 with initial void ratios $e_0 = 0.93$, 0.95, 0.97, 0.95, 0.95, and 0.95 respectively, as reported by Banerjee and Stipho.[2,3] The classical parameters M, κ, λ, ν, and R are taken from the above references (ν was changed to 0.15 from the suggested value 0.30), and m was put equal to 0.20. According to step 1 of the suggested calibration procedure of Section 10.7, the experimental curves for OCR =

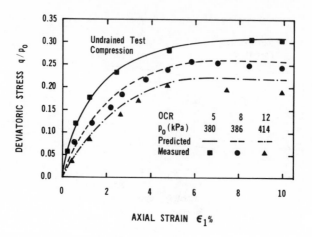

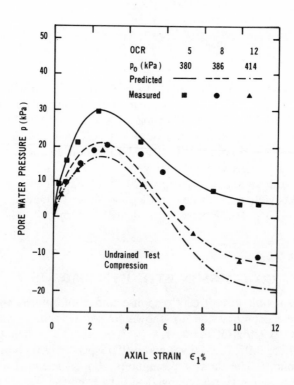

Figure 10.7 Predictions of the model for heavily overconsolidated clay (experimental data from Banerjee and Stipho[3])

1.2, Figure 10.6, were used to obtain $h = 44$. Observe the internal consistency of the model shown by the good fit of all three curves simultaneously corresponding to OCR = 1.2 in Figure 10.6. Subsequently, the curves for OCR = 5, Figure 10.7, were used to obtain $A = 0.30$ (again one curve would suffice). This yields the set No. 1 of material constants and the predictions and comparison with experimental data for all OCR are shown in Figures 10.6, 10.7 by curved lines.

The soil parameters set No. 2 with $h = 58$ and $e_0 = 1$ is used to predict the undrained kaolin response under cyclic deviatoric loading as reported experimentally by Wroth and Loudon.[28] Again the classical critical state parameters are taken from the above work. The comparison of calculated versus experimental behaviour for the undrained stress path is shown in Figure 10.8(a). The $q-\varepsilon_1$ and $p-\varepsilon_1$ curves are shown in Figures 10.8(b), 10.8(c), but no experimental data were available for comparison. In all the above, the values of h are different than the ones reported in Ref. 13. This is because equation (10.57) has been slightly modified and h is non-dimensionalized by introducing p_a.

10.10 CONCLUSION

In this chapter a general aspect of the concept of the bounding surface was presented, supplemented by a procedure to construct bounding surface models in general. The combined bounding/yield surface models[8,15,18] can be included within the framework of the general aspects presented in Section 10.3, and further elaboration on this can be found in Chapter 8 of this volume. The general equations of soil plasticity were developed within a bounding surface formulation. Attention was subsequently given in constructing within the framework of critical state soil mechanics a simple model in invariant stress space for clays. No yield/loading surface is explicitly introduced and a simple 'radial' rule associates stress points within the bounding surface with their 'images' on the surface. The state and the functioning of the model are defined only in terms of the stress and the plastic change of the void ratio. These properties provide certain attractive features of simplicity which can be very important from the point of view of the numerical analysis for large systems.

Despite its simplicity and with the exception of the capability to account for anisotropy, the model can describe realistically the soil response under different monotonic and cyclic loading conditions at any OCR, including unstable behaviour and cyclic mobility with further potential to include tension response. Comparison with experiments demonstrates these properties.

The present formulation introduces only two new material constants h and m associated with the general functioning of the model. A third material constant A aims at improving the shape of the bounding surface used, but it is not essential to the general concept (other shapes can be used). Methods for the calibration of h, A are proposed and applied.

As a final conclusion, perhaps the value of this presentation can be embodied in the demonstrated simple idea that any sound classical yield surface soil plasticity model can be easily transformed into a corresponding and more flexible bounding surface model.

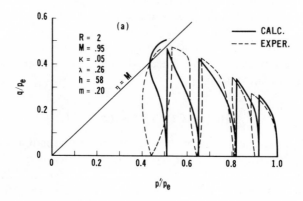

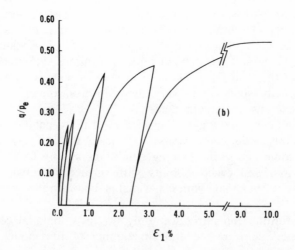

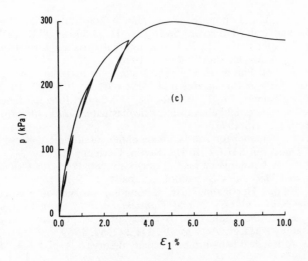

Figure 10.8 (a)(b)(c) Theory versus experiments for undrained cyclic loading (experimental data from Wroth and Loudon[28])

ACKNOWLEDGEMENT

The present work was conducted in part under the sponsorship of the University of California, Davis, U.S. Army Corps of Engineers WES Contract No. DACA 39-79-M-0059 and National Science Foundation Grant No. CME-79-10835.

REFERENCES

1. Al-Hussaini, M. M., and Townsend, F. C. *Investigation of Tensile Testing of Compacted Soils*, Miscellaneous paper S-74-10, U.S. Army Engineer Waterways Experiment Station, Vicksburg, Mississippi, 1974.
2. Banerjee, P. K., and Stipho, A. S. Associated and non-associated constitutive relations for undrained behaviour of isotropic soft clays, *Int. J. Num. Anal. Meth. Geomech.*, **2**, 35–56, 1978.
3. Banerjee, P. K., and Stipho, A. S. An elastoplastic model for undrained behaviour of heavily overconsolidated clays, *Int. J. Num. Anal. Meth. Geomech.* (short Communication), **3**, 97–103, 1979.
4. Bazant, Z. P., Ansal, A. M., and Krizek, R. J. Chapter 15 of this volume.
5. Dafalias, Y. F., and Popov, E. P. (1974, 1975) A model of nonlinearly hardening materials for complex loadings, *Proc. 7th U.S. National Congress of Applied Mechanics*, 149 (Abstract), Boulder, USA, June 1974 and *Acta Mech.*, **21**, 173–92, 1975.
6. Dafalias, Y. F. *On Cyclic and Anisotropic Plasticity: (i) A General Model Including Material Behaviour under Stress Reversals, (ii) Anisotropic Hardening for Initially Orthotropic Materials*, Ph.D. Thesis, University of California, Berkeley, 1975.

7. Dafalias, Y. F., and Popov, E. P. A simple constitutive law for artificial graphite-like materials, *Trans. 3rd SMiRT*, C 1/5, London, U.K., 1975.
8. Dafalias, Y. F., and Popov, E. P. Plastic internal variables formalism of cyclic plasticity, *J. Appl. Mech.*, **98**(4), 645–50, 1976.
9. Dafalias, Y. F., and Popov, E. P. Cyclic loading for materials with a vanishing elastic region, *Nuclear Engineering and Design*, **41**(2), 293–302, 1977.
10. Dafalias, Y. F., Ramey, M. R., and Sheikh, I. A model for rate-dependent but time-independent material behavior in cyclic plasticity, *Trans. 4th SMiRT*, L 1/8, San Francisco, U.S.A., 1977.
11. Dafalias, Y. F. A model for soil behavior under monotonic and cyclic loading conditions, *Trans. 5th SMiRT*, K 1/8, Berlin, Germany, 1979.
12. Dafalias, Y. F. A bounding surface plasticity model, *Proc. 7th Canadian Congress of Applied Mechanics*, Sherbrooke, Canada, 1979.
13. Dafalias, Y. F., and Herrmann, L. R. A bounding surface soil plasticity model, *Int. Symp. on Soils under Cyclic and Transient Loading*, Swansea, U.K., 1980.
14. Eisenberg, M. A., and Phillips, A. A theory of plasticity with non-coincident yield and loading surfaces, *Acta Mech.*, **11**, 247–60, 1969.
15. Krieg, R. D. A practical two-surface plasticity theory, *J. Appl. Mech.*, **42**, 641–6, 1975.
16. Mroz, Z. On the description of anisotropic work hardening, *J. Mech. and Physics of Solids*, **15**, 163–75, 1967.
17. Mroz, Z., Norris, V. A., and Zienkiewicz, O. C. An anisotropic hardening model for soils and its application to cyclic loading, *Int. J. Num. Anal. Meth. Geomech.*, **2**, 203–21, 1978.
18. Mroz, Z., Norris, V. A., and Zienkiewicz, O. C. Application of an anisotropic hardening model in the analysis of elastoplastic deformation of soils, *Géotechnique*, **29**(1), 1–34, 1979.
19. Mroz, Z., Norris, V. A., and Zienkiewicz, O. C. Chapter 8 in this volume.
20. Petersson, H., and Popov, E. P. Constitutive relations for generalized loadings, *Proc. ASCE*, **103**, EM4, 611–27, 1977.
21. Popov, E. P., and Ortiz, M. Macroscopic and microscopic cyclic metal plasticity, *Proc. 3rd ASCE/EMD Specialty Conference*, Austin, U.S.A., 1979.
22. Prevost, J. H. Plasticity theory for soil stress–strain behavior, *Proc. ASCE*, **104**, EM, 1177–96, 1978.
23. Roscoe, K. H., and Burland, J. B. On the generalized stress–strain behaviour of 'wet' clay, in Heyman, J., and Leckie, F. A. (eds.), *Engineering Plasticity*, Cambridge University Press, pp. 535–609, 1968.
24. Sangrey, D. A., Henkel, D. J., and Espig, M. I. The effective stress response of a saturated clay soil to repeated loading, *Can. Geotech. J.*, **6**(3), 241–52, 1969.
25. Schofield, A. N., and Wroth, C. P. *Critical State Soil Mechanics*, McGraw-Hill, London, 1968.
26. Valanis, K. C. A theory of viscoplasticity without a yield surface, Part 1: General Theory, Part II: Application to mechanical behavior of metals, *Arch. of Mech.*, **23**, 517–51, 1971.
27. Valanis, K. C., and Read, H. E. Chapter 14 of this volume.
28. Wroth, C. P., and Loudon, P. A. The correlation of strains with a family of triaxial tests on overconsolidated samples of kaolin, *Proc. Geotechnical Conference*, Oslo, **1**, 159–63, 1967.
29. Zytynski, M., Randolph, M. R., Nova, R., and Wroth, C. P. On modelling the unloading–reloading behaviour of soils, *Int. J. Num. Anal. Meth. Geomech.* (short comm.), **2**, 87–94, 1978.

Soil Mechanics—Transient and Cyclic Loads
Edited by G. N. Pande and O. C. Zienkiewicz
© 1982 John Wiley & Sons Ltd

Chapter 11

A Model for the Cyclic Loading of Overconsolidated Soil

M. J. Pender

SUMMARY

The background to the author's critical state/work-hardening plasticity stress–strain model for overconsolidated soil is discussed and the performance of the model when applied to cyclic loading is reviewed. Closed hysteresis loops are generated by the model for two-way cyclic loading when equal proportions of the undrained strength in compression and extension are mobilized. However, the model does not work so well for cyclic loading which involves the mobilization of different proportions of the strength in compression and extension. In one-way cyclic loading the predicted strain accumulates rapidly with the number of cycles. Experimental evidence shows that generally a stable state is reached if the cyclic stress level is not too large. The model is extended so that this phenomenon can be included. This is achieved by modifying the hardening function so that after each change in the direction of shearing, from compression to extension and vice versa, there is an increase in the stiffness of the model and a reduction in the rate of pore pressure increase. Thus a steady state is eventually reached. This extension is achieved with the introduction of two cyclic hardening parameters.

11.1 INTRODUCTION

Pender[18] has presented a mathematical model for the stress–strain behaviour of overconsolidated soil. The model was developed within the framework of critical state soil mechanics using a constitutive relationship for a work-hardening plastic material. The purpose of this chapter is to review the performance of the model when applied to cyclic loading and to present some recent developments.

In its initial form the model gives a good representation of a wide range of soil behaviour. Four parameters are needed to specify a given soil and these are obtained from routine tests. With these data it is then possible to model drained and undrained behaviour for a range of overconsolidation ratios.

So successful was the initial phase of the model development, that phenomena not originally covered were examined more closely with a view to extending the model. The first of these extensions was into the field of cyclic stress–strain behaviour. The model describes the large strain variation of apparent shear modulus and equivalent viscous damping ratio. The details are given by Pender.[16] However, at very small strain amplitudes, soil is known to exhibit an elastic shear modulus. Thus, another parameter, the small strain shear modulus, was added. This affects the modelling when small strains are important or when there is some restraint on the strains, such as in the oedometer where the lateral strain is zero. The second of these extensions was to cover the normally consolidated region. This was achieved by the hypothesis that the stress space is divided into two regions: an 'inner' region in which the soil behaves in an overconsolidated fashion and an 'outer' region in which the soil deforms in a normally consolidated fashion. The details are given by Pender.[17] Plastic strains are assumed to occur in both of these regions. The boundary between the two regions is given by the undrained stress path for normally consolidated soil. When a drained stress path crosses this boundary there is a 'softening' in the stress–strain response of the soil because the volumetric and shear strain rates increase. This boundary is, in effect, a generalization of the concept of preconsolidation pressure. The oedometer test gives one point on the boundary between overconsolidated and normally consolidated behaviour, other stress paths give other points on the boundary.

11.2 BACKGROUND TO THE DEVELOPMENT OF THE MODEL

The author's original intention in developing the model was to provide a good qualitative description of the range of overconsolidated soil behaviour with relatively few and simple input data. The modelling of normally consolidated soil behaviour had achieved some success with the Cam clay (Schofield and Wroth[23]) and the modified Cam clay (Roscoe and Burland[22]) models. It thus seemed worthwhile to attempt a similar approach for overconsolidated behaviour. Despite the difficulties associated with the interpretation of many aspects of the stress–strain behaviour of overconsolidated soil, difficulties which are not nearly so significant for normally consolidated soil, the model was reasonably successful. Figures 6, 10, 13 and 15 in reference 18 provide evidence of this. The difficulties in interpreting overconsolidated stress–strain behaviour, such as the observation that heavily overconsolidated soil almost never reaches the critical state in either drained or undrained loading, calls into question the application of the critical state approach to overconsolidated behaviour. Parry[14] has provided an illuminating interpretation of the behaviour of overconsolidated soil at peak shear stress. He showed that, at the peak shear stress, the rate of change of pore pressure for undrained tests and the rate of volume change in drained tests is a function of how far the specimen is from the critical state.

In common with the Cam clay models, the author's model is not a predictive model in the sense that exploration of the model response will reveal new aspects of soil behaviour. Rather it is a descriptive model with the purpose of providing a compact numerical representation of many aspects of overconsolidated soil behaviour. The basic features of the soil behaviour that are to be modelled must be incorporated when the model is formulated. A measure of the success of such a model is the range of phenomena and accuracy of the modelling achieved in relation to the number of items of input data, and the difficulty associated with the measurement of these. In this regard the critical state framework is particularly useful because it provides a context for the development of the model which is based on soil behaviour, the parameters thus have a readily identified physical significance.

Although the terminology, starting points and structure of this model have much in common with the Cam clay models, there is one very important difference in viewpoint, namely, the manner in which the incremental stress–strain relationship is derived. Cam clay and modified Cam clay start with a work equation which postulates a mechanism for the dissipation of plastic work within the soil skeleton. The work equation provides the information from which the yield locus is determined. This procedure gives the impression that the Cam clay models are based on an understanding of the physical processes occurring within the soil. However, in reality, both the theory of plasticity and critical state view of soil behaviour are macroscopic rather than microscopic approaches to the behaviour of the material. Thus they imply no understanding about what is occurring within the soil at the microstructural level. The writer therefore started by making hypotheses which give mathematical functions to specify various aspects of the soil behaviour, these are: the yield locus, the ratio of the plastic strain increments and the shape of the undrained effective stress path. The second and third of these functions are very dependent on the critical state idealization.

11.2.1 Choice of constitutive relationship

The author's modelling is based on the assumption that soil can be regarded as a work-hardening plastic material. The constitutive equation for such a material has the form:

$$d\varepsilon_{ij}^{p} = h \frac{\partial g}{\partial \sigma_{ij}} df, \tag{11.1}$$

where $d\varepsilon_{ij}^{p}$ is the plastic strain increment tensor
  σ_{ij} is the stress tensor
  h is the hardening function
  g is the plastic potential
 and df is the differential of the function f which defines the
    yield locus.

This choice of constitutive relationship is a matter of convenience and does not imply anything fundamental about soil behaviour. Many other possibilities are available. Two very significant advantages of the plasticity approach are the diagrammatic interpretation that is provided as to how the constitutive relationship works, and the clear distinction between loading and unloading. The major disadvantage in applying plasticity concepts to soil behaviour lies in the concept of yielding. The classical experimental work of Taylor and Quinney[25] and the much more detailed work of Bertsch and Findlay[3] has shown that, for metals, the phenomenon of yielding is quite abrupt. The more complex behaviour of soil makes it rather difficult to pick up clearly when yield actually occurs, a problem which has been discussed by Lewin and Burland.[8] These difficulties have led the author to adopt an approach that is not based on verifying the fine details of the soil response, but rather on comparing the broad aspects of the response of soil with the model predictions. The same approach has been followed in extending the model to a wider range of phenomena. The yield locus concept, used by the author, is based additionally on the ideas that the size of the region within the yield locus is very small and almost all soil deformation is plastic. The hypothesis was made that the yield locus is reduced to a line. A discussion of similar ideas is given by Palmer and Pearce,[13] who explore the idea that the yield locus for soil is reduced to a point in the stress space.

As an example of an alternative stress–strain formulation to work harden-ing plasticity there is the endochronic model. It is of interest that the form of the stress–strain equation given by Valanis and Read[27] for the conventional triaxial test is of the same form as equation (29) in Pender[18] for the work-hardening plasticity model. This equation relates to the stress–strain curve for a lightly overconsolidated material having a constant p stress path, where p is equal to the critical state value for the current void ratio.

11.2.2 Overconsolidated stress–strain behaviour—elastic or plastic?

The basic idea of the author's model is the hypothesis that even for overconsolidated behaviour, plastic stress–strain behaviour is important. Recoverable strains are a significant part of the volumetric response, but are significant only at very small strain levels for the shear response.

The Cam clay models have been developed with the purpose of modelling the strains in the normally consolidated region. Consequently, they recog-nize recoverable volumetric strains for overconsolidated behaviour, but, in the first instance, assume that there are no recoverable shear strains. There have been two approaches to improving this situation. Roscoe and Bur-land[22] simply augment the shear strains calculated with modified Cam clay model for drained normally consolidated behaviour, with the shear strains observed in a normally consolidated undrained test. Alternatively, the overconsolidated behaviour is modelled as elastic. The view that overconsolidated soil can be described by an elastic stress–strain relationship is widespread, e.g. Morgenstern.[9] Wroth[29] introduces a development of this and makes the shear modulus and bulk modulus functions of the mean principal effective stress, Poisson's ratio being constant. Wroth and Zytynski[30] have illustrated the application of this version of modified Cam clay to finite element computations. For many applications this stress-dependent elastic model is a good and convenient idealization of overconsolidated soil behaviour. However, there are a number of features of the behaviour of overconsolidated soil which are contrary to elastic behaviour. Some of these are:

(1) Parry and Amerasinghe[15] performed undrained triaxial compression tests on heavily overconsolidated kaolin. They performed load–unload cycles for a series of increasing shear stress levels. They then compared the strain recovered during the unloading part of the test with the strain just prior to the unloading. Some of their data are reproduced in Figure 11.1, which also includes some data reported by others.

It is evident that the strain recovered on unloading is directly proportional to the strain at the commencement of unloading. Since the loading curves are nonlinear, the unloading curves are also nonlinear. Thus, the behaviour is not elastic because neither the loading or unloading curves are linear, nor are the strains generated during loading recovered on unloading.

If the strains were recovered on unloading it might be possible to assume that a stress-dependent elastic model would represent the soil behaviour well. Alternatively it could be argued that, for a continuous

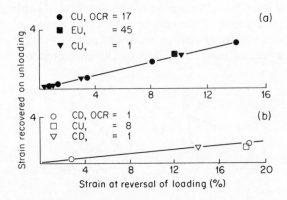

Figure 11.1 Strain recovered on unloading from triaxial compression on over-consolidated clay: (a) Kaolin; (b) Weald clay (after Parry and Amerasinghe[15])

loading process, the unloading behaviour is irrelevant, so that an elastic modulus representative of the loading behaviour would be suitable. However, the nonlinearity that is always observed for large stress changes means that such a description of the soil stress–strain behaviour is likely to be valid only for stress changes that are small in relation to the existing stress state.

(2) The assumption of elastic behaviour precludes the possibility of modelling hysteresis. As stated above, the strain on unloading is not, in general, recovered. Thus, during a load–unload–reload cycle, work is dissipated in deforming the soil skeleton. The amount of hysteresis is a function of the strain at the beginning of the unloading. It is observed at stress levels well below failure. Only when the stress path engages the yield locus during loading will the modified Cam clay model generate hysteresis. Hysteresis is observed at stress levels which are well inside the yield locus.

(3) Purely elastic behaviour cannot represent dilatancy. It is well-known that the onset of dilatancy occurs well before the peak strength is mobilized. Thus, a simple elastic model is unable to model this very important phenomenon.

(4) The modified Cam clay model assumes isotropic hardening, i.e. the yield locus is expanded uniformly, rather than locally, by a stress path which engages it. This implies that, on unloading, the behaviour is elastic as the soil is then, in effect, overconsolidated. In fact, the opposite occurs. Figure 11.2 gives some results of Ohmaki[12] for drained compression tests at constant mean principal stress on lightly overconsolidated clay. The fact that the volume of the specimen continues to decrease on unloading suggests that plastic deformation is occurring during unloading, as elastic behaviour demands that the direction of the strain reverses when the direction of loading reverses.

Similar observations have also been reported by others, for example, Namy[11] gives results of conventional drained tests. This experimental observation had an important effect on the development of the author's

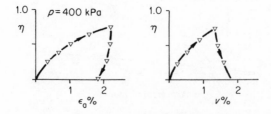

Figure 11.2 Drained constant mean principal effective stress tests on clay (after Ohmaki[12])

model. It is the reason why the yield locus is always carried along with the current stress point.

The above four comments outline the reasons for the decision to model overconsolidated stress–strain behaviour as work-hardening plastic, rather than elastic or pseudo-elastic.

One feature of soil behaviour that is not handled well by the plasticity approach is anisotropy. It is well known that many overconsolidated soils have anisotropic stiffness properties. In fact, certain aspects of soil stress–strain behaviour are clearly anisotropic, and cross anisotropic elastic models have been used to describe these by Atkinson[2] and Wesley.[28] The author's model and the Cam clay models assume isotropic behaviour. The modelling of anisotropic behaviour, even with an elastic model, is not a simple matter. The difficulty lies in the measurement of the values for the various moduli. A very high standard of experimental work is required. Thus the various models available have strengths and weaknesses. The choice of model depends on the particular application in question; there are no doubt cases in which an anisotropic elastic model furnishes more useful predictions than an isotropic work-hardening plasticity model. It is the belief of the author that the overall pattern of behaviour encompassed by the critical state/work-hardening plasticity approach is sufficiently broad to offer a useful description of soil behaviour.

11.2.3 Application of the model to static stress–strain behaviour

The model makes the hypothesis that the undrained stress paths are parabolic and 'seek' a critical state appropriate to the current void ratio. The form of the paths is given by

$$\left(\frac{\eta - \eta_0}{AM - \eta_0}\right)^2 = \frac{p_{cs}}{p}\left\{\frac{1 - p_0/p}{1 - p_0/p_{cs}}\right\}, \tag{11.2}$$

where η is the stress ratio q/p

q is the principal effective stress difference, $(\sigma_1' - \sigma_3')$

p is the mean principal effective stress, $(\sigma_1' + 2\sigma_3')/3$

M is the stress ratio, q/p, at the critical state

η_0 is the stress ratio at the start of the undrained path

p_0 is the value of p at the start of the undrained path

p_{cs} is the value of p for the point on the critical state line corresponding to the current void ratio

A is $+1$ for loading in compression and negative with a magnitude not necessarily unity, for loading in extension.

The incremental axial plastic strain for undrained loading is given by

$$d\varepsilon^p = \frac{2\kappa(p/p_{cs})(\eta - \eta_0)\,dn}{(AM)^2(1 + e)(2p_0/p - 1)\{(AM - \eta_0) - (\eta - \eta_0)p/p_{cs}\}}, \tag{11.3}$$

where $d\varepsilon^p$ is the increment in plastic distortion; for triaxial conditions,
$$\varepsilon^p = 2(\varepsilon^p_{axial} - \varepsilon^p_{radial})/3$$
$-\kappa$ is the slope of the line in the e, $\ln p$ plane for swelling under
    spherical stress conditions
e is the void ratio of the soil.

The derivation and discussion of these equations is given by Pender.[18] Equation (11.3) embodies two ways in which the stress path affects the stress–strain response of the soil. The position of the current stress point relative to the critical state, both with regard to the mean principal effective stress and the stress ratio η, provides one control. The position of the current stress point relative to initial stress conditions, or the stress conditions at the last reversal point for cyclic loading, provides the other control. Thus a type of scaling comes from the position of the current stress point relative to the initial stress conditions and the critical state condition towards which it is moving. The main innovation in the development of the incremental stress–strain equations is the utilization of the expression for the undrained stress path as a means for determining the hardening function. In work-hardening plasticity the determination of the hardening function usually presents the biggest problem. Specifying the form of the undrained stress path, in effect, specifies the plastic volumetric component of the strain. Thus, if the yield function and plastic potential are assumed to be known, the hardening function can be inferred.

At a later stage in this paper, the results of cyclic loading tests on Drammen clay presented by Andersen et al.,[1] will be discussed. At this point it is appropriate to compare the model predictions with static undrained tests on the Drammen clay for a range of overconsolidation ratios. The comparision for the stress–strain curves, stress paths and pore pressure response is given in Figure 11.3. The parameters for the model were determined from two of the figures in the paper by Andersen et al.[1] Their Figure 2 gives the results of oedometer and spherical compression tests, whilst Figure 7 gives the undrained stress path for a normally consolidated specimen. This is enough to estimate the values for the material parameters, the initial conditions of the overconsolidated specimens and the critical state point for each of the specimens. The predictions in Figure 11.3 were then obtained by the immediate application of equations (11.2) and (11.3). The model is seen to give a good representation of the stress paths and stress–strain curves for the three overconsolidation ratios. The differences between the model and real undrained stress paths, Figure 11.3(b), are shown in the form of pore pressure, u, response in Figure 11.3(c). Clearly this is a more sensitive way of comparing the shapes of the stress paths. It is of note that the one set of parameters ($M = 1.23$, $\kappa = 0.0185$, p_0/p_{cs} for normally consolidated undrained behaviour $= 2.15$ and $\lambda = 0.2200$) is sufficient to

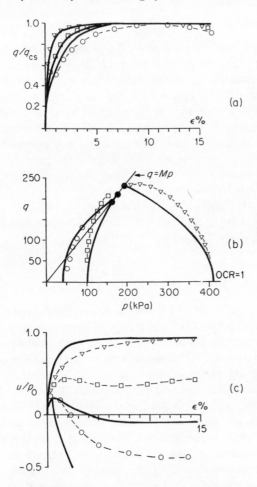

Figure 11.3 Static undrained tests on Drammen clay: (a) stress–axial-strain curves; (b) effective stress paths; (c) pore pressure response (∇, $\square$, $\bigcirc$ results Andersen *et al.*[1]—model predictions) $M = 1.23$, $\kappa = 0.0185$, $\lambda = 0.2200$, $(p_0)_{nc}/p_{cs} = 2.15$

model the behaviour for the three overconsolidation ratios considered. The initial conditions for each of the specimens differ but the same parameter values are used for each case. The model in the same form as discussed here is also capable of predicting drained response for overconsolidated material; once again the same four parameters, with appropriate initial conditions, are all that is required.

11.2.4 Discussion

The critical state/work-hardening plasticity model thus gives a good numerical description of the range of overconsolidated stress–strain behaviour. The model is such that values for soil parameters can be determined reasonably easily. Likewise the conceptual basis for the model is straightforward. Thus it yields a good return for the number of starting assumptions and the amount of data needed to apply the model. However, it is far from perfect; it has many defects in detail and there are also some phenomena that it cannot handle. The initial development of the model provides a basis for the description of other aspects of soil behaviour, or possibly the development of several variants to describe specific phenomena not covered in the earlier development. Two such variants have been mentioned above. The main purpose of this chapter is to present further developments for the modelling of cyclic stress–strain behaviour.

11.3 APPLICATION TO CYCLIC LOADING

The observation in Section 11.2.2(4), that on drained unloading, the volume of normally consolidated soil continues to decrease, suggests that the yield locus is carried along with the current stress point. Thus plastic deformation occurs whether the soil is being sheared in compression or extension, and the state of the soil moves continually towards the critical state appropriate to the current void ratio. The only information that the model is required to 'remember' is the state of the soil at the last reversal point. Each time the stress path changes from compression to extension, and vice versa, the starting conditions are reinitialized. This arises from the way in which the model is formulated; no new parameters are introduced to handle cyclic loading. This concept, that the yield locus is always carried along with the current stress point, has a considerable computational advantage. It is not then necessary to keep track of the position of the current yield locus relative to the current stress point. In common with the Cam clay models, the author's model does not consider time effects such as creep or increased stiffness resulting from very rapid loading. The cyclic stiffness of the soil model is thus not affected by the frequency of the loading.

It is of interest to note that this approach to cyclic loading provides stress–strain curves that are very similar in shape to those obtained by Pyke,[21] Figure 11.4. In effect, then, the model incorporates implicitly the scaling rules that Pyke suggested in applying a modified Masing's rule to hyperbolic cyclic loading curves.

As in the static loading case it is assumed that the state of the soil is always moving towards the critical state value of the mean principal effective stress. The question needs to be asked: does soil behave in this way? As

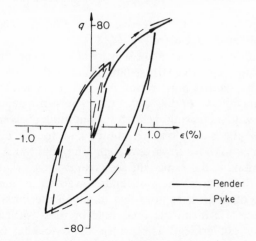

Figure 11.4 Comparison between the author's cyclic loading model and that of Pyke[21]

evidence in support, the experimental work of Taylor and Bacchus[26] is of interest. Laboratory prepared specimens of a silty clay were cyclically loaded under strain-controlled conditions. The strain control was such that for each cycle of loading the specimen was deformed to a set strain in compression and then to a strain of equal magnitude in extension. Thus the soil was subjected to two-way cyclic loading, although the maximum stress in the compression part of the cycle was greater than the maximum stress in the extension part of the cycle. Each specimen was subjected to 100 cycles at a particular strain amplitude. It was then loaded to failure in static undrained compression; the effective stress paths are shown in Figure 11.5. Also

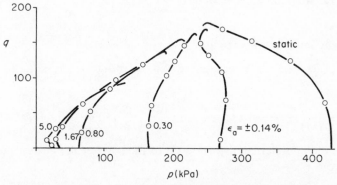

Figure 11.5 Static undrained effective stress paths after 100 cycles of strain-controlled cyclic loading (Taylor and Bacchus[26])

included in this figure is the effective stress path for static loading of a normally consolidated specimen.

Three features of Figure 11.5 are worthy of comment. Firstly, after cyclic loading the residual effective stress state has moved to the left of the initial condition, i.e. a positive pore water pressure is generated by the cyclic loading. The magnitude of the pore water pressure generated after 100 cycles increases with increasing cyclic strain amplitude. Secondly, the effective stress paths for the static loading, subsequent to the cyclic loading, all exhibit a tendency to move towards a common point. Since the void ratio of all these specimens is the same, this behaviour shows that, even after the cyclic loading, the state of the soil tends towards a common critical state point. It is also of interest to note that although the initial state of all the specimens is wet of critical, the final state of those with the larger strain amplitudes is dry of critical. The critical state seeking behaviour of the author's model makes it difficult to include this phenomenon. The state of the soil in cyclic undrained loading moves towards the critical state value of the mean principal effective stress, either from the wet side or the dry side. The undrained stress paths given by equation (11.2) cannot cross the line $p = p_{cs}$. Thirdly, the effective stress paths for the post cyclic loading of those specimens that have been subjected to the larger strain amplitudes do not reach the critical state point, the undrained shear strength is less than that of the specimen not subjected to cyclic loading. The total plastic work done during the 100 cycles can be thought of as causing an accumulation of damage with a consequent reduction in strength. The amount of plastic work increases with increasing strain amplitude, hence the strength reduction increases with increasing strain amplitude.

These data reported by Taylor and Bacchus[26] suggest that, even though the undrained cyclic loading response of soil is more complex than the static loading behaviour, the critical state idealization still offers a viable framework for modelling cyclic behaviour.

11.3.1 Cyclic loading when $p=p_{cs}$

Equation (11.3) can be integrated for a constant p stress path when $p = p_{cs}$:

$$\varepsilon_{p_{cs}}^{p} = \frac{2\kappa}{(AM)^2(1+e)}\left\{(AM-\eta_0)\ln\left(\frac{AM-\eta_0}{AM-\eta}\right)-(\eta-\eta_0)\right\}+\varepsilon_0^{p}, \quad (11.4)$$

where ε_0^{p} is the cumulative distortion up to the start of the current loading from η_0.

Stress–strain loops generated with equation (11.4) are given in Figure 11.6. Figure 11.6(a) shows how the strains accumulate when the reversal

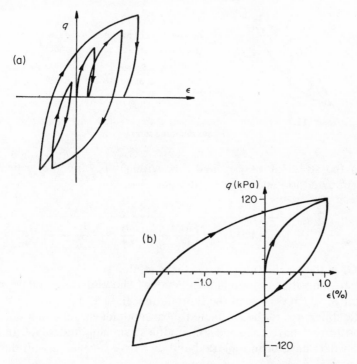

Figure 11.6 Stress–strain loops generated with equation (11.4): (a) accumulation of strains for irregular loading; (b) steady-state cyclic loading

points are not symmetrically placed relative to the critical state points. Figure 11.6(b) shows how closed hysteresis loops are possible when the same proportion of the critical state strength is mobilized in compression and extension.

11.3.2   Apparent shear modulus

From the closed stress–strain loops. of Figure 11.6(b), it is possible to calculate the apparent shear modulus and equivalent viscous damping ratio. These parameters, and particularly their variation with strain amplitude, are frequently used to illustrate the nonlinear behaviour of soil in steady state (i.e. when there is no accumulation of strain) cyclic loading. The two parameters are defined in Figure 11.7.

This diagram corrects a small error in the earlier paper on cyclic loading, Pender.[16] The apparent shear modulus defined there is not the true value but three times the apparent shear modulus. Using the definition of G given in Figure 11.7, the apparent shear modulus, for steady state cycling

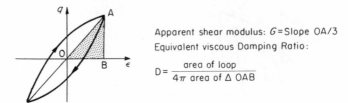

Figure 11.7 Definition of apparent shear modulus and equivalent viscous damping ratio

between the stress ratios $+\eta_j$ and $-\eta_j$ with $p = p_{cs}$ and the critical state stresses in extension and compression equal, is

$$G/q_{cs} = \frac{\eta_j(1+e)}{3\kappa\left\{\left(1+\dfrac{\eta_j}{M}\right)\ln\left(\dfrac{M+\eta_j}{M-\eta_j}\right) - \dfrac{2\eta_j}{M}\right\} + \dfrac{\eta_j q_{cs}(1+e)}{G_e}}, \quad (11.5)$$

where G_e is the small strain elastic shear modulus.

Comparision with equation (11) in Pender[16] shows that the difference is in the factor 3, which occurs on the bottom line. In fact, this correction makes very little difference to the calculated shear modulus–strain amplitude curve. Seed and Idriss[24] have collected soil moduli strain amplitude data. In Figure 11.8 their data have been replotted and the curve calculated with the aid of

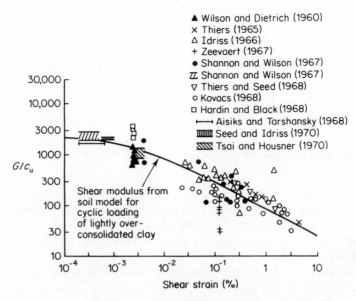

Figure 11.8 Apparent shear moduli data for saturated clays (after Seed and Idriss[24])

equation (11.4) and (11.5) plotted. The strain amplitude on the abscissa is the maximum shear strain in the specimen. It is apparent that the soil model gives a good representation of the deterioration of apparent shear modulus with increasing strain amplitude. It is emphasized that the curve in Figure 11.8 has been calculated for a given set of soil properties—$M = 0.90$, $e = 1.377$, $\kappa = 0.02$, $G_e = 200$ MPa—and initial conditions $e = 1.377$, $p_0 = p_{cs} = 220$ kPa. It is not intended as a best fit through the data points. Rather it represents the variation of apparent shear modulus for a particular initial condition and set of soil properties. Other curves for different initial conditions and soil properties can be calculated readily.

11.3.3 Equivalent viscous damping ratio

The corrected expression for the equivalent viscous damping ratio is:

$$D = \frac{(2/\pi)\left(1+\dfrac{\eta_i}{M}\right)\left\{\ln\left(\dfrac{M+\eta_i}{M-\eta_i}\right)-2\left(\dfrac{\eta_i}{M}\right)\right\}}{\left(\dfrac{\eta_i}{M}\right)\left\{\left(1+\dfrac{\eta_i}{M}\right)\ln\left(\dfrac{M+\eta_i}{M-\eta_i}\right)-\dfrac{2\eta_i}{M}\right\}+\dfrac{(1+e)p_{cs}(\eta_i)^2}{3\kappa G_e}} \cdot \tag{11.6}$$

The variation in equivalent viscous damping ratio with strain amplitude calculated with equations (11.4) and (11.6) is compared with the data collected by Seed and Idriss[24] in Figure 11.9. This shows that equations

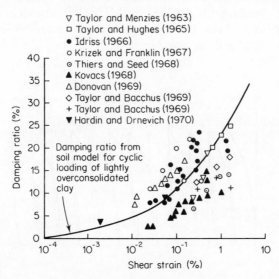

Figure 11.9 Equivalent viscous damping ratios for saturated clays (after Seed and Idriss[24])

(11.4) and (11.6) give a good representation of the variation in damping ratio with strain amplitude. The parameter values used in calculating the curve were the same as those for the curve in Figure 11.8.

11.4 EXTENSION TO NON-STEADY STATE CYCLIC LOADING

Figure 11.6(b) illustrates the concept of steady state cyclic loading: if $p = p_{cs}$ and equal proportions of the undrained strength in extension and compression are mobilized, the model generates closed stress–strain loops. It is not necessary that the strengths in extension and compression be the same, only that the same proportion of each is mobilized at the extremities of the loading cycle. On the other hand, the behaviour for irregular cyclic loading is shown in Figure 11.6(a). The application of equation (11.4) to this situation could not be expected to give closed stress–strain loops. Equation (11.4) can also be applied to the case when the soil is cycled between fixed stress limits which are not the same proportion of the strength of the soil. An example is shown for one-way cyclic triaxial loading (i.e. with the stress always compressive) in Figure 11.10; the Drammen clay parameters (Figure 11.3) are used.

It is clear from this diagram that the strains accumulate rapidly with the number of cycles. This is contrary to the observed behaviour. When the applied stress level does not mobilize too large a proportion of the soil strength, a shake-down phenomenon is observed, in which the increase in strain per cycle decreases with increasing number of cycles. Some results of Anderson et al.[1] and Brown, Lashine, and Hyde[4] presented in Figure 11.11, illustrate this point. Thus equation (11.4) may give reasonable modelling of irregular cyclic loading if the number of cycles is low, but it is not suitable

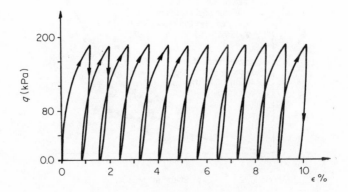

Figure 11.10   Accumulation of strain given by equation (11.4) for a one-way cyclic loading

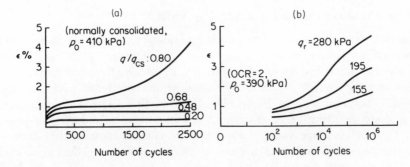

Figure 11.11   Strain accumulation in one-way cyclic loading: (a) Andersen *et al.*[1] on Drammen clay; (b) Brown, Lashine, and Hyde[4] on a silty clay

for large numbers of cycles. Similar problems emerge when equation (11.3) is applied to the case in which $p \neq p_{cs}$. The pore pressure response then becomes important. The application of equation (11.2) for the undrained effective stress path to cyclic loading gives too rapid a build up in pore water pressure with number of cycles. This means that the path moves towards p_{cs} too rapidly.

The extension of the model to remedy the above two defects is the purpose of the remainder of this chapter.

One approach that could be used to generate the cyclic hardening behaviour is the postulation of an elastic domain, the size of which increases during cyclic loading. Thus the rate of accumulation of strain would decrease as the number of cycles increased. Such an approach has been used by Mroz, Norris and Zienkiewicz.[10] A similar development is the postulation of a number of nested yield loci; see Prevost.[20] These approaches offer more flexibility than the author's approach at the expense of a greater computational task (mentioned in Section 11.3) and conceptual difficulty, not to mention the number of parameters required and the determination of values for these. As explained in Section 11.2.1, it is unlikely that the fine details of these approaches could be verified experimentally. Herein, the concept that the yield locus remains the same as before with an altered hardening function is explored. Thus, every change of stress ratio, regardless of the direction of loading, gives plastic strain. In this sense the author's model has something in common with the hyperbolic and Ramberg–Osgood models, Idriss, Dobry and Singh[7] and Pyke.[22] Figure 11.4 shows that the shape of the stress–strain loops generated with equation (11.4) is similar to that generated by Pyke's adaption of the hyperbolic model. However, the author's model is an effective stress, rather that total stress, approach and so the pore pressure response during the cyclic loading is also modelled.

11.4.1 Modification of the hardening function

The above comments reveal that both equations (11.2) and (11.3) need to be modified to model cyclic loading more satisfactorily. The fact that the undrained stress path does not approach p_{cs} as rapidly as equation (11.2) predicts means that the plastic volumetric strain is reduced with each cycle in just the same way as Figure 11.11 shows that the plastic distortional strains are reduced for each cycle. The constitutive relationship, equation (11.1), involves three functions f, g and h. The observation that both the plastic volumetric and distortional strain components are affected by cyclic loading suggests that it is the hardening function, h, which should be modified to account for cyclic loading. The yield locus, f, and the plastic potential, g, can, in the first instance, be left unchanged.

In presenting and interpreting experimental data, particularly for cyclic loading between fixed stress limits, the number of cycles is an obvious and convenient independent variable. However, equation (11.2) for the stress path and the stress–strain equations (11.3) and (11.4) suggest that the number of half cycles, or the number of times the direction of loading changes from compression to extension and vice versa, is more likely to be significant. Not only do the equations for the model direct attention to half cycles because the stress conditions at the previous turning point are 'remembered', but also from the physical point of view it is possible to envisage rearrangement of soil particles during each half cycle, such that the soil structure at each turning point is stiffer than it was before.

A modified hardening function is derived in the same way that the hardening function was derived for the original form of the model. An undrained stress path is possible when the plastic component of the volumetric strain is matched by an equal and opposite recoverable volumetric strain increment. As with the Cam clay models, the author's model takes the recoverable volumetric strain from the linear plot of the volume change in an e, $\ln p$ plot:

$$\mathrm{d}v^r = \frac{\kappa\,\mathrm{d}p}{p(1+e)},\tag{11.7}$$

where $\mathrm{d}v^r$ is the recoverable volumetric strain increment.

Thus, using equations (11.1) and (11.7), a differential expression for the undrained stress path is:

$$h\frac{\partial g}{\partial p}\,\mathrm{d}f + \frac{\kappa\,\mathrm{d}p}{p(1+e)} = 0.\tag{11.8}$$

An alternative differential form of the undrained stress path is obtained from equation (11.2):

$$\mathrm{d}p = \frac{2(1-p_0/p_{cs})(p^2/p_{cs})(\eta-\eta_0)\,\mathrm{d}\eta}{(AM-\eta_0)^2(2p_0/p-1)}.\tag{11.9}$$

Now, as explained in Section 11.4, during cyclic undrained loading the magnitude of dp appears to be rather smaller than given by equation (11.9). The following modification of the hypothesis that the undrained stress paths are parabolic is proposed:

$$dp_{\text{cyclic}} = dp_{\text{(equation 11.9)}} \left(\frac{\eta - \eta_0}{AM - \eta_0} \right)^{\xi},\qquad (11.10)$$

where ξ is the cyclic hardening index.

Combining equations (11.8), and (11.10) the following expression is obtained for the modified hardening function:

$$h = \frac{2\kappa(p_0/p_{cs} - 1)(\eta - \eta_0)^{(1+\xi)}}{(AM - \eta_0)^{(2+\xi)} p_{cs}(1+e)(2p_0/p - 1)\, \partial g/\partial p}.$$

Assuming the same plastic potential function as before:

$$\partial g/\partial p = \frac{(AM)^2(p_0/p_{cs} - 1)((AM - \eta_0) - (\eta - \eta_0)p/p_{cs})}{(AM - \eta_0)^2}.$$

On substitution of this the modified hardening function is given by

$$h = \frac{2\kappa(\eta - \eta_0)^{(1+\xi)}}{(AM)^2 p_{cs}(1+e)(2p_0/p - 1)(AM - \eta_0)^{\xi}((AM - \eta_0) - (\eta - \eta_0)p/p_{cs})}.$$

$$(11.11)$$

As with the earlier development of the model, $\partial g/\partial q = 1$ and $df = p\, d\eta$, on substitution of these and equation (11.11) into equation (11.1):

$$d\varepsilon^{\text{p}} = \frac{2\kappa(p/p_{cs})(\eta - \eta_0)^{1+\xi}\, dn}{(AM)^2(1+e)(2p_0/p - 1)(AM - \eta_0)^{\xi}((AM - \eta_0) - (\eta - \eta_0)p/p_{cs})}.$$

$$(11.12)$$

11.4.2 Cyclic hardening index, ξ

The objective is to arrive at a model which can be used for cyclic loading, regular and irregular, and for all stress levels up to failure. The ability to handle irregular loading is necessary for application to earthquake and offshore problems. Thus at the beginning of each half cycle the index ξ should be such that no information about the stress conditions at the next turning point is required. The use of half cycles rather than full cycles places restrictions on ξ. Experimental evidence shows that the soil specimen decreases in length during one-way compressive cyclic loading. If the index ξ has too large a value the strain recovered during a given unloading half cycle may be greater than the strain on the next loading half cycle.

A number of forms were tried for ξ. The first of these, ξ is a constant, is capable of very successful modelling of repeated loading between fixed stress limits, i.e. of one of the curves in Figure 11.11. However, the same value of ξ does not work for a different set of cyclic stress limits. If, for example, ξ is found to model the one-way cyclic loading curve for $q/q_{cs} = 0.68$ in Figure 11.11(a), and then that value of ξ applied to one-way cyclic loading with $q/q_{cs} = 0.20$, the problem mentioned above, in which the specimen increases in length, arises. Alternatively, if the value of ξ is applied to the case with $q/q_{cs} = 0.80$ the strains accumulate too rapidly. Thus ξ is related to the level of cyclic stress. It is proposed that the change in q in the previous half cycle, normalized with respect to p_{cs}, be part of ξ.

Figure 11.11(a) shows that, at least for modest levels of cyclic loading, the strain does not accumulate indefinitely. Rather the rate of increase of strain seems to decrease with increasing numbers of cycles and approach a stable state. Thus the cyclic hardening index, ξ, should increase as the number of half cycles increases. The hypothesis is made that ξ has the form:

$$\xi = (|q_p|/p_{cs})^\alpha (H^\beta - 1), \tag{11.13}$$

where q_p is the change in q in the previous half cycle

H is the number of the current half cycle

α, β are soil parameters for cyclic hardening.

The second part of the right-hand side of equation (11.13) is zero when $H = 1$. Thus for the first half cycle of loading equations (11.10) and (11.12) reduce to the same form as equations (11.9) and (11.3). It will be shown below that with this form of ξ equations (11.10) and (11.12) model many aspects of undrained cyclic loading behaviour.

11.4.3 Application to cyclic loading when $p = p_{cs}$

The result of applying equation (11.12) to two way cyclic loading, between fixed stress limits in which equal proportions of the undrained strength in extension and compression are mobilized, is given in Figure 11.12(a). The stiffening effect produced by equation (11.12), as opposed to the stable state closed loops in Figure 11.6(b), is apparent. As in Figure 11.6(b) the effect of starting with a compressive half cycle followed by a half cycle in extension leads to an increase in length of the specimen shown in Figure 11.12(a). Because the stiffening effect is based on half cycles, the result of the cyclic loading is a net increase in length. Similar behaviour is shown for Drammen clay in Figure 10(b) of Andersen *et al.*[1] For the calculations plotted in Figure 11.12(a), the magnitude of q_{cs} in extension is the same as that in compression. If the magnitude of q_{cs} in extension was modelled as being less than that in compression, the length increase in two-way cyclic loading,

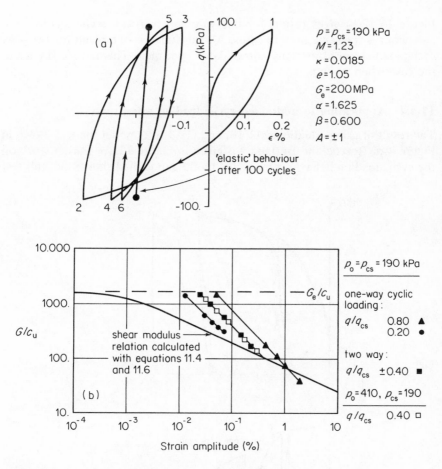

Figure 11.12   Cyclic loading when $p = p_{cs}$, using equation (11.12): (a) Stress–strain loops;
(b) Apparent shear modulus variation

when the first half cycle is compressive, would be accentuated. The result of using equation (11.12) rather than equation (11.3) to calculate the strains is a gradual increase in the range over which the small strain elastic shear modulus provides a significant contribution to the strains. This effect is presented in Figure 11.12(b), where the change in apparent shear modulus for the compressive half cycles is plotted. For the initial half cycles the apparent shear modulus lies close to the relationship calculated with equations (11.4) and (11.6). Because of the stiffening effect of the cyclic loading the apparent shear modulus migrates towards the elastic value. Thus the small strain elastic shear modulus and the curve calculated with equations (11.4) and (11.6) provide bounds on the apparent shear modulus behaviour.

Figure 11.12(b) gives results for one-way and two-way cyclic loading; the behaviour is the same in both cases. Also included is one result for two-way cyclic loading on a normally consolidated specimen. This is very similar to the case when $p = p_{cs}$.

11.4.4 Application to undrained cyclic loading when $p \neq p_{cs}$

The result of using equations (11.10) and (11.12) to model the test shown in Figure 4 of Roscoe and Burland[22] is given in Figure 11.13. Earlier work on the cyclic loading behaviour predicted by the model had shown that this test

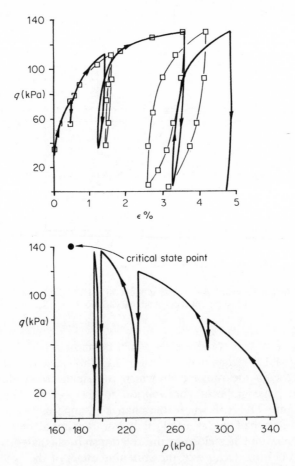

Figure 11.13 Comparison between the cyclic undrained test on kaolin reported by Roscoe and Burland[22] (□) and the model prediction (—) ($M = 0.81$, $\kappa = 0.035$, $e = 1.377$, $p_{cs} = 173$, $p_0 = 345$, $\alpha = 3.00$, $\beta = 1.40$, $G_e = \infty$)

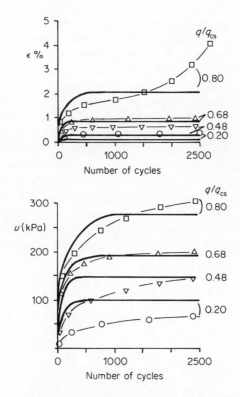

Figure 11.14 Comparison between predicted and measured (Andersen *et al.*[1]) behaviour of Drammen clay in a one-way cyclic triaxial loading (properties as for Figure 11.3, $\alpha = 1.625$, $\beta = 0.600$)

result was a rather demanding one to model, because at the end of the fifth and seventh half cycles the state of the soil is very close to the critical-state condition. There is thus a tendency for the model to predict large strains during these half cycles. The values of α and β used (3.00 and 1.40) have controlled this effect at the expense of too small a recovered strain during the sixth and eighth half cycles. Roscoe and Burland did not present the effective stress path; the calculated path is given in Figure 11.13(b).

The result of using equations (11.10) and (11.12) to predict the one-way cyclic loading behaviour of Drammen clay is compared with the results of Andersen *et al.*[1] in Figure 11.14. The parameters used are the same as those in Figure 11.3 with the addition of $\alpha = 1.625$ and $\beta = 0.600$. It is clear that these values of α and β have the effect of limiting the build-up in strain and pore pressure to the first few hundred cycles. In particular for $q/q_{cs} = 0.20$

the predicted strain is much smaller than the observed strain. Alternative values of α and β which allow a more gradual build-up in the strain were found to produce the same final pore pressure for each level of cyclic loading as the stress paths all reached the $p = p_{cs}$ condition. Anderson *et al.* found that when $q/q_{cs} = 0.80$ the strain kept increasing rather than reaching a stable state as predicted by the model. Presumably this is a type of fatigue phenomenon. Since the stress level is a significant fraction of the undrained strength the onset of fatigue occurs much earlier than would be expected for the lower stress levels. In Figure 11.11(b) the impression is gained that even for relatively small stress levels a fatigue phenomenon is manifested if enough load cycles are applied. The modelling of fatigue is not attempted here. The behaviour modelled in Figure 11.14, in which, for conditions wet of critical, the cyclic stress path and stress–strain curves reach a steady state condition such that $p > p_{cs}$, is consistent with the equilibrium line concept discussed by France and Sangrey[6] for cyclic un-drained loading.

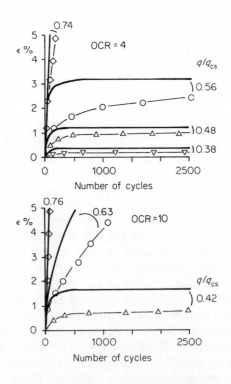

Figure 11.15 One-way cyclic undrained triaxial loading of Drammen clay for over-consolidion ratios 4 and 10

Using the same parameters, it is possible to use equations (11.10) and (11.12) to model undrained cyclic loading of Drammen clay when the initial state of the soil is dry of critical. The calculated results for the strain build up are given in Figure 11.15 along with the results of Andersen *et al*. It is clear that the general features of the strain build-up are modelled quite well, and in particular the fact that the strain accumulates more rapidly for cyclic loading when the initial state is overconsolidated rather than normally consolidated. It is of interest to note that Andersen *et al*.[1] give results from three different laboratories for specimens with an initial overconsolidation ratio of 4. There are considerable differences between the three sets of results. This serves to emphasize the folly of attempting to validate a model by concentrating on just one set of results. A good qualitative description of the general features of the behaviour is more important. The pore pressure response is not presented in Figure 11.15. The model predicts that for initial states dry of critical $p \rightarrow p_{cs}$, i.e. the pore pressure decreases as the number of cycles increases. However, the experimental evidence shows that the pore pressure initially increases, i.e. p moves away from p_{cs}, and eventually settles down to a steady value. It might be possible to model this effect by using the more complex form of parabolic stress path described by the author in reference 19, coupled with the cyclic hardening index idea.

11.5 DISCUSSION

Figures 11.14 and 11.15 show that the addition of the cyclic hardening index idea to the original formulation of the model provides a viable way of modelling the cyclic loading behaviour of Drammen clay. The cyclic hardening parameters α and β have the disadvantage of being non-standard soil parameters, unlike those used in the original development of the model. Furthermore, no simple procedure suggests itself for the determination of α and β. Accurate modelling of a large number of cycles places considerable demands on the method of determining the parameter values. In modelling static behaviour a small error in κ or M may not be too important. However, in cyclic loading these errors will be carried forward through each cycle, so that the net effect could be a considerable accumulation of error.

The model has a limiting apparent shear modulus determined by the small strain elastic shear modulus, G_e. The extended model developed herein predicts that the apparent shear modulus of the soil migrates towards this value during cyclic loading regardless of the level of cyclic stress, Figure 11.12(b). Thus, the range of strains over which the small strain elastic shear modulus operates eventually extends well beyond the 10^{-4} usually quoted. Closer examination of experimental results may reveal that the apparent shear modulus operating when the stable state is reached is less than G_e.

It is mentioned in Section 11.4.4 that a fatigue phenomenon may be the reason for the reduced undrained strength subsequent to cyclic loading shown in Figure 11.5, and the gradual increase in strain with increasing number of cycles for $q/q_{cs} = 0.80$ in Figure 11.11(a). A possible approach to modelling this might be to suggest that during cyclic loading there is a gradual decrease in p_{cs}. The magnitude of this decrease would be a function of the number of half cycles and some measure of the damage to the soil skeleton, such as the accumulated plastic work. This would give a gradual increase in strain and pore pressure as well as reduced strength. Although it would be relatively easy to modify the model further to include this effect, problems in the determination of the appropriate parameters would be considerable. At least the cyclic hardening parameters, α and β, can, in principle, be determined from the results of a few cycles of loading. This would not be the case with a fatigue phenomenon, test results with large numbers of cycles would be necessary. Thus, the possibility of modelling fatigue is not considered further here.

Final question which arises relates to the 'memory' of the soil. The model presented in this chapter implies that the soil can 'count' the number of times the direction of shearing is reversed. What happens at the end of the cyclic loading? Does the soil continue to 'remember' the value of H when the cyclic loading ceases? Alternatively, does the stiffening effect decay with time, or as the excess pore pressure dissipates? Careful examination of experimental results is needed to answer these questions.

11.6 CONCLUSIONS

With a relatively small adaptation of the earlier version of the author's stress–strain model for overconsolidated soil it has been possible to model, at least for the overall phenomena, soil subject to large numbers of loading cycles. Two additional soil parameters have been introduced: the cyclic hardening parameters α and β. Unlike the parameters for the initial development of the model, all of which were standard soil properties, α and β have the disadvantage of not having a readily identified physical meaning or standard procedure for their measurement.

This extension of the model is based on two ideas. Firstly, the yield locus always moves with the current stress point. Thus, the cyclic loading is not modelled as a process in which an elastic domain, the size of which changes during the cycling, is developed. Secondly, for each half cycle, the material gets progressively stiffer. This is achieved by modifying the hardening function part of the constitutive relationship. For undrained loading a reduction in the rate at which the strain builds up with increasing number of cycles and a reduction in the rate at which the pore water pressure rises results. Thus, with a sufficiently large number of cycles the strain and effective stress state reach an equilibrium condition.

In Figures 11.14 and 11.15 the predictions of the revised model are compared with the undrained cyclic triaxial test results on Drammen clay reported by Andersen *et al.*[1] It is evident that the model provides a reasonable approximation to the observed behaviour. Modelling ranging from the normally consolidated condition to overconsolidation ratios of 10 is done with the one set of parameters. The total number of parameters that must be provided to specify a particular soil for this cyclic loading development of the model is seven.

NOTATION

c_u	undrained shear strength
e	void ratio
f	yield function
g	plastic potential
h	hardening function
p	mean principal effective stress, $(\sigma_1' + 2\sigma_3')/3$
p_0	value of p at the beginning of a half cycle
p_{cs}	value of p at a critical state point
q	$(\sigma_1' - \sigma_3')$
q_{cs}	value of q at a critical state point
q_p	change in q in the previous half cycle
u	pore water pressure
v	volumetric strain
v^r	recoverable volumetric strain
A	a parameter which is $+1$ for loading in compression and negative with a magnitude not necessarily unity for loading in extension
D	equivalent viscous damping ratio
G	apparent shear modulus
G_e	small strain elastic shear modulus
H	half cycle number
M	critical state friction parameter
α	a parameter related to ξ
β	a parameter related to ξ
ε_{ij}	plastic strain tensor
ε^p	plastic distortion (for conventional triaxial conditions: $\varepsilon = 2\,(\varepsilon_{\text{axial}} - \varepsilon_{\text{radial}})/3$)
ε_0^p	plastic distortion at the beginning of the current half cycle
$\varepsilon_{p_{cs}}^p$	plastic distortion for a $p = p_{cs}$ stress path
κ	minus slope of the swelling line in the e, $\ln p$ plane
λ	minus slope of the compression line
η	stress ratio q/p
η_0	value of η at the beginning of a half cycle
η_i	value of η for steady state cyclic loading between fixed stress limits with $p = p_{cs}$
ξ	cyclic hardening index
σ_{ij}	stress tensor
$\sigma_1'\,\sigma_3'$	major and minor principal effective stress

REFERENCES

1. Andersen, K. H., Pool, J. H., Brown, S. F., and Rosenbrand, W. F. Cyclic and static laboratory tests on Drammen clay, *J. Geotech. Eng. Div., Proc. ASCE*, **106**, 499–529, 1980.
2. Atkinson, J. H. Anisotropic elastic deformations in laboratory tests on undisturbed London clay, *Géotechnique*, **25**, 357–74, 1975.
3. Bertsch, P. and Findlay, N. N. Experimental study of corners, normality and Bauschinger effect in subsequent yield surfaces, *Proc. 4th U.S. Nat. Congr. Applied Mechanics*, **2**, 893–907, 1962.
4. Brown, S. F., Lashine, A. K. F., and Hyde, A. F. L. Repeated load triaxial testing of a silty clay, *Géotechnique*, **25**, 95–114, 1975.
5. Eekelen, H. A. M. van, and Potts, D. M. The behaviour of Drammen clay under cyclic loading, *Géotechnique*, **28**, 173–96, 1978.
6. France, J. W. and Sangrey, D. W. Effects of drainage on repeated loading of clays, *J. Geotech. Eng. Div., Proc. ASCE*, **103**, 769–85, 1977.
7. Idriss, I. M., Dobry, R., and Singh, R. D. Nonlinear behaviour of soft clays during cyclic loading, *J. Geotech. Eng. Div., Proc. ASCE*, **104**, 1427–47, 1978.
8. Lewin, P. I., and Burland, J. B. Stress probe experiments on saturated normally consolidated clay, *Géotechnique*, **20**, 38–56, 1970.
9. Morgenstern, N. R. Stress–strain relationships for soils in practice, *Proc. 5th Panamerican Conf. Soil Mech. Fdn. Engng.* Buenos Aires, 1975.
10. Mroz, Z., Norris, V. A., and Zienkiewicz, O. C. An anisotropic hardening model for soils and its application to cyclic loading, *Int. J. Num. Anal. Meth. Geomech.*, **2**, 203–21, 1978.
11. Namy, D. L. *An Investigation of Certain Aspects of Stress–Strain Relationships for Clay Soils*, Ph.D. thesis, Cornell University, 1970.
12. Ohmaki, S. Elastic behaviour of normally consolidated clay, *Proc. 3rd Australia–New Zealand–Conference on Geomechanics*, Wellington, **2**, 127–132, 1980.
13. Palmer, A. C., and Pearce, J. A. Plasticity theory without yield surfaces, *Symposium on Plasticity and Soil Mechanics* (A. C. Palmer, editor), Cambridge University Engineering Dept., 188–200, 1973.
14. Parry, R. H. G. Discussion Roscoe, K. H., Schofield, A. N., and Wroth, C. P., On the yielding of soils, *Géotechnique*, **8**, 180–83, 1958.
15. Parry, R. H. G., and Amerasinghe, S. F. Components of deformation in clays, *Symposium on Plasticity and Soil Mechanics* (A. C. Palmer, editor), Cambridge University Engineering Dept., 108–126, 1973.
16. Pender, M. J. Modelling soil behaviour under cyclic loading, *Proc. 9th Int. Conf. Soil Mech. Fdn. Engng.*, Tokyo, **2**, 325–31, 1977.
17. Pender, M. J. A unified model for soil stress-strain behaviour, *Proc. 9th Int. Conf. Soil Mech. Fdn. Engng., Tokyo, Specialty Session No. 9: Constitutive Equations of Soils*, 213–22, 1977.
18. Pender, M. J. A model for the behaviour of overconsolidated soil, *Géotechnique*, **28**, 1–25, 1978.
19. Pender, M. J. Cyclic mobility—a critical state model, *Symposium on Soils under Cyclic and Transient Loading*, Swansea, 325–33, 1980.
20. Prevost, J. H. Mathematical modelling of soil stress–strain–strength behaviour, *3rd International Conference on Numerical Methods in Geomechanics*, Aachen, **1**, 347–61, 1979.
21. Pyke, R. M. Nonlinear soil models for irregular cyclic loadings, *J. Geotech. Eng. Div., Proc. ASCE*, **105**, 715–26, 1979.

22. Roscoe, K. H., and Burland, J. B. On the generalised stress–strain behaviour of 'wet' clay, in Heyman, J., and Leckie, F. A. (eds.), *Engineering Plasticity*, Cambridge University Press, 535–609, 1968.
23. Schofield, A. N. and Wroth, C. P. *Critical State Soil Mechanics*, McGraw-Hill, 1968.
24. Seed, H. B., and Idriss, I. M. *Soil Moduli and Damping Factors for Dynamic Response Analysis*, report EERC-70–10 Earthquake Engineering Research Centre, University of California, 1970.
25. Taylor, G. I., and Quinney, H. The plastic distortion of metals, *Phil. Trans. Roy. Soc.*, **A230,** 323–62, 1931.
26. Taylor, P. W., and Bacchus, D. R. Dynamic cyclic strain tests on clay, *Proc. 7th Int. Conf. Soil Mech. Fdn. Engng.*, Mexico City, **1,** 401–9, 1969.
27. Valanis, K. C., and Read, H. E. A theory of plasticity for hysteretic materials— 1: Shear response, *Computers and Structures*, **8,** 503–10, 1978.
28. Wesley, L. D. The nature of anisotropy in soft clays, *Proc. 3rd Australia–New Zealand Conference on Geomechanics*, Wellington, **1,** 219–24, 1980.
29. Wroth, C. P. Some aspects of the elastic behaviour of overconsolidated clay, in Parry, R. H. G. (ed.), *Stress–Strain Behaviour of Soils* (Roscoe Memorial Symposium), Foulis, 347–61, 1971.
30. Wroth, C. P., and Zytynski, M. Finite element computations using an elasto-plastic soil model for geotechnical problems in soft clay, *Proc. 9th Int. Conf. Soil Mech. Fdn. Engng., Tokyo, Specialty Session on Computers in Soil Mechanics: Present and Future*, 193–243, 1977.

Soil Mechanics—Transient and Cyclic Loads
Edited by G. N. Pande and O. C. Zienkiewicz
© 1982 John Wiley & Sons Ltd

Chapter 12

Modelling and Analysis of Cyclic Behaviour of Sands

J. Ghaboussi and H. Momen

SUMMARY

A material model is presented in this chapter which is capable of representing the cyclic behaviour of sands reasonably accurately. A combination of isotropic and kinematic hardening is used for a yield surface which remains bounded by a rigid plastic failure surface. A non-associative flow rule is used and the volumetric strains are determined by a semi-empirical rule. Several drained and undrained cyclic triaxial tests are simulated and the results are compared with the experimental results. A reasonable level of accuracy is demonstrated in various aspects of the cyclic behaviour of sands. The role of membrane penetration is also investigated and it was shown that the membrane penetration can significantly influence the results of undrained cyclic triaxial tests.

12.1 INTRODUCTION

The capabilities for realistic dynamic analysis of geotechnical systems are rapidly developing. The major components of these methodologies are constitutive laws for representing the behaviour of soils under cyclic stresses. Recently, a number of researchers have developed and proposed various material models for representing the cyclic behaviour of soils.[2,6,8,16,17,22] Different approaches to modelling of cyclic behaviour of soils have been used with varying degrees of success. However. there is still insufficient verification of models and a shortage of good cyclic test results to be used in verification.

A material model is presented here, with proper verification, which can be used in finite element analysis of transient behaviour of soil structures under dynamic loading. For such an analysis to be successful, the material model must accurately represent the following aspects of the soil behaviour.

(1) hysteretic energy dissipation;
(2) accumulated irreversible deformations;
(3) volumetric deformations.

Material hysteresis determines the amount of the effective damping in soils.

Energy dissipation through hysteresis is probably the major portion of the actual damping in the geotechnical systems. The accumulated irreversible deformations provide a measure of damage and the potential volumetric strains cause pore pressure generation and possibility of liquefaction in masses of saturated cohesionless soils. The material model presented in this paper will be compared to experimental results and evaluated in terms of its capability in representing these three aspects of the behaviour.

The material model presented here is for completely drained behaviour of sands and is formulated in terms of effective stresses. Thus, the proposed model will not directly provide any information regarding the pore pressures. Such an effective stress material model is intended for use in a two-phase model of saturated sands in which the volumetric strains of sand structure and the pore fluid are coupled. In such an approach, which was used in Refs. 4, 5, 7, 9, the migration of the pore fluid and the redistribution of pore pressures during and after the transient loading are taken into account. However, in a completely undrained condition, when uniform pore pressures throughout the soil can be assumed, as in cyclic triaxial tests, it is possible to compute the pore pressures directly from the proposed model. Several undrained cyclic triaxial tests are simulated in this paper by assuming that the volume of the sample remains constant and the compressibility of the pore water can be neglected.

The proposed material model consists of a failure surface and a yield surface. The failure surface forms an asymptotic envelope of the yield surface which undergoes a combination of isotropic and kinematic hardening. A plastic potential is so chosen as to give only deviatoric plastic strain increments from the flow rule. The volumetric strain increments are determined on the basis of plastic work. The formulation of various aspects of the proposed material model is given in the following sections.

12.2 FAILURE AND FAILURE SURFACE

Among many definitions of failure of soil,[11] the one most appropriate in this study is a definition of failure as a state corresponding to very large plastic strains. Theoretically, it can be approached only asymptotically. This definition is very similar to the concept of critical state.[18] It allows for unlimited plastic flow at failure without any change in the state of stress. Further, the condition of no change in void ratio at failure can be met by the use of a specific potential surface which will be described later.

The failure surface acts as an asymptote to the yield surface. Due to 'hardening' the yield surface can asymptotically approach, but can never cross, the failure surface. It must be pointed out that this latest requirement excludes any strain softening and post-peak behaviour.

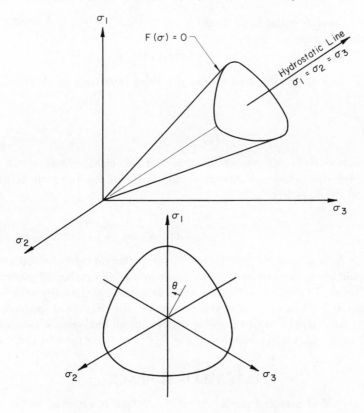

Figure 12.1   Failure surface in the principal stress space

Considerable research has been done regarding the actual shape of the failure surface. Most failure surfaces proposed in recent years[1,13,21] have more or less similar shapes. The failure surface used in this paper is of a conical shape with its axis coinciding with the space diagonal. The cross section of the failure surface on a plane normal to its axis is a round cornered hexagon of Mohr–Coulomb type. The failure surface is shown in Figure 12.1 and given by the following equation:

$$F(\sigma) = S_{ij}S_{ij} - \tfrac{2}{27}[MI_1 R(\theta)]^2 = 0, \tag{12.1}$$

in which

$$I_1 = \sigma_{kk},$$

$$S_{ij} = \sigma_{ij} - \tfrac{1}{3}\delta_{ij}I_1$$

and M is the failure material parameter.

The variable θ is the Lode angle in the octahedral plane, defined by the following equation.

$$\sin 3\theta = -(3\sqrt{3}J_3)/(2J_2^{1.5}),$$ (12.2)

where J_2 and J_3 are the second and the third invariants of the deviatoric stress:

$$J_2 = \tfrac{1}{2}S_{ij}S_{ij},$$ (12.3)

$$J_3 = \tfrac{1}{3}S_{ij}S_{jk}S_{ki}.$$ (12.4)

The function $R(\theta)$ determines the shape of the cross-section of the yield surface on the octahedral plane. A simple form for function $R(\theta)$ was proposed by Argyris et al.:[1]

$$R(\theta) = \frac{2K}{(1+K)-(1-K)\sin 3\theta},$$ (12.5)

in which K is a material parameter which is the ratio of the radii of the failure surface in compression and extension on the octahedral plane. This simple form of function $R(\theta)$ gives a concave yield surface for values of K less than 0.7. A more elaborate expression for $R(\theta)$ was proposed by Willem and Warnke[21] which remains convex for all the possible range of the parameter K. This latter expression for $R(\theta)$ has been used in this study.

12.3 YIELD SURFACE

In the proposed material model the yield surface is assumed to be of the same shape as the failure surface and the hardening rule is a combination of isotropic and kinematic hardening. Assuming the same shape for the yield surface as the failure surface is based on model requirements rather than experimental evidence; there is no concrete experimental evidence on the shape of the yield surface. However, since the failure surface acts as an asymptote for the hardening yield surface, it is required that the yield surface not intersect the failure surface. This requirement can be met by assuming that the yield is of the same shape as the failure surface.

Due to kinematic hardening the material develops an anisotropy. In the presence of this induced anisotropy, the state of the material is not determined solely on the basis of the state of stresses. It is also necessary to quantify the state of the induced anisotropy. To achieve this, a new tensor quantity α_{ij}, is introduced. This tensor is normalized so that:

$$|\alpha| = (\alpha_{ij}\alpha_{ij})^{1/2} = 1.$$ (12.6)

Geometrically, α_{ij} is directed along the axis of the yield surface, which in general does not coincide with the space diagonal. Thus, α_{ij} can be considered as the kinematic hardening parameter.

Any state of stress can be decomposed into a component, $\bar{I}\alpha_{ij}$, along the axis of the yield surface and a component, $\bar{S}_{ij}$, normal to the axis of the yield surface.

$$\bar{I} = \alpha_{ij}\sigma_{ij} \tag{12.7}$$

$$\bar{S}_{ij} = \sigma_{ij} - \bar{I}\alpha_{ij}. \tag{12.8}$$

It is evident that $\bar{I}$ and $\bar{S}_{ij}$ are anisotropic equivalents of the first invariant of stress, I_1, and the deviatroic stress, S_{ij}, in an isotropic condition. In fact, in an isotropic condition, when the exis of the yield surface coincides with the space diagonal, the tensor $\boldsymbol{\alpha}$ takes the form:

$$\alpha_{ij} = \frac{1}{\sqrt{3}} \delta_{ij}. \tag{12.9}$$

The yield surface in general is of the following form.

$$f(\sigma_{ij}, \alpha_{ij}. k) = 0. \tag{12.10}$$

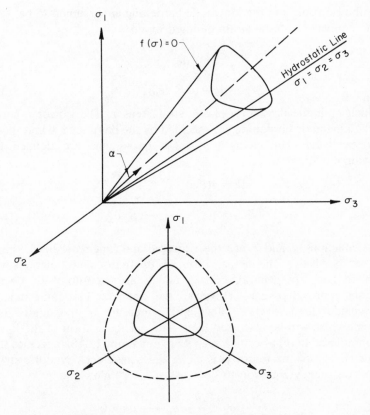

Figure 12.2 Yield surface in the principal stress space

In this equation α_{ij} and k are the kinematic and the isotropic hardening parameters, respectively. The specific form of the yield surface used in this paper is written in terms of the equivalent stress quantities defined earlier. The general shape of the yield surface is shown in Figure 12.2.

$$f = \bar{S}_{ij}\bar{S}_{ij} - [R(\bar{\theta})k\bar{I}]^2 = 0. \tag{12.11}$$

$R(\bar{\theta})$ is the same function as that used in the definition of the surface. However, $\bar{\theta}$ is defined in terms of the equivalent deviatoric stresses:

$$\sin 3\bar{\theta} = -(3\sqrt{3}\bar{J}_3)/(2\bar{J}_2^{1.5}), \tag{12.12}$$

$$\bar{J}_2 = \tfrac{1}{2}\bar{S}_{ij}\bar{S}_{ij} \tag{12.13}$$

$$\bar{J}_3 = \tfrac{1}{3}\bar{S}_{ij}\bar{S}_{jk}\bar{S}_{ki} \tag{12.14}$$

12.4 HARDENING RULE

Both the isotropic and the kinematic hardening are assumed to be governed by the equivalent plastic strain denoted by ξ.

$$\xi = \int d\xi, \tag{12.15}$$

$$d\xi = (de_{ij}^p \, de_{ij}^p)^{1/2}, \tag{12.16}$$

in which e_{ij}^p is the deviatoric plastic strain tensor. The isotropic hardening and the kinematic hardening are specified by the parameter k and the tensor α_{ij}, respectively. The changes in these parameters are defined by the following equations:

$$dk = g(\xi) \, d\xi, \tag{12.17}$$

$$d\alpha_{ij} = c(\xi) \frac{1}{(\bar{S}_{kl}\bar{S}_{kl})^{1/2}} \bar{S}_{ij} \, d\xi. \tag{12.18}$$

The functions g and c are the isotropic and kinematic hardening functions, respectively. The actual form of the hardening functions will be described later. In general, both g and c are monotonically decreasing functions approaching zero. However, the kinematic hardening function, c, also includes load reversal effect and a measure of the degree of load reversal, which will be discussed later. The particular form of the kinematic hardening used here, as can be seen from equation (12.18), specifies that the rate of change of the tensor $\boldsymbol{\alpha}$ is along the equivalent deviatoric stress $\bar{\mathbf{S}}$.

The consistency equation to be satisfied is as follows.

$$\frac{\partial f}{\partial \sigma_{ij}} \, d\sigma_{ij} + \frac{\partial f}{\partial \alpha_{ij}} \, d\alpha_{ij} + \frac{\partial f}{\partial k} \, dk = 0. \tag{12.19}$$

12.5 FLOW RULE

In the past associated flow rule has been used frequently in soil plasticity models. In a previous work, the present authors also used the associated flow rule.[8] However, it has been amply demonstrated in recent years that the use of the associated flow rule results in unrealistic volumetric strains. The strategy used in this paper is to define a potential function such that it yields only deviatoric strains. The volumetric strains are subsequently computed on the basis of plastic work. It is recognized that such a non-associated flow rule may result in violation of stability postulates of theory of plasticity for some possible, but unlikely, stress paths. However, as will be demonstrated in a later publication, it is possible to remedy this problem by introducing some restrictions on the hardening functions.

The flow rule is given by the following equation:

$$de_{ij}^{p} = \frac{1}{h}(n_{kl}\, d\sigma_{kl})n_{ij}',$$ (12.20)

in which h is the 'plastic modulus' which can be related to the hardening functions g and c. The unit tensors n and n' are normal to the yield surface and the potential surface, respectively:

$$n_{ij} = \frac{\partial f}{\partial \sigma_{ij}} \Big/ \left(\frac{\partial f}{\partial \sigma_{kl}} \frac{\partial f}{\partial \sigma_{kl}}\right)^{1/2},$$ (12.21)

$$n_{ij}' = \frac{\partial f'}{\partial \sigma_{ij}} \Big/ \left(\frac{\partial f'}{\partial \sigma_{kl}} \frac{\partial f'}{\partial \sigma_{kl}}\right)^{1/2}.$$ (12.22)

The potential function f' is not defined explicitly. It is only sufficient to define the unit normal vector to the potential function which is given by the following equation:

$$\frac{\partial f'}{\partial \sigma_{ij}} = \frac{\partial f}{\partial \sigma_{ij}} - \frac{1}{3}\left(\frac{\partial f}{\partial \sigma_{kl}} \delta_{kl}\right)\delta_{ij}.$$ (12.23)

Using equation (12.20), the increment of the equivalent plastic strain, $d\xi$, can be expressed as follows:

$$d\xi = \frac{1}{h}(n_{ij}\, d\sigma_{ij}).$$ (12.24)

Substitution of equations (12.17), 12.18), and (12.24) into the consistency equation (12.19) results in the following relation between the plastic modulus h and the hardening functions g and c.

$$h = \beta_1 g + \beta_2 c,$$ (12.25)

in which β_1 and β_2 are scalar values dependent on the state of stress

$$\beta_1 = \frac{2\,|\bar{S}|}{|\partial f/\partial \sigma|}\,\frac{|\sigma|^2}{\bar{I}}, \tag{12.26}$$

$$\beta_2 = \frac{2\,|\bar{S}|}{|\partial f/\partial \sigma|}\,R(\theta)\bar{I}. \tag{12.27}$$

Of the three material functions, h, g, and c, two must be specified and the third is computed from equation (12.25). In Ref. 9 for a cyclic bi-axial shear model, it was found convenient to specify the plastic modulus, h, and the kinematic hardening function, c. However, it is physically more meaningful to specify the plastic modulus and the isotropic hardening function and compute kinematic hardening function c from equation 12.25. The special form of the functions h and g, used here, are described in the next section.

12.6 HARDENING FUNCTIONS

The asymptotic nature of the failure condition, as defined earlier, requires that the plastic modulus h and the hardening functions c and g asymptotically approach zero with increasing equivalent plastic strain ξ. This requirement is, of course, strictly valid when the failure condition is approached by monotonically increasing stresses. Modelling of the cyclic behaviour requires that the influence of unloading on the plastic modulus and hardening functions be established through reasonable postulates. The following two postulates are introduced here:

(1) Isotropic hardening is independent of stresses.
(2) Upon unloading, the initial value of the plastic modulus and thus also the kinematic hardening function, depend on the 'degree of stress reversal' to be defined later.

An exponentially decaying function is used to define the isotropic hardening function as follows:

$$g(\xi) = G_0 \exp\left(-\frac{G_0}{k_f}\xi\right). \tag{12.28}$$

In this equation G_0 is a material parameter and k_f is the final value of k, which determines the final size of the yield surface at failure.

$$k_f = \int_0^\infty g(\xi)\,\mathrm{d}\xi. \tag{12.29}$$

There is some experimental evidence indicating that the final radius of the yield surface is approximately half the radius of the failure surface.

The plastic modulus is the product of two functions: a backbone decaying function h_1; and, a function h_2 which includes the effect of stress reversal.

$$h = h_1 h_2, \qquad (12.30)$$

$$h_1 = p' \left[1 - \frac{\alpha q}{MRp'} \right]^n \qquad (12.31)$$

$$h_2 = h_u + SR[H_0 f(\xi_u) - h_u], \qquad (12.32)$$

$$\alpha = \text{sign}\,(S_{ij} \bar{S}_{ij}),$$

in which

H_0 (initial plastic modulus) and n are material parameters
p' is the effective normal stress
$q = \sqrt{(3J_2)}$,
h_u = the value of h at the last unloading, and
ξ_u = the value of ξ at the last unloading.

The degree of stress reversal SR is given by the following equation:

$$SR = \tfrac{1}{2}[1 - \bar{S}_{ij}^r \bar{S}_{ij}^u / (|\bar{S}^r| \, |\bar{S}^u|)], \qquad (12.33)$$

in which

$\bar{S}_{ij}^u$ = the value of $\bar{S}_{ij}$ at the last unloading (transition from plastic to elastic state);
$\bar{S}_{ij}^r$ = the value of $\bar{S}_{ij}$ at the last reloading (transition from elastic to plastic state).

The symbol $|\cdot|$ denotes the vector length. It can be seen that the value of the 'degree of stress reversal' varies between zero and one. When $\bar{S}^r$ and $\bar{S}^u$ coincide, there is no stress reversal and $SR = 0$. When the angle between $\bar{S}^r$

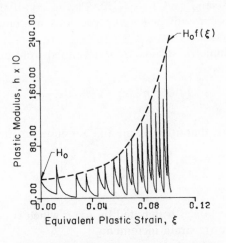

Figure 12.3 Plastic modulus in cyclic drained triaxial test with constant stress amplitude (cf. Figure 12.7)

and $\bar{S}^u$ is equal to π, there is complete stress reversal and $SR = 1$. The function $f(\xi)$ in equation (12.32) represents the general hardening of the effect of the stress reversal. The function $f(\xi)$ is a monotonically increasing function with an initial value of $f(0) = 1.0$.

The plastic modulus for a constant stress amplitude cyclic triaxial test is shown in Figure 12.3. In a cyclic triaxial test the stress reversals are complete and the peaks shown in Figure 12.3 occur at stress reversals.

12.7 VOLUMETRIC STRAINS

The flow rule defined earlier does not yield any plastic volumetric strains, since the outward normal to the potential function is also normal to the hydrostatic line. The volumetric strain increments are divided into two parts: those resulting from changes in deviatoric stresses (or dilatancy volumetric strains), $d\varepsilon_d^p$; and, the volumetric strain increments resulting from changes in the mean normal stress (or consolidation volumetric strains), $d\varepsilon_c^p$:

$$d\varepsilon^p = d\varepsilon_c^p + d\varepsilon_d^p. \tag{12.34}$$

The consolidation volumetric strains are computed from a standard equation as follows:

$$d\varepsilon_c^p = \frac{0.434(C_c - C_s)}{(1 + e_0)p'} \, dp', \tag{12.35}$$

in which e_0 is the initial void ratio, C_c and C_s are the virgin and rebound compressibilities of sand and $p' = \sigma'_{kk}/3$. Plastic consolidation volumetric strain increments occur only when the previous maximum mean normal stress, p', is exceeded.

The dilatancy volumetric strains are determined from plastic work which is expressed as follows:

$$dw^p = p' \, d\varepsilon_d^p + \sqrt{(\tfrac{2}{3})}q \, d\xi, \tag{12.36}$$

in which $q = \sqrt{(3J_2)}$.

It has been shown[15] that for a given sand a unique function can be defined as follows:

$$S(\xi) = \int \frac{1}{p'} \, dw^p. \tag{12.37}$$

The slope of the function S is denoted by $\mu(\xi)$, which is used to determine the dilatancy volumetric strain increments

$$d\varepsilon_d^p = \left[\mu(\xi) - \sqrt{\left(\frac{2}{3}\right)\frac{q}{p'}} \right] d\xi, \tag{12.38}$$

in which

$$\mu(\xi) = dS(\xi)/d\xi. \qquad (12.39)$$

In this equation $\mu(\xi)$ is treated as a material property function.

The equation (12.38) is only valid for the case of triaxial stress condition where the vectors S and de^p are in the same direction. Thus, the following relation is true for triaxial stress condition:

$$\sqrt{(\tfrac{2}{3})}q \, d\xi = |S| \, |de^p|. \qquad (12.40)$$

In a general stress condition, because of the shape of the cross-section of the yield surface, the two vectors S and de^p need not be in the same direction. Thus, for a general stress condition, the dilatancy volumetric strains are computed from the following relation:

$$d\varepsilon_d^p = \mu(\xi) \, d\xi - \frac{1}{p'} \, S_{ij} \, de_{ij}^p. \qquad (12.41)$$

An implied assumption is being made, that the function $\mu(\xi)$ determined from triaxial stress conditions is also valid for an arbitrary stress path. This assumption requires experimental verification. Experimentally determined forms of the function $\mu(\xi)$ will be discussed in the next section.

12.8 MATERIAL PARAMETERS

A brief discussion of the material parameters required for the proposed model is given in this section. The influence of each parameter on the response of the material model is discussed briefly. The material parameters for the proposed model can be classified into the following categories: elastic moduli, failure parameters, hardening (or deformation) parameters, and dilatancy parameters. These groups of parameters are discussed in the following.

Elastic moduli.   In the elastic region, the material is assumed to be isotropic. Thus two elastic parameters are required, such as Young's modulus E and Poisson's ratio v. The elastic moduli are computed from the slope of the unloading portion of the stress–strain curves; the initial slope of the stress–strain curves (initial moduli) include both elastic and plastic deformations. The unloading moduli generally increase with mean effective normal stress, p', and decrease with increasing equivalent plastic deformations, ξ (degradation). On the basis of experimental data the following equation is used to

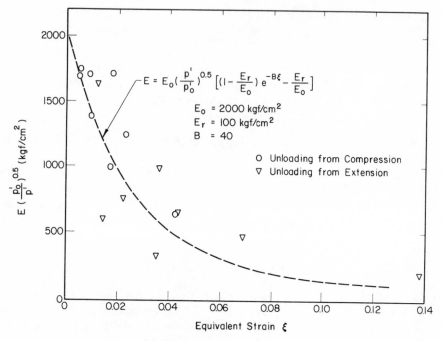

Figure 12.4   Degradation of elastic (unloading) modulus

represent the increase with p' and degradation of the elastic modulus:

$$E = E_0\left(\frac{p'}{p_0'}\right)^m\left[\left(1-\frac{E_r}{E_0}\right)\exp\left(-B\xi\right)+\frac{E_r}{E_0}\right], \qquad (12.42)$$

in which

E_0 = initial value of elastic modulus at effective mean normal stress p_0',
E_r = residual value of elastic modulus at effective mean normal stress p_0',
and, m and B are material parameters. A value of $m \approx 0.5$ in most cases provides a good fit of measured data. Thus the important parameters are E_0 and the degradation parameter B. Shown in Figure 12.4 are typical experimental data on degradation of elastic modulus compared with equation (12.42). The experimental data points represent unloading elastic moduli from compression and extension sides in a cyclic triaxial test.

Failure parameters.   Two parameters M and K define the failure surface. The parameter M defines the failure in triaxial compression. The parameter K is the ratio of shear strengths at triaxial extension and triaxial compression at the same value of the effective mean normal stress. The two parameters are related to the internal angle of friction through the following equations:

$$M = \frac{6 \sin \phi_u}{3 - \sin \phi_u},$$ (12.43)

$$K = \frac{\sin \phi_u'(3 - \sin \phi_u)}{\sin \phi_u(3 + \sin \phi_u')},$$ (12.44)

in which ϕ_u and ϕ_u' are the ultimate angles of friction in compression and extension, respectively.

Hardening parameters. The three parameters required for the isotropic hardening function and the plastic modulus are G_0, H_0, and n. It may be noted that for $n = 2$ a hyperbolic equivalent stress–strain relation is obtained as follows:

$$\frac{q}{p'} = \frac{MH_0\xi}{M + H_0\xi}.$$ (12.45)

However, in some cases a value of n slightly less than 2 results in a better fit of experimental data in a monotonic stress–strain relation. The initial plastic modulus is determined from the initial modulus E_i when loading starts from the hydrostatic stress condition.

$$H_0 = 1/(1/E_i - 1/E_0)$$ (12.46)

The parameter G_0 plays an important role in the cyclic behaviour of the material. It controls the general hardening and reduction of hysteresis with the increase of number of cycles. However, at present no direct method of determinning the parameter G_0 is available; it can only be determined by trial and error.

Dilatancy parameter. The material function controlling the dilatancy is $\mu(\xi)$, which is the slope of the plot of the normalized plastic work, S, versus the equivalent plastic strain, ξ. Such a plot for a number of standard triaxial tests[19] in drained condition is shown in Figure 12.5. It can be seen that for values of ξ greater than 2% S can be assumed to be a linear function of ξ, thus μ is constant. However, for values of ξ smaller than 2% the slope of S versus ξ is decreasing. In general, it is found satisfactory to represent the S versus ξ curve with three linear segments. Thus, μ will have three constant values for three ranges of ξ.

The experimental values of S given in Figure 12.5 are lower for triaxial extension than for triaxial compression; the ratio of μ in extension and compression is approximately equal to K, which is the ratio of the shear strength in extension and compression. Thus the appropriate values of the dilatancy parameter to be used is $R(\theta)\mu$ in which μ is for triaxial compression.

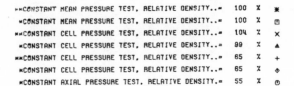

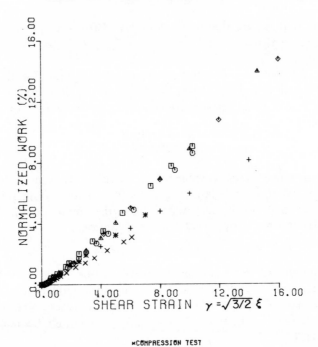

Figure 12.5   Normalized plastic work, S, for a number of tests reported in Ref. 19

12.9   SIMULATION OF DRAINED CYCLIC TRIAXIAL TESTS

The physical properties and the material parameters used in the simulation of cyclic drained tests are shown in Table 12.1. These parameters were estimated from the monotonic portion of the triaxial test in compression and extension. The experimental results are reported in Refs. 19 and 20. The stress–strain curves and volumetric strains for triaxial compression and extension computed from the proposed model are compared with the experimental curves in Figure 12.6. As can be seen, the model simulation results are reasonably close to measured values.

Table 12.1 Physical properties and material parameters for drained tests

Type of test	Comp.	Cyclic	Cyclic
Number of figure in which the simulation results are shown	12.6	12.7	12.8
Physical properties			
Void ratio e_0	0.722	0.750	0.740
initial relative density D_r	0.65	0.60	0.62
consolidation pressure P_0 (kg/cm^2)	2.0	2.0	2.0
Elastic parameters			
E_0 (kg/cm^2)	1800	1700	1700
E_r (kg/cm^2)	100	100	100
B	55	55	55
ν	0.20	0.20	0.20
Failure parameters			
M	1.55	1.55	1.55
K	0.65	0.71	0.71
Deformation parameters			
H_0	280	280	280
n	2.0	2.0	2.0
G_0	18	18	18
Compressibilities			
C_c	0.042	0.045	0.045
C_s	0.014	0.014	0.014
Dilatancy parameters			
$\mu \begin{cases} \xi < 0.005 \\ 0.005 < \xi < 0.020 \\ \xi > 0.02 \end{cases}$	$\begin{matrix} -0.25 \\ 0.85 \\ 0.95 \end{matrix}$	$\begin{matrix} -0.20 \\ 0.85 \\ 0.95 \end{matrix}$	$\begin{matrix} -0.20 \\ 0.85 \\ 0.95 \end{matrix}$

The results of the simulation of a cyclic triaxial test are summarized in Figure 12.7. In this test the lateral stress is kept constant and the axial stress is cycled between two constant stress levels. Figure 12.7(a) shows the stress ratio q/p' versus shear strain $\gamma = \varepsilon_a - \varepsilon_r$ as computed from the material model. It is interesting to note the general hardening of the material and the gradual decrease of the hysteresis with the number of cycles. The first and ninth cycle stress–strain curves are compared with the experiment in Figure 12.7(b). The model simulation results for the first cycle are reasonably close to the experimental results. However, the accumulation of errors in each cycle has caused a shift of the ninth cycle stress–strain curve away from the experimental curve. Although the amount of the hysteresis at the ninth cycle is represented adequately, the amount of accumulated irreversible deformations are significantly underestimated. The volumetric deformation from the model simulation are compared with the experimental results in

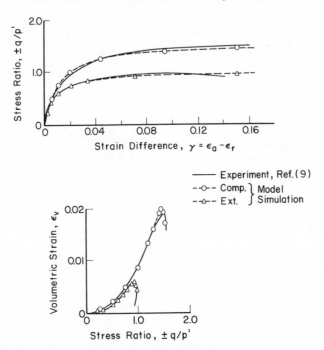

Figure 12.6 Model simulation of drained triaxial tests

Figure 12.7(c). Although the details differ from the measured values, overall it can be seen that the simulated volumetric strains are reasonably accurate.

The simulation results are compared in Figure 12.8 to experimental results for a cyclic test in which the axial stress amplitude is increased after every cycle. All the three important aspects of the cyclic behaviour—namely, hysteretic damping, accumulated irreversible deformations and volumetric strains—seem to be accurately represented in this test.

12.10 MEMBRANE PENETRATION EFFECTS IN UNDRAINED CYCLIC TRIAXIAL TESTS

Cyclic triaxial tests on saturated sands are often performed to study the liquefaction potential. In these tests each stress cycle causes a certain amount of pore pressure increase and the number of cycles required to cause 'liquefaction' of 'cyclic mobility' is used as an index of liquefaction potential. The validity of such a test result to some extent rests on the assumption that the sample is in undrained condition and the total volume of the sample remains constant, if the compressibility of the pore water is neglected and the sample is fully saturated. However, if no additional measures are taken to prevent membrane penetration, the total volume of the sample cannot be assumed to remain constant and the assumption of undrained condition is not strictly valid. This study is an attempt to investigate the influence and the relative significance of membrane penetration on the excess pore pressures in cyclic triaxial tests.

The effective stress path for a cyclic triaxial test reported in Ref. 10 is shown in Figure 12.12(a). Three distinct stages are evident from this figure.

Stage 1   This stage usually comprises the first cycle. Significant effective stress reduction and pore pressure increase occur at the first cycle. Upon loading from the isotropic stress condition, the sample exhibits permanent volume change tendency under shear stress (dilatancy).

Stage 2   After the first cycle, the pore pressure increase per cycle remains constant. A striking feature of the effective stress path at this stage is that most of the effective stress decrease occurs in unloading, while in reloading the effective stress remains constant. In these cycles the sample does not have a significant tendency for permanent volume change due to dilatancy. All the effective stress changes occur due to changes in the total mean pressure.

Stage 3   This stage starts when the effective stress paths has reached a 'near failure condition', or 'initial liquefaction'.[7] The stage after initial liquefaction is characterized by rapid changes in the effective stress.

Prior to application of any stress difference, $q = \sigma_1 - \sigma_3$, the membrane has penetrated into the sample and the amount of this penetration is dependent on the difference between the cell pressure and the pore pressure, $\sigma_3 - u$. As the axial stress is increased, the sample is at stage one and the tendency for volume decrease causes an increase in the pore pressure. Since the cell pressure is constant and the pore pressure is increasing, the membrane penetration tends to decrease from its initial value. The net effect is that the generated pore pressures are lower and the strength higher than the case in which the membrane penetration is prevented.

The influence of the membrane penetration in the second stage is the opposite of its effect in the first stage. In unloading, there is no tendency for volume deformation in the sample. Thus, assuming the pore water to be incompressible, if the membrane penetration was prevented, the decrease in the total mean pressure would be transferred to the pore pressure and the effective stress would remain constant. However, in standard tests, when no measures are taken to reduce the membrane penetration, as the axial stress is reduced, the pore pressure decreases and the difference between the cell pressure and the pore pressure increases, resulting in more membrane penetration into the sample. The net result is that only part of the total pressure decrease is transferred to the pore pressure and the remainder results in effective stress decrease.

On reloading, the pore pressure is increasing and the membrane tends to move out. Since the sample does not have any volume change tendency, as a result the pore pressure increase is equal to the total mean pressure increase and the effective stress remains constant.

In addition to the above effects in the second stage, there is some dilatancy volume change tendency in the sample towards the end of the unloading and reloading. The presence of this effect is evidenced by a slight aberration from the straight line in the effective stress path towards the end

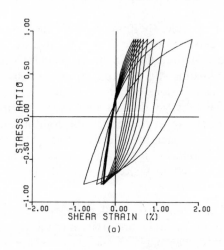

(a)

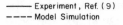

——— Experiment, Ref. (9)
– – – Model Simulation

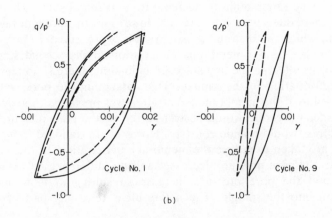

(b)

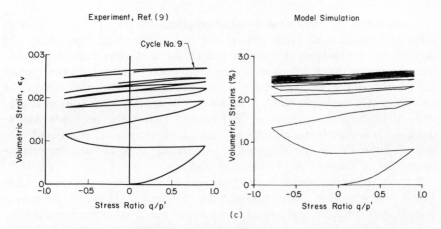

Figure 12.7 (a) Model simulation of cyclic drained triaxial test; (b) comparison of first and ninth cycle with experiment; (c) comparison of volumetric strains with experiment

of unloading and reloading. In these regions, as in the first stage, the pore pressure is generally underestimated.

Finally, in the third stage the effect of the membrane penetration is similar to those in the first and second stages. However, the actual rate of pore pressure change is so high that the influence of the membrane penetration on the effective stress path may be negligible.

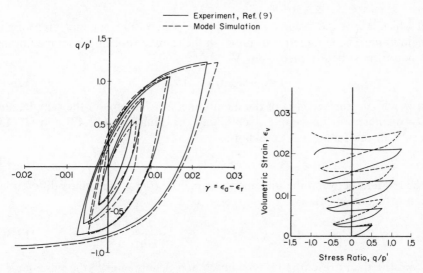

Figure 12.8 Comparison of model simulation and experiment for a drained cyclic triaxial tests

In summary, the membrane penetration affects the pore pressure development in an undrained cyclic triaxial test in two opposing ways. When the sample has a tendency for volume decrease and thus a pore pressure increase under shear deformation, the membrane penetration causes smaller pore pressure changes than if the member penetration was to be prevented. On the other hand, when there is no tendency for volume change upon unloading, the membrane penetration causes the pore pressure to decrease less than if the membrane penetration was to be prevented. The overall influence of the membrane penetration depends on the relative magnitude of the above-mentioned two effects.

To the authors' knowledge no experimental study of the effect of membrane penetration in cyclic triaxial tests is available. However, in triaxial compression tests the influence of membrane penetration has been studied experimentally. Of particular interest are two recent works. Lade and Hernandez[14] used brass platelets inside the membrane to reduce the membrane penetration. A similar effect was achieved by Kiekbusch and Schuppener[12] by injecting liquid rubber inside the membrane. Both these studies show that by reducing membrane penetration higher pore pressures develop in the triaxial compression test, similar to the first stage in cyclic triaxial test. The numerical values given in Ref. 12 are used here to estimate the influence of the membrane penetration in cyclic triaxial tests.

A normalized membrane penetration volume, S, is defined in Ref. 12 as follows:

$$S = V_m/\log \sigma_3', \tag{12.47}$$

in which V_m is the volume of membrane penetration per unit area of the membrane. The volumetric strain ε_m in the sample resulting from membrane penetration is determined from V_m:

$$\varepsilon_m = V_m A/V = 4V_m/D, \tag{12.48}$$

in which A = surface area of the membrane, V = Volume of the sample, and D = diameter of the sample. Combining the equations (12.47) and (12.48) results in the following expression:

$$\varepsilon_m = 4S \log \sigma_3'/D. \tag{12.49}$$

The relation between the increments of ε_m and σ_3' is obtained by differentiation of equation (12.49):

$$d\varepsilon_m = C_m \, d\sigma_3', \qquad C_m = \frac{1.737S}{D\sigma_3'}. \tag{12.50}$$

From the data presented by Kiekbusch and Schuppener,[12] the parameter S appears to be almost independent of σ_3' and strongly depends on the grain size.

The total volume of the sample remains constant if the compressibility of the pore water is neglected. Thus, the sum of the volumetric strain in sand, $d\varepsilon_s$, and the volumetric strain due to membrane penetration, $d\varepsilon_m$, must be zero:

$$d\varepsilon_s + d\varepsilon_m = 0. \tag{12.51}$$

The volumetric strain in the sand has two components. One is due to changes in the mean effective stress, $p' = (\sigma_1 + 2\sigma_3')/3$. The second component of volumetric strain, $d\varepsilon_d$, is due to changes in the octahedral shear stress, (dilatancy):

$$d\varepsilon_s = C\,dp' + d\varepsilon_d, \tag{12.52}$$

in which

$$C = \frac{0.434C_s}{(1 + e_0)\sigma_m'}. \tag{12.53}$$

C_s is the compressibility of sand and e_0 is the initial void ratio. Substituting equations (12.50) and (12.52) into (12.51) and using the effective stress relations $p' = p - u$ and $\sigma_3' = \sigma_3 - u$, leads to the following equation for the increment of pore pressure.

$$du = \frac{C\,dp + C_m\,d\sigma_3 + d\varepsilon_d}{C + C_m}. \tag{12.54}$$

In a standard triaxial test the cell pressure is kept constant; thus $d\sigma_3 = 0$ and the pore pressure is given by the following equation:

$$du = \frac{C\,dp + d\varepsilon_d}{C + C_m}. \tag{12.55}$$

In unloading there is no volume change tendency, thus $d\varepsilon_d = 0$. If the membrane penetration was prevented, $C_m = 0$, then $du = dp$, which implies no change in the effective stress. However, in the cyclic triaxial test result reported by Ishihara, Tatsuoka, and Yasuda[10] (Figure 12.12(a)), which apparently included membrane penetration effects, the slope of dq/dp' in unloading is approximately equal to $7:1$. The same phenomenon appears in tests performed by Castro,[3] but with a slope of approximately $6:1$ in unloading (Figure 12.13(a)). In both tests the reloading is almost vertical. Using equation (12.55) with $d\varepsilon_d = 0$, the unloading slope can be determined and is given by the following equation:

$$\frac{dq}{dp'} = 3\left(\frac{C}{C_m} + 1\right). \tag{12.56}$$

Table 12.2 Physical properties and material parameters for undrained tests

References where the test results are reported.	10	10	3	3
Type of test	comp. ext. cyclic	cyclic	comp.	cyclic
Simulation results shown in Figure	12.10, 12.11	12.12, 12.14, 12.15	12.9	12.13
Physical properties				
void ratio e_0	0.750	0.737	0.623	0.770
mean grain size D_{50} (mm)	0.44	0.44	0.18	0.18
sample diameter D (cm)	5.00	5.00	3.56	3.56
consolidation pressure p_0' (kg/cm^2)	3.0	2.1	4.0	4.0
Elastic parameters				
E_0 (kg/cm^2)	2500	2000	3500	2000
E_r (kg/cm^2)	100	100	100	100
B	40	40	40	40
ν	0.20	0.20	0.20	0.20
Failure parameters				
M	1.55	1.55	1.33	1.33
K	0.61	0.61	0.75	0.70
Deformation parameters				
H_0	510	420	1200	1400
n	2.0	2.0	2.0	2.0
G_0	30	30	30	30
Compressibilities				
C_c	0.042	0.042	0.042	0.042
C_s	0.014	0.015	0.012	0.007
Dilatancy parameters				
$\mu \begin{cases} \xi < 0.002 \\ 0.002 < \xi < 0.007 \\ \xi > 0.007 \end{cases}$	0.10 0.50 0.77	0.30 0.50 0.70	0.10 0.50 0.75	0.45 0.60 0.72
Membrane penetration Parameter S	0.0044	0.0044	0.0023	0.0023

The physical properties of the Fuji sand used by Ishihara, Tatsuoka, and Yasuda and the sand used by Castro have been extracted from Refs. 10 and 3 and given in Table 12.2. Also shown in this table are the values of the membrane penetration constant S determined from Ref. 12 for the D_{50} values in Table 12.2. Using these values the slope of the unloading effective stress path computed from equation (12.56) for Ishihara's and Castro's tests are 9 : 1 and 7 : 1 respectively. These values compare well with the actual measured slopes.

12.11 SIMULATION OF UNDRAINED CYCLIC TRIAXIAL TESTS

The purpose of the simulation of cyclic triaxial tests presented in this section is twofold; to demonstrate the capability of the proposed model in undrained condition and to investigate the influence of membrane penetration in undrained tests. The tests simulated are those reported by Ishihara, Tatsuoka, and Yasuda[10] and Castro.[3] The physical properties of sands are reported in the above references and material properties used in model simulation are determined from monotonic test results. The material parameters which could not be directly determined from test results are estimated. The physical properties and the material parameters are summarized in Table 12.2.

The results of the simulation of tests, in which the axial stress is increasing monotonically while the lateral stress is kept constant, are summarized in Figures 12.9 and 12.10. In the q, p' plane the total stress path has a slope of 3 to 1. The effective stress path and the stress–strain curves are compared with the experimental results. The model simulation is performed with and without the membrane penetration. It can be seen that the results of the

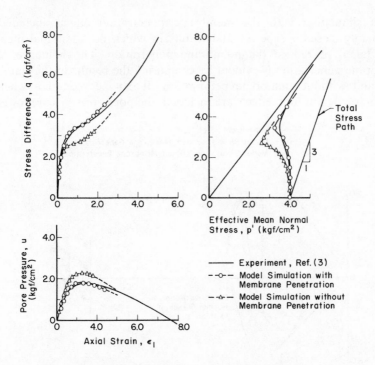

Figure 12.9 Simulation of undrained triaxial tests

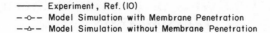

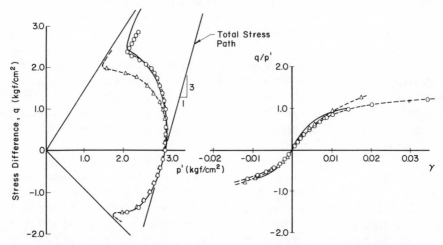

Figure 12.10 Simulation of undrained triaxial compression and extension test

model simulation with the membrane penetration effects included are reasonably close to experimental results in which no additional measures were taken to reduce the membrane penetration. The influence of the membrane penetration is evident by comparing the results of the simulation with and without membrane penetration. It can be seen that when the membrane penetration effects are included, the pore pressures are lower and

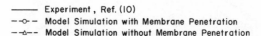

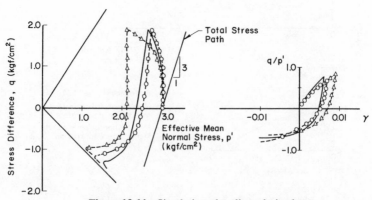

Figure 12.11 Simulation of cyclic undrained test

the shear strength is higher than those when the membrane penetration effects are excluded. It is also interesting to note from Figure 12.10 that the membrane penetration has much more significant influence in triaxial compression than in triaxial extension. This is due to the fact that in triaxial compression the pore pressure is increasing from the beginning of application of axial stress, where as in triaxial extension for up to $q = 1.0$ kg/cm^2 (65% mobilized strength) there is very little pore pressure increase; in fact pore pressure decreases before it increases.

Direct experimental investigation of the effect of membrane penetration in triaxial extension is not available at the present. However, the trend shown in the model simulations appears to be reasonable.

The results of simulation of one cycle of stress are shown in Figure 12.11.

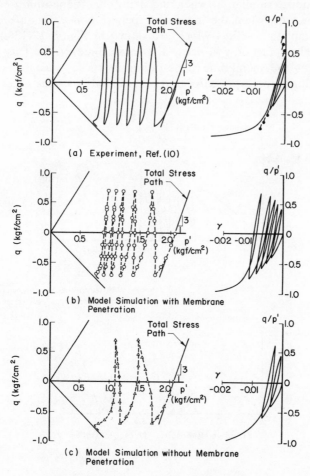

Figure 12.12 Simulation of cyclic undrained test

In this case also the result of simulation with the membrane penetration effects included is close to the experimental results. The two opposing effects of membrane penetration on cyclic tests, as discussed in the previous section, can be clearly seen from Figure 12.11. When the membrane penetration effects are excluded the pore pressure is higher in loading, but in unloading the pore pressure decreases with the same rate as the total stress and the effective pressure remains constant. However, over the whole cycle when the membrane penetration effects are excluded the pore pressure is higher than when the membrane penetration effects are included.

The results of model simulation of a cyclic triaxial test reported by Ishihara, Tasuoka, and Yasuda[10] are summarized in Figure 12.12. As can be seen, the initial liquefaction occurs after five cycles in the experiment and also in the model simulation when the membrane penetration effects are included. However, when the membrane penetration effects are excluded, the model simulation results which are shown in Figure 12.12(c) indicate that the initial liqufaction will occur after only three cycles. This is a clear indication that the membrane penetration in cyclic triaxial tests causes underestimation of pore pressures and overestimation of the cyclic strength

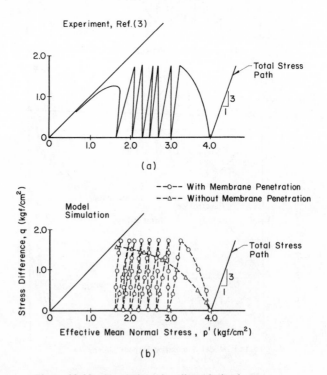

Figure 12.13 Simulation of cyclic undrained test

or the number cycles required to cause liquefaction. Direct experimental evidence on the effects of membrane penetration in cyclic triaxial tests is not available. However, indirect experimental evidence is available based on comparison of tests performed on samples with different diameters. As discussed earler, the membrane penetration effects are inversely proportional to the sample diameter. Thus, for smaller diameter samples the measured liquefaction strength will be higher. Experimental data to this effect has been presented by Lade and Hernandez.[14]

The results of the model simulation of a cyclic test reported by Castro[3] are shown in Figure 12.13. The sand in this test is in a very loose state, with a relative density of 0.33. Since the stress cycles are on the compression side, it can be expected that the influence membrane penetration will be more

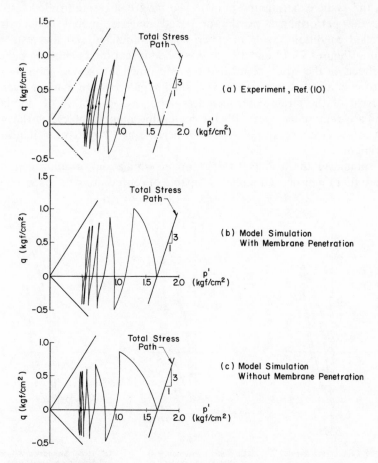

Figure 12.14 Effective stress path for strain-controlled cyclic undrained test

significant in this test than the previous cyclic test, which was stressed in compression and extension. This is due to the fact that, as observed earlier, the membrane penetration effects are more significant in compression than in extension. It is also observed here that when the membrane penetration effects are included the results of the model simulation are very similar to the experimental stress path, except when the stress path approaches the initial liquefaction, where the experimental results exhibit a significant softening—no doubt due to the very low density of the sand used in this test. When the membrane penetration effects are excluded, the initial liquefaction is reached in the first loading. This indicates that the true undrained shear strength (when the membrane penetration is prevented) is lower than the applied stress.

All the cyclic experiments simulated so far are stress-controlled tests. To verify the performance of the proposed material model under strain-controlled conditions, such an experiment is simulated and the results are shown in Figures 12.14 and 12.15. It is evident that the membrane penetration effects in the strain-controlled test are much less severe than those in the stress-controlled tests. However, similar patterns are evident over the first cycle of the strain-controlled test. The overall quality of simulation results are reasonable, but with the increasing number of cycles the amplitude of the stress ratio q/p' seems to increase at a higher rate than in the experiment.

In summary, the results of the model simulation of undrained tests indicate that the proposed model is capable of representing the experimental

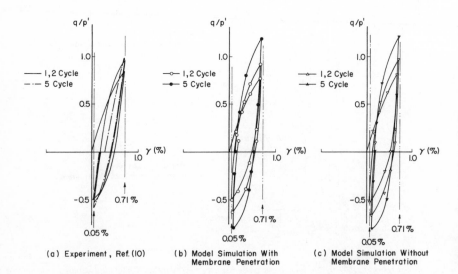

Figure 12.15 Stress–strain diagram for strain-controlled cyclic undrained test

results with a reasonable degree of accuracy only if the membrane penetration effects are included. It was also shown that membrane penetration can significantly influence the results of undrained tests. However, further direct experimental data on the influence of membrane penetration is needed before this effect can be properly quantified.

ACKNOWLEDGEMENT

This research was started, and a major part of it carried out, under National Science Foundation Grant No. ENV-76-00626. Part of the computer cost was provided by the Research Board of University of Illinois through the Civil Engineering Department. This support is gratefully acknowledged.

REFERENCES

1. Argyris, J. H., *et al. Recent Developments in Finite Element Analysis of PCRV*, 2nd International Conference, SMIRT, Berlin, 1973.
2. Bazant, E. P., and Krizek, R. J., Endochronic constitutive law for liquefaction of sand, *Proc ASCE*, **102** (EM2), 225–38, April 1976.
3. Castro, G., *Liquefaction of Sands*, Harvard Soil Mechanics Series, No. 81. Cambridge, Mass., 1969.
4. Ghaboussi, J., and E. L. Wilson, Variational formulation of dynamics of fluid saturated porous elastic soilds, *Proc. ASCE*, **98** (EM4), 947–63, August 1972.
5. Ghaboussi, J., and E. L. Wilson, Liquefaction analysis of saturated granular soils, *Fifth World Conference on Earthquake Engineering*, Rome, June 1973.
6. Ghaboussi, J., and Karshenas, K., On the finite element analysis of certain material nonlinearities in geomechanics, *Proc. Int. Conf. Finite Elements in Nonlinear Solid and Structural Mechanics*, Geilo, Norway, August 1977.
7. Ghaboussi, J., and Dikmen, U. S., Liquefaction analysis of horizontally layered sands, *J. Geotech. Eng. Div. ASCE*, **104** (GT3), 341–56, March 1978.
8. Ghaboussi, J., and Momen, H., Plasticity model for cyclic behavior of sands, *3rd International Conference on Numerical Methods in Geomechanics*, Aachen, April 1979.
9. Ghaboussi, J., and Dikmen, U. S., *LASS-III, Computer Program for Seismic Response and Liquefaction of Layered Ground Under Multi-Directional Shaking*, Report No. UILU-ENG-79-2012, Department of Civil Engineering, University of Illinois at Urbana-Champaign, Urbana, Illinois, July 1979.
10. Ishihara, K., Tatsuoka, F., and Yasuda, S., Undrained deformation and liquefaction of sand under cyclic stresses, *Soils and Foundations*, **15** (1), March 1975.
11. Karshenas, M., and Ghaboussi, J., *Modeling and Finite Element Analysis of Soil Behavior*, Report No. UILU-ENG-79-2020, Department of Civil Engineering, University of Illinois at Urbana-Champaign, Urbana, Illinois, December 1979.
12. Kiekbusch, M., and Schuppener, B., Membrane penetration and its effect on pore pressures, *J. Geotech. Eng. Div.*, *ASCE*, **103** (GT11), 1267–1279, November 1977.
13. Lade, P. V., and Duncan, J. M., Elasto-plastic stress–strain theory for cohesionless soil, *J. Geotech. Eng. Div.*, *ASCE*. **101** (GT10), pp. 1037–53, October 1975.
14. Lade, P., and Hernandez, S., Membrane penetration effects in undrained tests, *J. Geotech. Eng. Div.*, *ASCE*, **103** (GT2), 109–125, February 1977.

15. Moroto, N., A new parameter to measure degree of shear deformation of granular material in triaxial compression tests, *Soils and Foundations* **16** (4), 1–11, December 1976.

16. Mroz. Z., Norris, V. A., and Zienkiewicz, O. C., An anisotropic hardening model for soils and its application to cyclic loading, *Int. J. Num. Anal. Meth. Geomech.*, **2**, 203–21, 1978.

17. Sandler, I. S., and Baron, M. L. Recent developments in the constitutive modeling of geological materials, *3rd International Conference on Numerical Methods in Geomechanics*, Aachen, April 1979.

18. Schofield, A. N., and Wroth, C. P., *Critical State Soil Mechanics*, McGraw-Hill, London, 1968.

19. Tatsuoka, F. PhD. Thesis, University of Tokyo, 1972.

20. Tatsuoka, F., and Ishihara, K., Drained deformation of sand under cyclic stresses reversing direction, *Soils and Foundations*, **14**(3), December 1974.

21. Willem, K. J., and Warnke, E. P. Constitutive model for triaxial behavior of concrete, *Seminar on Concrete Structures Subjected to Tri-axial Stresses*, ISMES, Bergamo, Italy 1974.

22. Zienkiewicz, O. C., Chang, C. T., and Hinton, E. Nonlinear seismic response and liquefaction, *Int. J. Num. Anal. Meth. Geomech.*, **2**, 381–404, 1978.

Soil Mechanics—Transient and Cyclic Loads
Edited by G. N. Pande and O. C. Zienkiewicz
© 1982 John Wiley & Sons Ltd

Chapter 13

A Constitutive Model for Soil under Monotonic and Cyclic Loading

R. Nova

SUMMARY

An overview of the behaviour of soils in triaxial compression suggests that monotonic loading can be modelled by means of an elastic–plastic constitutive law allowing for isotropic hardening and hardening–softening transition. It is shown experimentally that the constitutive law should be non-associate. The hardening function is assumed to depend on the previous history of the material not only through the plastic volumetric strain but, for sands, even through the plastic deviatoric strain.

Several predictions of the mathematical model in the most common triaxial tests are given in closed form. It is shown that experimental results on normally consolidated clay and on virgin sand are well matched in drained and undrained tests. Even the phenomenon of liquefaction under monotonic loading can be described by the model.

The behaviour of soil in unloading–reloading, i.e. within the current yield locus, is reckoned to be 'paraelastic', i.e. path independent between suitably defined stress reversal points. The constitutive law in this range, its domain of validity and the memory of the material are discussed. It is shown that the model accounts for hysteresis loops, shear compaction and pore pressure build-up in cyclic undrained tests. The modelling of cyclic mobility is then possible.

13.1. INTRODUCTION

Consider a sample of virgin soil, i.e. of remoulded clay consolidated from a slurry or of sand deposited in the laboratory either in a loose or in a dense state. Imagine that the sample is loaded with cycles of isotropic pressures of varying amplitudes. Even for very small amplitudes, at the end of each cycle a permanent volumetric strain occurs. Upon reloading, after the completion of a hysteresis loop, the material tends to follow the curve of virgin loading, as if the unloading–reloading cycle had not been performed. All the loops have similar shape.

Similar results can be obtained for deviatoric strains, e.g. in a drained test in which a sample of virgin soil is sheared at constant isotropic pressure to which it was previously consolidated. Note that the preconsolidation pressure has a fundamental influence on the behaviour of the soil, since the shear strength and the rigidity of the sample increase almost linearly with it. Also, it has been demonstrated[29] that in a triaxial test the experimental curve which links the deviatoric strain normalized to the effective isotropic pressure is unique for a given stress path.

Finally, it is well known that dense sand exhibits a peak strength followed by a softening behaviour when sheared in compression.

These observations suggest, as first demonstrated in Ref. 6, that the behaviour of virgin soils can be modelled by means of an elastic–plastic model with isotropic hardening (without initial yield surface). The existence of hysteresis loops in unloading–reloading, however, suggests, that the behaviour within a yield surface cannot be considered purely elastic and that a model that allows for hysteresis should be adopted. Moreover, to describe the behaviour of dense sand a criterion for hardening–softening transition should be given.

In the following a model based on these experimental observations will be presented. It is composed of two parts. The first one concerns the behaviour of virgin soils, the second* deals with the modelling of hysteresis in unloading–reloading. These have a quite different structure. The former[15,21] is an elastic–plastic work-hardening model with a non-associated flow rule. The latter[11,19,21] employs a 'paraelastic' law, i.e. the constitutive relation is assumed to be path independent between subsequent stress reversal points, as defined in Section 13.3.

A rigorous presentation of the model would require a tensorial formulation from which eventually we would derive a simple constitutive law to interpret the behaviour of soils in standard triaxial tests. For the sake of clarity, the historical development of the model will instead be followed. This method of presentation is simpler, allowing a better understanding of the physical meaning of the constitutive parameters and taking the reader easily from the very simple modelling of a triaxial test under monotonic loading to the modelling of complex cyclic path in three dimensions:

It will be assumed in the following that the material is isotropic and that strains are small. These are indeed limitations of the model since actual soils are anisotropic to a certain extent and strains as high as 20% are of practical interest. Nevertheless, an extension of the model to cover these features is formally possible, but with a great increase in complexity, seldom justified in practice.

---

* The modelling of the behaviour in unloading–reloading is the result of a research programme still under development conducted in collaboration by Dr Tomasz Hueckel of the IPPT-PAN (Warsaw) and the writer.

Soil will be considered as a two-phase material, completely saturated, so that the effective stress principle applies. The constitutive relations will be always written in terms of effective stresses, often referred to simply as stresses. Strains and stresses will be taken positive in compression.

The material properties will be considered independent of time, i.e. viscous and rate effects will not be covered by the model. A superposed dot will indicate increment or rate, but in the latter time is intended only as an ordering factor in the manner generally used in theory of plasticity. Monotonic and cyclic loading will be intended as quasistatic.

13.2. THE MODELLING OF MONOTONIC LOADING

Consider a specimen of virgin soil in a standard triaxial apparatus. Assume that the state of stress and strain is uniform within the sample so that it can be considered as a material element. Assume further that no shear stress is imposed on the sample either by the end platens or by the rubber membrane. Only two sets of parameters are then necessary to completely define the states of stress and strain within the sample. The former is determined by two quantities linked to invariants of the stress tensor; namely, the mean effective pressure σ'_m and the deviatoric stress q defined as:

$$\sigma'_m = \frac{\sigma'_1 + 2\sigma'_3}{3}, \qquad q = \sigma'_1 - \sigma'_3, \tag{13.1}$$

where σ'_1 and σ'_3 are the axial stress and the cell pressure, respectively. The plane σ'_m, q will henceforth be called the triaxial plane.

Correspondingly, the latter is determined by the volumetric strain v and the deviatoric strain ε defined as:

$$v = \varepsilon_1 + 2\varepsilon_3, \qquad \varepsilon = \tfrac{2}{3}(\varepsilon_1 - \varepsilon_3) \tag{13.2}$$

A superscript p or e will indicate plastic or elastic strains respectively. Since the soil is virgin, the initial yield surface is assumed to be concentrated at the origin. The first application of loading will therefore cause plastic strain to occur. The yield function f will in general depend on a set of n hidden variables ψ_i, which depend in turn on the plastic history of the sample. Since the material is assumed to be isotropic, these will depend only on the plastic volumetric and deviatoric strains. Thus:

$$f = f(\sigma'_m, q, \psi_i) \leqslant 0, \tag{13.3}$$

$$\psi_i = \psi_i(v^p, \varepsilon^p). \tag{13.4}$$

Plastic strains occur only if $f = 0$ and at the same time $\dot{f} = 0$, i.e. there is no elastic unloading. Therefore

$$\frac{\partial f}{\partial \sigma'_m} \dot{\sigma}'_m + \frac{\partial f}{\partial q} \dot{q} + \sum_1^n \frac{\partial f}{\partial \psi_i} \dot{\psi}_i = 0, \tag{13.5}$$

From equation (13.4) we have:

$$\dot{\psi}_i = \frac{\partial \psi_i}{\partial v^p} \dot{v}^p + \frac{\partial \psi_i}{\partial \varepsilon^p} \dot{\varepsilon}^p. \tag{13.6}$$

Assume that a plastic potential exists, i.e. that it is possible to define a function g such that:

$$\dot{v}^p = \Lambda \frac{\partial g}{\partial \sigma'_m}, \qquad \dot{\varepsilon}^p = \Lambda \frac{\partial g}{\partial q}, \tag{13.7}$$

where Λ is a non-negative scalar. By substituting (13.7) in (13.6) and (13.6) in (13.5), it is possible to determine the plastic multiplier:

$$\Lambda = -\frac{\dfrac{\partial f}{\partial \sigma'_m} \dot{\sigma}'_m + \dfrac{\partial f}{\partial q} \dot{q}}{\displaystyle\sum_1^n \frac{\partial f}{\partial \psi_i} \left(\frac{\partial \psi_i}{\partial v^p} \frac{\partial g}{\partial \sigma'_m} + \frac{\partial \psi_i}{\partial \varepsilon^p} \frac{\partial g}{\partial q} \right)} \tag{13.8}$$

or

$$\Lambda = \frac{1}{H} \left(\frac{\partial f}{\partial \sigma'_m} \dot{\sigma}'_m + \frac{\partial f}{\partial q} \dot{q} \right). \tag{13.9}$$

The factor

$$H = -\sum_1^n \frac{\partial f}{\partial \psi_i} \left(\frac{\partial \psi_i}{\partial v^p} \frac{\partial g}{\partial \sigma'_m} + \frac{\partial \psi_i}{\partial \varepsilon^p} \frac{\partial g}{\partial q} \right) \tag{13.10}$$

is referred to as hardening modulus. Values of $H > 0$ characterize hardening, $H = 0$ perfectly plastic, $H < 0$ softening behaviour.

To determine the plastic incremental response of the soil to any stress increment, it is thus necessary to identify the appropriate expressions for f, g and ψ_i.

This will be done experimentally.

Consider first the expression of the plastic potential and define dilatancy as the ratio between the volumetric and the deviatoric plastic strain rates:

$$d = \frac{\dot{v}^p}{\dot{\varepsilon}^p}. \tag{13.11}$$

On the basis of the experimental work of Stroud[25] on sand it is possible to assume that the dilatancy depends only on the state of stress and not on the stress increment. A particular stress–dilatancy relationship can be written as

$$\eta = M - \mu d \tag{13.12}$$

where M and μ are positive material constants and η is the stress ratio, defined as:

$$\eta = \frac{q}{\sigma'_m}. \tag{13.13}$$

From the definition of plastic potential, we know that its projection onto the triaxial plane, given by the equation

$$q = q(\sigma'_m) \tag{13.14}$$

is such that at any point

$$\frac{dq}{d\sigma'_m} = -\frac{\dot{v}^p}{\dot{\varepsilon}^p}. \tag{13.15}$$

Therefore equation (13.12) becomes:

$$\frac{dq}{d\sigma'_m} = \frac{q}{\mu\sigma'_m} - \frac{M}{\mu} \tag{13.16}$$

Integrating above we get the explicit expression of Eq. (13.4) in the form:

$$q = \frac{M\sigma'_m}{1-\mu}\left[1 - \mu\left(\frac{\sigma'_m}{\sigma_g}\right)^{(1-\mu)/\mu}\right] \tag{13.17}$$

where σ_g is the value of the isotropic effective pressure when $\eta = M$ as shown in Figure 13.1. When $\mu = 1$ equation (13.17) reduces to the Cam Clay plastic potential.[24]

Equation (13.12) is however not completely satisfactory. From considerations of symmetry it would be expected that no shear strain should occur during isotropic consolidation, whilst equation (13.12) implies a finite dilatancy for $\eta = 0$. Actually, if we look at the data with greater care we see that interpolation by means of a straight line is fairly good except for low values of η, where the shear deformation is small and it is difficult to evaluate the dilatancy correctly. To ensure the fulfilment of the assumed isotropy requirements, it is convenient to assume that for low values of η the stress–dilatancy relationship is given by a hyperbola of equation:

$$\eta d = a \tag{13.18}$$

which is in agreement with some experimental findings of Namy[14] on a silty clay. The constant a can be determined by requiring that a smooth transition exists from equation (13.18) to equation (13.12). We find that $a = M^2/4\mu$ and that the transition value of the stress ratio is given by $\eta_{tr} = M/2$.

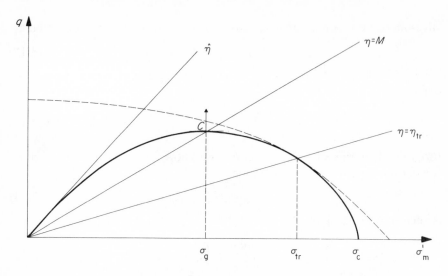

Figure 13.1   The plastic potential

Following the same path of reasoning that led to equation (13.17), it is possible to derive the equation of the plastic potential for $\eta \leqslant \eta_{tr}$ as

$$\frac{q^2}{a} + \sigma_m'^2 = \sigma_c^2, \tag{13.19}$$

where σ_c will be called the preconsolidation pressure. The final picture of the plastic potential in the triaxial plane is given in Figure 13.1. It is shown in Ref. 20 that a physical meaning can be associated with both parts of the expression of the plastic potential.

Assume now that the soil hardens isotropically. This implies that the yield loci, i.e. the projections of the yield function onto the triaxial plane, are represented by a set of closed curves all similar in shape but with increasing dimensions. Thus the yield function depends only on a single parameter and can be expressed by:

$$f = f(\sigma_m', q, \psi) = 0. \tag{13.20}$$

In equation (13.8), there are therefore only two partial derivatives to be evaluated. $\partial \psi / \partial v^p$, $\partial \psi / \partial \varepsilon^p$. They can in principle be determined by performing tests on which successive $\dot{\varepsilon}^p$ and $\dot{v}^p$ are zero. The first condition is fulfilled in a consolidation test on an isotropic material under purely isotropic pressure. Since actual soils are never truly isotropic, some shear strains can be measured even in such a test, but these are negligible if the degree of anisotropy of the specimen is low. Therefore an $\eta = 0$ process is adequate for finding $\partial\psi/\partial v^{\mathrm{p}}$, and ψ can be identified with the consolidation pressure σ_{c}. It is generally accepted that

$$\dot{v}\bigg|_{\eta=0} = \lambda\frac{\dot{\sigma}_{\mathrm{c}}}{\sigma_{\mathrm{c}}}. \tag{13.21}$$

It seems reasonable to assume that the recoverable strain rate is given by

$$\dot{v}^{\mathrm{e}}\bigg|_{\eta=0} = B_0\frac{\dot{\sigma}_{\mathrm{c}}}{\sigma_{\mathrm{c}}}, \tag{13.22}$$

where λ and B_0 are material constants: λ is given by the slope of the semilog plot of the isotropic stress–strain relation in a isotropic loading test, while B_0 is given by the initial slope of the unloading curve in the same test. The plastic volumetric strain rate is then given by

$$\dot{v}^{\mathrm{p}}\bigg|_{\eta=0} = (\lambda - B_0)\frac{\dot{\sigma}_{\mathrm{c}}}{\sigma_{\mathrm{c}}}, \tag{13.23}$$

and therefore

$$\frac{\partial\psi}{\partial v^{\mathrm{p}}} = \frac{\partial\sigma_{\mathrm{c}}}{\partial v^{\mathrm{p}}} = \frac{\sigma_{\mathrm{c}}}{\lambda - B_0}. \tag{13.24}$$

From the expression of the plastic potential it is possible to see that a process for which no plastic volumetric strain occurs is given by the equation $\eta = M$. Although we do not know yet the equation of the yield function, the hypothesis of isotropic hardening allows us to say that there is a constant ratio between σ_{c} and the value of the isotropic pressure, say σ_{y}, when we are consolidating along the $\eta = M$ line so that:

$$\sigma_{\mathrm{y}} = Y\sigma_{\mathrm{c}}. \tag{13.25}$$

It is reasonable to assume that:

$$\dot{\varepsilon}^{\mathrm{p}}\bigg|_{\eta=M} = \frac{\lambda - B_0}{D}\frac{\dot{\sigma}_{\mathrm{y}}}{\sigma_{\mathrm{y}}}, \tag{13.26}$$

where D is another material constant. In fact, it has been shown[14] that in η constant tests there exists a logarithmic relationship between the deviatoric strain and the isotropic pressure. In this type of test, elastic deviatoric strains are negligible, and then equation (13.26) holds.

Note that D is positive for sand, but is zero for kaolin. Even for sand, D is greater than zero only in the hardening range. It will be assumed that D drops suddenly to zero at the start of softening. D is strictly connected with the relative density of the sand. It increases with increasing density and may be taken as zero for very loose sands. Thus from (13.26) we get

$$\frac{\partial \sigma_y}{\partial \varepsilon^p} = \frac{D}{\lambda - B_0} \sigma_y,$$ (13.27)

and from (13.25)

$$\frac{\partial \psi}{\partial \varepsilon^p} = \frac{\partial \sigma_c}{\partial \varepsilon^p} = \frac{\partial \sigma_c}{\partial \sigma_y} \frac{\partial \sigma_y}{\partial \varepsilon^p} = \frac{D}{\lambda - B_0} \sigma_c.$$ (13.28)

Using equations (13.24), (13.28), it is possible to derive a method for the experimental determination of the yield locus as shown in Ref. 17. In fact, by integrating equations (13.24), (13.28), we get, after some algebras,

$$v^p + D\varepsilon^p = (\lambda - B_0) \ln \sigma_c + \text{const.}$$ (13.29)

Since by definition σ_c is constant for a given yield locus, we find that the contours for which

$$v^p + D\varepsilon^p = \text{const.}$$ (13.30)

give the family of yield loci.

Thus, to determine the yield function it is only necessary to evaluate the plastic strains occurring in simple drained tests such as those of Figure 13.2.

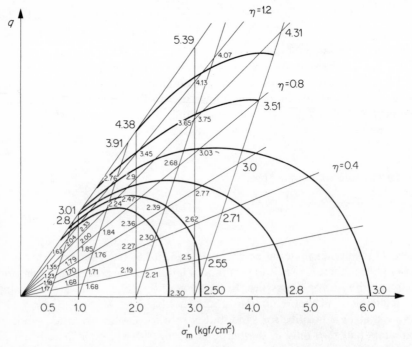

Figure 13.2 Determination of the yield function. Numbers indicate the computed values of the constant in equation (13.30)—data from Ref. 26

To do that we should deduct from the total strains the elastic part. It will be assumed that in an elastic plastic process (monotonic loading) the volumetric strain rate is given by equation 13.22 and the elastic shear strain rate by

$$\dot{\varepsilon}^e = \tfrac{2}{3} L_0 \dot{\eta} \tag{13.31}$$

where L_0 is a material constant. L_0 can be derived from the initial slope of the deviatoric stress strain curve in a σ'_m constant test. Further discussion on the elastic behaviour will be given in Section 13.3.

When a sufficient number of tests has been performed, a family of contours (13.30) can be drawn. A standard procedure for curve fitting can be used to give an analytical expression for the yield function. Note that strains must be recorded from the very beginning of the experiment, from the first application of the cell pressure, since plastic volumetric strains occur even under isotropic consolidation.

It has been shown in Ref. 17 that for high values of η the yield function can be well approximated by the equation:

$$\eta + m \ln \sigma'_m = \text{const.} \tag{13.32}$$

proposed in Ref. 22 and confirmed in Ref. 27.

Equation (13.32) is not applicable for low values of η as can be easily verified. It will be assumed in the following that for $\eta \leqslant M/2$ the yield function coincides with the plastic potential, i.e. is given by equation (13.19). For higher values of η the yield function is given by equation (13.32), which can also be written as

$$\eta = M - m \ln \frac{\sigma'_m}{\sigma_y}. \tag{13.33}$$

The comparison between the calculated and the experimental results for a normally consolidated kaolin, (data from Ref. 1) are shown in Figure 13.3.. From equations (13.17), (13.19), (13.24), (13.28), (13.33) it is then possible to derive the explicit expressions for Λ, which read:

$$\eta \leqslant \frac{M}{2}: \quad \Lambda = \frac{\lambda - B_0}{\sigma'_m(d+D)} \frac{d\dot{\sigma}'_m + \dot{q}}{\eta + d} \quad \equiv (\lambda - B_0) \frac{(\eta + d)\dot{\sigma}'_m/\sigma'_m + \dot{\eta}}{(\eta + d)(d + D)},$$

$$d = \frac{a}{\eta}, \tag{13.34}$$

$$\eta \geqslant \frac{M}{2}: \quad \Lambda = \frac{\lambda - B_0}{\sigma'_m(d+D)} \frac{(m - \eta)\dot{\sigma}'_m + \dot{q}}{m} \equiv (\lambda - B_0) \frac{m\dot{\sigma}'_m/\sigma'_m + \dot{\eta}}{m(d + D)},$$

$$d = \frac{M - \eta}{\mu}. \tag{13.35}$$

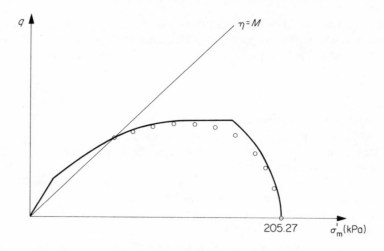

Figure 13.3 Comparison between calculated and experimental yield
locus—data from Ref. 1

Thus combining equations (13.22), (13.29) and equations (13.34) or (13.35), where applicable, the constitutive law in incremental terms is given by

$$\dot{v} = B_0 \frac{\dot{\sigma}'_m}{\sigma'_m} + \Lambda d, \tag{13.36}$$

$$\dot{\varepsilon} = \tfrac{2}{3} L_0 \dot{\eta} + \Lambda.$$

By integrating equation (13.36) it is therefore possible to determine the complete strain history of the sample for a given stress path.

Consider, for instance, a drained, constant cell pressure test on a normally consolidated clay $(D = 0)$. Since $\sigma'_3 = \text{const.}$, the stress path is given by the condition:

$$\dot{q} = 3\dot{\sigma}'_m. \tag{13.37}$$

By substituting (13.37) in (13.36) and integrating it is possible to derive the stress–strain equations. We get:

if $\eta \leqslant \dfrac{M}{2}$

$$\begin{cases} v = -\lambda \ln (1 - \eta/3) + \tfrac{1}{2}(\lambda - B_0) \ln (1 + \eta^2/a), \\ \varepsilon = \tfrac{2}{3} L_0 \eta + (\lambda - B_0) \left\{ -3/a \ln (1 - \eta/3) - \dfrac{1}{\sqrt{a}} \tan^{-1} (\eta/\sqrt{a}) \right\}, \end{cases} \tag{13.38}$$

if $\eta \geqslant \dfrac{M}{2}$

$$\begin{cases} v = (\lambda - B_0)/m \, (\eta - M/2) - \lambda \ln \left(\dfrac{3 - \eta}{3 - M/2} \right) + v \Big|_{\eta = M/2}, \\ \varepsilon = \tfrac{2}{3} L_0 (\eta - M/2) + \dfrac{\mu (\lambda - B_0)}{m (3 - M)} \left\{ (3 - m + M) \ln \dfrac{1}{2 (1 - \eta/M)} + m \ln \dfrac{3 - \eta}{3 - M/2} \right\} \end{cases} \qquad (13.39)$$

$$+ \, \varepsilon \Big|_{\eta = M/2}.$$

A comparison between calculated and experimental results for a normally consolidated kaolin (data from Ref. 28) are given in Figure 13.4. The

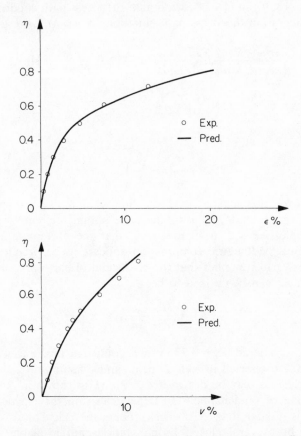

Figure 13.4 Constant cell pressure test on a normally consolidated kaolin-data from Ref. 28

material parameters used are $\lambda = 0.113$, $B_0 = 0.022$, $M = 0.9$, $L_0 = 0.0397$, $m = 0.53$, $\mu = 0.51$.

As a general procedure, the parameters λ, B_0, L_0 and m should be determined as previously outlined whilst M and μ could be found by performing a series of η constant tests and evaluating the value of the dilatancy. In practice, the value of M can be more easily derived from equation (13.39b) as the value to which the deviatoric stress–strain relation tends asymptotically, whilst μ can be simply determined by a best fitting procedure. The constant M appears then to be the critical state stress ratio.

Consider now an oedometric test starting from an isotropic state σ_0. Assuming that the oedometer ring is rigid, no lateral strain will occur. Therefore $\dot{\varepsilon}_3 = 0$ and consequently

$$\dot{v}/\dot{\varepsilon} = \tfrac{3}{2}. \tag{13.40}$$

Substituting (13.38) in (13.34), we obtain the stress path condition which yields after integration the stress path equation. If $\eta \leqslant M/2$, we get

$$\ln \frac{\sigma'_m}{\sigma_0} = \left(L_0 + \frac{B_0 d_0}{a + d_0^2}\right) \frac{a}{\lambda d_0} \ln \frac{1}{1 - \eta d_0/a} + \frac{1}{2}\left(\frac{a}{a + d_0^2}\frac{B_0}{\lambda} - 1\right) \ln \left(1 + \frac{\eta^2}{a}\right)$$

$$- \frac{d_0}{a + d_0^2}\frac{B_0}{\lambda}\sqrt{a}\,\tan^{-1}\frac{\eta}{\sqrt{a}}, \tag{13.41}$$

where

$$d_0 \equiv \frac{3}{2}\left(1 - \frac{B_0}{\lambda}\right). \tag{13.42}$$

It is easy to see that if $d_0 > 2a/M$, the line $\eta = a/d_0$ is an asymptote to which the stress path tends. By using the same parameters employed in the previous example, we see in Figure 13.5, that the asymptote is rapidly approached. Thus oedometric compression appears to be practically an η constant process, which implies that the coefficient of earth pressure at rest K_0 is constant. K_0 can be derived to be

$$K_0 = \frac{3d_0 - a}{3d_0 + 2a}. \tag{13.43}$$

In the case considered $K_0 = 0.73$, which compares favourably with the value of $K_0 = 0.70$ reported in Ref. 2 for a similar kaolin.

Similar conclusions may be drawn if $d_0 < 2a/M$, as shown in Ref. 18.

The hardening of sand happens to depend also on the plastic deviator strain, i.e. D is greater than zero. This renders the calculations of strains more difficult but several closed form solutions are available[15] for the most common test.

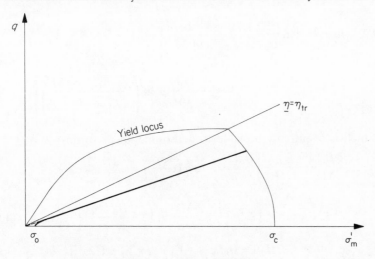

Figure 13.5 Effective stress path in a oedometric test

Note that if $D>0$, H can be greater than zero even if $d<0$. Therefore the stress ratio can be greater than M even in the elastoplastic range. Thus the critical stress ratio may be bypassed by a dense sand, as experimentally verified.

Consider an η constant test; it is easy to see that

$$\frac{\partial v}{\partial \varepsilon} = \frac{B_0 D + \lambda d}{\lambda - B_0} = \text{const.} \qquad (13.44)$$

Here the strain path in an η constant test is a straight line, since d is constant.

In an undrained test, the effective stress path is controlled by the condition of no volume change, i.e. $\dot{v}=0$. Thus from (13.36)

$$\Lambda d = -B_0 \dot{\sigma}'_\text{m}/\sigma_\text{m}.$$

By integrating equation (13.45) it is possible to obtain the equation of the effective stress path:

$$\eta \leqslant M/2: \quad \ln \frac{\sigma'_\text{m}}{\sigma_\text{c}} = -\frac{1-B_0/\lambda}{1+\dfrac{D^2}{a}\dfrac{B_0^2}{\lambda^2}} \left[\tfrac{1}{2}\ln\left(1+\frac{\eta^2}{a}\right) \right.$$

$$\left. +\frac{B_0 D}{\lambda\sqrt{a}}\tan^{-1}\left(\frac{\eta}{\sqrt{a}}\right) - \ln\left(1+\frac{B_0}{\lambda}D\frac{\eta}{a}\right) \right], \quad (13.46)$$

$$\eta \geqslant M/2: \quad \ln \frac{\sigma'_m}{\sigma_c} = \left(\ln \frac{\sigma'_m}{\sigma_c} \right)_{\eta = M/2} - \frac{1 - B_0/\lambda}{m/M} \left[\eta/M \right.$$

$$\left. - \frac{1}{2} + \frac{B_0 D}{\lambda M} \mu \ln \left(1 + \frac{1/2 - \eta/M}{1/2 + \frac{B_0 D \mu}{\lambda M}} \right) \right], \quad (13.47)$$

from which is eventually possible to determine shear strains:

$$\eta \leqslant \frac{M}{2}: \quad \varepsilon = \frac{B_0}{M} \frac{1 - B_0/\lambda}{1 + \frac{D^2 B_0^2}{\lambda^2 a}} \left[\frac{B_0 D}{\lambda \sqrt{a}} \sqrt{\mu} \ln \left(1 + \frac{\eta^2}{a} \right) \right.$$

$$\left. - 2\sqrt{\mu} \tan^{-1} \left(2\sqrt{\mu} \frac{\eta}{M} \right) + \frac{M\lambda}{DB_0} \ln \left(1 + \frac{B_0}{\lambda} \frac{D\eta}{a} \right) \right] + \frac{2}{3} L_0 \, \eta, \quad (13.48)$$

$$\eta \geqslant \frac{M}{2}: \quad \varepsilon = \varepsilon \bigg|_{\eta = M/2} + \frac{B_0}{M} \frac{1 - B_0/\lambda}{m/M} \mu \ln \left[1 + \frac{\eta/M - 1/2}{1 + \frac{B_0 D \mu}{\lambda M} - \frac{\eta}{M}} \right] + \frac{2}{3} L_0 \left(\eta - \frac{M}{2} \right).$$

$$(13.49)$$

The solution for a σ'_m constant test is also straightforward. It is shown in Ref. 15 that the shear strains can be directly derived from equations (13.48, 13.49) simply by taking D instead of DB_0/λ and by multiplying equations (13.48, 13.49) by λ/B_0. The volumetric strains can be derived from equations (13.46, 13.47) by taking D instead of DB_0/λ and by multiplying the two equations by $(-\lambda)$.

The analytical solution for a drained σ'_3 constant test is more complex, but strains can be easily found via numerical integration. The same holds for every conceivable stress path (without elastic unloading).

Before making a comparison of the theoretical results with actual experimental data, it is still necessary to define the criterion for hardening–softening transition. The model as it stands is unable to predict such behaviour. We shall postulate therefore that there exist a stress ratio $\eta = \eta_f$ at which hardening–softening transition occurs. This ratio is given by the asymptote toward which the stress path in an undrained test tends, i.e.

$$\eta_f = M + \frac{B_0}{\lambda} \mu D, \quad (13.50)$$

as can be easily derived by equation (13.47). In fact, in a drained test the hardening–softening transition implies a peak in the deviatoric stress–strain relation, i.e. failure as usually accepted. Experimentally the locus of the peaks in the triaxial plane is approximately a straight line not far from the asymptote to which the undrained tests experimentally tend. The assumption of the line $\eta = \eta_f$ as the locus at which hardening–softening transition occurs then appears justified. Note that since in the softening range D has been taken as zero, for large deformations the stress ratio tends asymptotically from above to M which does not depend on the density of the sand, as is experimentally verified. Since also $(B_0/\lambda)\mu$ is not very influenced by it, it is the term D that accounts for the dependency of the stress ratio at failure on the density index. It has been shown in Ref. 27 that m increases with the sand density.

Consider now a set of tests conducted by Tatsuoka[26] on the Fuji River sand. The comparison between experimental and calculated curves are shown in Figures 13(6–9). The parameters used in the calculations are $M = 1.34$, $\mu = 0.7$, $D = 0.561$, $\lambda = 0.01275$, $B_0 = 0.00544$, $m = 0.55$, $L_0 = 0.00715$. The constant D has been determined via equation (13.50). The general agreement between the two types of results is satisfactory and, as far as the effective stress path in the undrained test and volumetric strains are concerned, very good. Note that for $\eta = M$ the effective stress path in the undrained test has a direction change. The line for which this occurs has been defined as the line of phase transition,[27] and indicates the change from a compressive to a dilatant behaviour of sand. The constitutive law adopted accurately models this feature. The line of phase transition coincides with the critical state line. Since it is easier to determine the line of phase transition, the parameter M may be more conveniently derived from the undrained test results.

An interesting observation can be made on the expression of the effective stress path in the undrained test, equation (13.47). It is shown in Ref. 15 that when η is a solution of the equation

$$(1 - B_0/\lambda)\eta^2/M^2 - (m/M + 1 - B_0/\lambda)\eta/M + \left(1 + \frac{B_0 D\mu}{\lambda M}\right)\frac{m}{M} = 0,$$

$$(13.51)$$

then the stress path has a horizontal tangent. This happens when

$$\eta/M = A \pm \sqrt{(A^2 - B^2)}, \tag{13.52}$$

where:

$$\begin{cases} A = \tfrac{1}{2}\left(1 + \dfrac{m/M}{1 - B_0/\lambda}\right), \\[2mm] B^2 = \left(1 + \dfrac{B_0 D\mu}{\lambda M}\right)\dfrac{m/M}{1 - B_0/\lambda}. \end{cases} \tag{13.53}$$

Therefore if $A^2 < B^2$ no solution exists, i.e. the deviatoric stress q increases monotonically; if $A^2 = B^2$ there is a point of inflection with horizontal tangent; if $A^2 > B^2$ there are a maximum and a minimum, i.e. q increases up to $\eta = \eta_1 = A - \sqrt{(A^2 - B^2)}$ then decreases until $\eta = \eta_2 = A + \sqrt{(A^2 - B^2)}$ and eventually increases again.

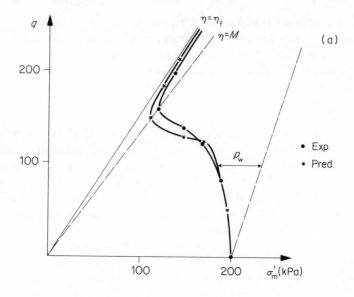

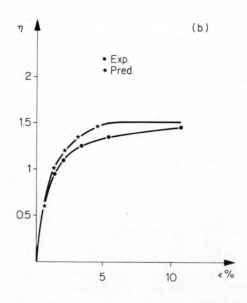

Figure 13.6 Undrained test on Fuji River sand—data from Ref. 26:
(a) effective stress path; (b) deviatoric stress–strain relationship

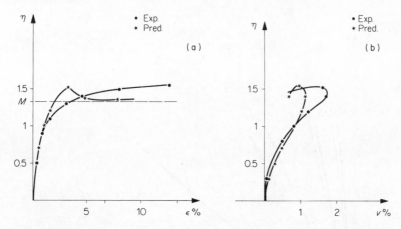

Figure 13.7 Constant cell pressure test on Fuji River sand—data from Ref. 26:
(a) shear strains; (b) volumetric strains

Consider the last case. Imagine now a load-controlled, undrained test. When the deviatoric stress reaches the value corresponding to $\eta = \eta_1$, a sudden failure of the specimen occurs: the load can be sustained only because the area of the specimen increases rapidly and consequently the deviatoric stress drops to a value that can be much smaller than that corresponding to $\eta = \eta_1$. This process is accompanied by large deformations and by an increase of the pore pressure that approaches the value of the confining pressure. This phenomenon, experimentally verified on loose

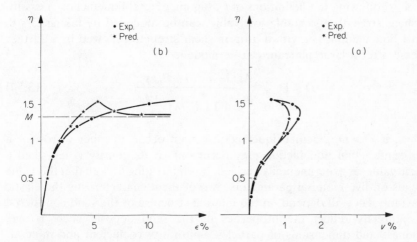

Figure 13.8 σ'_m constant test on Fuji River sand—data from Ref. 26: (a) volumetric strains; (b) shear strains

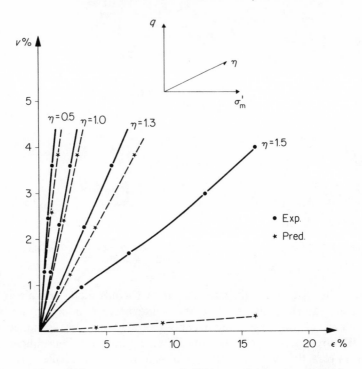

Figure 13.9 η constant test on Fuji River sand—data from Ref. 26

Banding sand by Castro,[4] has been defined as actual liquefaction by Casagrande.[3] Following the definitions of Casagrande, total liquefaction, i.e. with the shear strength at a stable low value, can be modelled by taking $D = 0$. Partial liquefaction, i.e. with a drop in shear strength followed by a further increase after a large plateau, can be modelled if

$$D < D_{\text{crit}} \equiv \frac{M}{\mu} \frac{(1 - m/M - B_0/\lambda)^2}{4(1 - B_0/\lambda)\dfrac{mB_0}{\lambda M}}. \tag{13.54}$$

Thus, if $D > D_{\text{crit}}$ actual liquefaction cannot occur. in other words for a given sand actual liquefaction may occur only if its density is less than a critical value, as experimentally verified. It is interesting to note that this value depends on five material parameters, two of them characterizing the elastic behaviour. They all depend on the internal structure of the sand considered and indirectly reflect the properties of the sand deposit, such as age, roundness and dimensions of particles, uniformity coefficient and mineralogy. From this viewpoint, actual liquefaction loses some of its mysterious aspect: it can be seen as a phenomenon that occurs when the relative density

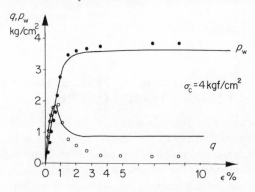

Figure 13.10 Undrained test on a sand exhibiting actual liquefaction: stress–strain relation and pore water pressures—data from Ref. 4

of a sand is less than a critical value that depends in an 'integral' way on the internal structure and on properties of the constituent material. As a comparison with experimental data, consider a test conducted by Castro.[4] It is shown in Figure 13.10 that the model gives good qualitative predictions of the behaviour at very low density. The parameters employed are $M = 1.2$, $\lambda = 0.00807$, $B_0 = 0.00151$, $L_0 = 0.00818$, $\mu = 1.0$, $m = 0.35$, $D = 0$.

Equation (13.44) is valid also for normally consolidated clay with $D = 0$. It is possible to show that:

$$\eta_1 = \frac{m}{1 - B_0/\lambda}.$$ (13.55)

For this value of η a peak occurs in the deviatoric stress–strain curve, as sometimes experimentally verified. Since a null value for D is appropriate for very loose sand for which liquefaction is observed, one may think that the model predicts liquefaction even for clays, which is obviously not true for two reasons. First, note that the whole discussion on liquefaction has been based on the hypothesis that equation (13.47) applies. Therefore η_1 must be greater than $M/2$. If $D = 0$ this implies that:

$$\frac{1}{2} \leqslant \frac{m/M}{1 - B_0/\lambda} \leqslant 1.$$ (13.56)

For clays the ratio m/M is much higher than in case of sands and often

$$\frac{m}{M} > 1 - B_0/\lambda,$$ (13.57)

so that no peak occurs. Even when conditions (13.56) are satisfied, since m/M is high, the stress path is flat and the peak is not pronounced. Large deviatoric strains are reached with pore pressure far away from the value of the confining pressure.

13.3 THE MODELLING OF SOIL BEHAVIOUR UNDER CYCLIC LOADING

The behaviour in unloading–reloading cannot be successfully described by an elastic constitutive law, even if non-linear. In fact, it is well known that wide hysteresis loops occur in unloading–reloading. Sometimes[13] the loop width is even greater than the plastic shear strain. Depending on the drainage conditions shear compaction or pore pressure build-up occur in cyclic loading. None of these phenomena can be taken into account by an elastic model. On the other hand, to limit the number of constitutive parameters it may appear convenient to preserve some degree of path independence.

The model that will be presented here, due to Hueckel and Nova[10–13,19,21] is based on the hypothesis that the stress–strain law is piecewise path independent. The points which delimit the validity of the single branch of the constitutive law will be called stress-reversal points. A proper definition of stress reversal point will be given later on in this section. The strains in this range will be called paraelastic strains.

Consider first the stress–strain law. It is well known that in a uniaxial test at constant isotropic pressure the hysteresis loop is roughly symmetric with respect to the segment that connects its extremities. This is the basis of the Ramberg–Osgood model.[23] It has been demonstrated by Hardin and Drnevich[7] that the shear compliance is linearly dependent on the shear strain.

The experimental results are not simply modelled if other tests are considered. For instance in a purely isotropic test the stress–strain loop does not have such a symmetry, which is not respected even by the deviatoric loops if triaxial tests different from the previous one are considered, e.g. constant cell pressure test. Nevertheless it has been shown in Ref. 19 that the loop symmetry is restored if a convenient set of stress variables is employed. These are the logarithmic measure of the isotropic pressure $\ln(\sigma'_m/\sigma_a)$ where σ_a is the atmospheric pressure, and the stress ratio η. In fact, it is well known that if experimental results of a purely isotropic unloading–reloading test are drawn in a semilog plot, the stress–strain loop regains a symmetry. In a constant cell pressure test the symmetry is respected in the plane η, ε. Since these variables appear to be very useful in simplifying the modelling of the behaviour of soil, not only in the hysteretic range but even in the elastoplastic one, they will be used extensively in the following and will be called geotechnical variables. Note that between σ'_m and q and the geotechnical variables there is a one-to-one correspondence.

The stress–strain loops do not close exactly at the stress reversal point, as postulated by the Ramberg–Osgood model, but a little below, so that under cyclic loading at constant shear amplitude the loops tends to shift to the right in the direction of increasing strains. This phenomenon is accompanied by a shear compaction effect, and conversely in an undrained test by a pore pressure build-up after each cycle. Since this latter phenomenon may lead to cyclic liquefaction and is the reason of the loss of undrained strength due to cyclic loading, to be successful, a model of soil under cyclic loading must include this fundamental effect.

To take account of all such experimental findings, it is assumed that in the triaxial plane the stress–strain law is given by:

$$\begin{bmatrix} \Delta^M v \\ \Delta^M \varepsilon \end{bmatrix} = \begin{bmatrix} B & \sqrt{\tfrac{2}{3}} B\theta \, \mathrm{sgn}\, \Delta^M \eta \\ 0 & \tfrac{2}{3} L \end{bmatrix} \begin{bmatrix} \ln \sigma'_m / \sigma'^M_m \\ \Delta^M \eta \end{bmatrix} \tag{13.58}$$

where the symbol $\Delta^M y$ means $\Delta^M y = y - y^M$, i.e. it indicates the difference between the current value of a variable and its value at the last, Mth, stress reversal. B and L are the bulk and shear compliances in terms of geotechnical variables and θ is a shear compaction parameter. B and L are assumed to be linear functions of a scalar, χ, that is a measure of the paraelastic strain difference, and that is called therefore the strain amplitude parameter:

$$B = B^M(1 + \omega_v \chi^M), \qquad L = L^M(1 + \omega_\varepsilon \chi^M), \tag{13.59}$$

$$\chi^M = (\tfrac{1}{3} \Delta^M v^2 + \tfrac{3}{2} \Delta^M \varepsilon^2)^{1/2}. \tag{13.60}$$

The parameters ω_v and ω_ε are positive material constants. ω_v may be determined in a isotropic unloading–reloading test ($\eta = 0$) by imposing that the calculated curve goes through the point at which stress reversal occurs. θ and ω_ε may be found from the stress path and the deviatoric stress–strain curve, respectively, in a cyclic undrained test ($\Delta^M v = 0$). B^M and L^M are equal to B_0 and L_0 respectively at the first stress reversal, i.e. when the virgin specimen strained monotonically to reach a certain deformed state is unloaded for the first time. Assume the performance of a 'radial' unloading–reloading path from the virgin condition, i.e. assume that the stress point initially on the yield locus moves back and forth on a straight line. We shall postulate that at the start of reloading the compliances B and L are restored to the initial values B_0 and L_0. The new stress–strain law is formally identical to that employed in the unloading path but stresses and strains must be henceforth referred to the stress reversal point, until yielding is reached again. In this way the observed loops symmetry is ensured. Moreover in a σ'_m constant test if one disregards the small contribution to χ given by the shear compaction effect, the constitutive law predicts results similar to those obtained experimentally in Ref. 7.

If the stress path is not radial, the same concept of unloading is no longer self-evident and a criterion distinguishing between unloading and reloading (or vice versa) must be postulated. It is experimentally observed that, in terms of geotechnical variables, the material tends to become softer as the strain difference from the state at the last stress reversal point increases. This feature has been called elastic softening,[8] as opposed to the elastic hardening for materials exhibiting locking behaviour. It is evident from equation (13.59) that the condition for elastic softening is given by:

$$\dot{\chi}^M > 0. \tag{13.61}$$

For a given state of the material, there is a locus in the stress space, corresponding through (13.58)(13.60) to the equation

$$\chi^M = \text{const.} \tag{13.62}$$

Apply now a stress rate to the material. If the vector of the stress rate is directed outwards the locus, χ^M increases, and if it is directed inwards, χ^M decreases. Since this latter possibility is in contradiction with the observed softening, we shall assume that an inward directed stress rate gives rise to a stress reversal, so that the constitutive law must be updated by referring stresses and strains to this last stress reversal point.

The values of the compliance moduli just after the stress reversal are assumed to depend on the orientation of the stress rate with respect to the vector connecting the Mth and the $(M+1)$th stress reversal points. It will be postulated[9] that for complete stress reversal, i.e. for radial paths, the compliances will be restored to the value they had just after the Mth stress reversal, whilst for neutral paths, i.e. $\dot{\chi}^M = 0$, they assume the values given by equation (13.59) as if $\dot{x}^M > 0$. For rates inclined in between, a linear interpolation rule will be adopted.

The locus for which a stress reversal occurs is called dead locus and is stored in the memory of the material. Investigations of simple uniaxial cycles make it possible to infer that the memory of soils concerning the hysteretic range is hierarchically organized. It is apparent, in fact, that after the completion of a uniaxial loop the material has no memory of it, i.e. it behaves as if the loop did not exist. To generalize this feature to multidimensional loading conditions, it has been assumed that when a dead locus is reached again by the stress point, the material forgets all the 'younger' dead loci and the current locus. The dead locus is reactivated and the origin to which stresses and strains should be referred is the stress reversal point pertinent to the dead locus. For the sake of continuity of strains, at the reactivation point the strain difference between the state at this point and the state at the origin of the dead locus given by (13.59) must be complemented by the strains which have arisen along the path within the reactivated dead locus.[12] The yield locus acts as the oldest dead locus.

Summing up, the structure of the constitutive law is elastic in sections. The history dependence of the material behaviour within the yield locus has been lumped in the stress reversal points. Thus the material enjoys a discrete memory instead of a continuous one as in the theory of plasticity.

Having defined the constitutive law and its limits of validity even in the hysteretic range, we can now determine the stress–strain relations for cyclic tests. Consider first an undrained test on a normally consolidated clay under cyclic loading with variable amplitude. Assume the programme of loading to be that imposed in Ref. 29 on a normally consolidated kaolin. The first loading yields elastoplastic strains and the effective stress path is given by equation (13.46). The unloading–reloading paths within the yield locus are given by

$$\sigma'_m = \sigma'^M_m \exp\{-\sqrt{(\tfrac{2}{3})}\theta \, |^{\Delta M}\eta|\}. \tag{13.63}$$

Since for this particular loading programme, the yield locus is reached before the end of each cycle, the hysteretic memory is cleared after every cycle. From the second cycle ahead, yielding occurs for $\eta > M/2$. Then from equation (13.47), we get that the stress path is given by

$$\sigma'_m = \sigma'^i_m \exp\left\{\frac{1 - B_0/\lambda}{m}(\eta - \eta_i)\right\}, \tag{13.64}$$

where σ'^i_m and η_i are the stress coordinates of onset of yielding. Figure 13.11 shows the comparison between calculated and experimental stress paths and the calculated pore pressures p_w. The constitutive parameters employed are $M = 0.96$, $\lambda = 0.113$, $B_0 = 0.022$, $\mu = 0.67$, $m = 0.7$, $\omega_v = 23$, $\omega_\varepsilon = 150$, $L_0 = 0.00397$, $\theta = 0.1$.

The movement of the effective stress point towards the critical state is satisfactorily matched and the pore pressure predictions are reasonable. This result suggests that the constitutive law adopted may be able to model even the liquefaction of sand under cyclic loading. However, it must be emphasized that in common laboratory liquefaction tests the sand specimens undergo an extension phase. This is a range in which the model has not been tested so far. This problem will be the subject of further research and will not be treated here.

The 'paraelastic' behaviour concerns also the behaviour of overconsolidated clays, which are clays naturally or artificially unloaded from a virgin condition. Unfortunately, the model has not yet been tested extensively in this area. Nevertheless it is shown in Ref. 19 that it is possible to predict the behaviour of an artificially overconsolidated kaolin under undrained monotonic loading. Consider a set of samples of the kaolin used in Ref. 29, normally consolidated to an isotropic pressure σ_c that will be referred to as the preconsolidation pressure. Suppose now that each sample is unloaded to a different isotropic pressure σ'^i_m, so that the overconsolidation ratios are

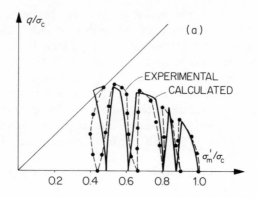

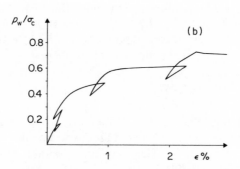

Figure 13.11 Cyclic undrained test on a normally consolidated kaolin—data from Ref. 29: (a) effective stress path; (b) calculated pore pressures

given by $OCR_i = \dfrac{\sigma_c}{\sigma_m^i}$. Suppose now that the specimens are sheared to failure in undrained conditions. Since stress reversal occurs within the yield locus neither during isotropic unloading nor during the undrained part of the test, the volumetric strain is constrained to be constant in this latter part and equal to:

$$^{\Delta M}v^{i*} = -\frac{B_0 \ln OCR_i}{1 - (\omega_v/\sqrt{3}) \ln OCR_i}. \tag{13.65}$$

By virtue of (13.59) it is possible to determine $^{\Delta M}\varepsilon^i$ as

$$^{\Delta M}\varepsilon^i = \frac{1 + [1 - (1 - \tfrac{2}{3}L_0^2\omega_\varepsilon^{2\Delta M}\eta^2)(1 - \tfrac{1}{3}\omega_\varepsilon^{2\Delta M}v^{i*}v^{i*2})]^{1/2}}{1 - \tfrac{2}{3}L_0^2\omega_\varepsilon^{2\Delta M}\eta^2}(\tfrac{2}{3}L_0^{\Delta M}\eta). \tag{13.66}$$

Thus χ^M can be determined through (13.60) and the stress path can be eventually found:

$$\sigma'_m = \sigma_c \exp \left\{ \frac{\Delta M_v^{i*}}{B_0(1+\omega_v\chi^M)} - \sqrt{(\tfrac{2}{3})}\theta \, |{}^{\Delta M}\eta| \right\}. \tag{13.67}$$

When the yield locus is reached the stress path equation is ruled by the elastoplastic model.

The comparison between theoretical and experimental data is shown in Figure 13.12. Although the predictions are not as good as for the case of normally consolidated clays, the agreement is satisfactory. In particular the change in shape of the stress path with increasing OCR is predicted.

It is worth noting that for a heavily overconsolidated clay, the portion of the yield locus for $\eta > M$ constitutes a limiting curve that cannot be crossed by any stress path. In fact when the stress point reaches this curve a softening process starts, since $D = 0$ and the dilatancy is negative. Thus this part of the yield locus coincides with the failure locus. For lightly overconsolidated clays the yield locus is reached in the hardening range—positive dilatancy—and the failure locus is given by the straight line $\eta = M$. The critical stress ratio is approached asymptotically by the deviatoric stress–strain curve. The failure locus for a given preconsolidation pressure is then given by the curve shown in Figure 13.13. It may be noted that the theoretical curve is qualitatively in very good agreement with the experimental failure locus for clay.

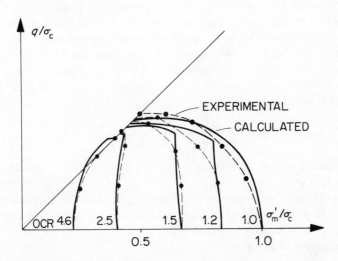

Figure 13.12 Effective stress paths in undrained tests on overconsolidated specimens of kaolin—data from Ref. 29

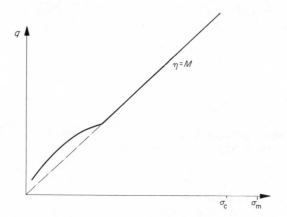

Figure 13.13   Failure locus for a kaolin with a given
preconsolidation pressure

It is worth noting that the behaviour of dense, virgin sand is different from
that of overconsolidated clays, as demonstrated for instance by the shape of
the effective stress path in undrained tests or by the value of K_0. The model
consequently treats them differently: the straining of dense sands is considered an elastoplastic process whilst that of overconsolidated clays is considered paraelastic, at least for moderate loading. *In situ*, it is however
possible to find dense sands, overconsolidated because of the removal of a
pre-existing load or because of the fluctuation of the water table. This kind
of material must be modelled as an overconsolidated clay.

Another very important test that can be described by this model is the
confined compression test with cycles of unloading–reloading. For the sake
of simplicity, consider a virgin kaolin. By imposing the kinematic condition
of no lateral strain we found equation (13.41) for the virgin loading. Assume
now an unloading of the specimen. It is possible to show[19] that the equation
of the stress path is given by

$$\sigma'_m = \sigma_m^M \exp \left\{ \frac{L_0^{\Delta M}\eta}{B_0(1+(\omega_v - \omega_\varepsilon)L_0 \,|^{\Delta M}\eta|)} - \sqrt{(\tfrac{2}{3})}\theta \,|^{\Delta M}\eta| \right\}, \quad (13.68)$$

where σ_m^M is the effective pressure at the unloading point. A similar equation
holds for reloading, but one should have care to change the starting values
of the compliances after the stress reversal. Since the direction of the stress
path is not known *a priori* at the stress reversal point, an iterative procedure
is necessary to determine B^1 and L^1.

Shear strains are given by:

$$^{\Delta M}\varepsilon = \frac{\tfrac{2}{3}L_0 \,|^{\Delta M}\eta|}{1 - L_0\omega_\varepsilon \,|^{\Delta M}\eta|} \quad (13.69)$$

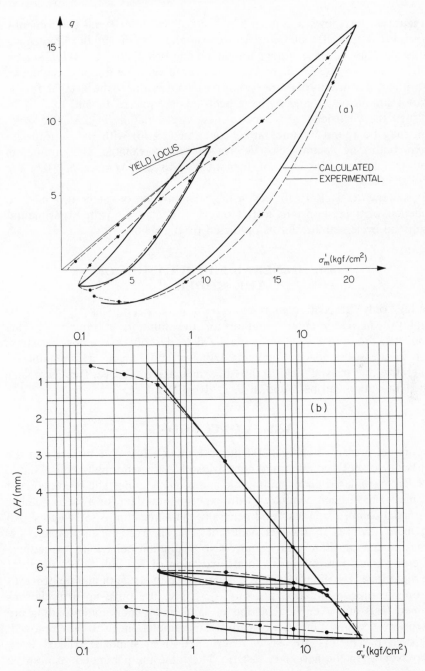

Figure 13.14 Oedometric test on a silty clay—courtesy of Studio Geotecnico Italiano: (a) effective stress path; (b) vertical strains

Figure 13.14 shows a comparison between calculated and experimental curves for a silty clay at Gioia Tauro (supplied by the Studio Geotecnics Italiano). The values of the parameters employed are $B_0 = 0.0028$, $L_0 = 0.0021$, $\omega_\varepsilon = 230$, $\omega_v = 125$, $\theta = 0.1$. The overall agreement looks acceptable. The different concavity of the reloading paths may be due to the fact that, for the sake of simplicity, B_1 and L_1 have been taken equal to B_0 and L_0.

When the yield locus is reached a new virgin loading begins. The stress path may be found by integrating equation (13.40) with the appropriate initial conditions. Again, a closed-form solution is available. The parameters used to calculate this part of the stress path are $M = 1.2$, $m = 0.6$, $\lambda = 0.66$, $\mu = 0.54$.

It is shown in Ref. 10 that even the phenomenon of ratchetting may be modelled in a σ'_m constant test. However, in contrast with experimental results no cycle stabilization is predicted by the theory.

13.4. GENERAL FORMULATION OF THE MODEL

So far, only 'triaxial' compression tests have been considered. To be employed in a computer program for the solution of boundary value problems, the model must be generalized to describe every possible stress path in the stress space. Following Drucker and Prager[5] it is possible to generalize the 'triaxial' field to the full principal stress space. Details of such a generalization can be found in Refs. 10, 12 and 21.

13.5. CONCLUSIONS

The model presented is a general-purpose model. It is able to interpret the behaviour of both granular and cohesive soils, either normally consolidated or overconsolidated under monotonic and cyclic loading. The predictions of the model have been compared with experimental data for a large number of triaxial compression tests and generally the agreement is satisfactory. All the main features of the behaviour of soils in triaxial compression, i.e. critical state, transition from compressive to dilatant behaviour, hardening–softening transition, hysteresis loops, variation of K_0 with OCR, shape of the yield locus for different OCRs, dependence of strength and deformability on the preconsolidation pressure, pore pressure build-up in cyclic un-drained tests and even liquefaction of sand under monotonic loading are taken into account. Only ten constitutive parameters need to be employed. All such parameters are easily determined from standard triaxial tests. Moreover when the necessary tests to determine the parameters cannot be performed the available closed form solutions make it easy to calibrate a required set of parameters. Note that (apart from the σ_3 constant test for sand) all the calculated results presented have been determined with a hand calculator. As far as cycling loading is concerned, the limited number of parameters and the possibility of closed-form solutions are due to the assumption of a paraelastic law which greatly simplifies the constitutive modelling.

Finally, the model enjoys a 'Chinese box' structure. By setting to zero some constants, the effect of strain accumulation in the hysteresis range, the hardening dependence on the second invariant of plastic strain deviator, etc., may be dropped in successive simplifications of the theory to obtain in the limit the simple 'Granta gravel' model. This allows us to treat problems with any desired accuracy using a unified numerical algorithm.

It is evident that apart from the limitations to isotropic, inviscid soils discussed in the introduction, the model has several flaws. The assumed sharp transition from hardening to softening for dense sands leads to prediction of a sharp peak which is not observed in experiments. Further, for this kind of materials the failure locus has been postulated, instead of being determined from a more rigorous failure criterion, linked perhaps with the change of the internal structure of the material.

Moreover, no attempt has been made to correlate the constitutive parameters with the initial state of the sample, i.e. its initial void ratio or density. Also, the model applies to somewhat artificial soils, without any natural cementation and without initially locked-in stresses. The overconsolidation of clays has been induced artificially by consolidating and then allowing these to swell with no memory of their past history.

Perhaps the most important limitation of the model lies in that only behaviour in triaxial compression tests has so far been checked. Unfortunately, we can be sure that the behaviour of the model in extension or in plane strain is not as good as for compression. This is due to the assumed Drucker–Prager type of three-dimensional formulation. It must be emphasized, however, that other more complex extensions of the model are possible to get a better approximation of the real soil behaviour even under these conditions. However, the model can be implemented in a finite element program in its present form and used for solving boundary value problems. The engineer may often know *a priori* whether the condition of loading of a certain part of a soil mass can be approximated better by a compression, extension or plane strain test and evaluate the parameters in accordance with this prediction.

Another limitation common to all constitutive models comes from their relation to experimental tests. We always model the combination of the behaviour of soil coupled to the experimental apparatus. This argument is often raised to deny validity to constitutive modelling in general. It is the opinion of the author that it is necessary to make a distinction between quantitative and qualitative errors which may be induced by 'imperfections' of the testing device. The former do not alter the essence of the model. It is only necessary to adjust the constitutive parameters. The better the experimental technique employed the better the predictions of the behaviour of the prototype will be. On the contrary the latter make a revision of the constitutive model compulsory. Fortunately, in general these kind of revisons are not frequent.

Finally a constitutive model is useful not only because it can be used in a computer program, but also because it gives a mental picture, essential in the understanding of the behaviour of soils.

ACKNOWLEDGEMENT

The author is indebted to his friend and colleague Dr. Tomasz Hueckel of the IPPT/PAN Warsaw without whose contribution the theory concerning the behaviour in unloading–reloading could never have been written. The author is also grateful to the Italian National Research Council (C.N.R.) and to the University of Cambridge for financial support.

REFERENCES

1. Balasubramaniam, A. S. *Some Factors Influencing the Stress–Strain Behaviour of Clay*, Ph.D. Thesis, University of Cambridge, 1969.
2. Burland, J. B. *Deformation of Soft Clay*, Ph.D. Thesis, University of Cambridge, 1967.
3. Casagrande, A. Liquefaction and cyclic deformation of sands. A. Critical review, *Harvard SM Series No. 88*, presented at the 5th Panam. Conf. SMFE Buenos Aires (November 1975), 1976.
4. Castro, G. Liquefaction of sand, *Harvard SM Series No. 81*, 1969.
5. Drucker, D. C., and Prager, W. Soil mechanics and plastic analysis or limit design, *Quart. Appl. Math.*, **10**, 157–65, 1952.
6. Drucker, D. C., Gibson, R. E., and Henkel, D. J. Soil mechanics and workhardening theories of plasticity, *Trans. ASCE* **122**, 338–46, 1957.
7. Hardin, B. O., and Drnevich, V. P. Shear modulus and damping in soils: measurement and parameter effect, *Proc. SMFE-ASCE*, **98**, 603–24, 1972.
8. Hueckel, T. Coupling of elastic and plastic deformations of bulk solids, *Meccanica* **4**, 227–35, 1976.
9. Hueckel, T., and Nova, R. A mathematical model of hysteresis of soils, in *Proc. Matériaux et structures sous chargement cyclique*, Palaiseau, 101–10, 1978.
10. Hueckel, T. On a neutral path uniqueness in a piecewise description of cyclic behaviour of inelastic solids, *J. de Mécanique* **20**(2), 365–78, 1981.
11. Hueckel, T., and Nova, R. On paraelastic hysteresis of soils and rocks, *Bull. Acad. Pol. des Sciences*, Sec. Sc. Techn. **27**(1), 49–55, 1979.
12. Hueckel, T., and Nova, R. 'Some hysteresis effects of the behaviour of geologic media' *Int. J. of Solids and Structures*, **15**, 625–42, 1979.
13. Hueckel, T., and Nova, R. Hysteresis behaviour of soils and rocks, *Proc. 5th SMiRT* K 1/7, Berlin, 1979.

14. Namy, D. *An Investigation of Certain Aspects of Stress–Strain Deformation for Clay Soils*, Ph.D. Thesis, Cornell University, 1970.
15. Nova, R. *Theoretical Studies of Constitutive Relations for Sands*, M.Sc. Thesis, University of Cambridge, 1977.
16. Nova, R. On the hardening of soils, *Archiwum Mechaniki Stosowanej*, **29**(3) 445–58, 1977.
17. Nova, R., and Wood, D. M. An experimental programme to define the yield function for sand, *Soils and Foundations* **18**(4) 77–86, 1978.
18. Nova, R. Un modello costitutivo per l'argilla, *Rivista Italiana di Geotecnica* **13** (1) 37–54, 1979.
19. Nova, R., and Hueckel, T. An engineering theory of soil behaviour in unloading and reloading, *Meccanica* **16**(3), 1981.
20. Nova, R., and Wood, D. M. A constitutive model for sand in triaxial compression, *Int. J. Num. Analyt. Meth. Geomech.* **3**(3) 255–78, 1979.
21. Nova, R., and Hueckel, T. A "geotechnical" stress variables approach to cyclic behaviour of soils, *Int. Symp. on Soils under Cyclic and Transient Load*, Swansea, **1**, 301–14, 1980.
22. Poorooshasb, H. B. Deformation of sand in triaxial compression, *4th Asian Conf. SMFE*, Bangkok, **1**, 63–6, 1971.
23. Ramberg, W., and Osgood, W. R. *Description of Stress Strain Curves by Three Parameters*, Tech. note 902, Nat. Adv. Comm. for Aeron. Washington D.C., 1943.
24. Schofield, A. N., and Wroth, C. P. *Critical State Soil Mechanics*, Wiley, 1968.
25. Stroud, M. A. *The Behaviour of Sand at Low Stress Levels in the Simple Shear Apparatus*, Ph.D. Thesis, University of Cambridge, 1971.
26. Tatsuoka, F. *Shear Tests in a Triaxial Apparatus. A Fundamental Research on the Deformation of Sand*, Ph.D. Thesis, University of Tokyo (in Japanese), 1972.
27. Tatsuoka, F., and Ishihara, K. Yielding of sand in triaxial compression, *Soils and Foundations*, **12**(2) 63–76, 1974.
28. Walker, A. F. *Stress Strain Relationships for Clay*, Ph.D. Thesis, University of Cambridge, 1965.
29. Wroth, C. P. The predictions of shear-strains in triaxial tests on normally consolidation clays 6th *ICSMFE*, Montreal, **2**, 417–20, 1965.
30. Wroth, C. P., and Loudon, P. A. The correlations of strains with a family of triaxial tests on overconsolidated samples of Kaolin, *Proc. Geot. Conf. Oslo*, **1**, 159–63, 1967.

Soil Mechanics—Transient and Cyclic Loads
Edited by G. N. Pande and O. C. Zienkiewicz
© 1982 John Wiley & Sons Ltd

Chapter 14

A New Endochronic Plasticity Model for Soils

K. C. Valanis and H. E. Read

14.1 INTRODUCTION

Current seismic design procedures for nuclear power plants and other large civil systems rely primarily on linear methods. Under some conditions, linear methods appear to provide satisfactory predictions for engineering purposes, whereas in other instances the need for more sophisticated methods has become apparent. Furthermore, the linear methods may result in inefficient designs which are very conservative for loads in the low frequency range of interest to seismic design. The potential reduction in the present design conservatism that may be achievable through the utilization of advanced nonlinear methods could lead to very substantial savings in design and construction costs.

Recent advances in computer technology have made multidimensional non-linear analyses economically feasible for use in seismic design. The lack of suitable mathematical models for describing the response of soils to earthquake-type loading has, however, inhibited the use of these advanced methods. While notable progress has been made during the past decade in formulating mathematical models which can describe the response of soils to relatively simple deformation histories, such as occur in the standard laboratory tests, progress has been particularly slow in developing models that can describe soil behaviour under the general type of loading produced by earthquakes; such loading is highly irregular and complex, being characterized by near-cyclic motions involving numerous reversals in stress rate. When soils are subjected to such loading, they exhibit substantial non-coincidence of unloading and reloading paths and strong dependence of these paths on the history of deformation.

In 1976, the Electric Power Research Institute (EPRI) initiated a research effort, called the Soil-Structure Interaction Program, which was established to develop advanced non-linear methods for seismic design and analysis of nuclear power plants. One of the key tasks in this program was to develop advanced non-linear soil models suitable for representing the complex,

375

hysteretic behaviour soils exhibit during earthquake loading. While there were several one-dimensional soil models available which had been developed for this purpose, none of the existing multidimensional soil models were felt to be adequate.

In 1971, a new approach for describing the rate-independent yet history-dependent response of inelastic materials was proposed.[18,19] This theory, called endochronic, was different from the earlier theories of plasticity in that it did not require the concept of a yield surface for its development. In the early efforts to apply the new theory to real materials, the emphasis was on metals.[20-2] It soon became evident that the theory was able to predict not only the salient features of the plastic behaviour of metals, but also a number of observed features of metal plasticity that lay beyond the scope of the existing plasticity theories.

Systems, Science and Software (S³) as part of the EPRI Soil-Structure Interaction Program, undertook a research effort to develop an endochronic theory for soils. A particularly attractive feature of the theory for modelling soil behaviour is that it does not require a yield surface; the notion that plastic flow takes place abruptly after a yield point has been reached is more suited to metals than soils, which develop plastic strain in a gradual, continuous manner immediately upon application of loading. The Endochronic theory is three-dimensional, and its ability to describe soil response to earthquake-type loadings was demonstrated in the initial investigation S³ conducted for EPRI,[11,25] which considered the response of soils to numerous cycles of shear, as well as the three-dimensional behaviour of soil, including cyclic triaxial loading.

Despite the fact that the original Endochronic theory was able to describe most of the important features of soil response, it was nevertheless unable to predict closed hysteresis loops for small unload–reload processes under one-dimensional conditions. For such deformations, the theory predicted a slope at the reloading point that was smaller than the unloading slope at the same point. This feature of the theory did not appear to be in agreement with the observed behaviour of most materials, including soils, although it must be admitted that there are very few data available on the response of materials to such processes to either verify or invalidate this feature of the theory. Nonetheless, it seemed reasonable to assume that most materials exhibit elastic behaviour at, and in a small neighbourhood of, the point of unloading, and this was not the case with the original endochronic theory. Furthermore, some suspicions were raised that the small unload–reload feature of the theory could lead to numerical difficulties if the model was used in conjunction with computer codes to analyse wave propagation problems,[14] although this was never demonstrated numerically in a well-defined wave propagation problem.

Since the endochronic soil model is being developed to provide an advanced description of soil behaviour which will be used in computer codes to analyse soil-structure interaction of nuclear power plants, it was imperative that the questions regarding the small unload–reload features of the theory be thoroughly investigated and resolved before a further model development was undertaken. To address this issue, an investigation was performed by S^3 under EPRI support, and has resulted in the development of a new endochronic model[28] which not only is free of the undesirable small unload–reload feature that typifyed the earlier version of the theory but provides more extensive modelling capability with remarkably few material parameters. The new model has the capability to describe most of the features of soil behaviour important under seismic loading, including densification/dilatancy, strain hardening (softening), hysteresis, and creep under cyclic loading (ratchetting*). It is our position that the endochronic theory now rests on sound physical, thermodynamic and analytical grounds, and that it can be used with confidence in the numerical solution of wave propagation problems.

In this Chapter, the new endochronic model is described, with particular attention to its application for describing soil behaviour. A numerical scheme for treating the one-dimensional version of the model is presented and incorporated into a small computer program, which is used to perform numerical studies to illustrate various features of the new model. Comparisons are given between model predictions and experimental observations for several simple types of deformations of soils to demonstrate the modelling capabilities; these include multiple loading–unloading of sand under hydrostatic compression, cyclic simple shear of sand over numerous cycles of deformation, and densification of sand under repeated cycles of simple shear. An analytical proof is given to show that, under one-dimensional conditions the new model produces hysteresis loops which always close, no matter how small the magnitude of the unload–reload process.

14.2 A BRIEF REVIEW OF THE ENDOCHRONIC THEORY

In the late 1960s, the formulation of constitutive theories of viscoelastic materials from concepts of irreversible thermodynamics and internal state variables reached an advanced level of development.[17] On the basis of this success, it was natural to inquire if a similar approach could be used to establish a theory of plasticity, and the attempt by Valanis to explore this question led to the development of the endochronic theory in 1971.[18,19]

---

* Editors' note: see Chapter 18 for fuller explanation of this term.

14.2.1 Early concept of intrinsic time

In the rate-independent version of the original theory,[20,21,23] the intrinsic time ζ is defined as the distance along a path in the strain space E with metric* $\mathbf{P}$. If E and E' are two adjacent strain states in this space, the distance, $\mathrm{d}\zeta$, between them is given by the expression:

$$\mathrm{d}\zeta^2 = P_{ijkl}\,\mathrm{d}\varepsilon_{ij}\,\mathrm{d}\varepsilon_{kl}, \tag{14.1}$$

where the fourth-order symmetric positive-definite tensor, $\mathbf{P}$, is a material property. The distance, $\mathrm{d}\zeta$, between two adjacent strain states may, therefore, vary from material to material, even though the strain coordinates may be the same in all cases.

The basis of the endochronic theory is the concept of the intrinsic time scale. By hypothesis, this is the 'natural' scale with respect to which the memory of a material of its past deformation history should be measured. The intrinsic time scale, z, is related to the intrinsic time measure, ζ, by the relation:

$$\mathrm{d}z = \frac{\mathrm{d}\zeta}{f} \tag{14.2}$$

where f is a function of the *history* of strain. In applications, f has generally been taken as a function† of ζ. The function f is of thermodynamic origin and is related proportionally to the degree of internal friction in a material. If a material hardens, $f(\zeta)$ increases with ζ and is a constant otherwise.

14.2.2 The original endochronic theory

In the original version of the endochronic theory,[18,19] the governing equations for isotropic materials may be decomposed into deviatoric and hydrostatic components. In that theory, the deviatoric stress $\mathbf{s}$ is related to the history of the deviatoric strain $\mathbf{e}$ by the linear functional relation:

$$\mathbf{s} = \int_0^z \mu(z - z')\frac{\partial \mathbf{e}}{\partial z'}\,\mathrm{d}z' \tag{14.3}$$

where $\mathbf{s}$ is zero in the reference configuration, $z = 0$, and the shear modulus, $\mu(z)$, is given by a Dirichlet series, i.e.

$$\mu(z) = \lambda_\infty + \sum_{r=1}^{n} \lambda_r e^{-\rho_r z}, \tag{14.4}$$

---

* The origins of this idea lie in earlier works of Ilyushin[6] and Pipkin and Rivlin,[8] in which the metric was Euclidean and the same for all materials. Their theories, therefore, were not endochronic.

† In an equivalent manner, it may be expressed as a function of z, since ζ and z are functionally related through equation (14.2)

where κ_∞, λ_r and ρ_r are positive constants. The hydrostatic stress, σ, is related to the history of volumetric strain, ε, in a similar fashion by the linear functional relation:

$$\sigma = 3 \int_0^z K(z - z') \frac{\partial \varepsilon}{\partial z'} \, dz', \tag{14.5}$$

where $\sigma = \sigma_{kk}/3$ and $\varepsilon = \varepsilon_{kk'}$ in the usual notation where the summation convention is employed, and the bulk modulus, $K(z)$, is given by a Dirichlet series of the form of equation (14.4) Note again that $\sigma = 0$ in the reference configuration.

14.2.3 Criticisms of original theory

The original version of the endochronic theory was recently attacked[14] on the basis of a conjecture that the theory might give rise to numerical instabilities in the solution of wave propagation problems. This conclusion was arrived at on the basis of a construction of wave propagation solutions for infinitesimal amplitude waves, using the unloading–reloading characteristics of the original theory mentioned earlier. It was claimed that a family of small amplitude wave propagation solutions in a doubly infinite bar could be constructed with quiescent initial conditions.

A careful inspection of Ref. 14 reveals, however, that the initial conditions referred to were not quiescent because the velocity did not vanish everywhere but was, in fact, *double valued* at the centre of the bar at time zero. This, in our view, is the cause of the non-uniqueness of solution, and not the endochronic model. The above point notwithstanding, it was also asserted that small numerical perturbations would propagate throughout the solution region causing serious cumulative errors. This last assertion was never demonstrated numerically in a well-defined wave propagation problem.

The theory was also attacked in Ref. 14 on the basis of its prediction of unloading–reloading behaviour which violates Drucker's postulate of material stability.[5] Since this postulate is not of thermodynamic origin and can be violated by standard frictional physical systems, as demonstrated by Mandel,[7] this characteristic of the theory is not, in itself, particularly disturbing. In fact, for soils, Drucker's postulate is generally not considered relevant.

One other point, however, merited attention. On the basis of equation (14.3), the predicted shear response for small unload–reload processes gives rise to hysteresis loops which are as shown in Figure 14.1 by the dotted line. The limited experimental data available, however, indicate that small hysteresis loops are closed for such materials as metals and rate-independent soils, as indicated by the solid line in Figure 14.1.

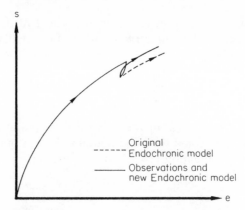

Figure 14.1 Stress paths for small unload–reload processes under one-dimensional conditions

14.2.4 An endochronic theory with a finite yield surface

It was shown in a recent work[26] that the openness of the loops is thermodynamic in nature and has to do with the fact that the intrinsic time rate of dissipation (i.e. with respect to z) at the onset of unloading is equal to the intrinsic time rate of dissipation upon continuation of loading. Because, in general, rate-insensitive materials initially unload in an elastic manner and, therefore, with essentially zero rate of dissipation the discrepancy between model prediction and observation, shown in Figure 14.1, is bound to arise.

In Ref. 26, it was shown that if the measure of intrinsic time is redefined in terms of the increment of *plastic* strain, the rate of dissipation at the onset of loading, unloading and reloading is, in fact, zero. To develop this new concept of intrinsic time more fully, consider that in the process of deformation of a material element, the succession of strain states traces two paths, namely, one ζ_D in the six-dimensional deviatoric plastic strain space and another, ζ_H, in the one-dimensional volumetric plastic strain space. A two-dimensional space $S(\zeta_D, \zeta_H)$ may be constructed into which the six-dimensional space is mapped by the transformations:

$$d\zeta_D = \|d\theta\|, \tag{14.6}$$

$$d\zeta_H = |d\Theta|, \tag{14.7}$$

where the double vertical bars enclosing a symbol denote its norm and the single vertical bar signifies absolute value.

A two-dimensional intrinsic time space $Z(z_D, z_H)$ is then defined as one into which the previous space S is mapped by the transformations:

$$dz_D^2 = k_{00} \, d\zeta_D^2 + k_{01} \, d\zeta_H^2, \tag{14.8}$$

$$dz_H^2 = k_{10} \, d\zeta_D^2 + k_{11} \, d\zeta_H^2, \tag{14.9}$$

where the k_{rs} are material-dependent parameters which may depend on the intrinsic time measures, i.e.

$$k_{rs} = k_{rs}(\zeta_D, \zeta_H) \qquad (r, s = 0, 1). \tag{14.10}$$

Note that the form of equations (14.8) and (14.9) provides the theory with the provision for coupling between the deviatoric and volumetric components of deformation.

If, for certain materials, it is found that $k_{00} = k_{10}$ and $k_{01} = k_{11}$, it follows from equations (14.8) and (14.9) that

$$dz_D = dz_H = dz \tag{14.11}$$

and

$$dz^2 = k_{00} \, d\theta_{ij} \, d\theta_{ij} + k_{01}(d\Theta)^2. \tag{14.12}$$

Equation (14.12) may be written alternately in the form:

$$dz^2 = P_{ijkl} \, d\varepsilon_{ij}^P \, d\varepsilon_{kl}^P, \tag{14.13}$$

where

$$P_{ijkl} = \left(k_{01} - \frac{k_{00}}{3}\right)\delta_{ij} \, \delta_{kl} + k_{00} \, \delta_{ik} \, \delta_{jl}. \tag{14.14}$$

In essence, the concept of intrinsic time defined by equations (14.8) and (14.9) affords greater generality in the description of coupled phenomena, such as cross hardening, densification and dilatancy, and shear travel effects, and makes the rate of dissipation vanish at the onset of loading, unloading or reloading.

As an illustrative example, consider the case of pure shear. In this instance, we can write

$$d\theta = de - \frac{ds}{2\mu_0}, \tag{14.15}$$

where de is the increment of total shear strain, ds represents the increment of shear stress, and μ_0 denotes the elastic shear modulus. The increment of intrinsic time is given by equation (14.8), i.e.

$$dz = |d\theta|, \tag{14.16}$$

where the subscript D has been omitted and we have set $k_{00} = 1$ for convenience. With dz defined in this manner, it was shown[26] that equation (14.3) leads to the expression

$$s = s_y^0 \frac{\partial \theta}{\partial z} + \int_0^z \rho(z - z') \frac{\partial \theta}{\partial z'} dz', \tag{14.17}$$

where s_y^0 denotes a yield stress and the kernel function $\rho(z)$ is defined

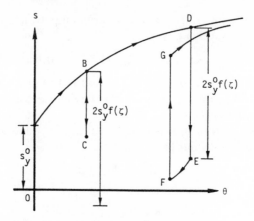

Figure 14.2 Shear stress versus shear strain as predicted by equation (14.8) for various points of unloading and reloading

according to the following equation

$$\rho(z) = \sum_{1}^{n} R_r e^{-\beta_r z}, \tag{14.18}$$

with $R_r > 0$, $\beta_r > 0$.

Under conditions of uniaxial loading, equation (14.17) gives a stress response depicted in Figure 14.2. Certain important features of the above response are worth pointing out. Unloading from a point D to a point F, below E (where DE is twice the current yield stress) and reloading, gives a closed loop as shown, where $GF = DE$. Unloading at another typical point, such as B, to a point C (where BC is less than twice the yield stress at B) and reloading gives rise to a path which returns along CB back to the original curve at B and on to D.

The response at $\theta = 0$ is indeterminate from equation (14.17). However, a process for which $\theta \equiv 0$ implies from equation (14.15) that

$$e = \frac{s}{2\mu}, \tag{14.19}$$

i.e. that the response at $\theta = 0$ is, in fact, elastic. Note that the features of this version of the endochronic theory are akin to classical plasticity theory.

The full three-dimensional version of the theory derived in Ref. 26 is summarized below. The constitutive equations governing the deviatoric stress $\mathbf{s}$ and the hydrostatic pressure σ are:

$$\mathbf{s} = s_y^0 \frac{\partial \boldsymbol{\theta}}{\partial z_D} + \int_0^{z_D} \rho(z_D - z') \frac{\partial \boldsymbol{\theta}}{\partial z'} \, dz', \tag{14.20}$$

$$\sigma = \sigma_y^0 \frac{\partial \Theta}{\partial z_H} + \int_0^{z_H} \phi(z_H - z') \frac{\partial \Theta}{\partial z'} \, dz', \tag{14.21}$$

where the plastic deviatoric strain increment $d\boldsymbol{\theta}$ and the plastic volumetric strain increment $d\Theta$ are given by the expressions:

$$d\boldsymbol{\theta} = d\mathbf{e} - \frac{d\mathbf{s}}{2\mu_0}, \tag{14.22}$$

$$d\Theta = \varepsilon - \frac{d\sigma}{3K_0}, \tag{14.23}$$

and the increments in the intrinsic time scales, dz_D and dz_H, are defined by equations (14.8) and (14.9).

It was pointed out in Ref. 26 that equations (14.20) and (14.21) embody the kinematic-cum-isotropic hardening model of classical plasticity. For instance, with equation (14.20) in mind, one can write

$$\mathbf{r} = \int_0^{z_D} \rho_1(z_D - z') \frac{\partial \boldsymbol{\theta}}{\partial z'} \, dz'. \tag{14.24}$$

Equations (14.20) and (14.4) then require that

$$\|\mathbf{s} - \mathbf{r}\|^2 = (s_y^0)^2 \left(\frac{d\zeta_D}{dz_D}\right)^2. \tag{14.25}$$

Of course, equation (14.25) is the equation of a hypersphere with centre r and radius $s_y^0(d\zeta_D/dz_H)$. The hypersphere translates with the history deformation and the vector of translation $\mathbf{r}$ is given by equation (14.24). The radius of the sphere expands or contracts, depending on whether $d\zeta_D/dz_D$ is an increasing or decreasing function of z_D.

It is interesting to note that in the case where the deviatoric and hydrostatic response are not coupled* $d\zeta_D/dz_D$ is, in fact, f_D. Thus, if the material hardens, $f_D(z_D)$ is an increasing function of z_D and the yield surface expands. If the material softens, then $f_D(z_D)$ is a decreasing function of z_D and the yield surface contracts. Similar conclusions can be drawn in relation to the hydrostatic response.

14.3 A NEW ENDOCHRONIC THEORY

14.3.1 Formulation of the new theory

The new endochronic theory is derived from the constitutive framework described in Section 14.2.4. To show how this can be done, let us consider

* In this case, $k_{rs} = 0$, $r \neq s$; $k_{00} = 1/f_D(\zeta_D)$.

the stress–strain behaviour in shear under monotonic loading conditions. We shall initially suppose that a finite yield point does exist. We now address ourselves to the case where the slope of the stress–strain path is continuous at the yield point. In this particular case, it follows that a plot of stress versus plastic strain θ yields a curve which has an infinite slope at $\theta = 0$.

With the above notions in mind, we return to equation (14.8) and differentiate it with respect to θ to obtain after some manipulation the result:

$$\left.\frac{ds}{d\theta}\right|_0 = s_y^0\left(\frac{d^2\zeta}{dz^2}\frac{1}{f(\zeta)}\right)_0 + \rho_1(0), \qquad (14.26)$$

where the suffix D has been dropped. We now wish to make the slope given in equation (14.26) infinite irrespective of the form of the hardening (softening) function $f(\zeta)$ of the material. This can be accomplished by setting

$$\rho_1(0) = \infty. \qquad (14.27)$$

If we further require that plastic deformation takes place upon initiation of loading, this may be achieved by setting

$$s_y^0 = 0. \qquad (14.28)$$

The notions described above, in regard to the shear component of deformation, can be applied equally as well to the hydrostatic component of deformation. When this is done, the three-dimensional response of isotropic endochronic materials, which yield immediately upon application of loading, is governed by equations (14.8), (14.9), (14.22) and (14.23) given earlier, together with the following equations:

$$\mathbf{s} = \int_0^{z_D} \rho(z_D - z')\frac{\partial\boldsymbol{\theta}}{\partial z}, dz', \qquad (14.29)$$

$$\sigma = \int_0^{z_H} \phi(z_H - z')\frac{\partial\Theta}{\partial z'}dz', \qquad (14.30)$$

where the kernel functions $\rho(0) = \phi(0) = \infty$. The above system of equations summarize the new endochronic theory.

---

*The singularities in the kernels $\rho(z)$ and $\phi(z)$ must be integrable in the sense that the integrals

$$\int_0^z \rho(z')\,dz', \qquad \int_0^z \phi(z')\,dz'$$

must exist for all finite z.

14.3.2 Interpretation of the new theory in terms of a mechanical model

In the discussion of the original endochronic theory given earlier in Section 14.2.2, it was noted that $\rho(z)$ and $\phi(z)$ were expressible as finite sums of exponentially decaying terms, i.e.,

$$\rho(z) = \sum_{r=1}^{n} R_r e^{-\beta_r z}, \tag{14.31}$$

$$\phi(z) = \sum_{r=1}^{m} P_r e^{-\lambda_r z}. \tag{14.32}$$

However, in the new theory, we require that $\rho(0) = \phi(0) = \infty$, so n or m cannot be finite. Thus, it becomes necessary to set both equal to infinity, in which event we have:

$$\rho(z) = \sum_{r=1}^{\infty} R_r e^{-\beta_r z}, \tag{14.33}$$

$$\phi(z) = \sum_{r=1}^{\infty} P_r e^{-\lambda_r z}. \tag{14.34}$$

These series must be convergent for all values of $z > 0$, but must diverge at $z = 0$.

Let us consider an example to demonstrate that the above conditions can be satisfied. For this purpose, we focus attention on the deviatoric kernel function, $\rho(z)$, and let

$$\rho(z) = \sum_{r=1}^{\infty} \frac{R_0}{r} e^{-\beta_0 r z}. \tag{14.35}$$

For $z = 0$, equation (14.35) becomes

$$\rho(0) = \sum_{r=1}^{\infty} \frac{R_0}{r}, \tag{14.36}$$

which is a well-known divergent series. On the other hand, when $z > 0$, equation (14.35) is convergent, as may be shown by the ratio test. For this purpose, consider the series:

$$X = \sum_{r=1}^{\infty} R_0 e^{-\beta_0 r z} \qquad (z > 0), \tag{14.37}$$

which is convergent, and has the sum

$$X = \frac{R_0}{e^{\beta_0 z} - 1}. \tag{14.38}$$

Since every term ρ_r in equation (14.35) is equal to, or is smaller than, the

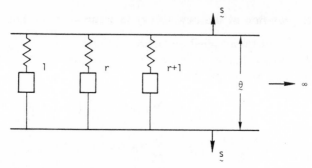

Figure 14.3 An infinite array of simple endochronic elements
in parallel

corresponding term X_r in equation (14.37), it follows that $\rho < X$. Consequently, ρ is a convergent series. Similar arguments can be made for the series ϕ. Thus, both kernels ρ and ϕ may be represented by infinite series of simple endochronic mechanical elements in parallel, as shown in Figure 14.3 for the case of shear.

Let us now examine the response of a typical simple endochronic element from the array shown in Figure 14.3. The details of such an element are depicted in Figure 14.4

The classical endochronic constitutive equation for the response of the element applies. Thus,

$$\mathbf{Q}_r = b_r \frac{d\mathbf{q}_r}{dz}, \tag{14.39}$$

$$\mathbf{Q}_r = (\boldsymbol{\theta} - \mathbf{q}_r) R_r, \tag{14.40}$$

where b_r is the resistance of the dissipative element r, and R_r is the spring

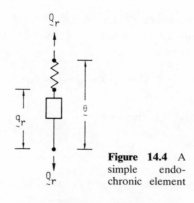

Figure 14.4 A simple endo-chronic element

stiffness. By combining equations (14.39) and (14.40), one finds that

$$\frac{d\mathbf{Q}_r}{dz} + \beta_r \mathbf{Q}_r = \frac{d\mathbf{\theta}}{dz} R_r, \tag{14.41}$$

where

$$\beta_r = \frac{R_r}{b_r}. \tag{14.42}$$

From equation (14.41),

$$\mathbf{Q}_r = R_r \int_0^z e^{-\beta_r(z-z')} \frac{\partial \mathbf{\theta}}{\partial z'} dz'. \tag{14.43}$$

But from Figure 14.3,

$$\mathbf{s} = \sum_1^\infty \mathbf{Q}_r. \tag{14.44}$$

Hence,

$$\mathbf{s} = \sum_{r=1}^\infty R_r \int_0^z e^{-\beta_r(z-z')} \frac{\partial \mathbf{\theta}}{\partial z'} dz', \tag{14.45}$$

thus recovering equation (14.29) in view of equation (14.31).

One feature of the constitutive relation given by equation (14.45) is that the overall strain is the *plastic* strain. However, we recall that

$$\mathbf{e} = \mathbf{\theta} + \frac{\mathbf{s}}{2\mu_0}. \tag{14.46}$$

This implies that the complete model, in terms of the total strain $\boldsymbol{\epsilon}$, is the infinite series model, itself in series with a spring of stiffness, $2\mu_0$. The complete model is then as shown in Figure 14.5.

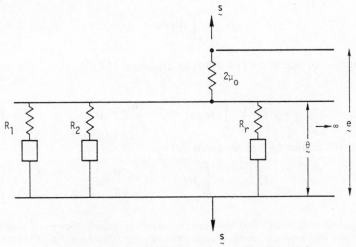

Figure 14.5 The complete infinite element model

A similar mechanical model can be constructed for the hydrostatic component of response.

14.3.3 Continuously distributed spectrum of elements

For the purposes of analysis and application, it is convenient, at this stage, to introduce a continuously distributed spectrum of elements. The corresponding mechanical model is depicted in Figure 14.6. To this end, we write

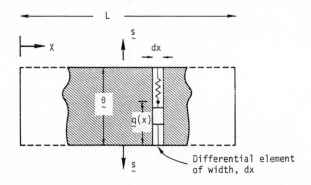

Figure 14.6 The continuously distributed element model

equations (14.31) and (14.32), respectively, in the form*

$$\rho(z) = \int_0^\infty R(r)e^{-rz}\,dr, \tag{14.47}$$

$$\phi(z) = \int_0^\infty P(r)e^{-rz}\,dr. \tag{14.48}$$

Substitution of equation (14.47) into equation (14.29) yields the relationship:

$$s = \int_0^\infty \int_0^z R(r)e^{-r(z-z')}\frac{\partial\boldsymbol{\theta}}{\partial z}\,dz'\,dr \tag{14.49}$$

Thus, if we define $Q(r, z)$ as given below:

$$\mathbf{Q}_r(r, z) = \int_0^z R(r)e^{-r(z-z')}\frac{\partial\boldsymbol{\theta}}{\partial z'}\,dz', \tag{14.50}$$

---

*The functions $R(r)$ and $P(r)$ must satisfy certain constraints. They must both be non-negative since they represent springs of distributed elements; their integral in the interval $(0, \infty)$ must diverge; finally, their Laplace transform must exist.

it follows from equation (14.49) that

$$\mathbf{s} = \int_0^\infty \mathbf{Q}_r(r, z)\, \mathrm{d}r. \tag{14.51}$$

Let us consider now a differential element of the continuously distributed element model shown in Figure 14.6. Let the spring stiffness distribution be $R(x)$ and the friction coefficient be $b(x)$ and let the stress in the element be $\mathrm{d}\mathbf{s}$. The spring constant of the element is, therefore, $R(x)\,\mathrm{d}x$ and the friction coefficient is $b(x)\,\mathrm{d}x$. Then, since we are dealing with an endochronic element, equations (14.39) and (14.40) apply, in which case,

$$\mathrm{d}\mathbf{s} = [\boldsymbol{\theta} - \mathbf{q}(x)]R(x)\,\mathrm{d}x, \tag{14.52}$$

$$\mathrm{d}\mathbf{s} = \frac{\mathrm{d}\mathbf{q}(x)}{\mathrm{d}z}\, b(x)\,\mathrm{d}x. \tag{14.53}$$

Let

$$\mathbf{Q}_x = \frac{\mathrm{d}\mathbf{s}}{\mathrm{d}x}. \tag{14.54}$$

Then we recover exactly equation (14.41), where $\beta(x)$ is defined by

$$\beta(x) \equiv \frac{R(x)}{b(x)}. \tag{14.55}$$

Thus,

$$\mathbf{Q}_x = \int_0^z R(x) e^{-\beta(x)(z - z')} \frac{\partial \boldsymbol{\theta}}{\partial z'}\, \mathrm{d}z' \tag{14.56}$$

Using equation (14.51) or equation (14.54), it follows that

$$\mathbf{s} = \int_0^z \rho(z - z') \frac{\partial \boldsymbol{\theta}}{\partial z'}\, \mathrm{d}z', \tag{14.57}$$

where

$$\rho(z) = \int_0^L R(x) e^{-\beta(x)z}\, \mathrm{d}x. \tag{14.58}$$

To obtain equation (14.47), we introduce the transformation

$$\beta(x) = r, \tag{14.59}$$

together with the conditions $\beta(0) = 0$, $\beta(L) = \infty$. In view of equation (14.59), x is a function of r, i.e.

$$x = x(r). \tag{14.60}$$

Thus, as a result of equation (14.58),

$$\rho(z) = \int_0^\infty R^*(r)e^{-zr}\,\mathrm{d}r, \qquad (14.61)$$

where

$$R^*(r) = R[x(r)]\frac{\mathrm{d}x}{\mathrm{d}r}. \qquad (14.62)$$

Dropping the star in $R^*(r)$ in equation (14.61), we obtain equation (14.47).

14.3.4 An alternative approach to incorporation of hardening

The approach described in the preceding sections to account for hardening or softening behaviour in the new endochronic theory is based on the use of an intrinsic time scale z which depends on the intrinsic time measure ζ through the relationship:

$$\mathrm{d}z = \frac{\mathrm{d}\zeta}{f(z)}. \qquad (14.63)$$

The form of the function $f(z)$ is selected to provide the desired hardening or softening behaviour.

An alternative way of incorporating hardening/softening into the model has been found, however, which appears to provide some advantages from a numerical standpoint. To formulate this alternative approach, let us consider the deviatoric response, and set:

$$\mathrm{d}z = \mathrm{d}\zeta. \qquad (14.64)$$

Let us recall that the deviatoric kernel function $\rho(z)$ was obtained by summing the response of an infinite array of simple endochronic elements assembled in parallel, as in Figure 14.3. Let us consider now a constituent element from this array, as depicted in Figure 14.4. Its mechanical response is governed by equations (14.39) and (14.40), i.e.

$$\mathbf{Q}_r = b_r \frac{\mathrm{d}\mathbf{q}_r}{\mathrm{d}z}, \qquad (14.65)$$

$$\mathbf{Q}_r = R_r(\boldsymbol{\theta} - \mathbf{q}_r). \qquad (14.66)$$

If b_r and R_r are constants, one obtains equation (14.57). However, for materials that exhibit hardening or softening, b_r and R_r may depend on a variable, or variables, which characterize the hardening.

Let us suppose that both b_r and R_r depend, for example, on z through the same hardening (softening) function $H_\mathrm{D}(z)$, i.e.

$$b_r = b_r^0 H_\mathrm{D}(z), \qquad (14.67)$$

$$R_r = R_r^0 H_\mathrm{D}(z). \qquad (14.68)$$

By eliminating Q_r from equations (14.65) and (14.66), we obtain the following differential equation in q_r:

$$\frac{d\mathbf{q}_r}{dz} + \beta_r \mathbf{q}_r = \beta_r \boldsymbol{\theta}, \tag{14.69}$$

where

$$\beta_r = R_r/b_r. \tag{14.70}$$

Note that β_r is a constant because of relations (14.67) and (14.68).* The solution of equation (14.69) with the initial conditions $\mathbf{q}_r(0) = \boldsymbol{\theta}(0) = 0$, is the following:

$$\boldsymbol{\theta} - \mathbf{q}_r = \int_0^z e^{-\beta_r(z-z')} \frac{\partial \boldsymbol{\theta}}{\partial z'} dz' \tag{14.71}$$

Upon substitution into equation (14.66), we obtain for Q_r

$$Q_r = H_D(z) R_r^0 \int_0^z e^{-\beta_r(z-z')} \frac{\partial \boldsymbol{\theta}}{\partial z'} dz', \tag{14.72}$$

and by summing forces,

$$\mathbf{s} = \sum_{r=1}^{\infty} \mathbf{Q}_r. \tag{14.73}$$

Therefore, from equations (14.72) and (14.73), we obtain

$$\mathbf{s} = H_D(z) \int_0^z \rho(z-z') \frac{\partial \boldsymbol{\theta}}{\partial z'} dz', \tag{14.74}$$

where

$$\rho(z) = \sum_{r=1}^{\infty} R_r^0 e^{-\beta_r z}. \tag{14.75}$$

Using reasoning similar to that above, it can be shown that expressions corresponding to equations (14.74) and (14.75) also apply to hydrostatic behaviour, whenever the hydrostatic material parameters b_r and P_r depend on the same hardening function H_H, i.e.,

$$b_r = b_r^0 H_H, \tag{14.76}$$

$$P_r = P_r^0 H_H. \tag{14.76}$$

When this is the case, the hydrostatic response is given by the equation:

$$\sigma = H_H \int_0^{z_H} \phi(z_H - z') \frac{\partial \Theta}{\partial z'} dz', \tag{14.77}$$

---

* Note also that β_r is a constant so long as b_r and R_r depend on the same function H_D, regardless of the argument of this function.

where

$$\phi(z) = \sum_{r=1}^{\infty} P_r^0 e^{-\lambda_r z}. \tag{14.78}$$

In the case of soils, hydrostatic hardening occurs due to volumetric compaction and, consequently, it would seem reasonable for such materials to assume that the function H_H depends on the plastic volumetric strain, i.e. $H_H = H_H(\Theta)$.

14.3.5 Mathematical form of the new theory for arbitrary deformation histories

Below, we consider in some further detail the mathematical form which the new theory takes for arbitrary deformation histories. To simplify the discussion, the developments given below are derived only for the case of simple shear. However, the basic form of the results so obtained apply equally well to the more general model, including the representations of hydrostatic response and of more general deviatoric behaviour.

Restricting attention now to the case of simple shear, the governing endochronic constitutive relation is, from equation (14.74),

$$s = H_D \int_0^{z_D} \rho(z - z') \frac{d\theta}{dz'} dz'. \tag{14.79}$$

where H_D denotes the hardening (softening) function, for deviatoric response, θ is the plastic component of the shear strain, and z_D represents the deviatoric intrinsic time.

Upon letting $z_1, z_2, \ldots, z_n$ denote the values of the deviatoric intrinsic time z_D at which reversals in the shear strain rate have occurred, equation (14.79) may be written in expanded form as

$$s = H_D \left[\lambda_1 \int_0^{z_1} p(z - z') \, dz' + \lambda_2 \int_{z_1}^{z_2} \rho(z - z') \, dz' + \cdots \right.$$
$$\left. + \lambda_{n+1} \int_{z_n}^{z} \rho(z - z') \, dz' \right], \tag{14.80}$$

where, if $d\theta/dz' = +1$ for the initial part of the deformation, we have

$$\lambda_r = \begin{cases} +1, & \text{if } r \text{ is odd,} \\ -1, & \text{if } r \text{ is even.} \end{cases} \tag{14.81}$$

Equation (14.80) may alternatively be rewritten in the more compact form

$$s = H_D \left\{ \sum_{r=1}^{n} \left[\lambda_r \int_{z_{r-1}}^{z_r} \rho(z - z') \, dz' \right] + \lambda_{n+1} \int_{z_n}^{z} \rho(z - z') \, dz' \right\} \tag{14.82}$$

If we now set

$$\int_0^z \rho(x')\,dx' \equiv F(z) \tag{14.83}$$

and let

$$x' = z - z', \tag{14.84}$$

then

$$\int_{z_{r-1}}^{z_r} \rho(z-z')\,dz' = F(z-z_{r-1}) - F(z-z_r) \tag{14.85}$$

and

$$\int_{z^n}^z \rho(z-z')\,dz' = F(z-z_n). \tag{14.86}$$

Substituting these results into equation (14.82), we obtain the following expression for s:

$$s = H_D\left\{ F(z) - 2\left[F(z-z_1) - F(z-z_2) + \cdots + (-1)^{n+1}F(z-z_n) \right] \right\} \tag{14.87}$$

or

$$s = H_D\left[F(z) - 2\sum_{r=1}^{n} (-1)^{r+1}F(z-z_r) \right]. \tag{14.88}$$

Although the above result has been developed for the case of simple shear, the general form applies to more general deviatoric behaviour as well as to hydrostatic response. In the latter case, the pressure σ is given by an equation of the form

$$\sigma = H_H\left[G(z_H) - 2\sum_{r=1}^{m} (-1)^{r+1}G(z_H - z_r^H) \right], \tag{14.89}$$

where H_H is the hardening function for hydrostatic response, G is defined similarly to F, i.e.

$$G(z_H) \equiv \int_0^{z_H} \phi(x')\,dx', \tag{14.90}$$

and $z_1^H, z_2^H, \ldots, z_n^H$ denote the values of the hydrostatic intrinsic time z_H at which the volumetric strain rate $\dot{\varepsilon}$ changes sign.

14.3.6 An illustrative example: pure shear response

To illustrate some features of the new model described in the preceding sections, let us consider the case of pure shear and adopt the following form

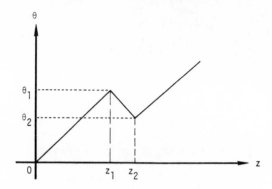

Figure 14.7 Typical history of $\theta(z)$

for the kernel function $\rho(z)$:

$$\rho(z) = \rho_0/z^\alpha, \tag{14.91}$$

where α is a constant which satisfies the inequality $0 < \alpha < 1$. Introducing this kernel function into equation (14.29) and ignoring the suffix D leads to the expression

$$s = \rho_0 \int_0^z \frac{1}{(z-z')^\alpha} \frac{d\theta}{dz'} dz'. \tag{14.92}$$

An arbitrary pure shear deformation involving loading, unloading, reloading, etc. will produce the history of θ versus z depicted in Figure 14.7. Loading takes place in the range $0 \leq z \leq z_1$; unloading occurs in the range $z_1 < z \leq z_2$; while reloading takes place for $z_2 < z < \infty$. Let us now consider the stress–strain relations which result from equation (14.92) for these three cases.

(1) Loading: $\dfrac{d\theta}{dz} = 1$

In this case, equation (14.92) gives

$$s = \left(\frac{\rho_0}{\beta}\right)\theta^\beta, \tag{14.93}$$

where $\beta = 1 - \alpha$; $0 < \beta < 1$. It then follows that

$$\theta = \left(\frac{s^\beta}{\rho_0}\right)^{1/\beta}. \tag{14.94}$$

Since, from equation (14.46),

$$e = \theta + \frac{s}{2\mu_0}, \tag{14.95}$$

it follows that

$$e = \frac{s}{2\mu_0} + \left(\frac{s^\beta}{\rho_0}\right)^{1/\beta}.$$ (14.96)

Interestingly enough, this is the Ramberg–Osgood equation for one-dimensional stress–strain response, introduced in 1943, which is frequently used to describe the behaviour of soils under shear loading (see Refs. 12 and 16, for example).

(2) Unloading:

$$\frac{d\theta}{dz} = +1, \quad 0 \leq z \leq z_1; \quad \frac{d\theta}{dz} = -1, \quad z_1 < z \leq z_2.$$

Again, using equation (14.92), we find that at any z such that $z_1 < z \leq z_2$,

$$s(z) = R_0\{z^\beta - 2(z - z_1)^\beta\},$$ (14.97)

where

$$z = 2\theta_1 - \theta, \quad z_1 = \theta_1 \quad \text{and} \quad R_0 = \rho_0/\beta.$$ (14.98)

Upon setting

$$x = \theta_1 - \theta,$$ (14.99)

equation (14.97) takes the form

$$s_U(x) = R_0\{(\theta_1 + x)^\beta - 2x^\beta\}$$ (14.100)

where the suffix U specifically designates the functional dependence of s on x during unloading.

Remark 1: The slope $ds/d\theta$ of the stress–plastic strain curve, at the onset of unloading, $\theta = \theta_1$ ($x = 0$), is infinite. Therefore, the slope of the stress–strain curve at unloading is the elastic slope.

Combining equations (14.99) and (14.100), we obtain

$$s_U(\theta) = R_0\{(2\theta_1 - \theta)^\beta - 2(\theta_1 - \theta)^\beta\},$$ (14.101)

which can be differentiated to give

$$\frac{ds_U}{d\theta} = R_0\left\{\frac{2\beta}{(\theta_1 - \theta)^\alpha} - \frac{\beta}{(2\theta_1 - \theta)^\alpha}\right\},$$ (14.102)

where $\alpha = 1 - \beta$. Note that at $\theta = \theta_1$, $ds_U/d\theta = \infty$.

(3) Reloading:

$$\frac{d\theta}{dz} = +1, \quad 0 \leq z \leq z_1; \quad \frac{d\theta}{dz} = -1, \quad z_1 < z \leq z_2; \quad \frac{d\theta}{dz} = +1, \quad z_2 \leq z \leq \infty.$$

Using equation (14.92) in conjunction with the above values of $d\theta/dz$, one obtains the following equation for s during reloading:

$$s = R_0\{z^\beta - 2(z - z_1)^\beta + 2(z - z_2)^\beta\}, \tag{14.103}$$

where

$$z_1 = \theta_1; \qquad z_2 = 2\theta_1 - \theta_2$$
$$z - z_1 = \theta + \theta_1 - 2\theta; \qquad z - z_2 = \theta - \theta_2.$$

Setting

$$y = \theta - \theta_2, \quad \Delta = \theta_1 - \theta_2 \qquad (\theta \geqslant \theta_2), \tag{14.104}$$

one obtains the following expression for s:

$$s_R(y) = R_0\{(y + \Delta + \theta_1)^\beta - 2(y + \Delta)^\beta + 2y^\beta\}, \tag{14.105}$$

where the suffix R specifically designates the functional dependence of s on y during reloading.

Remark 2. The slope of the stress–plastic strain curve at the onset of reloading, $\theta = \theta_2$, is infinite. Therefore, the slope of the stress–strain curve at reloading is the elastic slope.

The conclusion is arrived at on the basis of equation (14.105) following a calculation similar to that of Remark 1. It is also evident that the reloading slope is higher than the unloading slope at reloading.

A proof of hysteresis loop closure is given in Ref. 29.

14.4 APPLICATION OF NEW ENDOCHRONIC THEORY TO SOILS

In this section, the ability of the new endochronic theory to describe important features of the response of soils is demonstrated. Endochronic models are developed for several simple types of deformation, including simple shear and hydrostatic compression, and, where appropriate, applied to existing soil data. Particular attention is given to the ability of the model to describe observed hysteretic behaviour of soils during multiple cycles of unloading and reloading.

14.4.1 Shear of cohesionless soils

The mechanical response of cohesionless soils subjected to shear depends on the initial relative density of the soils. If the initial relative density is less than critical, i.e., a loose soil, the shear stress typically increases monotonically with increasing shear strain to a limiting shear stress, the value of which depends on the confining pressure. If the initial relative density is greater than critical, the stress–strain curve first increases to a peak stress and then decreases toward a finite limiting stress, whose value also depends upon the confining pressure.

The soil response features described above can be easily represented by the endochronic theory. To illustrate this, consider equation (14.29), which for the special case of simple shear reduces to the form

$$s = \int_0^z \rho(z - z') \frac{\mathrm{d}\theta}{\mathrm{d}z'} \mathrm{d}z', \tag{14.106}$$

where, to account for the hardening–softening behaviour exhibited by dense cohesionless soils, we adopt the expressions

$$\mathrm{d}z = \mathrm{d}\zeta / f(z) \tag{14.107}$$

and

$$\mathrm{d}\zeta = |\mathrm{d}\theta|. \tag{14.108}$$

Equation (14.106) may then be rewritten as

$$s = \int_0^z \rho(z - z') f(z') \frac{\mathrm{d}\theta}{\mathrm{d}\zeta} \mathrm{d}z'. \tag{14.109}$$

We have found that the response of soils to simple shear can be described very well by the following forms for $\rho(z)$ and $f(z)$:

$$\rho(z) = \rho_0 \frac{e^{-kz}}{z^\alpha}, \tag{14.110}$$

$$f(z) = 1 + \beta^* e^{-k_1 z}, \tag{14.111}$$

where ρ_0, α, β^*, k and k_1 are constants. Upon substituting equations (14.110) and (14.111) into equation (14.109) and performing the integration, it can be placed in the form:

$$s = s_\infty [F(z) - 2F(z - z_1) + 2F(z - z_2) - \cdots$$
$$+ (-1)^{n+1} 2F(z - z_n)], \tag{14.112}$$

where

$$F(z) = \gamma(a)^{-1} [\gamma(a, kz) + \beta^* (k/q)^a e^{-k_1 z} \gamma(a, qz)] \tag{14.113}$$

and

$$s_\infty = \rho_0 k^{-a} \gamma(a), \tag{14.14}$$

$$1 - \alpha = a, \tag{14.15}$$

$$k - k_1 = q,$$

and $z_1, z_2, \ldots, z_n$ denote the values of the intrinsic time scale at which reversal in the sign of the shear strain rate have occurred. Also, $\gamma(a)$

denotes the complete Gamma function and $\gamma(a, \ldots)$ is the incomplete Gamma function.

The system of equations that completely define simple shear for arbitrary unloading–reloading thus consists of equation (14.112)–(14.115) together with the elastic relation:

$$ds = 2G(de - d\theta), \tag{14.116}$$

where G denotes the shear modulus and e is the shear strain.

Our experience to date, in applying the above model to soil data, has shown that good fits to such data can be obtained for values of a (or α) close to 0.5. For the special case of $\alpha = 0.5$, equation (14.113) reduces to the following form:

$$F(z) = \text{Erf}\,(\sqrt{kz}) + \beta e^{-k_1 z}\,\text{Erf}\,(\sqrt{qz}), \tag{14.117}$$

where Erf ( ) denotes the error function and

$$\beta = \beta^* \sqrt{\left(\frac{k}{k - k_1}\right)} \tag{14.118}$$

Figures 14.8–14.11 depict various features of the simple shear behavior predicted by the model described above . To determine to what extent the hysteretic features of the above model during unloading and reloading processes are sensitive to the parameter a (or equivalently α), two values of a were selected and corresponding values of k were found which gave close

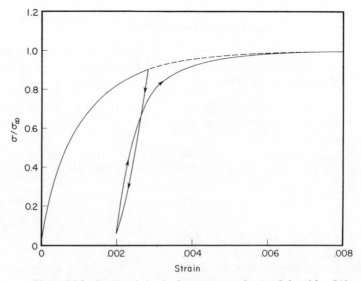

Figure 14.8   Predicted simple shear response for $\alpha = 0.6$ and $k = 540$

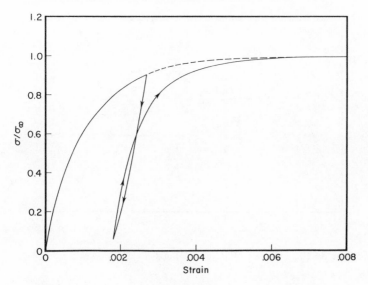

Figure 14.9 Predicted simple shear response for $\alpha = 0.5$ and $k = 700$

fits of the model to a prescribed initial loading curve with $\beta^* = 0$. The curves shown in Figure 14.8 were calculated with $a = 0.6$ and $k = 540$, while the curves given in Figure 14.9 correspond to $a = 0.5$ and $k = 700$. Furthermore, the shear modulus G was assigned the value 19.75 MPa in both cases. These figures illustrate the ability of the model to describe, without the use of yield surfaces, the general inelastic response features of soil under shear. Furthermore, they depict another important feature of the new theory, namely, that the initial loading path (the continuation of which is indicated by the dashed lines) is 'remembered' by the model; after unloading, the reloading path returns to the continuation of the initial loading curve. Comparison of the predicted unloading–reloading paths shown in these figures reveals that they are virtually identical. Consequently, it appears that the hysteretic features of this particular model are set, once the initial loading curve has been prescribed.

Figures 14.10 and 14.11 illustrate some of the features of the model when β is nonzero. Such values of β typify the response of dense cohesionless soils, and evidently β depends in some manner on the initial relative density. Figure 14.10 shows initial loading curves for several different values of β that appear to bracket the general type of simple shear response exhibited by cohesionless soils. Similar behaviour is also exhibited by overconsolidated clays. Figure 14.11 shows the hysteretic feature of the model for the case in which unloading and reloading occur after the peak stress has been reached. The curve shown in the figure was obtained with $\beta = 2$.

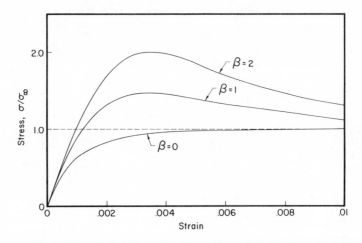

Figure 14.10 Influence of β on the initial loading curves for simple shear

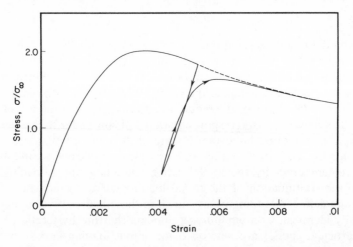

Figure 14.11 Hysteretic unloading–reloading response of simple shear model with $\beta = 2$

14.4.2 Cyclic shear of cohesionless soils

The simple shear version of the model, described in Section 14.3.4, has been applied to the cyclic shear data for drained sand reported in Ref. 4. The data were obtained in a cyclic shear device that is a modified version of the type initially designed by the Norwegian Geotechnical Institute. The tests were performed on cylindrical samples of sand enclosed in wire-reinforced rubber membranes. The end surfaces of the cylindrical samples were subjected to cyclic relative displacements normal to the cylinder axis, while maintaining a fixed distance between these surfaces. Moreover, each test was conducted with a dead load applied to the specimen along the direction of the cylinder axis, and the frequency of the cyclic relative displacement was 1 Hz.

Although the experimental device described above is commonly referred to as a 'simple shear' device, the states of stress that it produces in a soil specimen are not strictly simple shear. In order to have simple shear, uniform shear stresses would have to be applied over the curved lateral surface of the cylindrical specimen and this is not, of course, the boundary condition applied by the wire-reinforced flexible membrane. Furthermore, while such tests are typically performed under a constant vertical load, the pressure in the soil does not remain constant during a test, but increases due to the build-up of the lateral stresses from hardening.

A detailed analysis of the response of sand in the cyclic shear device described above has recently been performed with a dynamic 3-D finite element code (Ref. 2). The results of this analysis reveal that the cyclic response of sand predicted with the sophisticated approach is similar to that calculated under the assumption of uniform simple shear.

On this basis, we applied the simple shear version of the new endochronic model described in Section 14.3.5 to the set of cyclic shear data given in Ref. 4 corresponding to a relative density of 45%, an applied vertical stress of 0.192 MPa, and a peak shear strain of 0.3%. The initial state from which intrinsic time z is measured was taken to be the state prior to the initiation of shearing. A more precise approach would adopt the state of the material prior to the application of the vertical load as the initial state. However, no information is available on the deformation that occurred during vertical load application, and consequently the corresponding increment in intrinsic time could not be determined.

The appropriate governing system of equations for simple shear is, from Section 14.3.5,

$$s = H\{F(z) - 2[F(z - z_1) - F(z - z_2)$$
$$+ \cdots + (-1)^{n+1}F(z - z_n)]\}, \tag{14.119}$$

$$ds = 2G(de - d\theta), \tag{14.120}$$

$$dz = |d\theta|, \tag{14.121}$$

where, for the present application, we take:

$$F(z) = \mathrm{Erf}\,(\sqrt{kz}). \tag{14.122}$$

Here, Erf ( ) denotes the error function and k is a material constant. Furthermore, for materials that exhibit cyclic hardening, such as the loose dry sand considered in Ref. 4, it is assumed that the hardening function H

depends on z as follows:

$$H = H_0 + (H_\infty - H_0)(1 - e^{-\gamma z}).$$ (14.123)

The integration of the non-linear system of equations given by equations 14.119 to 14.123 requires the use of numerical methods. For this purpose, a small computer program was developed* to calculate the shear stress for a prescribed cyclic shear strain history; the numerical scheme adopted for this purpose utilized an iteration scheme based on the secant method.

One important feature of the model described above should be noted, since it greatly increases the computing speed. Consider equation (14.119), the right-hand side of which contains an increasing number of terms as the number of cycles (and therefore the number of reversals in the sign of the shear strain rate) increases. In general, the right-hand side of equation (14.119) will contain $2n + 1$ terms after n cycles of deformation. Furthermore, the values of the $2n$ quantities $z_1, z_2, \ldots z_{2n}$ must also be retained. This means that, for the present case of interest, which involves up to 300 cycles of deformation, there will be about 600 terms present during the final cycle and 600 terms of z_i to be remembered, and this could prove computationally unattractive. If, however, the function has a form which saturates, with suitable rapidity, to a limiting value as z increases, some important advantages occur from a computing standpoint. First of all, the function adopted in equation (14.122) saturates to unity for sufficiently large values of the argument. When this occurs, there is no further need to calculate its value, since for all future deformation, the value of z continues to increase, and the function remains at its limiting value. Similar comments obviously apply to the general term $F(z - z_i)$ also. Secondly, if a term $F(z - z_i)$ saturates, it no longer becomes necessary to retain in memory the value of z_i, or for that matter any of the z_j which are less than z_i. Consequently, we can say that when the function F has a form that saturates, the corresponding endochronic model exhibits 'fading memory'.

The notions discussed above were incorporated into the computer program listed in Appendix C of Ref. 29 to increase its computing efficiency. Using this program, numerical calculations were performed relative to the data of interest from Ref. 4. Excellent agreement between the data and the calculations was obtained for the following values of the model parameters:

$$G = 19.75 \text{ MPa} \qquad H_\infty = 74.7 \text{ kPa}$$
$$H_0 = 28.73 \text{ kPa} \qquad \gamma = 30.6 \qquad\qquad (14.124)$$
$$k = 750$$

Moreover, a uniform strain increment of $\Delta e = 0.01\%$ was used throughout the calculation.

---

* A listing of this computer program is given in Appendix C, Ref. 29.

The calculated stress–strain paths for cycles 1, 2, 10 and 300, based on the parameters listed above, are shown in Figure 14.12(a)–(d), together with the data from Ref. 4, which is denoted by open circles. The dashed curves that appear in these figures are an attempt to pass smooth curves through the data. The close agreement between the predicted and observed response of the sand over 300 cycles of deformation illustrates the powerful predictive capability of the new theory. Note that the model requires only five parameters.

14.4.3 Hydrostatic compression of soils

There are three essential features that characterize the hydrostatic response of a wide variety of soils, namely:

(1)   The pressure–volumetric strain loading path exhibits a concavity at moderately high pressures, such as depicted in Fig. 14.13.
(2)   During hydrostatic unloading–reloading processes, soils exhibit hysteresis loops similar to that shown in the figure.
(3)   The unloading slope, at the onset of unloading, increases with volumetric compaction.

The observed hysteresis indicates that significant intergrain sliding accompanies the process of compression, leading to a more compact grain configuration. This, in turn, leads to a greater number of intergrain friction surfaces, a result which increases the amount of frictional resistance that must be overcome to produce further compression. Furthermore, as more grains come into contact, during compression, the bulk modulus increases accordingly. These observations lead us to conclude that the frictional as well as elastic resistance to compression increases concomitantly with compaction. Mathematically, we interpret this to mean that the frictional and the elastic resistances to compression increase with the plastic volumetric strain Θ.

It has proven difficult to describe *all* of the above phenomena using existing theories. In some cases, highly non-linear theories of the hypoelastic type have been used to represent the above phenomena, but not without difficulty. For example, in developing a hypoelastic description for hydrostatic compression of soils, Romano[13] assumed constitutive functions for unloading that differed from those adopted during reloading. Also, while the earlier version of the endochronic theory for soils[11] could describe most of the observed features of hydrostatic compression of soils, it was incapable of portraying hysteretic response during unloading and reloading.

In that which follows, it is shown that the new endochronic theory provides a framework wherein *all* of the essential features of the hydrostatic compression of soils can be described with simplicity—and with constitutive

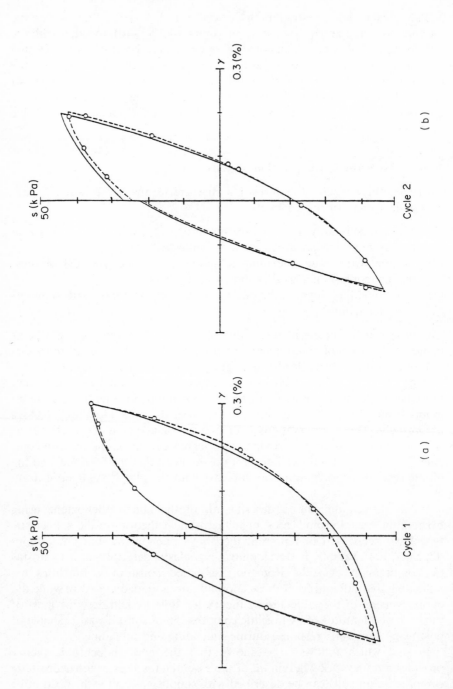

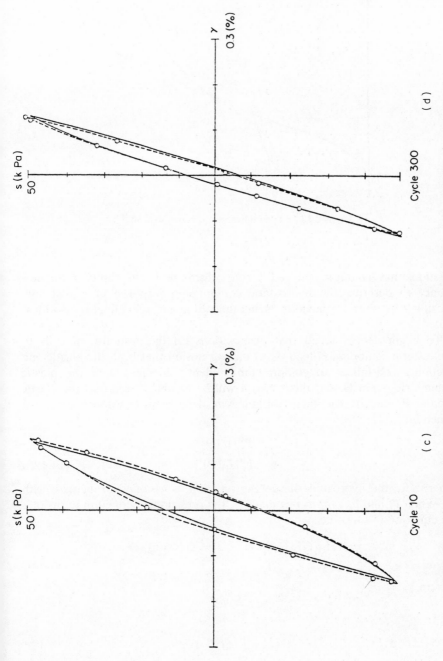

Figure 14.12 Response of sand to cyclic shear under an applied vertical stress of 192 kPa (data from Ref. 4)

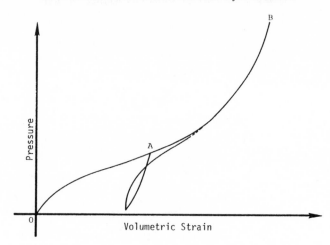

Figure 14.13 Typical hydrostatic response of soils

relations that *remain unchanged* during deformation. The ability of the new theory to describe the hydrostatic compression response of a real soil, including hysteretic behaviour during unloading and reloading, is also illustrated.

To begin, let us recall that observations on the response of soils to hydrostatic compression lead us to the conclusion that both the elastic and frictional resistances to compression evidently increase with the plastic volumetric strain Θ. On this basis, it will be assumed here that the elastic moduli P_r and the frictional resistances b_r depend on Θ through the *same* function, i.e.,

$$P_r = P_r^0 H_H(\Theta),$$
$$b_r = b_r^0 H_H(\Theta), \tag{14.125}$$

where H_H, the hydrostatic hardening function, is assumed to increase with Θ. When P_r and b_r vary in this manner, the pressure is defined in terms of z by equation (14.89), i.e.

$$\sigma = H_H\{G(z) - 2[G(z - z_1) - G(z - z_2) + \cdots + (-1)^{n+1}G(z - z_n)]\}, \tag{14.126}$$

where

$$dz = |d\Theta| \tag{14.127}$$

and

$$G(z) = \int_0^z \phi(x')\,dx'. \tag{14.128}$$

To describe the hydrostatic compression of real soils, we adopt the following forms of ϕ and H_H:

$$\phi = \phi_0 z^{-\alpha}$$
$$H_H = (\Theta_m - \Theta)^{-\beta}, \tag{14.129}$$

where α and β are constants such that $0 < \alpha < 1$, $\beta > 1$, and Θ_m denotes the maximum plastic volumetric strain that can be achieved in compression. Substitution of equation (14.129) into equation (14.128) leads to the result

$$G(z) = c_0 z^m, \tag{14.130}$$

where we have set $c_0 = \phi_0/(1-\alpha)$ and $m = 1 - \alpha$. To the above equations, we must adjoin the incremental elastic relation for hydrostatic compression, namely:

$$d\sigma = K(d\varepsilon - d\Theta). \tag{14.131}$$

Here, K denotes the bulk modulus which, in view of the apparent increase in the elastic resistance with plastic strain during compression, will be assumed, as a first approximation, to depend linearly on Θ, i.e.,

$$K = K_0 + m_0\Theta, \tag{14.132}$$

where K_0 and m_0 are constants. The complete system of equations that describe hydrostatic response therefore consist of equations (14.126), (14.127) and (14.130)–(14.132), which contain six material parameters, namely, c_0, m, β, Θ_m, K_0 and m_0. A small computer program, similar to that given in Appendix C of Ref. 29, was developed to solve this set of equations for a prescribed volumetric strain history.

To illustrate the modeling capability of the new endochronic theory, the hydrostatic compression data for the test identified as HCU-2 in Ref. 3 were selected for consideration, and the model described above was applied to it. A fit to the data was obtained with the values of the parameters listed below:

$$c_0 = 0.72 \qquad \Theta_m = 0.397$$
$$m = 0.45 \qquad K_0 = 11.72\ \text{MPa} \tag{14.133}$$
$$\beta = 3.76 \qquad m_0 = 2.76\ \text{GPa}$$

A comparison between the data and the model prediction based on the parameters listed above is given in Figure 14.14. An inspection of this figure reveals the excellent modelling capability provided by the model, including the hysteresis during unloading and reloading. It should be kept in mind that the model which provided these results contains constitutive relations that remain unchanged during the entire deformation and, in addition, requires only six material parameters.

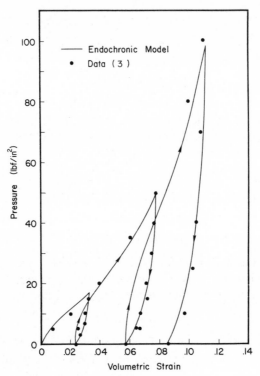

Figure 14.14 Hydrostatic compression of McCor-
mick Ranch soil

14.4.4 Densification under cyclic shear

Densification and dilatancy in isotropic materials result from coupling between the hydrostatic and deviatoric components of deformation. Densification occurs if applied shear produces a decrease in volume, while dilatancy
takes place if there is an increase in volume due to applied shear. In this
section, we describe how coupling between hydrostatic and shear components is easily achieved in the new endochronic theory through the
generalized concept of intrinsic time, first presented in Ref. 26. A relatively
simple endochronic model for treating densification due to cyclic simple
shear is developed and applied to the densification data for dry sand
reported by Silver and Seed.[15]*

To begin, let us recall equations (14.16) and (14.17), which provide
coupling between the deviatoric and volumetric modes of deformation.

---

* The justification for assuming the states of stress produced by the so-called 'simple shear'
device[15] to be states of simple shear was discussed earlier in Section 14.4.2, and will not be
repeated here.

These equations may be generalized to the forms:

$$dz_D^2 = k_{00}^2 \left(\frac{d\zeta_D}{f_D}\right)^2 + k_{01}^2 \left(\frac{d\zeta_H}{f_H}\right)^2, \tag{14.134}$$

$$dz_H^2 = k_{10}^2 \left(\frac{d\zeta_D}{f_D}\right)^2 + k_{11}^2 \left(\frac{d\zeta_H}{f_H}\right)^2, \tag{14.135}$$

through the addition of the hardening (or softening) functions f_D and f_H. The k_{rs} are new parameters, not equal to those in equations (14.16) and (14.17), which reflect the sensitivity of coupling between deviatoric and volumetric deformational modes.

On inspection of equation (14.135), we notice that the hydrostatic intrinsic time scale z_H is coupled to its deviatoric counterpart z_D through the coupling parameter k_{10}. During cyclic shearing, densification occurs, which means that both $d\zeta_D$ and $d\zeta_H$ increase simultaneously. Therefore, in developing a model for densification under cyclic shear, both have to be considered in the evaluation of dz_H.

From experimental observation, however, it is known that the rate of densification $d\zeta_H/d\zeta_D$ is small for sands and, accordingly, one may, as a first approximation, ignore the second term in the right-hand side of equation (14.135). In this case, we obtain

$$dz_H = k_{10} \frac{d\zeta_D}{f_D}, \tag{14.136}$$

where $d\zeta_D$ is the increment in the deviatoric intrinsic time measure, and f_D is proportional to the internal frictional resistive force in shear. On physical grounds, we expect this force to be essentially proportional to the applied pressure, σ_0. We therefore set

$$f_D = c_1 \left(\frac{\sigma_0}{\sigma_y}\right), \tag{14.137}$$

where c_1 is a dimensionless constant and σ_y is a constant having the units of stress. Without loss of generality, we may set $c_1 = 1$, in which case equation (14.136) becomes

$$dz_H = k_{10} \left(\frac{\sigma_y}{\sigma_0}\right) d\zeta_D. \tag{14.138}$$

We consider now the hydrostatic constitutive relation, which we take in the form:

$$\sigma = H_H \int_0^{z_H} \phi(z_H - z_H') \frac{\partial \Theta}{\partial z_H'} dz_H', \tag{14.139}$$

where dz_H is defined according to equation (14.138). For the purpose of

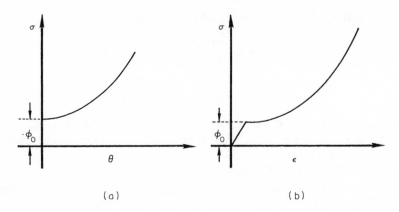

Figure 14.15 Dependence of σ on ε and θ for hydrostatic model assumed in densification analysis

illustrating the densification capability of the new theory, it is sufficient to consider a simple, yet physically realistic, description of hydrostatic behaviour. With this in mind, let us consider an endochronic hydrostatic model that has a yield point, such as is shown in Figure 14.15. The analytic representation of this model is achieved by setting

$$\phi(z) = \phi_0 \delta(z),$$
$$H_H = e^{k\Theta}, \tag{14.140}$$

where ϕ_0 and k are constants, and $\delta(z)$ denotes the Dirac delta function. Upon substituting equations (14.140) into equation (14.139), we obtain

$$\sigma = \phi_0 e^{k\Theta} \frac{d\Theta}{dz_H}. \tag{14.141}$$

At $\Theta = 0$, $\sigma = \phi_0$; therefore ϕ_0 is the pressure at which yield first takes place.

In the densification experiments reported in Ref. 15, the sand was subjected to various vertical loads before cyclic shearing was initiated and these loads remained constant thereafter. No information is available on the deformations that occurred during the applications of the vertical loads, and so the contributions made to the intrinsic time due to these initial compressions of the sand cannot be determined. In view of this, we will assume that the initial state of the sand is the state that exists just after the application of the vertical load.

If we denote the pressure produced by the initial compression by σ_0, where $\sigma_0 > \phi_0$, equation (14.141) can be written in the form

$$e^{k\Theta} d\Theta = \left(\frac{\sigma_0}{\phi_0}\right) dz_H. \tag{14.142}$$

Substituting from equation (14.138) into this equation leads to the following expression:

$$e^{k\Theta} \, d\Theta = k_0 \, d\zeta_D, \tag{14.143}$$

where we have set

$$k_0 = k_{10}\left(\frac{\sigma_y}{\phi_0}\right). \tag{14.144}$$

Upon integration of equation (14.143), we obtain the densification equation

$$\Theta = \frac{1}{k} \ln\left[1 + kk_0\zeta_D\right], \tag{14.145}$$

where the initial condition $\Theta = 0$ at $\zeta_D = 0$ has been observed. Equation (14.145) gives the desired relation between the plastic volumetric strain Θ due to cyclic shearing and the deviatoric intrinsic time ζ_D, which we recall to be given by the expression

$$\zeta_D = \int |d\theta|, \tag{14.146}$$

where $d\theta$ is the increment in the plastic shear strain If the elastic shear strain is neglected compared with the total shear strain e, and we introduce the engineering definition of shear strain, γ, where $d\gamma = 2de$. then

$$d\theta = \frac{d\gamma}{2}. \tag{14.147}$$

Substitution of equation (14.147) into equation (14.146), and performing the integration over N cycles of shear strain with amplitude γ_0, leads to the result

$$\zeta_D = 4N\gamma_0, \tag{14.148}$$

which may be combined with equation (14.145) to give the following expression for Θ:

$$\Theta = \frac{1}{k} \ln\left[1 + kk_0(4N\gamma_0)\right]. \tag{14.149}$$

The ability of this equation to describe densification data for sand[15] is shown in Figure 14.16, where the dependence of the densification Θ on shear strain amplitude γ_0 is depicted after 10 cycles for three different values of initial relative density, D_r. To obtain the predicted results, the material parameters k and k_0 were allowed to vary with the relative density, D_r, which appears to be physically plausible.

Several important points in regard to equation (14.149) and its derivation,

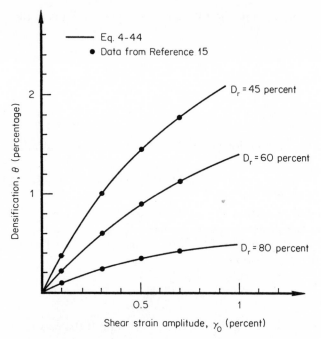

Figure 14.16 Densification of sand under cyclic shear after 10 cycles for several different values of initial relative density

should be emphasized. First of all, equation (14.143) is not an assertion but a consequence, in part, of the assumed form for the hydrostatic response; consequently, the form adopted for hydrostatic behaviour will influence the analytical form of the densification. This point is not present in the densification considerations presented in Ref. 4. Secondly, densification occurs through coupling between the hydrostatic and deviatoric intrinsic times, as indicated by equation 14.143. If there is no coupling, $k_{10} = k_0 = 0$ and no densification due to shear can take place. Again, this point is absent in Ref. 4. Finally, equation (14.149) implies that the densification is independent of the applied hydrostatic pressure, σ_0, which is in essential agreement with experimental data for sands.[15]

14.5 SHEAR TRAVEL

In this section, it is shown that the endochronic theory predicts an effect which has not been identified or investigated as yet.* What we are predicting

---

* This may not be the only instance but it is a rare instance in mechanics when a theory has predicted a phenomenon which, hitherto, has not been observed.

here is a new physical phenomenon, which we shall call *shear travel*, and which is the following:

A soil layer is subjected to a shear stress τ which is kept constant with time at some value τ_0 with a corresponding shear strain θ_0. While τ is kept constant, a cyclic pressure σ is applied to the soil layer to bring about a cyclic variation in θ. Then, if k_{01} is not zero, the shear deformation will increases with the number of cycles of θ where τ is kept constant at τ_0.

We shall show that this phenomenon can be predicted on the basis of equation (14.134). The constitutive equation for the shear response is equation (14.29), i.e.,

$$\tau = \int_0^{z_D} \rho(z_D - z') \frac{d\theta}{dz'} dz'. \tag{14.150}$$

For the purposes of the Laplace transform, the lower limit of integration is understood to be $0-$.

We take the Laplace transform of equation (14.150) to find that

$$\bar{\tau} = \bar{\rho} p \bar{\theta}, \tag{14.151}$$

where p is the Laplace transform parameter. This equation can be rewritten as

$$\bar{\theta} = \bar{\omega} \bar{\tau}, \tag{14.152}$$

where

$$\bar{\omega} = \frac{1}{\bar{\rho} p}. \tag{14.153}$$

Equation (14.152) implies that θ can be expressed in terms of τ in the form

$$\theta = \int_0^{z_D} \omega(z_D - z')\tau(z') \, dz', \tag{14.154}$$

where $\omega(z)$ is related to $\rho(z)$ by means of equation (14.153).

To describe shear travel, we assume that the part of z_D due to τ is negligible relative to the cumulative shear travel brought about by the cyclic variation of θ. In this event, since τ is constant and equal to τ_0 during travel, equation (14.154) can be placed in the form

$$\theta = \tau_0 \int_0^{z_D} \omega(z') \, dz'. \tag{14.155}$$

At this point, we recall equation (14.134). For the sake of simplicity, we set $f_D = f_H = 1$.

Again, presuming that the inequality $k_{00} \, d\zeta_D \ll k_{01} \, d\zeta_H$ holds in this particular experiment, we may set as a first approximation

$$dz_D = k_{01} \, d\zeta_H, \tag{14.156}$$

or

$$z_D = k_{01}\zeta_H, \tag{14.157}$$

where

$$\zeta_H = \int |d\theta| \tag{14.158}$$

If Δ is the amplitude of oscillation of θ and N is the number of cycles, it follows from equation (14.158) that

$$\zeta_H = 4N\Delta. \tag{14.159}$$

Since $\omega(z)$ is a positive function, the integral of $\omega(z)$ is a monotonically increasing function. Hence, θ increases monotonically with the number of cycles of θ.

14.6 CYCLIC CREEP

The phenomenon of cyclic creep, also termed 'ratchetting',[1] is a common feature of the response of many materials, including soils, concrete and metals, to asymmetric cyclic deformation. Cyclic creep is not a time-dependent effect, but arises because unloading and reloading paths for real materials generally are not coincident.

The endochronic model has the capability to describe cyclic creep, and to illustrate this, the simple shear model for sand, described in Section 14.4.2, was cycled between stress limits of 28.73 kPa and 2.39 kPa for 30 cycles, after initial loading to the upper stress limit. The calculations were performed with the computer subroutine ENDO, listed in Appendix C of Ref. 29. The calculated responses of the model for cycles 1–3, 10 and 30 is shown in Figure 14.17, and illustrates the growth in strain that occurs with increasing cycles of deformation. Note that the predicted response approaches elastic behaviour with increasing cycles of deformation due to the hardening present in the model.

CONCLUSION

The present paper is an exposition of the latest developments in the application of the endochronic theory to the mechanical response of soils to general types of deformation histories. Specifically, the theory is applied to (1) one-dimensional repetitive, loading–unloading (cyclic) histories and (2) to two-dimensional histories in which the normal stress is kept constant

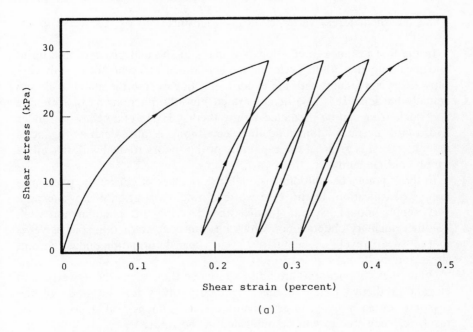

(a)

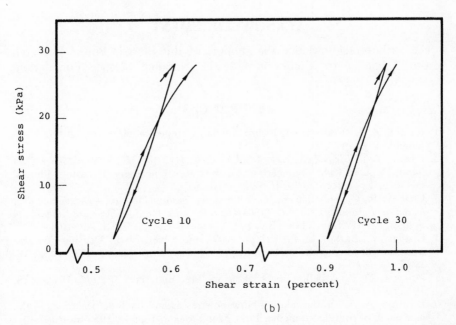

(b)

Figure 14.17 Response of endochronic shear model to asymmetric cyclic deformation, illustrating the phenomenon of cyclic creep (ratchetting)

while the shear stress varies cyclically (densification) or vice versa (shear travel).

In the first instance, predictions are made of the material response up to and including the 300th cycle, under conditions of cyclic shear with very satisfactory agreement with the experiment. In this case the material chosen actually hardens. It is also demonstrated how the phenomena of softening and ratchetting can be predicted by the theory. Hysteretic behaviour under hydrostatic cyclic loading–unloading conditions is also dealt with and a demonstration is given of the capability of the theory to deal with this effect simply and elegantly.

In the category of multidimensional histories lies of course the phenomenon of densification. In this case we give a demonstration of how, using a very simple model, we can predict and describe this phenomenon well. Another ancillary phenomenon which has not been studied in soils as yet is that of 'shear travel', i.e. cumulative cyclic shear deformation under constant shear and cyclic *normal* load.

While these applications are limited in scope they reveal the powerful and elegant predictive potential that the theory offers for the study of the response of soils under more complex three-dimensional histories. The results of such studies will be published in the future.

ACKNOWLEDGEMENT

The authors acknowledge the support of the Electric Power Research Institute, Palo Alto, California and the National Science Foundation, Washington, D.C.

REFERENCES

1. Bazant, Z. P. "Endochronic inelasticity and incremental plasticity," *Int. J. Solids Structures*, **14**, 691, 1978.
2. Bazant, Z. P., Krizek, R. J., and Shieh, C.-L. *Hysteretic Endochronic Theory for Sand*, Dept. of Civil Engineering, Northwestern University, Evanston, Illinois, Geotech. Engr. Report 79–4/654h, April 1979.
3. Chaney, R. C., and España, C. *Laboratory Testing Program—Development of Material Parameters for SIMQUAKE Site Soils for Soil-Structure Interaction Studies*, FUGRO, Inc., Long Beach, CA, Final Report to EPRI, August 1979.
4. Cuellar, V., Bazant, Z. P., Krizek, R. J., and Silver, M. L. Densification and hysteresis of sand under cyclic shear, *J. Geotech. Engr. Div.*, ASCE, **103**, 918, 1977.
5. Drucker, D. C. A definition of stable inelastic materials, *J. Appl. Mechs.*, **25**, 101, 1959.
6. Il'yushin, A. A. On the relation between stresses and small deformations in the mechanics of continuous media, *Prikl. Mat. i Mekh.*, **18**, 641, 1954 (in Russian).
7. Mandel, J. Conditions de Stabilité et Postulat de Drucker, *Proc. Intl. Union Theoret. Appl. Mech. Symp. on Rheology and Soil Mechs.*, Grenoble, France, 58, 1964.

8. Pipkin, A. C., and Rivlin, R. S. Mechanics of rate-independent materials, *Zeits. für Matem. Phys.*, **16,** 313, 1965.
9. Prager, W. in Appendix to Lindholm, U. S. (ed.) *Mechanical Behavior of Materials under Dynamic Loads*, Springer-Verlag, New York, 1968.
10. Ramberg, W., and Osgood, W. T. *Description of Stress–Strain Curves by Three Parameters*, NACA, Tech. Note 902, 1943.
11. Read, H. E., and Valanis, K. C. An endochronic constitutive model for general hysteretic response of soils, *Systems, Science and Software*, La Jolla, CA, Final Report to EPRI, Report NP-957, January 1979.
12. Richart, F. E. Some effects of dynamic soil properties on soil-structure interaction, *J. Geotech. Engr. Div.*, ASCE, **101** (GT12) 1197, 1975.
13. Romano, M. A continuum theory for granular media with a crtical state, *Arch. of Mechs.*, **26,** 1011, 1974.
14. Sandler, I. S. On the uniqueness and stability of endochronic theories of material behavior, *J. Appl. Mechs.*, **45,** 263, 1978.
15. Silver, M. L., and Seed, H. B. Volume changes in sands during cyclic loading, *J. Soil Mechs. Founds. Div.*, ASCE, **97,** 1171, 1971.
16. Streeter, V. L., Wylie, E. B., and Richart, F. E. Soil motion computations by characteristics method, *J. Geotech. Engr. Div.*, ASCE, **100,** 247, 1974.
17. Valanis, K. C. The viscoelastic potential and its thermodynamic foundations, *J. of Math. and Phys.*, **47,** 262, 1968.
18. Valanis, K. C. A theory of viscoplasticity without a yield surface, Part I: general theory," *Arch. of Mech.*, **23,** 517, 1971.
19. Valanis, K. C. A theory of viscoplasticity without a yield surface, Part II: Application to the mechanical behavior of metals, *Arch. of Mech.*, **23,** 535, 1971.
20. Valanis, K. C. Effect of prior deformation on cyclic response of metals, *J. Appl. Mechs.*, **41,** 441, 1974.
21. Valanis, K. C., and Wu, H.-C. Endochronic representation of cyclic creep and relaxation of metals," *J. Appl. Mechs.*, **42,** 67, 1975.
22. Valanis, K. C. An energy probability theory of fracture (an endochronic theory), *J. de Mécanique*, **14,** 843, 1975.
23. Valanis, K. C. On the foundations of the endochronic theory of viscoplasticity, *Arch. of Mechs.*, **27,** 857, 1975.
24. Valanis, K. C. Some Recent Developments in the Endochronic Theory of Plasticity—The Concept of Internal Barriers In Valanis, K. C. (ed.) *Constitutive Equations in Viscoplasticity: Phenomenological and Physical Aspects*, ASME, AMD-**21,** 1976.
25. Valanis, K. C., and Read, H. E. A theory of plasticity for hysteretic materials— I: shear response, *J. Computers and Structures*, **8,** 503, 1978.
26. Valanis, K. C. *Fundamental Consequences of a New Intrinsic Time Measure: Plasticity as a Limit of the Endochronic Theory*, The University of Iowa, Materials Engineering Report G224-DME-78-001, April 1978.
27. Valanis, K. C. *The Application of the Endochronic Theory of Plasticity to the Thermomechanical Behavior of Zircaloy*, Systems, Science and Software, La Jolla, CA, Final Report to EPRI, Report SSS-R-78-3764, August 1978.
28. Valanis, K. C. (1979) Endochronic theory with proper hysteresis loop closure properties, *Systems, Science and Software*, La Jolla, CA, Report No. SSS-R-80-4182, August 1979.
29. Valanis, K. C., and Read, H. E. *A New Endochronic Plasticity Model for Soils*, Systems, Science and Software Report SSS-R-80-4294, December 1979.

Soil Mechanics—Transient and Cyclic Loads
Edited by G. N. Pande and O. C. Zienkiewicz
© 1982 John Wiley & Sons Ltd

Chapter 15

Endochronic Models for Soils

Z. P. Bažant, A. M. Ansal and R. J. Krizek

15.1 INTRODUCTION

In practical field situations an element in a soil deposit is normally subjected to highly non-proportional stress paths and a generally applicable inelastic material model is therefore necessary. One rather effective model which has been recently developed for soils is based on endochronic theory. The simplest form of the theory which is today called endochronic was suggested by Schapery[24] who considered the viscosity coefficient of viscoplasticity appearing in the definition of reduced time to depend on the tangential octahedral component of the strain rate tensor, which leads to a reduced time of viscoplasticity that is completely equivalent to the definition of intrinsic time (given subsequently in equation (15.1)). The term "endochronic" was coined by Valanis[26] who was first to discover the tremendous usefulness and flexibility of the theory for modelling the response of materials to unloading, cyclic loading, and non-proportional loading. In this initial work Valanis[26] derived the general form of the theory from certain plausible thermodynamic hypotheses and applied it to describe the plastic-hardening behaviour (including the cross-hardening effect) and cyclic response of certain metals.

Motivated by the earlier work by Bažant[5,7,8] to characterize the constitutive relations of concrete, an endochronic formulation applicable to soils was first developed by Bažant and Krizek[4,6] and their coworkers who combined the endochronic formulation with a two-phase medium concept of the Biot[13] type. Later, the endochronic formulation was used in a one-dimensional finite difference scheme to model the liquefaction of infinite horizontal sand layers.[18] A finite element one-dimensional model of a sand layer was set up by Blázquez[14] and used to undertake an extensive parametric study of the different variables affecting liquefaction. At the same time, endochronic theory was applied to describe the quasi-static and cyclic behaviour of normally consolidated cohesive soils.[1,2,10,19] Further refinements of the endochronic model for sands were made[11] and this refined formulation was utilized to model by one and the same constitutive law the results of extensive cyclic simple shear tests and monotonic triaxial tests on a given sand. In the latest work Sener[25] applied an endochronic formulation to describe the sequence of the hysteresis loops leading to liquefaction of sand in a cyclic triaxial compression test.

15.2.   SUMMARY OF BASIC EQUATIONS

The basic concept of endochronic theory is that of the intrinsic time scale which is formulated to account for the dissipative effects of inelastic strain. Intrinsic time is defined as a monotonically increasing scalar function of strain, and one suitable definition, due to Valanis,[26] is:

$$dz^2 = \left(\frac{d\zeta}{z_1}\right)^2 + \left(\frac{dt}{\tau_1}\right)^2 \qquad (15.1(a))$$

where

$$d\zeta = F(\boldsymbol{\varepsilon}, \boldsymbol{\sigma}, \zeta)\, d\xi \qquad (15.1(b))$$

and

$$d\xi = \sqrt{(\tfrac{1}{2} de_{ij}\, de_{ij})} \qquad (15.1(c))$$

Here ε_{ij} is the strain tensor in Cartesian coordinates x_i ($i = 1, 2, 3$); z_1 and τ_1 are constant material parameters; $e_{ij} = \varepsilon_{ij} - \delta_{ij}\varepsilon = $ deviator of ε_{ij}; $\varepsilon = \tfrac{1}{3}\varepsilon_{kk} = $ volumetric (mean) strain; $\delta_{ij} = $ Kronecker delta; F is a strain-hardening–softening function; and ξ is the path length in deviatoric strain space (also called the distortion measure). In the above definition it is assumed that the inelastic behaviour is produced by the deviatoric strain increments;[5] this is because the primary source of inelasticity in soils may be attributed to the irreversible rearrangement of grain configurations associated with deviatoric strains. In both quasi-static and cyclic tests on soils it has been observed that the irreversible rearrangement of grains diminishes as inelastic straining progresses and intensifies as the strain magnitude increases. This fact is modelled by the strain hardening–softening function, F. Since hardening depends mainly on cumulative strains, while softening depends mainly on the stress and strain magnitudes, it is suitable to introduce

$$d\zeta = \frac{d\eta}{f(\eta)}, \qquad (15.2(a))$$

where

$$d\eta = F_1(\boldsymbol{\varepsilon}, \boldsymbol{\sigma})\, d\xi. \qquad (15.2(b))$$

The function $F_1(\boldsymbol{\varepsilon}, \boldsymbol{\sigma})$ is called the softening function[6] and the function $f(\eta)$, first introduced by Valanis,[26] is called the hardening function. This function limits the inelastic strain and thus serves a purpose similar to 'critical state' theory.

Soils also manifest significant inelastic volumetric strains, termed densification or dilatancy, which in seismic loadings are due primarily to shear.

The inelastic volume changes due to shear and to hydrostatic stress have different microstructural sources, and they are best treated separately. The inelastic volume change due to shear is expressed by the variable, λ, which is called densification or dilatancy, depending on the sign, and may be expressed as a function of the stress and strain invariants and the current accumulated value of λ:

$$d\lambda = L(\boldsymbol{\varepsilon}, \boldsymbol{\sigma}, \lambda)\, d\xi \tag{15.3}$$

The rate dependence of λ, which is important in the case of clays, can be introduced by generalizing equation (15.3) in a manner similar to equation (15.1(a)).

$$d\lambda^2 = [L(\boldsymbol{\varepsilon}, \boldsymbol{\sigma}, \lambda)\, d\xi]^2 + (\sigma_c\, dt/\tau_2)^2, \tag{15.4}$$

where τ_2 is a constant material parameter and σ_c is the consolidation stress.

For isotropic soils the stress–strain relations are conveniently written in terms of separate deviatoric and volumetric components:

$$de_{ij} = \frac{ds_{ij}}{2G} + de''_{ij}, \qquad de''_{ij} = \frac{s_{ij}}{2G}\, dz, \tag{15.5}$$

$$d\varepsilon = \frac{d\sigma}{3K} + d\varepsilon'', \qquad d\varepsilon'' = d\lambda, \tag{15.6}$$

in which $e_{ij} = \varepsilon_{ij} - \delta_{ij}\varepsilon$ are the deviatoric components of the strain tensor ε_{ij}; $\varepsilon = \frac{1}{3}\varepsilon_{kk}$ is the volumetric strain; $s_{ij} = \sigma_{ij} - \delta_{ij}\sigma$ are the deviatoric components of the stress tensor σ_{ij}; $\sigma = \frac{1}{3}\sigma_{kk}$ is the volumetric (mean) stress; e''_{ij} is the inelastic deviatoric strain; ε'' is the inelastic volume change due to distortion; and K and G are the bulk and shear moduli which depend on stress and strain states.

Inelastic volume changes due to hydrostatic stress transmitted between the soil particles (i.e., due to the effective stress) must now be taken into account. For this purpose, a second intrinsic time variable must be introduced:[8,25]

$$(dz')^2 = \left(\frac{d\zeta'}{z_2}\right)^2 + \left(\frac{dt}{\tau_1}\right)^2, \tag{15.7(a)}$$

where

$$d\zeta' = \frac{d\eta'}{h(\eta')}, \qquad d\eta' = H(\boldsymbol{\sigma})\, d\xi' \tag{15.7(b)}$$

and

$$d\xi = \sqrt{[I_1(d\boldsymbol{\varepsilon})]^2} = |d\varepsilon_{11} + d\varepsilon_{22} + d\varepsilon_{33}|. \tag{15.7(c)}$$

Here $I_1(d\boldsymbol{\varepsilon})$ is the first invariant of the incremental strain tensor; z_2 and τ_1 are material constants; $h(\eta')$ is the compaction hardening function; $H(\boldsymbol{\sigma})$ is the compaction softening function; and ξ' is a non-decreasing scalar variable called the compaction measure. The complete volumetric stress–strain relations may now be written as

$$d\varepsilon = \frac{d\sigma}{3K} + d\varepsilon'', \qquad d\varepsilon'' = d\lambda + \frac{\sigma}{3K} dz'. \qquad (15.8)$$

15.3   ADVANTAGES OF ENDOCHRONIC THEORY AND COMPARISONS WITH OTHER THEORIES

Since endochronic theory represents a substantial departure from previous thinking, it is not surprising that it has recently been the subject of intense debate and criticism.[22,23] These criticisms have proven to be very useful, because they have led to further refinements of the theory and an improved understanding of the fundamental hypotheses used in modelling inelastic behaviour. It appears that the criticisms which have been advanced thus far can either be met by certain extensions of the original formulation or reduced to requirements which represent hypotheses rather than a physical necessity.

The most substantial criticism has been violation of Drucker's stability postulate in the small for unload–reload infinitesimal loading cycles. In the case of a frictional material, violation of this postulate cannot, however, be generally a cause for alarm, because, as is known, frictional materials can violate the postulate and yet remain stable.[9,12] Nevertheless, the postulate should be satisfied for soils subjected to cycles of deviatoric strain at zero or constant hydrostatic stress, as in this case isotropic materials do not exhibit frictional effects. Moreover, violation of Drucker's postulate prevents closure of the hysteresis loops for small unload–reload cycles superimposed on a large static stress, and this is contrary to experimental evidence.

Therefore, the theory has been refined by the inclusion of jump-kinematic hardening[9] to satisfy Drucker's postulate and achieve closure of the hysteresis loops. This leads to the following stress–strain relations:

$$de''_{ij} = (s_{ij} - \alpha_{ij})\rho \, dz/2G, \qquad (15.9(a))$$

$$d\varepsilon'' = d\lambda + (\sigma - \alpha)\psi \, dz'/3K, \qquad (15.9(b))$$

in which α_{ij} and α represent the current centres of the deviatoric and volumetric loading surfaces in stress space, and ρ and ψ are coefficients (less than unity) which reduce the rate of accumulation of inelastic strain in cases of unloading and reloading and must satisfy a certain inequality so as to assure closing of the loops.[9] The centres of the loading must jump into the extreme stress point whenever loading changes to unloading or unloading to reloading. For this purpose one needs a three-way loading–unloading–reloading criterion, which may be conveniently expressed in terms of (1) the sign of the increment of the energies stored in the material, and (2) a check of whether the currently stored energy is maximum or less than some previously stored energy.

This refined formulation has been developed in detail[11,25] and extensive experimental data have been fitted. In particular, the latest version of this theory allowed for the first time the use of one and the same constitutive relation to represent data from cyclic simple shear tests with various amplitudes and many loading cycles, cyclic triaxial tests, and monotonic triaxial tests involving dilatancy rather than densification. The formulation is, however, substantially more complicated than the original simple version,[16,17] which nevertheless appears to be satisfactory as a crude approximation for seismic loading histories that involve large-amplitude cycles superimposed on a static stress.

Another criticism of endochronic theory was concerned with the continuity of the response for cyclic loading histories with vanishing amplitude of the cycles.[23] For example, if a cyclic stress of small amplitude is superimposed on a static stress and the amplitude tends to zero, then the response after an infinite number of cycles does not approach the response for the static stress. However, if one first fixes the number of cycles (for example, 10^7 cycles) and the amplitude is allowed to shrink to zero subsequently, then the response does approach that for the static load. Thus, although the material does not exhibit strong stability response in Liapunov's sense, it does exhibit a weaker stability or continuity. These stability properties are exactly the same as those exhibited by classical viscoplasticity,[9] and so there appears to be no strong basis for criticism on this point.

A further criticism[22] deals with the response to a loading path in the form of a regular staircase in strain space. When the number of stairs approaches infinity and the size of the stairs approaches zero, the staircase path tends to a straight-line path. So it may seem that the response to the staircase path should converge to the response for the straight-line path. However, this is not the case for the ordinary endochronic theory. Nevertheless, this at first seemingly objectionable feature does not appear entirely unreasonable if one considers the physics of inelastic behaviour; namely, the alternating segments of the staircase path produce damage (microcracks and microslips) that occurs predominantly on different planes than the damage produced by the smooth loading path. Since each path, even in the limit, leads to different damage, the responses can hardly be expected to approach each other.[12]

A similar situation is, in fact, obtained in classical plasticity if the path in stress space coincides with the movement of a corner of the loading surface and if hardening is described by the length of the path of plastic strain in strain space.[12] It should also be noted that a staircase path is, in the limit, non-differentiable, and it would not be unreasonable to exclude such paths from the range of validity of the theory. Finally, it is possible to modify the definition of intrinsic time by introducing a finite resolution in such a way that the length of the staircase path does approach the length of the smooth path, thereby achieving a full continuity of response. The full continuity may often be desirable for mathematical reasons.

From the conceptual point of view, it should be realized that classical plasticity implies a number of hypotheses which do not represent physical requirements (in fact, as many as six hypotheses). One of them is Drucker's stability postulate, and it may be shown that endochronic theory can be also obtained as a consequence of Drucker's postulate if some other hypotheses are altered in a certain manner that is not unreasonable.[12] In particular, the use of Drucker's postulate is questionable when dealing with the frictional aspects of soil.

As it appears today, endochronic theory is no less sound than any other theory of inelastic behaviour. However, it is also dubious to claim that the theory has more solid foundations than some other theories. The advantages of the endochronic approach must be viewed from the practical point of view, and they may be summarized briefly as follows:

(1) The fact that the reversibility at unloading, the salient feature of inelastic behaviour, is represented mathematically without any unloading criterion (even for the simplest endochronic theories) lends the theory considerable effectiveness and flexibility in modelling cyclic behaviour.

(2) In contrast to the increment of the plastic proportionality coefficient in plasticity, the intrinsic time parameter is independent of the loading surface and its evolution; this fact endows the theory with a great degree of versatility in fitting test data.

(3) The time-like interpretation helps to provide an intuitive understanding of the response.

(4) In the case of loading increments to the side of a proportional loading path (or tangential to the loading surface), the theory gives a softer response than plasticity theories without a vertex. This is certainly more correct, and it also is a safer model.[9] In fact, classical plasticity with normality represents the most unconservative or least safe assumption with regard to failures due to material instability when no experimental information on the response to such loading increments to the side exists, as is usually the case.

(5) Finally, the strain rate effect is modelled by endochronic theory in an automatic and relatively simple manner.

15.4 PRACTICAL APPLICATIONS AND PARAMETRIC STUDIES

During the past few years endochronic theory has been extensively used to fit experimental data from tests on cohesive soils subjected to cyclic and monotonic loadings and to investigate the influence of various parameters on the response of soil systems. Some typical results obtained from the latest of these studies will now be described.

15.4.1 Constitutive behaviour of cohesionless soils

The initial applications of endochronic theory several years ago were made to model the behaviour of saturated[6] and dry[16,17] sands. The first use of the theory to describe the build-up of pore pressure and subsequent liquefaction[6] illustrated the potential of the theory, but contained only limited fits of experimental data. The first extensive set of experimental data that were fit by the theory[16,17] resulted from about fifty strain-controlled cyclic simple shear tests with up to 300 cycles on an angular dry quartz sand at relative densities of 45%, 60%, and 80%. Vertical stresses ranged from 24 to 192 kN/m² and strain amplitudes ranged from 0.0001 to 0.005. The densification and hysteretic behaviour were described quite well in virtually all cases. This earlier model was refined[11] to include jump-kinematic hardening and two Maxwell elements, and it was then applied with even greater success to fit the data mentioned above; one typical set of hysteresis loops is shown in Figure 15.1. In the most recent work along this line, Sener[25] employed a refined version of this model combined with a Biot-type inelastic two-phase medium concept[4] to describe the sequence of hysteresis loops and the build-up of pore pressures leading to liquefaction in cyclic triaxial tests on sand.

15.4.2 State of stress in test specimen

To assess the importance of the inevitable stress and strain nonuniformity in cylindrical specimens subjected to cyclic simple shear tests, a three-dimensional endochronic finite element analysis was conducted.[11] The finite element grid, consisting of prismatic elements with nodes in the corners only, is depicted in Figure 15.2. Stresses and deformations were calculated in small time steps using the endochronic constitutive relation previously established under the assumption that stresses and strains were uniform. Not surprisingly, the distributions of stresses and strains were found to exhibit significant nonuniformity, as illustrated in Figure 15.3, although not as strong as elastic solutions would indicate. However, when the hysteresis loops calculated by this finite element analysis were compared with those obtained

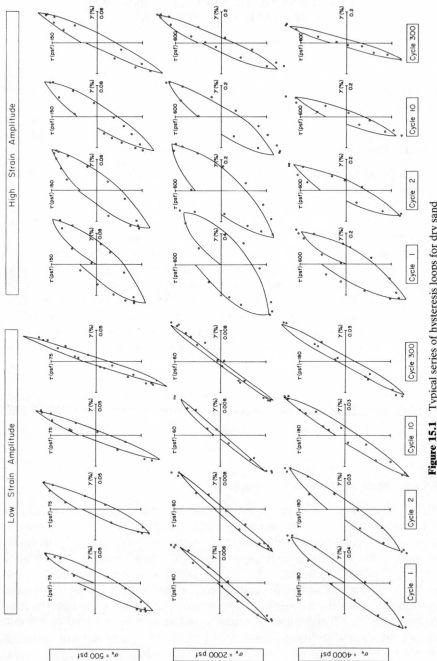

Figure 15.1   Typical series of hysteresis loops for dry sand

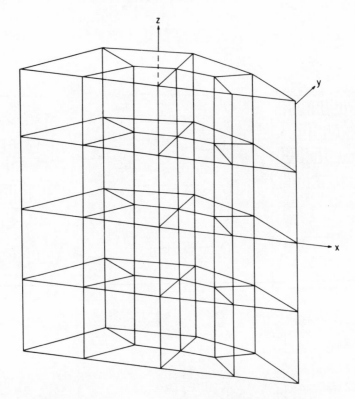

Figure 15.2 Finite element grid used in three-dimensional analysis
of cylindrical specimen subjected to cyclic simple shear

under the assumption of uniform stress and strain, the discrepancies were
within acceptable limits, although still significant. The inelastic behaviour of
soil tends to even out stress and strain nonuniformities and to mitigate their
effect on the overall response of the specimen.

15.4.3 Constitutive behaviour of cohesive soils

The two-phase medium concept (fluid and solid) is a natural choice for
analysing the undrained behaviour of saturated cohesive soils. The inelastic
generalization of this concept[4] was combined with an endochronic constitutive law for the solid phase to describe the stresses, strains, and pore
pressures observed for various clays. In one study,[2] data from seven different
clays that cover a relatively wide range of composition and behavioural
patterns were selected from the literature to constitute a data base for
determining the material parameters and their correlations with various soil
properties (such as the liquidity index, void ratio, etc). Of the fifteen model

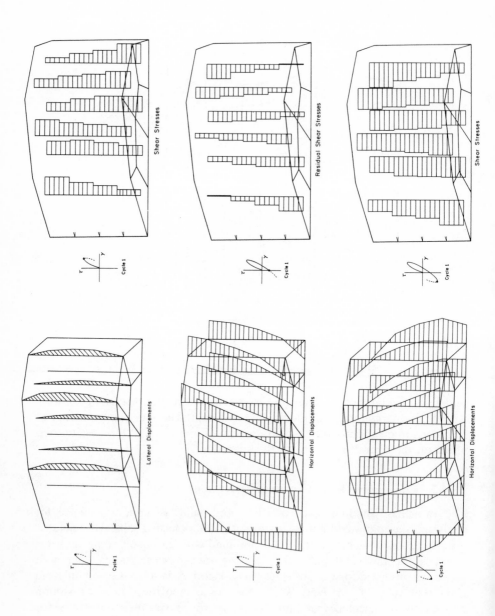

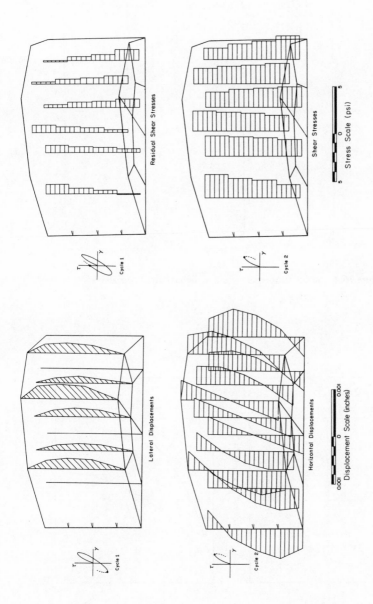

Figure 15.3 Typical distributions of stresses and displacements determined from endochronic analysis of cylindrical specimen subjected to cyclic simple shear

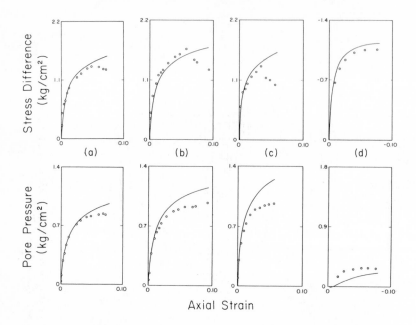

Figure 15.4   Stress–strain–pore pressure behaviour of cohesive soils

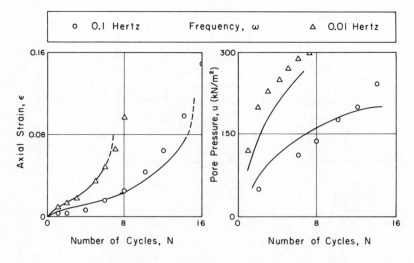

Figure 15.5   Cyclic triaxial response of kaolinite

parameters involved in this constitutive model, ten remained essentially constant for all of the normally consolidated clays investigated, while correlations with two readily measurable soil properties were developed for the other five. This model was then applied to describe test results from three-dimensional consolidated–undrained tests,[20] and the results are shown in Figure 15.4(a)–(c) for B equal to 0.2, 0.4, and 0.7, where $B = (\sigma_2 - \sigma_3)/(\sigma_1 - \sigma_3)$. Figure 15.4(d) shows the prediction made for a triaxial extension test in which the total confining stress was held constant and the axial stress was decreased.[21]

In a companion study concerned with describing the rate-dependent cyclic behaviour of cohesive soils,[19] the endochronic constitutive relationship was applied to model the results from two cyclic triaxial tests (one at a frequency of 0.01 Hz and the other 0.1 Hz) on specimens of kaolinite.[15] The resulting fits are illustrated in Figure 15.5. However, at this stage of development it was necessary to change the values of some of the material parameters from those obtained for the quasi-static tests. With further refinement of the model this should be unnecessary and one set of parameters should suffice to model the response for both quasi-static and cyclic tests.

15.4.4 Response of an earth dam

A two-dimensional finite element program using triangular plane strain elements and implicit step-by-step integration in time was developed to study the dynamic response of an earth dam subjected to a seismic excitation.[3] Both linear elastic and endochronic stress–strain laws for two-phase media were used to calculate the stresses and displacements caused in the dam by either (1) the modified version of the N–S component of the El Centro accelerogram from the 1940 earthquake (for both constitutive relations) or (2) the deconvoluted and enriched version of the Pacoima accelerogram from the 1971 San Fernando earthquake (for the endochronic model only). In both accelerograms the maximum values of the acceleration were scaled to be approximately equal, so only the cycle sequences of the records differ.

Figure 15.6 shows a cross-section of the dam and the finite element grid used in the analysis. The shell and foundation materials were treated as cohesionless soils, and the core and debris materials were considered to be cohesive soils. The seepage forces due to the head difference across the dam were neglected relative to the prevailing static and dynamic stresses, and the soil above the water table was treated as a one-phase medium. Typical time histories of horizontal displacements and shear stresses at points near the crest and mid-height of the dam are given in Figures 15.7 and 15.8 for both endochronic and linear elastic solutions. As is seen, the responses for the inelastic two-phase solution are considerably different from those of the

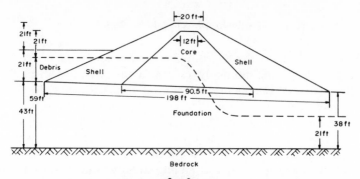

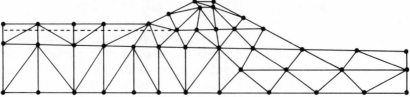

Figure 15.6 Cross-section of dam

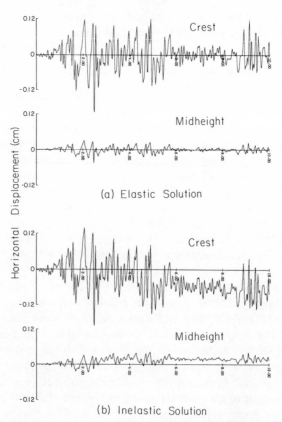

Figure 15.7 Time-dependent variation of typical horizontal displacements

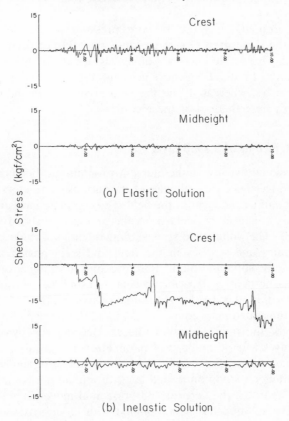

Figure 15.8 Time-dependent variation of typical shear stresses

elastic solution. In particular, they exhibit somewhat different peaks and amplitudes. In contrast to the elastic solution, the displacement and shear stress oscillations for the inelastic solution are not centred around the initial values (namely, the values for the static case), but rather about a mean value which wanders away from the initial values, reflecting the fact that permanent displacements and residual stresses are accumulated during the seismic excitation.

Comparisons of results calculated with the El Centro and Pacoima accelograms (both scaled to approximately the same maximum acceleration) showed that peak shear stresses obtained from the Pacoima record were approximately one-half of those obtained with the El Centro record. Similarly, the displacement at the crest of the dam is smaller for the Pacoima record than for the El Centro record. In addition, with the Pacoima record there was a dramatic increase in pore pressure in the debris material upstream of the dam after 11 s of shaking; this did not occur with the El Centro record. These differences are obviously due to the different nature of the two records. In the El Centro accelerogram the major peaks occur during the first 6 s, whereas in the Pacoima accelerogram the major peaks occur primarily near the end of the record.

15.4.5   Liquefaction of a sand layer

A two-dimensional dynamic finite element program has been developed by Blázquez[14] to investigate the influence of the various parameters that control the liquefaction of a sand layer in the field. The complete inelastic two-phase medium theory enhanced by inelastic strains and using a Biot-type approach was employed. The volumetric stress–strain relations include coupling between the strains and stresses in the fluid and solid phases and inelastic strains in both the solid and fluid.[4] Pore pressure increments were obtained from the inelastic densification increments by using the densification compliance deduced in this work. The equations of motion of the medium include both the added mass coupling effect and the viscosity for the relative displacement of the fluid and solid phases. Because the two-dimensional program involves a large number of parameters and requires considerable computer time, a one-dimensional version of the program was used to perform a parameter study on a sand layer of infinite horizontal extent and underlain by bedrock, the vertical and horizontal movements of which are prescribed. The solution has been coupled with an approximate analytical solution for the dissipation of pore pressure (that is, consolidation of the layer).

The profiles of the excess pore pressure at various times are shown in Figure 15.9 for the case where the bedrock is excited by a sinusoidal horizontal motion. Computational results are shown both for the case where a layer of dry sand exists above the water table and for the case where this layer is replaced by a rigid slab of equal mass, a simplification often made in practice. This simplification is seen to have a distinct effect on the time and location of liquefaction (defined as the point where the curve depicting the excess pore pressure touches the curve of the initial vertical effective stress). The replacement of the overlying sand layer by a rigid slab shortens the time to liquefaction and moves the point of initial liquefaction downward. In the extreme situation where the water table is at the ground surface, initial liquefaction occurs sooner and at a shallower depth. In addition, the acceleration profiles exhibit a decline in the amplitudes as the pore pressure increases and a complete vanishing of the oscillations in acceleration as liquefaction is approached. Obviously, this is because liquefied soil does not transmit shear waves.

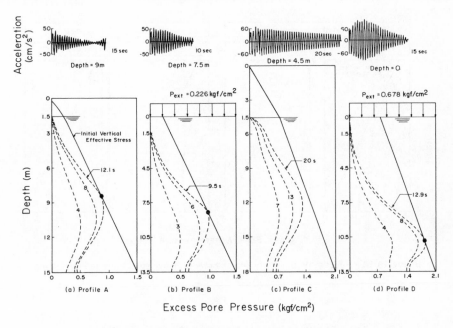

Figure 15.9 Effect of overburden condition on liquefaction

The effect of drainage is illustrated in Figure 15.10. Drainage at the top reduces the rate of increase of pore pressure, but the results also depend strongly on the permeability of the layer, higher permeability yielding a lower rate of increase in the pore pressure. Comparing cases with a draining top surcharge and an impervious top surcharge, we see that the latter gives shorter times to liquefaction (this is enhanced as the permeability increases) and shallower locations of the initially liquefied zone. Moreover, these calculations show that the value of the consolidation coefficient is extremely important. In addition, other calculations have demonstrated that the time to liquefaction decreases and the location of the initially liquefied zone becomes deeper as the value of the densification compliance, which relates the pore pressure increment to the inelastic densification, increases.

The effect of acceleration history of the excitation is exemplified in Figure 15.11. Compared are histories in which the acceleration amplitude decreases with time and those where it remains essentially constant with time. The shortest time to liquefaction is obtained in the former case, and the location of the initially liquefied zone is the deepest. For the case of a gradually increasing acceleration amplitude, the location of the liquefied zone is the shallowest and the rise in pore pressure occurs at an increasing rate, as contrasted with a decreasing rate in the former case.

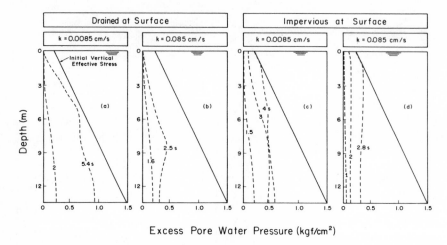

Figure 15.10   Effect of permeability and external drainage conditions on liquefaction

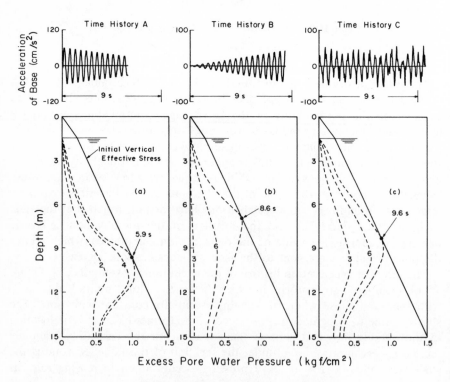

Figure 15.11   Effect of loading history on liquefaction

In the final part of this study, the layer was subjected to the El Centro 1940 accelerogram, including both horizontal and vertical accelerations of the bedrock. In order to assess the importance of using a complete two-phase model, computer solutions were also obtained for the assumption of equal displacements of the fluid and solid. This condition was found to have a drastic effect on the response and caused the location of the liquefied zone to move (in presence of vertical accelerations) all the way to the bedrock.

ACKNOWLEDGEMENT

The support provided by the National Science Foundation under Grant No. ENG-7807777 is gratefully acknowledged.

REFERENCES

1. Ansal, A. M., *An Endochronic Constitutive Law for Normally Consolidated Cohesive Soils*, Ph.D. Dissertation, Civ. Eng. Dept., Northwestern University, Evanston, Illinois, 166, 1977.
2. Ansal, A. M., Bažant, Z. P., and Krizek, R. J. Viscoplasticity of normally consolidated clays, *Proc. Amer. Soc. Civ. Eng., J. Geotech. Engng. Div.*, **105** (GT4), 519–37, 1979.
3. Ansal, A. M., Krizek, R. J., and Bažant, Z. P. *Seismic Analysis of an Earth Dam Based on Endochronic Constitutive Law*, Tech. Report, Dept. Civ. Eng., Northwestern University, Evanston, Illinois, 1979.
4. Bažant, Z. P., and Krizek, R. J. Saturated sand as an inelastic two-phase medium, *Proc. Amer. Soc. Civ. Eng., J. Eng. Mech. Div.*, **101** (EM4), 317–32, 1975.
5. Bažant, Z. P., and Bhat, P. D. Endochronic theory of inelasticity and failure of concrete, *Proc. Amer. Soc. Civ. Eng., J. of Eng. Mech. Div.*, **102** (EM4), 701–22, 1976.
6. Bažant, Z. P., and Krizek, R. J. Endochronic constitutive law for liquefaction of sand, *Proc. Amer. Soc. Civ. Eng., J. of Eng. Mech. Div.*, **102** (EM2), 225–38, 1976.
7. Bažant, Z. P. Endochronic and classical theories of plasticity in finite element analysis, *Int. Conf. on Finite Elements in Nonlinear Solid and Structural Mechanics*, Geilo, Norway, **1**, 1–15, 1971.
8. Bažant, Z. P., and Shieh, C. L. Endochronic model for nonlinear triaxial behavior of concrete, *Nuclear Eng. and Design*, **47**, 305–15, 1977.
9. Bažant, Z. P. Endochronic inelasticity and incremental plasticity, *Int. J. of Solids and Structures*, **14**(9), 691–714, 1978.
10. Bažant, Z. P., Ansal, A. M., and Krizek, R. J. Viscoplasticity of transversely isotropic clays, *Proc. Amer. Soc. Civ. Eng., J. Eng. Mech. Div.*, **105** (EM4), 549–65, 1979.
11. Bažant, Z. P., Shieh, C. L., and Krizek, R. J. *Hysteretic Endochronic Theory for Sand*, Tech. Report, Dept. of Civ. Eng., Northwestern University, Evanston, Illinois, 1979.
12. Bažant, Z. P. Work inequalities for plastic fracturing materials, *Int. J. of Solids and Structures* **16,** 873–901, 1980.

13. Biot, M. A. Theory of propagation of elastic waves in a fluid-saturated porous solid. I—Low frequency range, *J. of Acc. Soc. Amer.*, **28**(2), 168–78, 1956.
14. Blázquez, R. *Endochronic Model for Liquefaction of Sand Deposits as Inelastic Two-Phase Media*, Ph.D. Dissertation, Civ. Eng. Dept., Northwestern University, Evanston, Illinois, 201, 1978.
15. Brewer, J. H. *The Response of Cyclic Stress in Normally Consolidated Saturated Clay*, Ph.D. Dissertation, North Carolina State University, Rayleigh, N.C., 1972.
16. Cuellar, V. *Rearrangement Measure Theory Applied to Dynamic Behavior of Sand*, Ph.D. Dissertation, Civ. Eng. Dept., Northwestern University, Evanston, Illinois, 94, 1954.
17. Cuellar, V., Bažant, Z. P., Krizek, R. J., and Silver, M. L. Densification and Hysteresis of Sand Under Cyclic Shear, *Proc. Amer. Soc. Civ. Eng., J. Geotech. Div.*, **103** (GT5), 399–416, 1977.
18. El Zaroughi, A. A. *Application of Endochronic Constitutive Law to One-Dimensional Liquefaction of Sand*, Ph.D. Dissertation, Civ. Eng. Dept., Northwestern University, Evanston, Illinois, 158, 1978.
19. Krizek, R. J., Ansal, A. M., and Bažant, Z. P. Constitutive Equation for Cyclic Behavior of Cohesive Soils, *Proc. Amer. Soc. Civ. Eng., Geotech Eng. Div., Spec. Conf. on Earthquake Eng. and Soil Dynamics*, **2**, 557–68, 1978.
20. Lade, P. V., and Musante, H. M. *Three-dimensional Behavior of Normally Consolidated Cohesive Soil*, Report No. UCLA-ENG-7626, School of Eng. and Appl. Sciences, University of California, Los Angeles, California, 1976.
21. Parry, R. H. G. Triaxial compression and extension tests on remoulded saturated clay, *Geotech*, **10**(4), 166–80, 1960.
22. Rivlin, R. S. *Some Comments on the Endochronic Theory of Plasticity*, Report, Center for the Application of Mathematics, Lehigh University, Bethlehem, PA, October 1979.
23. Sandler, I. S. The uniqueness and stability of endochronic theories of material behavior, *J. of Appl. Mech, Transactions, ASME, Series E*, **45**, 263–6, 1978.
24. Schapery, R. A. On a Thermodynamic Constitutive Theory and Its Application to Various Nonlinear Materials, *Proc. of Int. Union of Theor. and Appl. Math. Symp.*, East Kilbride, (ed. B. A. Boley), Springer-Verlag, New York, N.Y., 259–85, 1968.
25. Sener, C. *An Endochronic Nonlinear Inelastic Constitutive Law for Cohesionless Soils Subjected to Dynamic Loading*, Ph.D. Dissertation, Civ. Eng. Dept., Northwestern University, Evanston, Illinois, 1979.
26. Valanis, K. C. A theory of viscoplasticity without a yield surface; Part I. General theory; Part II. Application to mechanical behavior of metals, *Arch. of Mech.* (*Archiwum Mechaniki Stosowanej*), **23**, 517–55, 1971.

Soil Mechanics—Transient and Cyclic Loads
Edited by G. N. Pande and O. C. Zienkiewicz
© 1982 John Wiley & Sons Ltd

Chapter 16

On Dynamic and Static Behaviour of Granular Materials

S. Nemat-Nasser

16.1 INTRODUCTION

It is reasonable to expect that constitutive relations developed to cover a wide variety of material behaviour under diverse loading conditions tend to be either so general that they cannot yield specific needed information, or they become too complicated to be of practical use. Therefore, it is both natural and necessary to attempt to develop for specific classes of materials under specific loading conditions simple constitutive descriptions which capture the essential physical features of the problem, while at the same time involving a minimum number of measurable material parameters, preferably amenable to physically meaningful identifications. On the other hand, while in science as contrasted to religion, old theories are continually discarded in favour of simpler, more effective new doctrines, no great advantage is gained by abandoning established theories which can still be effective, but require minor modifications to achieve the aesthetic appeal of intraconsistency.

In this chapter this point of view is illustrated by means of three examples that stem from some recent work at Northwestern University on static and dynamic behaviour of granular materials.

The first is the small deformation version of finite deformation plasticity theory with plastic volumetric change and internal friction, that has been formulated for application to finite plastic flows of granular materials, as well as porous metals.[10] This theory provides a systematic and consistent basis for critical state soil mechanics[15] which has been criticized for some of its inconsistencies. For example, the concepts of normality and associative flow rule used in the theory do not accord with the notion of internal friction and plastic compressibility. The introduction of the nonassociative flow rule in the manner developed by Nemat-Nasser and Shokooh[10] removes this inconsistency and at the same time provides a theory which contains parameters with clear physical definitions. The present kinematically linearized version of this theory is presented in notation common in soil mechanics in order to reach as wide an audience as possible.

The second example involves the phenomena of densification of granular masses and their possible liquefaction (when saturated and undrained) under cyclic shearing. Here, first by means of a simple dimensional analysis the essential features of the problem are brought into focus. Then several physically motivated assumptions are used in line with the work of Nemat-Nasser and Shokooh,[6,8] in order to obtain explicit expressions for the changes of the void ratio and the pore water pressure as functions of the number of cycles and other relevant parameters.

The final example concerns the methodology for a fundamental statistical approach to the description of the macroscopic response of granular masses on the basis of an examination of microstructural changes. This part is essentially a progress report on work that is being continued, although already some specific encouraging results have been obtained.[2,5,9,12]

No attempt will be made to give extensive references, as the cited few list a large number of relevant papers.

16.2 A PLASTICITY THEORY

16.2.1 Notation

A fixed rectangular Cartesian coordinate system with coordinate axes x_i, $i = 1, 2, 3$, is used. The macroscopic stress tensor (linearized), elastic strain tensor (linearized), and the plastic part of the strain tensor (linearized) are respectively denoted by σ_{ij}, e_{ij}, and ε_{ij}, where the stress denotes only the *effective* part (i.e. the normal stresses are reduced by the pore pressure) of the stress, if the material is saturated with nonzero pore pressure; note that for simplicity, we *do not* use a prime for this effective stress, nor a superposed p for the plastic part of the strain. Henceforth we do not consider the elastic part of the strain and, therefore, all kinematical quantities will refer to the *plastic part* of the deformation. Also, all stresses refer to the effective part of the stress.

16.2.2 General theory

The stress tensor is split into deviatoric (denoted by S_{ij}) and spherical parts, as

$$\sigma_{ij} = S_{ij} + p\delta_{ij}, \quad p = \tfrac{1}{3}\sigma_{kk}, \quad i, j, k = 1, 2, 3, \tag{16.1}$$

where repeated indices are summed and, as commonly used in soil mechanics, compression is viewed positive, p being the mean pressure.

We now introduce a yield function and a flow potential which are obtained by a minor but significant amendment to the commonly used J_2 (second invariant of deviatoric stress) flow potential:

$$\begin{aligned} f &\equiv \bar{\sigma} - F(p, \theta, \bar{\varepsilon}), \qquad \text{yield function;} \\ g &\equiv \bar{\sigma} + G(p, \theta, \bar{\varepsilon}), \qquad \text{flow potential,} \end{aligned} \qquad (16.2)$$

where

$$\bar{\sigma} = (\tfrac{1}{2} S_{ij} S_{ij})^{1/2}, \qquad \theta = \varepsilon_{kk}, \qquad \bar{\varepsilon} = \int_0^t (2\dot{\varepsilon}'_{ij} \dot{\varepsilon}'_{ij})^{1/2} \, dt; \qquad (16.3)$$

here, θ is plastic volumetric strain, $\dot{\theta} = \dot{v}/v$ is the rate of volumetric strain per unit current volume, $\bar{\varepsilon}$ represents the effective* distortional plastic strain, $\dot{\varepsilon}'_{ij}$ is the deviatoric part of the plastic strain, and t is a monotonically increasing load parameter; superposed dot stands for the rate. The rate of plastic strain is

$$\dot{\varepsilon}_{ij} = \lambda \frac{\partial g}{\partial \sigma_{ij}} = \lambda \left(\frac{S_{ij}}{2\bar{\sigma}} + \frac{1}{3} \frac{\partial G}{\partial p} \delta_{ij} \right). \qquad (16.4)$$

This and the consistency relation for the yield function, i.e. $\dot{f} = 0$, now give

$$\dot{\varepsilon}_{ij} = \frac{1}{H} \left(\frac{S_{ij}}{2\bar{\sigma}} + \frac{1}{3} \frac{\partial G}{\partial p} \delta_{ij} \right) \left(\frac{S_{kl}}{2\bar{\sigma}} - \frac{1}{3} \frac{\partial F}{\partial p} \delta_{kl} \right) \dot{\sigma}_{kl} \qquad (16.5)$$

which defines the plastic strain rate in terms of the corresponding stress rate. In (16.5) H is the work-hardening parameter,

$$H = \frac{\partial G}{\partial p} \frac{\partial F}{\partial \theta} + \frac{\partial F}{\partial \bar{\varepsilon}} \qquad (16.6)$$

which consists of two parts:

(1) work-hardening associated with volumetric strain,

$$h_1 = \frac{\partial G}{\partial p} \frac{\partial F}{\partial \theta} \qquad (16.7)$$

which may be positive or negative and which will be referred to as *density-hardening*; and

(2) work-hardening associated with plastic distortion,

$$h = \frac{\partial F}{\partial \bar{\varepsilon}} \qquad (16.8)$$

which is non-negative.

---

* In plasticity theory, $\bar{\sigma}$ and $\bar{\varepsilon}$ as defined by equations $(16.3)_{1,3}$ are called the 'effective' stress and strain, respectively; see Hill.[4] The term 'effective' is also used in soil mechanics to denote stresses when the pore water pressure is subtracted from the normal stresses. The context should make it clear which usage is being implied in the present paper.

To gain insight into the implication of the density-hardening parameter, h_1, observe from equation (16.4) that

$$\frac{\partial G}{\partial p} = \left(\frac{\bar{\sigma}}{S_{ij}\dot{\varepsilon}_{ij}}\right)\dot{\theta} = \frac{\dot{\theta}}{\dot{\bar{\varepsilon}}} = \frac{d\theta}{d\bar{\varepsilon}}, \tag{16.9}$$

where the fact that $\bar{\sigma}\dot{\bar{\varepsilon}} = S_{ij}\dot{\varepsilon}_{ij}$ is used. Hence, the sign of $\partial G/\partial p$ follows the sign of $\dot{\theta}$, because $\bar{\sigma}$ and the rate of distortional plastic work, $S_{ij}\dot{\varepsilon}_{ij}$, are both positive.

We refer to $\partial G/\partial p$ as the *dilatancy factor*, and observe that since $\partial F/\partial\theta$ is always positive (hardening due to compaction), h_1 has the same sign as $\dot{\theta}$: hardening during densification, softening during dilatancy, and being zero at the *critical* state.

The quantity $\partial F/\partial p$ in (16.5) represents the overall macroscopic friction factor, and is always positive. This can be seen by keeping θ and $\bar{\varepsilon}$ instantaneously fixed in the yield function which, in view of the consistency relation $\dot{f} = 0$, then gives

$$\frac{d\bar{\sigma}}{dp} = \frac{\partial F}{\partial p} > 0, \tag{16.10}$$

where $\bar{\sigma}$ can be viewed as the octahedral effective shear stress. Note that, in this theory, the overall friction factor, $\partial F/\partial p$, need not be a constant, as it probably is not, especially under large hydrostatic pressures.[1]

Equations (16.9) and (16.10) bring into focus the need for a non-associative flow rule for consistent plasticity theories of dilatant frictional granular materials. An associative flow rule demands $F \equiv G$, which then results in $\partial G/\partial p = \partial F/\partial p$, which is a contradiction, as the dilatancy factor $\partial G/\partial p$ is essentially a geometrical quantity, being positive, negative, or zero depending on the rate of volumetric strain per unit rate of distortional strain, whereas the friction coefficient, $\partial F/\partial p$, is a strictly positive quantity. The two quantities are, however, related because of the condition of energy requirement: for rigid granules the total rate of plastic work must equal the total rate of frictional loss, both measured per unit volume.

The total rate of plastic work per unit volume is

$$\sigma_{ij}\dot{\varepsilon}_{ij} = S_{ij}\dot{\varepsilon}_{ij} + p\dot{\theta}, \tag{16.11}$$

whereas the rate of frictional loss may be taken to be proportional to the coefficient of friction, the pressure, and the rate of effective distortional strain, i.e.

$$\dot{w}_f \approx \alpha\frac{\partial F}{\partial p}p\dot{\bar{\varepsilon}}, \tag{16.12}$$

where α is a parameter depending on the state of stress, e.g. $\alpha = 1$ for simple shear, and $\alpha \approx \sqrt{3}$ in the triaxial case; this last choice of α is consistent with equation (4.25) of Nemat-Nasser and Shokooh.[10] From (16.1) and (16.12) it now follows that

$$\frac{\partial G}{\partial p} = \alpha \frac{\partial F}{\partial p} - \frac{\bar{\sigma}}{p},$$

(16.13)

where (16.9) is used.

The critical state is defined by $\dot{\theta} = \partial G/\partial p = 0$. Hence, (16.13) yields, for $\dot{\theta} = 0$,

$$\alpha \frac{\partial F}{\partial p} - \frac{\bar{\sigma}}{p} = 0$$

(16.14)

which is a curve in the $\bar{\sigma},p$-plane, characterizing the critical states; see Figure 16.1. Points above this curve correspond to states with a negative dilatancy factor (dilation), and points below this curve are associated with positive $\dot{\theta}$ (compaction).

From the definition of the void ratio, $e = v_v/v_s$, $v = v_v + v_s$, it follows that

$$\dot{\theta} = -\frac{\dot{e}}{1+e},$$

(16.15)

where the minus sign indicates that the void ratio decreases in compaction for which $\dot{\theta}$ is regarded positive. Substitution from (16.9) and integration with respect to $\bar{\varepsilon}$ now results in

$$e = (1+e_0) \exp \left\{ -\int_0^{\bar{\varepsilon}} \frac{\partial G}{\partial p} \, d\bar{\varepsilon} \right\} - 1$$

$$= (1+e_0) \exp \left\{ -\int_0^{\bar{\varepsilon}} \left[\alpha \frac{\partial F}{\partial p} - \frac{\bar{\sigma}}{p} \right] d\bar{\varepsilon} \right\} - 1,$$

(16.16)

where e_0 is the initial void ratio; this equation is consistent with equation (4.19) of Nemat-Nasser and Shokooh.[10]

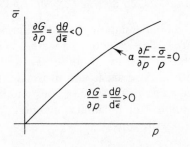

Figure 16.1 The critical curve in the $\bar{\sigma},p$-plane

16.2.3 General qualitative results

The theory presented above includes the general behaviour of loose as well as dense sand in simple shearing as well as in triaxial tests (monotonic loading only). Both cases can be discussed simultaneously in terms of the effective stress, $\bar{\sigma}$, and the effective distortional strain, $\bar{\varepsilon}$, which, however, will have a different meaning in each case.

For pure shear, the state of stress is defined by

$$\sigma_{11} = \sigma_{22} = \sigma_{33} = p,$$
$$\sigma_{12} = \sigma_{21} = \tau, \qquad \sigma_{23} = \sigma_{31} = 0. \tag{16.17}$$

Hence $\bar{\sigma} = \tau$. The strain rates, moreover, are defined by volumetric strain rate, $\dot{\theta}$, and by the only non-zero shear strain rate, $\dot{\varepsilon}_{12}$. Hence, $\dot{\bar{\varepsilon}} = 2\dot{\varepsilon}_{12}$, which is the rate of engineering shear strain.

For the triaxial state of stress, on the other hand,

$$\sigma_{11} = \sigma_1, \qquad \sigma_{22} = \sigma_{33} = \sigma_2, \tag{16.18}$$

with all other stress components being zero. The corresponding strain rates are

$$\dot{\varepsilon}_{11} = \dot{\varepsilon}_1, \qquad \dot{\varepsilon}_{22} = \dot{\varepsilon}_{33} = \dot{\varepsilon}_2, \tag{16.19}$$

with all other strain rate components being zero. In this case, therefore,

$$\bar{\sigma} = \frac{\sqrt{3}}{3}(\sigma_1 - \sigma_2), \qquad \text{and} \qquad \dot{\bar{\varepsilon}} = \frac{2\sqrt{3}}{3}(\dot{\varepsilon}_1 - \dot{\varepsilon}_2).$$

Since $\bar{\varepsilon}$ is a monotone increasing parameter, it is used as the 'time parameter' and therefore all superposed dots denote differentiation with respect to $\bar{\varepsilon}$. In particular, it follows that $\dot{\lambda} = \dot{\bar{\varepsilon}} = 1$.

Consider the yield function and note, from $\dot{f} = 0$, that

$$\frac{\mathrm{d}\bar{\sigma}}{\mathrm{d}\bar{\varepsilon}} - \frac{\partial F}{\partial p}\frac{\mathrm{d}p}{\mathrm{d}\bar{\varepsilon}} = H = \frac{\partial G}{\partial p}\frac{\partial F}{\partial \theta} + \frac{\partial F}{\partial \bar{\varepsilon}} = \left(\alpha\frac{\partial F}{\partial p} - \frac{\bar{\sigma}}{p}\right)\frac{\partial F}{\partial \theta} + \frac{\partial F}{\partial \bar{\varepsilon}} \tag{16.20}$$

which is a general differential equation giving the variation of stress in terms of the effective strain $\bar{\varepsilon}$. This equation, together with equation (16.16), models very nicely the behaviour of loose, as well as densely packed, granular materials, both in pure shear and in triaxial tests.

To illustrate this fact, consider the case of constant pressure test; the other cases can easily be studied in a similar manner. Then, $\mathrm{d}p/\mathrm{d}\bar{\varepsilon} = 0$, and equation (16.20) can be written as

$$\frac{\mathrm{d}\bar{\sigma}}{\mathrm{d}\bar{\varepsilon}} = (aM + h) - \frac{a}{p}\bar{\sigma}, \tag{16.21}$$

where the following notation is used:

$$a = \frac{\partial F}{\partial \theta} > 0, \qquad M = \alpha \frac{\partial F}{\partial p} > 0, \tag{16.22}$$

and where h is defined by (16.8).

Since h denotes distortional hardening, it may be assumed to depend only on the distortional strain $\bar{\varepsilon}$,

$$h = \hat{h}(\bar{\varepsilon}). \tag{16.23}$$

Then, depending on how fast the function $\hat{h}$ decays to zero with increasing $\bar{\varepsilon}$, differential equation (16.21) displays in the $\bar{\sigma}, \bar{\varepsilon}$-plane the loosely or densely packed granular material behaviour. Figure 16.2(a) illustrates two possible cases for the variation of $\hat{h}$, and Figure 16.2(b) gives the corresponding stress–strain relations.

It is seen that when the material is very loose, little distortional hardening exists, as the particles have more freedom to move relative to each other. In this case the distortional hardening is insignificant as compared with the density hardening. This results in Curve 1 of Figure 16.2(b). From equation (16.16) the corresponding variation of the void ratio is easily seen to be as shown by Curve 1 in Figure 16.2(c).

When the material is densely packed, on the other hand, the distortional hardening remains a dominant factor. Its variation with $\bar{\varepsilon}$ may take on a form schematically shown by Curve 2 of Figure 16.2(a). In this case, the critical state is first reached when $\bar{\sigma}/p = M$, at which point the right-hand side of (16.21) is still positive, and therefore $\bar{\sigma}$ continues to increase to a peak value, and then decreases as the right-hand side of (16.21) becomes negative. This continues, and the curve approaches asymptotically the limiting value,

$$\lim_{\bar{\varepsilon} \to \infty} \left\{ M + \frac{h}{a} \right\} = M^*; \tag{16.24}$$

see Curve 2 of Figure 16.2(b). The corresponding variation of the void ratio is shown by Curve 2 of Figure 16.2(c).

The model can easily be modified to include the effect of cohesion, as discussed in Ref. 10. Also, it is well known that the strain rate tensor, in general, is non-coaxial with the stress tensor, for dilatant frictional materials. This fact is examined from a more fundamental basis in Section 16.4. Here, however, it should be noted that one may modify equation (16.5) to read

$$\dot{\varepsilon}_{ij} = \frac{1}{H} \left\{ \frac{S_{ij}}{2\bar{\sigma}} + \frac{1}{3} \frac{\partial G}{\partial p} \delta_{ij} \right\} \left\{ \frac{S_{kl}}{2\bar{\sigma}} - \frac{1}{3} \frac{\partial F}{\partial p} \delta_{kl} \right\} \dot{\sigma}_{kl} + D_{ijkl}(\bar{\varepsilon}, \theta) \dot{\sigma}_{kl}, \tag{16.25}$$

where D_{ijkl} may be viewed as an 'effective secant compliance' tensor

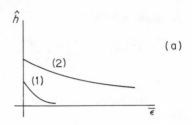

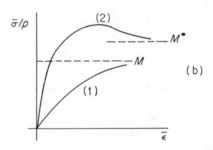

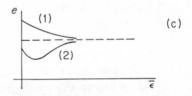

Figure 16.2 (a) Possible variations of the distortional hardening parameter, $\hat{h}$; (b) stress–strain relations; (c) variation of void ratio with distortional strain; *Note*: Curves marked (1) are for loose sands, and those marked (2) are for dense sands

depending on the effective distortional strain, as well as on the volumetric strain. For example if it is required that this additional term should not affect the definition of dilatancy, (16.9), and have no contribution to the rate of distortional work, then it may be taken as

$$D_{ijkl} = A\left[\delta_{ik}\delta_{jl} - \frac{1}{2\bar{\sigma}^2} S_{ij}S_{kl}\right],$$ (16.26)

where A is a function of $\bar{\varepsilon}$ and θ. It has been argued in the literature,[13,14] that modifications of this kind represent the vertex yield function.

16.2.4 Unloading and change of fabric

In the above discussion, no unloading has been considered. It is an experimental fact that there is always some strain recovery during unloading. This strain recovery may be partly elastic unloading and partly inelastic unloading due to (dissipative) rearrangement of grains, when cohesionless elastic granules are involved. Even if the granules are assumed to be rigid, so that there is no elasticity involved, a small incremental load reduction (unloading) is accompanied by some strain recovery. Traditionally, it has been convenient to regard this as elastic unloading; however, in reality, it stems from microstructural changes due to the changes of the 'fabric', and in fact, it involves a dissipative, thermodynamically irreversible process. In fact, for a collection of rigid granules, any deformation is plastic, be it during continuous loading, or as a result of strain recovery in unloading.

This is a fundamental question relating to the basic micromechanical behaviour of granular masses. It has been recently examined by this writer and co-workers in the context of a statistical approach, where the notion of a 'fabric tensor' has been introduced and related to the macroscopic stress tensor.[5,12]

16.3 LIQUEFACTION AND DENSIFICATION OF COHESIONLESS SAND

16.3.1 Notation

Attention is focused on cyclic shearing at constant confining pressure. Therefore, the state of stress is as defined by equation (16.17). Hence, for simplicity, the applied stress and the (constant) confining pressure are denoted by $\hat{\tau}$ and σ_c, respectively, and the corresponding shear strain (engineering) is denoted by γ. The pore water pressure is designated by $\hat{p}_w$, and the following dimensionless quantities are used:

$$\tau = \hat{\tau}/\sigma_c, \qquad p_w = \hat{p}_w/\sigma_c. \tag{16.27}$$

16.3.2 General discussion

When a sample of cohesionless sand is subjected to cyclic shearing under a fixed confining pressure, the grains are moved relative to each other, and because of the action of confining pressure, the sample tends to densify after each cycle. If the sample is saturated and undrained, the tendency toward densification causes an increase in pore water pressure, and therefore a decrease in the interparticular contact forces. When the pore water pressure increases to a value close to that of the confining pressure, the interparticular contact forces become very small, and therefore, the corresponding frictional resistance is rendered small: the overall shear resistance of the material tends to become negligibly small, i.e. the sample tends to liquefy.

If, on the other hand, drainage is provided, and the cyclic shearing is performed at low frequencies, the sample densifies with the increase in the number of cycles, resulting in an increase in its shear resistance.

Here we shall illustrate how a simple model can be generated, which would encompass the basic physical features of the phenomenon, and therefore can have effective predictive value.

16.3.3   Dimensional analysis

Let e_0 and e_m be, respectively, the initial and the minimum values of the void ratio for the considered sample. It is clear that, as the *relative void ratio*, $e_0 - e_m$, increases, the potential for liquefaction increases. In fact, if the grains are assumed to be rigid, no liquefaction can take place if $e_0 = e_m$.

With this in mind, we observe that the basic dimensionless parameters relevant for a first-order approximation theory of liquefaction in cyclic shearing, are: dimensionless applied shear stress amplitude, τ_0; pore pressure (measured at the end of each cycle), p_w; the relative void ratio, $e_0 - e_m$; and the number of cycles, N. Therefore, it is reasonable to expect that

$$F(\tau_0, p_w, e_0 - e_m, N) = C \qquad (16.28)$$

should define the relation between the stated dimensionless parameters, where the constant C may, however, depend on the initial void ratio e_0. The simplest form for this equation would be

$$\tau_0^a(e_0 - e_m)^b N = A_0 P(1 + p_w), \qquad a, b > 1, \qquad (16.29)$$

where positive exponents a and b must exceed 1, so that if a larger shear stress amplitude is used, or when initially looser sand is involved, then a fewer number of cycles would be required to attain the same pore pressure under the same confining pressure. A_0 on the right-hand side is a parameter which could only depend on the initial void ratio e_0. The function $P(1 + p_w)$ must be such that

$$P(1) = 1, \qquad P' \geq 0, \qquad P'' \leq 0. \qquad (16.30)$$

The first condition corresponds to the initial state $p_w = 0$ at $N = 0$, and the second two conditions guarantee that positive work is required to increase the pore pressure and that this increase results in less resistive materials (this is discussed further in the sequel). Note that, since A_0 is an as yet unspecified parameter, without a loss in generality, the exponent of N in equation (16.29) is chosen to be 1.

When a drained condition is involved, then the densification can be characterized by the following modified version of equation (16.29):

$$\tau_0^a(e - e_m)^b N = A_0, \tag{16.31}$$

where e now is the void ratio after N cycles of cyclic shearing at constant dimensionless shear stress amplitude τ_0.

From equation (16.29), liquefaction occurs when $p_w = 1$, so that $P(2) = P_l$ at $N = N_l$. From equation (16.31), on the other hand, the change in void ratio e can be calculated in response to a given number of cyclic shearings. In this case, most experiments are performed for cyclic shearing at constant strain amplitude, γ_0, rather than at constant stress amplitude. If we assume that

$$\tau_0 = A_1\gamma_0^{a_1} + A_2\gamma_0^{a_2} + \cdots, \tag{16.32}$$

then we can substitute for τ_0 into (16.31). We note that the exponents a_1 and $a_2, \ldots$, must all be positive and odd so that the shear stress amplitude versus the shear strain amplitude will be a centrally symmetric curve in the τ_0, γ_0-plane.

The above approach, although quite straightforward, does not bring the physics of the process to bear heavily on the mathematical model. In the sequel a physical approach is presented, from which equations similar to (16.29) and (16.31) emerge in a natural manner.

16.3.4   An energy approach

Based on an energy consideration, a unified densification and liquefaction theory for cohesionless sand in cyclic shearing has been developed by this writer and his associate.[6–8] Here, a brief summary of the theory is presented and results are compared with the above development.

To this end we first make the following physically obvious observations:

(1)   To change the void ratio from its current value e to $e + de$, an increment of energy dW is required for rearranging the sand grains (microstructural rearrangement);

(2)   This increment of energy dW increases as the void ratio approaches e_m;

(3)   Since an increase in the excess pore water pressure p_w results in a decrease in the intergranular forces, the required incremental energy dW decreases with increasing p_w.

Based on these we may write

$$dW = -\tilde{\nu}\frac{de}{f(1 + p_w)g(e - e_m)}, \tag{16.33}$$

where $\tilde{\nu}$ is a positive parameter, and the functions f and g are such that

$$f(1) = 1, \quad f' \geqslant 0, \quad g(0) = 0, \quad g' \geqslant 0. \tag{16.34}$$

Since

$$de = -\frac{e\sigma_c}{\kappa_w} dp_w, \tag{16.35}$$

where κ_w is the bulk modulus of water, equation (16.33) can also be expressed as

$$dW = \frac{\nu}{\kappa_w} \frac{e\, dp_w}{f(1+p_w)g(e-e_m)}, \tag{16.36}$$

where $\nu = \tilde{\nu}\sigma_c = \nu(\sigma_c)$, i.e. a parameter which depends on the confining pressure.

For densification of a drained sample, $p_w = 0$, and equation (16.32) becomes

$$dW = -\tilde{\nu}\frac{de}{g(e-e_m)} \tag{16.37}$$

which upon integration yields

$$\Delta W = -\tilde{\nu}\int_{e_0}^{e} \frac{dx}{g(x-e_m)}, \tag{16.38}$$

where ΔW is the total energy required to change the void ratio from e_0 to its current value e.

For liquefaction, we first integrate (16.35) to obtain

$$e = e_0 \exp\left\{-\frac{\sigma_c}{\kappa_w} p_w\right\}, \tag{16.39}$$

where it is assumed that initially $p_w = 0$. With the aid of (16.39), equation (16.36) can now be integrated. But since σ_c/κ_w, for pressures of several bars, is so small that $e \approx e_0$ in (16.39), we may simplify the calculations considerably (but without a loss in accuracy) by using e_0 instead of e in equation (16.36). Then upon integration, it follows that

$$\Delta W = \frac{\nu e_0}{g(e_0 - e_m)} \int_0^{p_w} \frac{dx}{f(1+x)}. \tag{16.40}$$

To obtain explicit results, we use approximate expressions for functions g and f in such a manner that conditions (16.4) are satisfied. Simplest functions of this kind are

$$\begin{aligned} g(e-e_m) &\equiv (e-e_m)^n, &&n > 1,\\ f(1+p_w) &\equiv (1+p_w)^r, &&r > 1. \end{aligned} \tag{16.41}$$

With these expressions, equations (16.38) and (16.40) upon integration respectively yield

$$e = e_{\mathrm{m}} + [(e_0 - e_{\mathrm{m}})^{1-n} + \bar{\nu} \, \Delta W]^{1/(1-n)}, \qquad n > 1, \qquad (16.42)$$

and

$$\Delta W = \frac{\hat{\nu} e_0}{(e_0 - e_{\mathrm{m}})^n} [1 - (1 + p_{\mathrm{w}})^{1-r}], \qquad r > 1, \qquad (16.43)$$

where we have set $\bar{\nu} = (n-1)/\tilde{\nu}$, and $\hat{\nu} = \nu/(r-1)\kappa_{\mathrm{w}}$, respectively.

To complete the solution, we must now estimate the energy loss ΔW. In a cyclic shearing, ΔW can be related to the area of hysteretic loop and the number of cycles.[8] Here two cases may be distinguished: (1) large amplitude shearing; and (2) small amplitude shearing, each either stress- or strain-controlled tests.

When the amplitude of shearing is large enough, all the particles are mobilized during each cycle, and hence the energy loss in each cycle is not very much dependent on the previous cycles. In this case we may assume that ΔW is proportional to the number of cycles, N. For very small amplitude shearing, on the other hand, the particles are only partially mobilized in each cycle, and hence the energy loss changes from cycle to cycle as the particles gradually take on new (more stable) positions relative to each other.

A series of experimental results on cyclic shearing of cohesionless sands has been carefully examined by Nemat-Nasser and Shokooh,[8] and because of the observed symmetry of the hysteretic loop, it has been concluded that the area of hysteretic loop in the ith cycle may be approximated by $A_i = h_i \tau_0^{1+\alpha}$, where α must be an even positive integer (because of the symmetry), and h_i is an increasing function of the number of cycles. Hence, for stress-controlled tests, it follows that

$$\Delta W = \sum_{i=1}^{N} \lambda_i h_i \tau_0^{1+\alpha} = \tau_0^{1+\alpha} \sum_{i=1}^{N} \lambda_i h_i = \hat{h}(N) \tau_0^{1+\alpha}. \qquad (16.44)$$

Moreover, for large stress amplitudes, we may assume that

$$\Delta W \simeq h N \tau_0^{1+\alpha}. \qquad (16.45)$$

From (16.45) and (16.43) it now follows that

$$\tau_0^{1+\alpha}(e_0 - e_{\mathrm{m}})^n N = \hat{\hat{\nu}} e_0 [1 - (1 + p_{\mathrm{w}})^{1-r}], \qquad (16.46)$$

where $\hat{\hat{\nu}} = \hat{\nu}/h$. Since α is even, it can easily be fixed by inspection of experimental results. For cohesionless samples considered by De Alba, Chan, and Seed,[3] it is immediately seen that $\alpha = 4$. Moreover, this result is not very sensitive to small variations in the other parameters. For example for De Alba *et al.*'s[3] experiments, n and r in equation (16.46) may be chosen in the ranges of 3 to 4 and 2 to 3, respectively, and hence $n = 3.5$

and $r = 2.5$ were selected for comparison with the considered experimental results. The parameter $\hat{\nu}$ then turned out to be a constant which did not vary much from test to test over a wide range of sample densities.[8]

A loss of total bearing capacity occurs momentarily, when the pore pressure equals the confining pressure, so that $p_w = 1$ in equation (16.46). If the number of cycles to this liquefaction is denoted by N_l, then equation (16.46) yields

$$\tau_0^{1+\alpha} = \frac{\eta e_0}{N_l(e_0 - e_m)^n},$$ (16.47)

where η is a constant. This equation has been compared with experimental results,[6,8] and it has been verified that over a wide range of densities, η indeed appears to be a constant which, of course, would depend on the material, the grain distribution, and the confining pressure. Figure 16.3 is a typical result.[6,8]

If α is taken to equal 4 (or any other positive even number), and since $n > 1$, it follows from equation (16.47) that, for the same number of cycles to liquefaction, the dimensionless shear stress amplitude increases as the initial void ratio approaches its minimum value; in fact, for void ratios close to the minimum (very dense send) a very large shear stress amplitude (approaching infinity as e_0 approaches e_m) is required. Moreover, since τ_0 is normalized

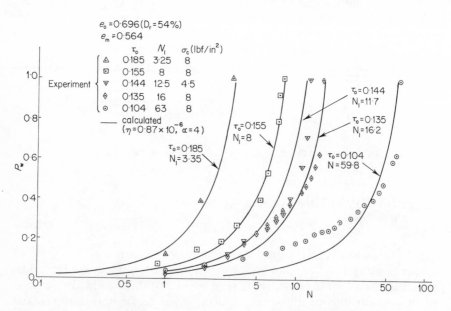

Figure 16.3 Normalized pore water pressure, p_w, in terms of the number of cycles, N, for indicated values of the normalized shear stress amplitude

with respect to the confining pressure, equation (16.47) again shows that, for the same number of cycles to liquefaction, the dimensional shear stress amplitude is proportional to the confining pressure. These are results which are confirmed both experimentally and by field investigations.

It is possible to relate the shear strain amplitude to the number of cycles and the shear stress amplitude during liquefaction of undrained sand. This is discussed by the present writer and his associates,[7] where comparisons with experimental results are also made.

For strain-controlled tests, one may use equation (16.32) in (16.45). However, as a first-order approximation, it has been found adequate to take

$$\tau_0 = K\gamma_0^{1/\beta}, \tag{16.48}$$

where β must be an odd positive integer. This then gives

$$\Delta W \simeq \hat{k}(N)\gamma_0^{(1+\alpha)/\beta} \tag{16.49}$$

Extensive comparison with experimental results for densification has been made[6,8] employing equation (16.49) with $\hat{k}(N) = k_0 N$ for large strain amplitudes ($\gamma_0 > 0.1\%$), and $\hat{k}(N) = k_0\sqrt{N}$ for small strain amplitudes ($\gamma_0 < 0.1\%$), where $n = 3.5$, $\alpha = 4$, and $\beta = 5$ have been used.

16.4 MICROMECHANICAL APPROACH

If one assumes that individual grains are rigid and their mutual interaction involves only friction over the contacting regions, then it is reasonable to expect that the overall macroscopic quantities, such as stress, strain, and their rates should be expressible in terms of the relevant microscopic quantities by means of a systematic averaging process. A fundamental program of this kind has been recently initiated by the writer and his associates, where several preliminary but significant results have been obtained.[2,5,9,11,12] In particular, explicit results for the overall stress tensor, the strain rate and spin tensors, and the macroscopic equations for the dilatancy phenomenon and the non-coaxiality of the stress and strain rates have been developed. In the sequel some of the results are briefly summarized, where attention is focused on the two-dimensional (plane strain) case; for simplicity, two-dimensional, rod-like granules may be envisaged.

16.4.1 Overall stress tensor

To develop overall stresses in terms of contact forces, the principle of virtual work can be used as an effective tool. To this end, let there be N contacts in a typical, suitably large volume, V, of sample (chosen as unit volume), denote the contact forces at a typical contact α by f_i^α, and let the vector which connects the centroid of, say, grain A to the centroid of grain B that are contacting at α, be denoted by $\mathbf{l}^\alpha$. Then, following the development outlined by Christoffersen, Mehrabadi, and Nemat-Nasser,[2] it is easy to show that the average stress tensor is describable as

$$\bar{\sigma}_{ij} \equiv \int_V \sigma_{ij}\, dV = \sum_{\alpha=1}^N f_i^\alpha l_j^\alpha. \qquad (16.50)$$

If, moreover, it is assumed that the rotation of individual grains coincides with the overall macroscopic rotation field evaluated at the corresponding contact point, then the stress tensor becomes symmetric and one has

$$\bar{\sigma}_{ij} = \tfrac{1}{2}N \langle f_i l_j + f_j l_i \rangle, \qquad \langle f_i l_j - f_j l_i \rangle = 0, \qquad (16.51)$$

where the notation

$$\frac{1}{N} \sum_{\alpha=1}^N f_i^\alpha l_j^\alpha \equiv \langle f_i l_j \rangle \qquad (16.52)$$

is used.

In Ref. 5 the implications of equations (16.51) and their relation with average tractions defined in terms of the resultant contact forces transmitted across an imagined plane, are examined in detail. In particular, it is shown that the definition of overall stress in terms of the virtual work method coincides with that developed with the aid of average tractions transmitted across imagined planes. Moreover, the stress tensor is related to the fabric tensor which characterizes the microstructure of the grain configuration.[2,5,12]

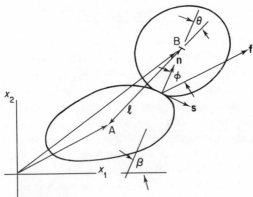

Figure 16.4 Two contacting granules (A and B)

At each contact, introduce the unit contact normal, $\mathbf{n}$, the unit vector, $\mathbf{s}$, normal to $\mathbf{n}$ as shown in Figure 16.4, and consider the following notation:

$$\mu = \tan \phi, \qquad b = \tan \theta, \qquad t = fl \cos \phi \sin \theta, \qquad (16.53)$$

where f and l are the magnitudes of the contact force and the vector $\mathbf{l}$, respectively; the superscript α is dropped for simplicity. Then, it is easy to show that

$$\bar{\sigma}_{ij} = N\langle t[\mu b \delta_{ij} + (1 - \mu b) n_i n_j + \tfrac{1}{2}(\mu + b)(n_i s_j + n_j s_i)]\rangle. \tag{16.54}$$

Moreover, the symmetry of the stress tensor results in

$$\langle fl \sin (\phi - \theta)\rangle = 0. \tag{16.55}$$

From equation (16.54) it follows that

$$\bar{\sigma}_{kk} = N\langle fl \cos (\theta - \phi)\rangle. \tag{16.56}$$

16.4.2 Overall kinematics

To obtain the overall kinematics, we observe that the rate of change of vector $\mathbf{l}$ corresponding to a typical contact α (the superscript α is dropped for simplicity) can be written as

$$\dot{l}_i = \left(\frac{\dot{l}}{l} \delta_{ij} + r_{ij}\right) l_j, \tag{16.57}$$

where the rotation matrix, r_{ij}, may be expressed as

$$r_{ij} = \dot{\Omega}(s_i n_j - s_j n_i), \qquad \dot{\Omega} = \dot{\nu} + \dot{\theta}, \tag{16.58}$$

where ν is the angle formed by the contact unit normal, $\mathbf{n}$, with the x_2-axis. Equation (16.57) can be interpreted as $\dot{l}_i = L_{ij} l_j$, where L_{ij} is the local deformation rate tensor. To solve this expression for L_{ij}, we note that this quantity can be defined only to within a vector perpendicular to the vector $\mathbf{l}$. This is an important observation, since the additional undefined quantity is exactly needed in order to provide a compatible overall deformation rate. Hence, we obtain for, say, the contact α,

$$L_{ij} = \frac{\dot{l}}{l} \delta_{ij} + r_{ij} + \dot{\gamma} k_i (\cos \theta \, s_j - \sin \theta \, n_j), \tag{16.59}$$

where $\mathbf{k}$ is any unit vector, and $\dot{\gamma}$ is a yet unspecified rate.

The overall deformation rate tensor (the velocity gradient) is then obtained by averaging, i.e.

$$\bar{L}_{ij} = \langle L_{ij}\rangle, \tag{16.60}$$

so that the macroscopic overall strain rate and the spin tensors are given by

$$\dot{\bar{\varepsilon}}_{ij} = \tfrac{1}{2}(\bar{L}_{ij} + \bar{L}_{ji}), \qquad \dot{\bar{\omega}}_{ij} = \tfrac{1}{2}(\bar{L}_{ij} - \bar{L}_{ji}); \tag{16.61}$$

the detailed results are given in Ref. 2 and will not be presented here.

16.4.3 Energy balance

Since the only source of dissipation is friction, the rate of stress-work, $\bar{\sigma}_{ij}\dot{\bar{\varepsilon}}_{ij}$, can be expressed by

$$\bar{\sigma}_{ij}\dot{\bar{\varepsilon}}_{ij} = N\langle(\mathbf{f}\cdot\mathbf{s})(\mathbf{l}^*\cdot\mathbf{s})\rangle, \tag{16.62}$$

where $\mathbf{l}^*$ is the rate of change of $\mathbf{l}$ observed in a coordinate system corotational with it, i.e.

$$\mathbf{l}^* = \frac{\dot{l}}{l}\mathbf{l}. \tag{16.63}$$

In terms of notation introduced in previous subsections, the rate of dissipation becomes

$$\bar{\sigma}_{ij}\dot{\bar{\varepsilon}}_{ij} = N\langle f\dot{l}\sin\theta\sin\phi\rangle. \tag{16.64}$$

It has been observed experimentally that, as granular masses deform, they momentarily form clusters consisting of instantaneously mutually immobile granules. This is true even for granular masses consisting of, say, spherical granules. In general, clusters are not spherical, and therefore, even for spherical granules anisotropic behaviour exists. Equation (16.64) shows this fact, because, if the clusters were spherical, then $\dot{l}$ would be zero, and there would be no energy dissipation. Note that, when clusters of mutually immobile grains constitute a unit granule, then $\mathbf{l}$ is the vector connecting the centroid of two adjacent contacting clusters.

16.4.4 Dilatancy and noncoaxiality

The plasticity theory presented in Section 16.2 gives a dilatancy equation which is in good accord with observation. However, it yields a strain rate coaxial with stress unless this is excluded by the additional term in the manner given in equation (16.25). The microscopic development, however, provides us not only with a dilatancy equation, but also with another equation which clearly shows noncoaxiality of the stress and the strain rate. These are obtained from the observation that the quantities $\langle n_i n_j\rangle$ and $\langle n_i s_j + n_j s_i\rangle$ are not independent. The detailed development is presented in Ref. 2. The final equations are:

$$(\sin^2\phi - R^2)\frac{\bar{\sigma}_{ij}\dot{\bar{\varepsilon}}'_{ij}}{\tau} = R\cos^2\phi\,\dot{\bar{\varepsilon}}_{kk}, \tag{16.65}$$

$$\bar{\sigma}_{ij}\overset{\triangledown}{\bar{\sigma}}_{kj} - \overset{\triangledown}{\bar{\sigma}}_{ik}\bar{\sigma}_{kj} = \frac{2\tau}{R}(\bar{\sigma}_{ik}\dot{\bar{\varepsilon}}_{kj} - \dot{\bar{\varepsilon}}_{ik}\bar{\sigma}_{kj}),$$

where

$$\overset{\triangledown}{\bar{\sigma}}_{ij} = \dot{\bar{\sigma}}_{ij} - \Omega_{ik}\bar{\sigma}_{kj} - \Omega_{jk}\bar{\sigma}_{ki},$$

$$\Omega_{12} = -\Omega_{21} = \dot{\bar{\omega}}_{12} - \langle\dot{\nu} + \dot{\theta}\rangle - \dot{\psi}, \tag{16.66}$$

$$R = \tau/p, \qquad \tan 2\psi = \frac{2\bar{\sigma}_{12}}{\bar{\sigma}_{11} - \bar{\sigma}_{22}},$$

and where τ is the maximum shear stress; see Figure 16.4 for definition of β and θ; $\nu = \pi/2 - \beta$.

16.4.5 Concluding comments

It is clear that the development summarized in this section is fundamental and should provide guidance for the formulation of constitutive relations for granular materials. It can be generalized to include the elasticity of the granules, the cohesive forces that may exist between the grains, the possibility of rate-dependent inelastic deformations of clusters of granules (e.g. in clay), and other relevant physical features. The author and his associates are pursuing these questions at this time. However, one of the most important outstanding problems that requires careful attention is the calculation of the stress rate in terms of the strain rate. This requires a knowledge of the manner by which new contacts are formed and some existing contacts are lost, as the deformation proceeds. In other words, we need to know the evolution of the distribution of contacts and their unit normals. While some theoretical conclusions can be obtained on the basis of reasonable simple assumptions, experiments must provide general guidance. The main problem is to identify the manner by which the 'fabric' changes in the course of deformation.[2,5,11,12]

ACKNOWLEDGEMENT

This work has been supported in part by the U.S. Geological Survey under Contract No. 14-08-0001-17770, and by U.S. Air Force Office of Scientific Research, Grant No. AFOSR-80-0017 to Northwestern University.

REFERENCES

1. Byerlee, J. D. The fracture strength and frictional strength of Weber sandstone, *Int. J. Rock Mech. Min. Sci. & Geomech. Abstr.*, **12,** 1–4, 1975.
2. Christoffersen, J., Mehrabadi, M. M., and Nemat-Nasser, S. A micromechanical description of granular material behavior, Earthquake Research and Engineering Laboratory Tech. Rep. No. 80-1-22, Dept. Civil Engrg., Northwestern University, Evanston, Ill., 1980; *J. Appl. Mech.*, **48,** 339–44, 1981.
3. De Alba, P. S., Chan, C. K., and Seed, H. B. *Determination of Soil Liquefaction Characteristics by Large-scale Laboratory Tests*, NUREG-0027, NRC-6, Shannon and Wilson, Inc., and Agbabian Assoc., Seattle, WA. Prepared for the U.S. Nuclear Regulatory Commission under Contract AT(04-3)-954. See also EERC Rep. 75–14, University of California, Berkeley, Calif., 1975.
4. Hill, R. *The Mathematical Theory of Plasticity*, Oxford University Press, 1950.
5. Mehrabadi, M. M., Nemat-Nasser, S., and Oda, M. On statistical description of stress and fabric in granular materials, Earthquake Research and Engineering Laboratory Tech. Rep. No. 80-4-29, Dept. Civil Engrg., Northwestern University, Evanston, Ill., 1980; to appear in *Int. J. for Numerical & Analytical Methods in Geomechanics*, **6,** No. 1, 1982.

6. Nemat-Nasser, S., and Shokooh, A. *A Unified Approach to Densification and Liquefaction of Cohesionless Sand*, Earthquake Research and Engineering Laboratory Tech. Rep. No. 77-10-3, Dept. Civil Engrg., Northwestern University, Evanston, Ill., 1977.

7. Nemat-Nasser, S., and Shokooh, A. A new approach for the analysis of liquefaction of sand in cyclic shearing, *Proc. Second Int. Conf. on Microzonation*, **2**, 957–69, San Francisco, Calif., 1978.

8. Nemat-Nasser, S., and Shokooh, A. A unified approach to densification and liquefaction of cohesionless sand in cyclic shearing, *Canadian Geotechnical J.*, **16**, 659–78, 1979.

9. Nemat-Nasser, S. On behavior of granular materials in simple shear, Earthquake Research and Engineering Laboratory Tech. Rep. No. 79-6-19, Dept. Civil Engrg. Northwestern University, Evanston, Ill., 1979; *Soils and Foundations*, **20**, 59–73, 1980.

10. Nemat-Nasser, S., and Shokooh, A. On finite plastic flows of compressible materials with internal friction, Earthquake Research and Engineering Laboratory Tech. Rep. No. 79-5-16, Dept. Civil Engrg., Northwestern University, Evanston, Ill., 1979; *Int. J. Solids & Struct.*, **16**, 495–514, 1980.

11. Oda, M., Konishi, J., and Nemat-Nasser, S. *Index Measures for Granular Materials*, Earthquake Research and Engineering Laboratory Tech. Rep. No. 80-3-26, Dept. Civil Engrg., Northwestern University, Evanston, Ill.; Some experimentally based fundamental results on the mechanical behavior of granular materials, *Géotechnique*, **30**, 479–95, 1980.

12. Oda, M., Nemat-Nasser, S., and Mehrabadi, M. M. A statistical study of fabric in a random assembly of spherical granules, Earthquake Research and Engineering Laboratory Tech. Rep. No. 80-4-28, Dept. Civil Engrg., Northwestern University, Evanston, Ill., 1980; to appear in *Int. J. for Numerical & Analytical Methods in Geomechanics*, **6**, No. 1, 1982.

13. Rice, J. R. The localization of plastic deformation. In Koiter, W. T. (ed.), *Theoretical and Applied Mechanics*, North-Holland, pp. 207–20, 1977.

14. Rudnicki, J. W., and Rice, J. R. Conditions for the localization of deformation in pressure-sensitive dilatant materials, *J. Mech. Phys. Solids*, **23**, 371–94, 1975.

15. Schofield, A., and Wroth, P. *Critical State Soil Mechanics*, McGraw-Hill, London, 1968.

Soil Mechanics—Transient and Cyclic Loads
Edited by G. N. Pande and O. C. Zienkiewicz
© 1982 John Wiley & Sons Ltd

Chapter 17

Fatigue Models for Cyclic Degradation of Soils

H. A. M. van Eekelen

17.1 INTRODUCTION

Most theories for soil behaviour are based on yield surfaces, flow rules and hardening laws, and are primarily meant to describe the effects of monotonic loading. In some models, kinematic hardening is included to describe unloading and damping and, with some luck, shakedown or incremental collapse may be achieved for a soil structure under inhomogeneous repeated loading.

Recently, some further modifications and complications have been introduced into these models in an effort to better describe some of the characteristic phenomena associated with persistent cyclic loading of clay or sand. Some of these models use sets of shifting yield surfaces within an isotropic yield surface, others are basically modified versions of the Cam Clay model. Although some success has been achieved in this way, agreement with experiment is less than perfect. Also, the nature of these models requires that in actual application, for earthquake or wave loading analysis, a stress–strain calculation is performed for each individual cycle, which makes the procedure impractical when large numbers of cycles are involved.

Cyclic loading changes the state of a soil sample. The stress–strain relations are influenced, the ultimate strength is affected, cumulative strains may develop, and cumulative pore pressures or volume strains are induced. In most models which have been constructed specifically to describe the behaviour of soils under cyclic loading, an extra state variable or memory parameter k is introduced to describe these changes. In the following, this memory parameter will be called 'fatigue'. In such a model, one only needs to know the initial state of a soil sample in terms of its response properties, and the current value of k, to predict behaviour under further cyclic or monotonic loading; one does not need details such as number, order, intensity or direction of past stress or strain cycles. It is also commonly assumed that the parameter k is monotonically increasing. This excludes materials with a 'fading memory', for which k could decrease during a period of relative quiet. Fatigue models for cyclic degradation of soils differ in the complexity of the soil model used; but what they have in common is the use of one (and in one case two) memory parameter(s) which are added on to the soil model to account for the cumulative effects of cyclic loading.

17.2 BASIC THEORY OF FATIGUE MODELS FOR CYCLIC DEGRADATION OF SOILS

Apart from a description of the initial state of the soil in terms of a static soil model or, in the simplest case, of a stress–strain relation for cycle number one, the basic elements of a fatigue theory for cyclic loading of soils are the following:

(1) a relation which gives the increase of the fatigue parameter k per cycle, dependent on the nature of the loading cycle and on the current value of k, and
(2) a set of relations which indicate the changes in soil response characteristics with increasing value of k.

Assuming that the parameter k accounts for all effects of past cyclic loading and that it is monotonically increasing, the first relation may be written in the form[6]

$$\frac{dk}{dN} = Q(k, \tau_c) \qquad \text{or} \qquad \frac{dk}{dN} = \hat{Q}(k, \gamma_c), \tag{17.1}$$

where the nature of a loading cycle is specified either by a generalized stress amplitude τ_c or by a generalized strain amplitude γ_c. The function Q or $\hat{Q}$ will depend on the initial state of the sample; for a given initial state it provides a yardstick to compare the effect of cycles with different intensities or loading geometries (if any). In many models, the function Q or $\hat{Q}$ is not given explicitly; it then has to be derived from expressions for the increase of k with number of cycles N in constant stress or strain amplitude tests, by taking the derivative with respect to N and substituting for N in terms of k.

Once the function Q or $\hat{Q}$ has been specified, one needs to indicate how the soil response to monotonic or cyclic loading changes with increasing value of k. In the simplest models[7,8,11] the change in cyclic response properties is given as a relation between stress and strain amplitudes

$$\gamma_c = F(k, \tau_c) \qquad \text{or} \qquad \tau_c = \hat{F}(k, \gamma_c). \tag{17.2}$$

In addition, a relation may be given for the ultimate strength after cycling, in terms of the initial strength and the value of k.

In more elaborate models, the state variable k is introduced as a parameter in a complete mechanical model for soil response (including damping), which may be a hardening elasto-plastic model,[15] or an endochronic theory of soil behaviour.[3]

An interesting special case arises when the function Q is separable, $Q(k, \tau_c) = g(k)h(\tau_c)$, with $g(k) > 0$ because the parameter k is monotonically increasing. In that case, the integral

$$K = \int_0^k \frac{1}{g(k')}\, dk'$$

is a monotonically increasing function of k, with inverse $k(K)$. For this state variable K we have $dK/dN = h(\tau_c)$. For experiments at constant τ_c one obtains $K = Nh(\tau_c)$, and for a more general process

$$K = \sum_{n=1}^{N} h(\tau_c^{(n)}), \tag{17.3}$$

where $\tau_c^{(n)}$ is the stress amplitude in the nth cycle. Hence, if the function Q is separable Miner's rule[13] of superposition applies: the cumulative effect of a number of stress cycles of varying intensity may be obtained by linear superposition, as in equation (17.3), independent of the order in which individual cycles occur. The same is true of strain cycles if the function $\hat{Q}$ is separable. It should be noted that the two versions of Miner's rule, on superposition of stress cycles and on superposition of strain cycles, are not equivalent: if one is applicable, then the other is not, except in a few special cases.[6]

In many empirical models for the behaviour of sands during earthquakes, Miner's rule is used to establish a simple conversion from the erratic actual stress history to an equivalent number of uniform stress cycles, from which the likelihood of soil liquefaction is estimated. These models have been reviewed by Seed.[19]

17.3 FATIGUE MODELS FOR CYCLIC DEGRADATION OF SOILS

Existing fatigue models differ in the complexity of the soil model used, and in the manner in which the parameter k is related to observables of the system. The simplest procedure is to identify k with an observable, such as the plastic volume strain for drained loading, or the cumulative pore pressure u for undrained loading. The function Q or $\hat{Q}$ may then be obtained by monitoring this variable in a number of constant amplitude tests. In some models, however, the fatigue parameter k is not directly related to any observable of the system, and the function Q or $\hat{Q}$ has to be obtained by a more elaborate process of curve fitting.

The most complex fatigue models are those in which the state variable k is introduced as a parameter in a complete mechanical model for soil response. Prévost[15] uses a complicated elasto-plastic model, with both isotropic and kinematic hardening, much like the models by Mroz[14] and Dafalias.*[4,5] He introduces a set of nested yield surfaces f_m, with radii r_m and associated plastic moduli H_m. Both r_m and H_m are specific functions of a memory parameter k. This state variable k appears to be the total length of the deviatoric strain path, but a closer analysis of the formulae involved reveals that this is not in fact the case (a similar situation arises in a paper by Zienkiewicz *et al.*[21]). The parameter k is directly related to the shrinking of the yield surfaces:

$$r_m = r_{m0} - \tau_{hf}(1.10 - 1.34e^{-k})$$

where r_{m0} is the value of r_m for monotonic loading, τ_{hf} is the measured shear strength for slow monotonic loading, and the initial value of k is zero. In cyclic loading at constant strain amplitude, the increase of k is given by

$$k = A \ln (4N - 3)[2D + \ln (4N - 3)],$$

where A is a given function of strain amplitude, and D is a numerical constant. Taking the derivative with respect to N and substituting for N in terms of k, one obtains the function $\hat{Q}$

$$\hat{Q} = CA\sqrt{(1 + k/AD^2)} \exp [-D\sqrt{(1 + k/AD^2)}]$$

with $C = 8De^D$. The function $\hat{Q}$ is clearly not separable, so that Miner's rule does not apply. The model also gives the form of the stress–strain cycles, and hence the damping. Strictly speaking, it is not a single parameter fatigue model, because the current geometry of the system of nested yield surfaces (i.e. the kinematic hardening) cannot be derived from the current value of k: it must be worked out separately from the details of the cyclic loading history. In an application to Drammen clay[16] reasonable agreement with cyclic loading data is achieved, but the number of material parameters involved is large (approximately 50).

Another rather complicated fatigue model is the one proposed by Cuellar, Bazant, and Krizek,[3] which is based on endochronic theory (it should be noted that some objections against endochronic models—non uniqueness and instability—have recently been raised by Sandler[20]†). The model is exceptional in that it contains *two* fatigue parameters k_1 and k_2, governing deviatoric strain and volumetric strain (for drained loading), respectively.

---

* *Editors' note:* See Chapters 8 and 10.
  † *Editor's note:* See Chapter 15 for comments on these objections and Chapter 14 for a new Endochronic theory.

For simple shear, the increase per cycle of these parameters is given by

$$\frac{dk_1}{dN} = \hat{Q}_1 = \gamma_c \quad \text{and} \quad \frac{dk_2}{dN} = \hat{Q}_2 = (\gamma_c)^q,$$

where q is a positive constant. Since both $\hat{Q}_1$ and $\hat{Q}_2$ are separable (they do not depend on k at all), the effect of a series of strain cycles is independent of the order of the cycles. The volumetric strain e is given in terms of the fatigue parameter k_2 by

$$e = \frac{1}{a} \ln (1 + ak_2),$$

and the deviatoric strain increments are given by an empirical relation which contains k_1

$$d\gamma = \frac{d\tau}{G} + \frac{\tau}{G} \frac{1}{Z(1 + Xk_1)^r} |d\gamma|. \tag{17.4}$$

Here G is shear modulus, and a, r, X, and Z are constants. Relation (17.4) also describes damping. As in the model by Prévost, generalized strain amplitudes are used to compute $\hat{Q}_1$ and $\hat{Q}_2$ for more general strain cycles. The model again contains a rather large number of empirical constants; another disadvantage is that the stress–strain law (17.4), which in more general form contains the second invariant $J_2(d\varepsilon)$ instead of $|d\gamma|$, is not linear in the strain increments.

If the increase of k_1 within each cycle is neglected, equation (17.4) may be integrated to give a relation between stress and strain amplitudes in drained simple shear cycling:

$$\tau_c = \hat{F}(k_1, \gamma_c) = G\gamma_c \left(Y \tanh \frac{1}{Y} \right),$$

with $Y = Z(1 + Xk_1)^r / \gamma_c$. With increasing number of cycles the stiffness increases (and the damping decreases), the ultimate stiffness being given by G.

Martin, Finn, and Seed[11,22] take the plastic volume decrease (e^p) induced by drained cyclic loading as fatigue parameter. From experiments by themselves and by Silver and Seed[20] on drained simple shearing of sand, they derive the functions $\hat{Q}$ and $\hat{F}$; their results may be represented as follows:

$$k = e^p = a\gamma_c \ln (1 + N/N_0) \quad \text{for constant } \gamma_c, \tag{17.5}$$

$$\tau_c = \hat{F}(k, \gamma_c) = G(k)\gamma_c/(1 + b\gamma_c). \tag{17.6}$$

$G(k)$ is a function of k that will be specified below (equation (17.16)). For crystal silica sand $N_0 = 2.15$, $b = 7$ (with γ_c as a percentage), and a is a numerical constant which depends on the relative density of the sand[22];

$a = 1.87$ for $D_r = 45\%$. Formulae (17.5) and (17.6) are approximations, with less than 2% error, to the more complicated expressions given by Martin, Finn, and Seed.[11]

The function $\hat{Q}$ is obtained by taking the derivative of k with respect to N in equation (17.5), and substituting for N in terms of k. The function $\hat{Q}$ is found not to be separable, and Miner's rule does not apply.

The constants a and N_0 in equation (17.5) are independent of vertical effective stress. Consequently, the model may be applied to undrained cyclic loading by introducing an elastic rebound modulus $E_r = du/de^{p1}$, where u is the pore pressure. With $E_r = \sigma_v/C$ one obtains, for instance

$$u/\sigma_{v0} = 1 - e^{-k/C}, \qquad \sigma_v = \sigma_{v0}e^{-k/C} \tag{17.7}$$

with k or $\hat{Q} = dk/dN$ as given above; σ_{v0} is the initial vertical effective stress. However, Martin, Finn, and Seed[11] use a more complicated expression for E_r, which results in

$$\sigma_v = \sigma_{v0}(1 - k/d\sigma_{v0}^n)^r, \tag{17.8}$$

where d, n, and r are numerical constants. It should be noted that this expression, as opposed to equation (17.7), predicts liquefaction after a finite number of cycles ($\sigma_v = 0$ at a finite value of k). Since with equation (17.7) or (17.8) the pore pressure u is a monotonically increasing function of k, the model for undrained loading is equivalent to one in which the pore pressure generated by cyclic loading is used as memory parameter.

A rather similar model has been proposed by Idriss *et al.*[8-10] for soft clays under (undrained) earthquake loading conditions. The model may be written as follows:[6]

$$k = t(\gamma_c)\ln(1+N) \qquad \text{for constant } \gamma_c, \tag{17.9}$$

$$\tau_c = \hat{F}(k, \gamma_c) = G(k)f(\gamma_c), \tag{17.10}$$

where $G(k) = Ge^{-k}$, with G a shear modulus. The function $f(\gamma_c)$, which defines the stress–strain relation in the first quarter cycle ($k = 0$) is called the 'backbone curve', and is given by a Ramberg–Osgood equation

$$\gamma_c = \tau_c/G + \alpha(\tau_c/G)^R. \tag{17.11}$$

The function $t(\gamma_c)$ in equation (17.9) is not linear as in equation (17.5). For Icy Bay clay from Alaska, for instance,[10] a modified hyperbolic equation is used:

$$t(\gamma_c) = \gamma_c/(0.4 + 7.3\gamma_c^{0.61}).$$

Damping is introduced by using Masing's rules[12] to construct hysteresis loops; some difficulties inherent in this procedure were recently pointed out by Pyke.[17] Although no attempt was made to identify k with an observable of the system, it has been suggested[9] that k may be related to the excess pore pressure u^+ generated by cyclic loading, possibly by a relation similar to equation (17.7). It should also be noted that for this model, again, Miner's rule of superposition does not apply.

Some other models in which the fatigue parameter k in undrained cyclic loading is identified with the cumulative pore pressure are summarized elsewhere.[6] More recently, van Eekelen and Potts[7] analysed extensive cyclic loading data on Drammen clay[1] in this same way. They obtained

$$k' = u^+$$
$$Q' = du^+/dN = A \exp (J_c/B\langle J_f \rangle), \qquad (17.12)$$

where u^+ is the part of the pore pressure due to cyclic loading, excluding 'static' pore pressure effects as occur in monotonic loading (some authors fail to make this distinction between cyclic and static pore pressures). J_c is the second invariant of the stress amplitude, and $\langle J_f \rangle$ is a suitably defined failure strength of the sample. Since Q' does not depend on $k = u^+$, Miner's rule for stress cycles applies.

For equation (17.2), a hyperbolic relation is adopted between the second invariant of the strain amplitude and a suitably normalized, k-dependent stress amplitude. The model is meant to apply to arbitrary loading geometries. It includes an expression for the decrease in ultimate strength due to cyclic loading, which may be interpreted as a shift of the 'virgin consolidation line' of the clay.

The pressure A in equation (17.12) was found to have the same value for samples with the same preconsolidation pressure p_0 and different consolidation pressures p_c (OCR = p_0/p_c = 1, 4, and 10). Although the tests on Drammen clay were all on samples with the same p_0, it seems reasonable to assume that the pressure A may be written as $A = ap_0$, with a a numerical constant. Switching to a normalized fatigue parameter $k = u^+/p_c$, equation (17.12) gives for the case of undrained simple shear cycling

$$Q = dk/dN = a(\text{OCR}) \exp (\tau_c/B\tau_f), \qquad (17.13)$$

with τ_f the initial shear strength of the sample. Again for simple shear, the relation between stress and strain amplitude may be written as

$$\tau_c = G(k)\gamma_c/(1 + 2.1\gamma_c), \qquad (17.14)$$

with γ_c as a percentage, and the function $G(k)$ as specified in equation (17.17). Substituting (17.14) into (17.13) and using the appropriate expression for the shear strength, one may integrate Q to obtain an expression for k in simple shear which is comparable with equations (17.5) and (17.9):

$$k = \beta \ln (1 + N/N_0) \qquad \text{for constant } \gamma_c, \qquad (17.15)$$

where β and N_0 are both functions of γ_c and overconsolidation ratio.

It is interesting to note that equations (17.5), (17.9), and (17.15), although independently arrived at and not stated in this form in the original papers, all predict a logarithmic increase of k with number of cycles in constant shear stress amplitude cycling. The relations between stress and strain amplitude are also very similar; omitting the labels on τ_c and γ_c, they may be written as follows:

$$\gamma/\gamma_0 = (\tau/\tau_0')(1 + 0.5\,|\tau/\tau_0'|^{2.5}) \qquad \text{Idriss } et\ al.[9] \text{ San Francisco Bay Mud}$$

$$\gamma/\gamma_0 = (\tau/\tau_0')\frac{1}{1 - 0.14\tau/\tau_0'} \qquad \text{Martin } et\ al.[11] \text{ Crystal silica sand}$$

$$\gamma/\gamma_0 = (\tau/\tau_0')\frac{1}{1 - 0.042\tau/\tau_0'} \qquad \text{van Eekelen and Potts[7] Drammen clay}$$

where γ_0 is a reference strain amplitude of 0.02%, and $\tau_0' = \gamma_0 G(k)$. The function $G(k)$ represents the cyclic degradation effect and is given by

$$G(k) = Ge^{-k} \qquad \text{Idriss } et\ al.[9]$$

$$G(k) = G\frac{1 + 1.87k}{1 + 0.54k}\left(\frac{\sigma_v}{\sigma_{v0}}\right)^{1/2} \qquad \text{Martin } et\ al.[11] \qquad (17.16)$$

$$G(k) = G(1 - k - u_{st}/p_c) \qquad \text{van Eekelen and Potts[7]} \qquad (17.17)$$

The constant G is the initial shear modulus, which depends on the consolidation pressure or vertical effective stress at the start of the test. In equation (17.16), σ_v/σ_{v0} is given in terms of k by equation (17.8); in equation (17.17) u_{st} is the 'static' pore pressure generated in monotonic loading—see Figure 17.1.

The dilemma of identifying a fatigue parameter can be circumvented by formulating instead a hypothesis on the tangential compliance $L_t = d\gamma_c/d\tau_c$, the change in strain amplitude resulting from a sudden change in stress amplitude. In general, of course, cyclic degradation will affect the value of L_t at a given stress amplitude, but Andersen et al.[1,2] assume that 'the immediate change in shear strain due to a change in cyclic shear stress is the same as it would have been in the first load cycle'. This means that, despite cyclic degradation, the material stiffness with respect to variations in stress amplitude remains unchanged. This seems somewhat improbable; it also does not agree with the predictions of the three models discussed above. For instance, both the model by Martin, Finn, and Seed[11] and the model by van Eekelen and Potts[7] predict (from equations (17.6) and (17.14)) that, in a constant stress amplitude test, the tangential stiffness has decreased to 36% of its original value when the strain amplitude has doubled, and to 11% when the strain amplitude has quadrupled, as compared to cycle number one.

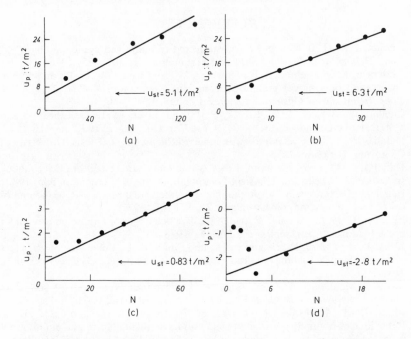

Figure 17.1 Observed pore pressures in cyclic loading; (a) simple shear, OCR = 1, $\tau_c/\tau_f = 0.67$; (b) two-way triaxial, OCR = 1, $q_c/q_f = 0.67$; (c) one-way triaxial, OCR = 4, $q_c/q_f = 0.39$; (d) simple shear, OCR = 10, $\tau_c/\tau_f = 0.65$. Reproduced by permission of The Institution of Civil Engineers, London

17.4 CONCLUSIONS

To describe quantitatively the cumulative effects of large numbers of loading cycles on the mechanical response properties of a soil, one may need to introduce a memory parameter to take account of the effects of past cyclic loading. If this memory parameter is identified with one of the observables of the system, the basic functions Q and F or $\hat{Q}$ and $\hat{F}$ of the 'fatigue' model may be determined relatively easily. The most obvious choice for this memory parameter is the plastic volume strain or (for undrained loading) the cumulative pore pressure generated by cyclic loading. A somewhat less convenient alternative is to use the total length of the deviatoric plastic strain path.

When properly analysed, there turns out to be a substantial degree of similarity between several seemingly very different theories for the behaviour of soil under cyclic loading.

Unless experimental data clearly indicate otherwise, it may be useful to formulate a fatigue model in such a way that Miner's rule of superposition applies, either for strain cycles or for stress cycles.

REFERENCES

1. Andersen, K. H. Behaviour of clay subjected to undrained cyclic loading. *Proc. Conf. Behaviour of Offshore Structures*, Trondheim, **1**, 392–403, 1976.
2. Andersen, K. H., Selnes, P. B., Rowe, P. W., and Craig, W. H. Prediction and observation of a model gravity platform on Drammen clay. *Proc. Second Int. Conf. Behaviour of Offshore Structures*, London, **I**, 427–46, 1979.
3. Cuellar, V., Bazant, Z. P., Krizek, R. J., and Silver, M. L. Densification and hysteresis of sand under cyclic shear. *J. Geot. Eng. Div., ASCE,* **103,** 399–416, 1977.
4. Dafalias, Y. F. *On Cyclic and Anisotropic Plasticity*, Ph.D. Thesis, University of California, Berkeley, 1975.
5. Dafalias, Y. F., and Herrmann, L. R. A bounding surface soil plasticity model. In *Soils under Cyclic and Transient Loading*, Balkema, Rotterdam, p. 335, 1980.
6. Eekelen, H. A. M. van. Single parameter models for progressive weakening of soils by cyclic loading. *Géotechnique*, **27,** 357–68, 1977.
7. Eekelen, H. A. M. van, and Potts, D. M. The behaviour of Drammen clay under cyclic loading. *Géotechnique*, **28,** 173–96, 1978.
8. Idriss, I. M., Dobry, R., Doyle, E. H., and Singh, R. D. Behaviour of soft clays under earthquake loading conditions. *Offshore Techn. Conf.*, OTC 2671, Dallas, Texas, 1976.
9. Idriss, I. M., Dobry, R., and Singh, R. D. Nonlinear behaviour of soft clays during cyclic loading. *J. Geot. Eng. Div., ASCE,* **104,** 1427–47, 1978.
10. Idriss, I. M., Moriwaki, Y., Wright, S. G., Doyle, E. H., and Ladd, R. S. Behaviour of normally consolidated clay under simulated earthquake and ocean wave loading conditions. In *Soils Under Cyclic and Transient Loading*, Balkema, Rotterdam, p. 437, 1980.
11. Martin, G. R., Finn, W. D. L., and Seed, H. B. Fundamentals of liquefaction under cyclic loading. *J. Geot. Eng. Div., ASCE,* **101,** 423–38, 1975.
12. Masing, G. Eigenspannungen und Verfestigung beim Messing. *Proc. Sec. Int. Congr. of Applied Mechanics*, 332–5, 1926.
13. Miner, M. A. Cumulative damage in fatigue. *Trans. ASME*, **67,** A159–64, 1945.
14. Mroz, Z., Norris, V. A., and Zienkiewicz, O. C. Application of an anisotropic hardening model in the analysis of elasto-plastic deformation of soils. *Géotechnique*, **29,** 1–34, 1979.
15. Prévost, J. H. Mathematical modelling of monotonic and cyclic undrained clay behaviour. *Int. J. for Num. Anal. Meth. in Geomech.*, **1**, 195–216, 1977.
16. Prévost, J. H., and Hughes, T. J. R. Finite element solution of boundary value problems in soil mechanics. In *Soils under Cyclic and Transient Loading*, Balkema, Rotterdam, p. 263, 1980.
17. Pyke, R. Nonlinear soil models for irregular cyclic loading. *J. Geot. Eng. Div., ASCE*, **105,** 715–26, 1979.
18. Sandler, I. S. On the uniqueness and stability of endochronic theories of material behavior. *J. Appl. Mech.*, **45,** 263–6, 1978.
19. Seed, H. B. Soil liquefaction and cyclic mobility evaluation for level ground during earthquakes. *J. Geot. Eng. Div., ASCE*, **105,** 201–55, 1979.
20. Silver, M. L., and Seed, H. B. Volume changes in sand during cyclic loading. *J. Soil Mech. Fdn Div., ASCE*, **97,** 1171–82, 1971.
21. Zienkiewicz, O. C., Chang, C. T., and Hinton, E. Non-linear seismic response and liquefaction. *Int. J. Num. Anal. Meth. Geomech.*, **2**, 381–404, 1978.
22. Finn, W. D. L., Byrne, P. M., and Martin, G. R., Seismic response and liquefaction of sands, *J. Geot. Eng. Div., ASCE*, **102,** 841–56, 1976.

Soil Mechanics—Transient and Cyclic Loads
Edited by G. N. Pande and O. C. Zienkiewicz
© 1982 John Wiley & Sons Ltd

Chapter 18

Shakedown of Foundations Subjected to Cyclic Loads

G. N. Pande

18.1 INTRODUCTION

A number of models for the behaviour of soils under cyclic loads have been proposed in preceding chapters (for example see Chapters 8–15 and Chapter 17). These models capture to varying degrees the important features of soil behaviour such as pore pressure generation, cyclic weakening, fatigue and degradation characteristics and cyclic hysteresis, etc., and can conceivably be incorporated in finite element codes for solution of boundary value problems. However, for a practical problem, the response of a soil structure can be evaluated for only a very limited number of cycles due to constraints on computing costs. In view of this, procedures for the prediction of shakedown or lack of it when the structure is subjected to a set of loads fluctuating arbitrarily within the given bounds is of considerable importance. Unfortunately such procedures for problems of interest to geotechnical engineers are lacking at present, even for simpler models of ideally elasto-plastic behaviour.

Limit theorems of collapse of elastic-perfectly plastic bodies occupy a distinct place in the soil mechanics literature.[4,28,30] This is partly due to the fact that they are closely related to the engineering solution of the stability problems of soil masses under simple loading programs. However, under relatively complex loading programs of unlimited duration, as occur in the foundations of marine structures, boundedness of overall plastic deformations cannot be ensured[11,29] even under loads lower than monotonic collapse at each stage in the loading program. Shakedown theorems, of which the collapse theorems are particular cases, have remained obscure in soil mechanics literature, and until recently few attempts have been made to develop engineering solutions for shakedown problems of soil masses subjected to repeated-cyclic loading.

The objectives of this chapter are twofold: firstly to introduce the concepts of shakedown-related bound theorems and survey the progress in the field of computation of shakedown loads for elasto/plastic continua; secondly to propose and develop some practical rules for shakedown problems of interest to geotechnical engineers. The emphasis, in this chapter, is on simplicity, physical explanations and practical application rather than on mathematical proof and rigour. Numerical examples have been included to illustrate the use of practical rules for evaluation of shakedown loads and the application of the displacement bounding theorems of shakedown to geotechnical situations.

It is necessary perhaps to make some comments on the relevance of the shakedown phenomenon in the research and development of constitutive relations of soils on the one hand and design and practice on the other. The stress state in soil samples tested in the laboratory is generally assumed to be homogeneous for monotonic loading tests. Experience with cyclic tests, although very limited, indicates that frictional constraints at the sample boundaries can set up stress gradients as the number of cycles increases.[14] The sample at this stage becomes a highly indeterminate structure. Similarly, in cyclic shear tests and all other cyclic tests, particularly when carried to a large number of cycles, what is observed is the response of a structure and evidently caution is necessary in developing constitutive models from the experimental structural response of test specimens.

18.2 STATEMENT OF THE PROBLEM AND SOME DEFINITIONS

Consider a general initial boundary value problem (Figure 18.1). We have to find the displacements $(\mathbf{u}(t), \mathbf{u} = (u, v, w)^{\mathrm{T}})$ and stresses $(\boldsymbol{\sigma}(t))$ for all points on the solid $(\mathbf{x} = (x_1, x_2, x_3)^{\mathrm{T}} \in \Omega)$ and at all instants of time $(t \in (0, \mathrm{T}))$, where $T > 0$ is the time representing the life of the structure. The symbol $\in$ stands

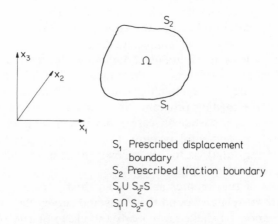

S_1 Prescribed displacement boundary
S_2 Prescribed traction boundary
$S_1 \cup S_2 = S$
$S_1 \cap S_2 = 0$

Figure 18.1   General initial-value problem

for 'included in'. The conditions to be satisfied by $\mathbf{u}(t)$ and $\boldsymbol{\sigma}(t)$ are

(a) equilibrium;
(b) compatibility of strains;
(c) constitutive relations of the material of the body;
(d) boundary conditions, i.e.

$$\mathbf{u}(\mathbf{x}, t) = \mathbf{g}(\mathbf{x}, t) \text{ on the boundary } S_1, \qquad (18.1)$$

$$\mathbf{n}^{\mathrm{T}}\boldsymbol{\sigma}(\mathbf{x}, t) = \mathbf{h}(\mathbf{x}, t) \text{ on the boundary } S_2, \qquad (18.2)$$

where $\mathbf{n}_i$ is the unit normal to the boundary at $(\mathbf{x}, t)$ and S_1 and S_2 are the portions of the boundary subjected to displacement and traction boundary conditions respectively. Obviously S_1 and S_2 are such that

$$S_1 \cup S_2 = S,$$

and $\qquad\qquad\qquad\qquad\qquad\qquad\qquad\qquad\qquad\qquad (18.3)$

$$S_1 \cap S_2 = 0.$$

Here S represents the total boundary of the solid.

We shall restrict ourselves to a class of this general problem in which the following assumptions are made:

(i) We consider only a quasi-static small deformation problem. Thus all inertia effects are neglected.
(ii) The constitutive relations are of linear elasto/plastic (not necessarily ideal plastic) type with an associated flow rule.
(iii) The functions $\mathbf{g}$ and $\mathbf{h}$ have maxima and minima within which they fluctuate arbitrarily.

Without loss of any generality the following additional assumptions are made mainly for clarity of presentation and discussion:

(1) The time t can be replaced by any monotonically increasing function as in theory of plasticity.
(2) The function $\mathbf{g}(\mathbf{x}, t)$ can be taken as identically zero and the function $\mathbf{h}(\mathbf{x}, t)$ can be taken to represent repeated application and removal of the cyclic component of boundary tractions. This function is diagrammatically represented in Figure 18.2.

Let a, b be the extreme values of the function $\mathbf{h}(\mathbf{x}, t)$; then the behaviour of the body depending on the values of a, b could fall in any of the following three categories:

(I) (a, b) such that no part of the body yields in the first cycle; plastic strains as well as plastic strain rates are zero at all the points on the body. Consequently the body behaves elastically and the number of repetitions of the load cycle have no influence. The displacements

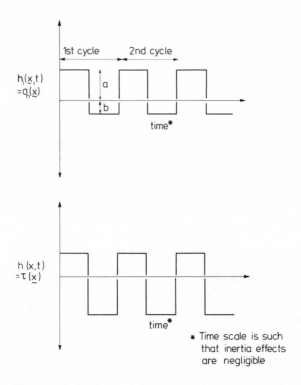

Figure 18.2 Typical variation of boundary tractions with
time

obtained during the first cycle are repeated in subsequent cycles. We
shall label such values of (a, b) as $(a, b)^e$.

(II) (a, b) such that collapse takes place in the first application of the load
cycle. Here it is assumed that in case of strain hardening plasticity,
there is a limiting value of material yield parameters. This implies that
in case of isotropic hardening, the hardening modulus varies with
strains and finally approaches zero. This is also the implication of
kinematically hardening models where a bounding or limiting yield
surface is specified.[6,21] Thus collapse in strain hardening/softening soil
plasticity is related to ultimate/residual parameters. We shall label
such values of (a, b) as $(a, b)^c$.

(III) If $(a, b)^e < (a, b) < (a, b)^c$, three distinct possibilities exist, viz.

(a) *Ratchetting* Ratchetting is the term used to describe progressive deformation. The body continually deforms under repeated cycles of loading
and after some number of cycles a limit state of serviceability may be
reached.

(b) *Elastic shakedown* Elastic shakedown implies that structure would adapt itself to cyclic loading in the sense that after a finite number of cycles, the structure behaves purely elastically for subsequent load cycles.

(c) *Plastic shakedown* Plastic shakedown is said to have taken place if after a finite number of cycles, the components of stress and strain form a hysteretic loop. There is no progressive deformation in each cycle since plastic strains of opposing nature are set up. This phenomenon is well known to metallurgists and structural engineers and is also called 'alternating plasticity' and is associated with low cycle fatigue failure.

Figure 18.3 shows diagrammatically the phenomena of shakedown and ratchetting.

A few remarks are appropriate here on the considerations involved in the design of geotechnical structures subjected to cyclic load. It is obvious that peak loads in the cycle should be such as to cause no ratchetting. (That ratchetting can take place in foundations subjected to cyclic loads was demonstrated by Rowe[26,27] on physical models and by Zienkiewicz *et al.*[33,34]

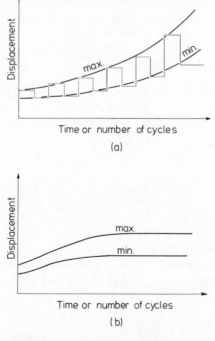

Figure 18.3 Displacement of a point on the body on repeated loading: (a) ratchetting; (b) elastic or plastic shakedown

on numerical models. Note that in some of these publications ratchetting was referred to as 'shakedown'). The factors of safety in such situations should be worked out with respect to elastic/plastic shakedown loads and not collapse loads. Unfortunately, in some cases, the deformations associated with elastic/plastic shakedown may be inadmissible from the serviceability point of view. In such cases the limiting value of deformation during a fairly large or unlimited loading programme under loads even less than the shakedown load will be of interest to geotechnical engineers.

18.3 SHAKEDOWN THEOREMS

If in addition to the assumptions outlined in Section 18.2 a further assumption is made, viz. that material behaviour is ideal elasto/plastic, two theorems relating to shakedown can be proved. These theorems provide an upper and lower bound on the shakedown load and are in fact more general than collapse theorems. As these theorems are not found in soil mechanics literature, they will be stated here without proof for clarity of discussion. The first shakedown theorem was proved by Melan in 1936[18-20] and second by Koiter[10] in 1956. It was only in 1956 that Koiter[10] pointed out the link between the shakedown theorems and collapse theorems.

(a)   *1st Theorem or Melan's Theorem*

If any time independent distribution of residual stress (f_{ij}) can be found such that the sum of these residual stresses and the elastic stresses (f_{ij}^e) (stress based on purely elastic behaviour) is a safe state of stress, i.e.

$$f_{ij} + f_{ij}^e = f_{ij}^s,\tag{18.4}$$

f_{ij}^s being such that

$$F(f_{ij}^s) \leq 0,\tag{18.5}$$

where F represents a yield function, at every point on the body and for all possible load combinations within the prescribed bounds of the loads, then the structure has an elastic shakedown. Conversely, if no time-independent distribution of residual stresses can be found satisfying the condition stipulated above, the structure will not shakedown and will eventually collapse.

(b)   *2nd theorem or Koiter's theorem*

The body will not shakedown, i.e. it will ultimately fail by cyclic plastic deformations if any admissible plastic strain rate cycle $\dot{\varepsilon}_{ij}^p(t)$ and any external loads $X_i(t)$, $p_i(t)$ within the prescribed limits can be found for which

$$\int_0^T \left[\int_\Omega X_i \dot{u}_i \, d\Omega + \int_{S_2} p_i \dot{u}_i \, ds \right] dt > \int_0^T \int_\Omega Q(\dot{\varepsilon}_{ij}^p) \, d\Omega \, dt, \qquad (18.6)$$

where $Q(\dot{\varepsilon}_{ij}^p)$ represents the plastic energy dissipation function and thus the right-hand side of equation (18.6) represents the total plastic energy dissipated over the whole body during the entire strain rate cycle. X_i, p_i and $\dot{u}_i$ represent the body forces, surface tractions and velocities respectively.

An admissible plastic strain rate cycle is defined as a cycle for which $\Delta\varepsilon_{ij}^p$ given by

$$\Delta\varepsilon_{ij}^p = \int_0^T \dot{\varepsilon}_{ij}^p \, dt \qquad (18.7)$$

constitutes a kinematically admissible strain distribution, which means that $\Delta\varepsilon_{ij}^p$ can be derived from

$$\Delta\varepsilon_{ij}^p = \tfrac{1}{2}(u_{i,j} + u_{j,i}), \qquad (18.8)$$

where u_i are displacements satisfying the displacement boundary condition.

The proof of both the theorems, as in the case of the collapse theorem, hinges on the associativity of the flow rule and convexity of the yield function. Further, they are valid only for ideal elasto/plastic materials. Nevertheless they can provide useful guidance for strain-hardening materials simply by adopting suitable modified values for yield parameters.

The two theorems have the identical role as that of lower and upper bound theorems of collapse. A number of applications of the first shakedown theorem have been made[12,13,24] in structural context to obtain lower bound shakedown loads.

18.4 MATHEMATICAL PROGRAMMING IN SHAKEDOWN PROBLEMS OF CONTINUA

In shakedown problems, the loading is conveniently defined by a set of load factors ($\boldsymbol{\mu}$) such that the vector of applied loads ($\mathbf{g}$) at any instant is a linear combination of load factors, i.e.

$$g_k = \mu_k q_k, \qquad (18.9)$$

where $\mathbf{q}$ is an arbitary reference load vector denoting various types of loads. Values of $\boldsymbol{\mu}$ are not known *a priori*. However, bounds on the load factors are known, i.e.

$$\mu_k^- < \mu_k < \mu_k^+. \qquad (18.10)$$

Some of the loading modes may have a correlation, e.g. the moment acting on the footing is related to horizontal shear loads if they are caused by the same force acting at some height above the footing on the structure which is assumed rigid. In general the bounds represented by equation (18.10) can be represented by

$$[A]\mu \leq a \qquad (18.11)$$

where matrix $[A]$ and vector a are known *a priori*. The classical shakedown problem can be stated as follows:

Let the load factor limits μ^-, μ^+ (or the vector a in equation (18.11)) be given an unknown value

$$\alpha\mu_k^- \leq \mu \leq \alpha\mu_k^+ \qquad \text{or} \qquad [A]\mu \leq \alpha a; \qquad (18.12)$$

then find the maximum value α_s of α allowing for shakedown. This can be stated as the following mathematical programming problem using Melan's theorem:

$$\text{Maximize } \alpha(\mu, f_{ij})$$

subject to

(1) $F(f_{ij}^e + f_{ij}) \leq 0$,
(2) f_{ij} is self-equilibrating residual stress,
(3) $[A]\mu \leq \alpha a$,
(4) $\alpha \geq 0$. $\qquad\qquad\qquad\qquad\qquad\qquad\qquad\qquad$ (18.13)

If f_{ij}^e and f_{ij} are both represented in a discrete manner as in a finite element procedure, f_{ij}^e is obtained from the linear combination of load factors, and a system of linear equations is obtained by applying the principle of virtual work for the residual stress field, f_{ij}, the form of the mathematical programming problem can be identified.

Formal procedure for computation of shakedown loads based on finite element discretization and piecewise linearization of the yield function was proposed by Maier.[15] His formulation[16] leads to a linear programming problem. He also extended[17] the shakedown theorems to allow for work hardening and second-order geometric effects. However, first numerical results for shakedown problem were given by Belytschko and Hodge[2] for the plane stress situation of a plate with a central circular hole subjected to an independently varying state of bi-axial stress. He used an equilibrium finite element formulation and used the continuous field description of the Von Mises yield function, which leads to a non-linear programming problem. The same problem has been solved by Corradi and Zavelani[5] using a displacement formulation of finite elements and piecewise linearization of the Von Mises yield function, which leads to a linear programming problem. Piecewise linearization of the yield function leads to a considerable increase in number of constraint equations in the optimization problem, which in turn increases the computational time. Hung and Konig[8] have proposed a non-linear programming formulation for the shakedown based on the 'yield criterion of the mean'; the yield criterion being checked only at one point in the element, thus reducing the number of constraint equations.

Aboustit and Reddy[1] have adopted the dual of the linear programming problem and solved for a plane strain representation of a footing subjected to an oscillating inclined load on drained soil obeying the Mohr–Coulomb yield criterion.

Progress in the field of numerical solution for shakedown problems has been very slow, although formal procedures for solution have been available for more than a decade or so. This is basically due to the fact that for two- or three-dimensional continua, even with crude finite element discretization, the mathematical programming problem becomes too large and excessive computing times are involved.

18.5 AN ENGINEERING APPROACH TO SHAKEDOWN PROBLEMS

From the above it is clear that some practical rules would be very useful to judge if shakedown would take place under a given set of bounds of loads. Such practical rules have been developed by Zarka and Casier[32] in the context of some simple problems related to nuclear pressure vessels involving thermal ratchetting in structures of kinematically hardening materials. For geotechnical problems, such as the loading of rigid or flexible mats on purely cohesive ideal elastoplastic foundations, two empirical procedures[23] are proposed for finding approximate lower bound values of shakedown loads.

(a) First cycle shakedown: As a first approximation we shall examine a first cycle shakedown (FCS), i.e. we will try to determine the maximum possible values of loads for which shakedown will take place after the first cycle of loading. It is particularly simple[23] to find first cycle shakedown load when an incremental load elastoplastic analysis is carried out. Experience on a limited number of problems indicates that if loads are higher than first cycle shakedown load, even by a few percent, the number of cycles to reach shakedown is so large as to make calculations economically impractical.

(b) Search for an optimal residual stress state by trial and error. It was shown in Section 18.4 that the determination of shakedown load reduces to an optimization problem. It is a common method in engineering design to optimize by trial and error. Of course a true optimum cannot be ensured by this procedure, but useful practical guidance can be obtained. A simple general procedure of search for an optimal residual stress state for any boundary value problem would be as follows:

(1) Perform an elastoplastic analysis incrementing the sets of loads (applied in any arbitrary manner) to any arbitrary value so as to induce plastic strains. Unload the structure and catalogue the residual stresses (f_{ij}).

(2) Compute elastic stresses (stresses assuming material behaviour as purely elastic) for unit or some reference values of all the sets of loads acting at the same location in the problem. Catalogue these stress fields $(f_{ij}^{e1}, f_{ij}^{e2}, \ldots, f_{ij}^{en})$.

(3) Scan: Let μ_1, μ_2, $\mu_3, \ldots$ be the shakedown load factors to be determined. Scan all finite elements and all Gauss points with $\mu_1, \mu_2, \mu_3, \ldots$ such that

$$F\left(f_{ij} + \sum_{k=1}^{n} \mu_k f_{ij}^{ek}\right) \leqslant 0, \qquad (18.14)$$

also noting that all μ's may not be independent.

(4) With $\mu_1, \mu_2, \mu_3, \ldots$, plot an interaction surface in load space. The domain enclosed by this surface represents shakedown domain, as according to Melan's theorem, a self-equilibrating residual state of stress (f_{ij}) has been found which, when superimposed on the elastic stresses due to loads within the shakedown domain, does not produce yielding, equation (18.14).

(5) Repeat steps 1, 3, and 4 with another excursion in the plastic domain along another arbitrarily chosen load path.

(6) Draw an enveloping curve/surface which can be taken as an approximate lower bound of the shakedown domain.

18.6 APPLICATION OF ENGINEERING APPROACH TO THE PROBLEMS OF FOOTINGS

We shall discuss the application of the engineering approach in the context of finite element computations of the shakedown of flexible footings in plane strain on weightless purely cohesive soils.

First, a very crude finite element mesh was adopted, the idea being that with a crude mesh, it might be possible to actually compute for many hundreds of cycles and draw conclusions from such computations for real structure. Studies with other finer meshes were also made. The loading adopted on the footing of width 'b' was vertically downwards only. This is a one parameter problem in which, theoretically, the collapse load is equal to the shakedown load. Figure 18.4 shows the variation of computed collapse load and FCS load with the number of degrees of freedom. It is seen that as the number of degrees of freedom increases, as is well known, the computed

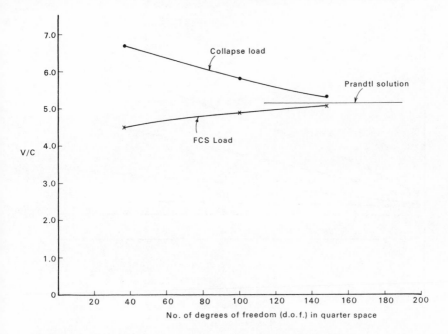

Figure 18.4 Variation of computed collapse load and FCS load with the number of degrees of freedom in the finite element discretization

collapse load approaches the theoretical value of 5.14C (C being cohesion) from above. It is interesting to note that the FCS load approaches the theoretical value of 5.14C from below. As a matter of fact an average (purely arbitrary) of computed collapse and computed FCS load always gives an improved value of theoretical collapse load. It may be noted that FCS load is found by rather trivial computations. To check the validity of the FCS criterion as a practical rule computations were made with a coarse mesh (37 degrees of freedom) for 400 cycles and with fine mesh (148 degrees of freedom) for 50 cycles at loads greater than first-cycle shakedown load. Figure 18.5 shows the variation of the non-dimensionalized displacement of the centre of the footing (δ/b) with the number of cycles at various intensities of loads. It is seen that even at an intensity of load slightly (0.2%) higher than first-cycle shakedown load, shakedown did not take place within a reasonable number of cycles. Thus it appears reasonable to assume that the first-cycle shakedown load can be taken as a practical guide in shakedown computations.

The FCS criterion and the procedure of search for optimal residual stress state by trial and error can be used judiciously to obtain the shakedown domain in a particular problem. Some such results have been presented by

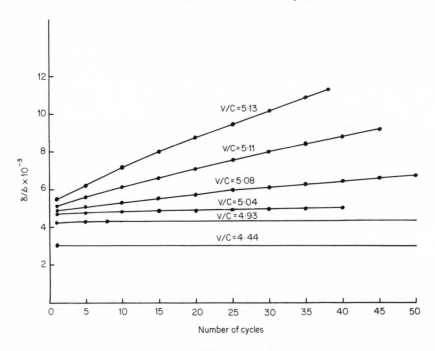

Figure 18.5 Variation of computed non-dimensionalized vertical displacement at the centre of footing with number of cycles at different intensities of uniformly distributed load

Pande, Davies, and Abdullah.[23] Figure 18.6 shows the collapse and FCS domains of the flexible footing subjected to cyclic vertical pressure (V) and cyclic horizontal pressure (H). Figure 18.7 shows a similar diagram for vertical pressure and moments for adhering and non-adhering footings in non-dimensional form.

Assuming horizontal loads and vertical loads to be inter-related, i.e.

$$\frac{H}{V} = \text{constant}, \tag{18.15}$$

and moment cycles to be symmetrical, the collapse and first-cycle shakedown domains are plotted in non-dimensional form in Figure 18.8 for various vertical and horizontal pressures and moments.

The results of Figures 18.7 and 18.8 are used as an illustration to compute the factor of safety (F.S.) against collapse and FCS of a typical gravity platform shown in Figure 18.9. The values of the forces have been adopted from a problem used for illustration by Hoeg.[7] Since the problem is axi-symmetric while design graphs (Figures 18.7 and 18.8) are for plane

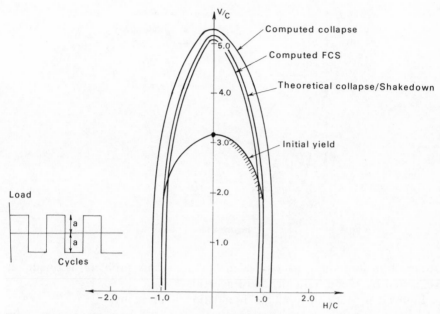

Figure 18.6 Non-dimensional interaction diagram for a flexible footing subjected to cyclic vertical and cyclic horizontal uniformly distributed loads

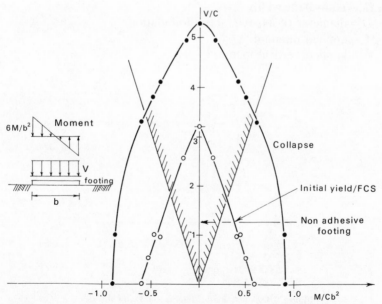

Figure 18.7 Non-dimensional interaction diagram for a flexible footing subjected to cyclic uniformly distributed vertical load and cyclic moments

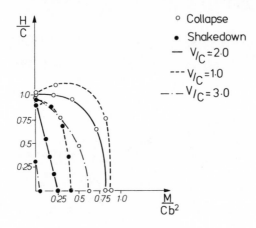

Figure 18.8

strain, transformation to equivalent plane strain problem is made as suggested by Janbu, Grande, and Eggereide.[9]

Equivalent width of a square foundation is given by:

$$B_0 = \frac{D}{2} - 2\frac{M_a}{V_a},$$ (18.16)

where B_0 = equivalent width
D = diameter of axi-symmetric foundation
M_a = applied moment
V_a = applied vertical load

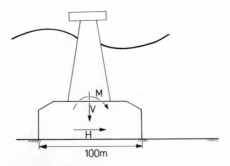

V=20,000 kN/m
M=±200,000 kN-m/m
H=±500,000 kN/m

Figure 18.9 Schematic diagram showing the design forces acting on a typical off shore gravity platform

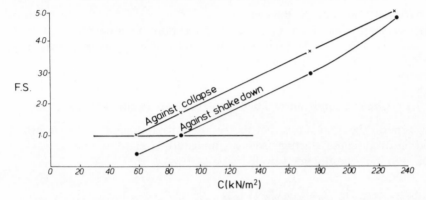

Figure 18.10   The computed variation of factors of safety against collapse and FCS with cohesion

Assuming the platform rests on an ideal elastoplastic half space of purely cohesive material (relevant to undrained total stress analysis), factors of safety (F.S.) against collapse and FCS can be computed for various values of cohesion (C) and are plotted in Figure 18.10. It is noted that considerable difference between the two values exists and could be very important for low values of cohesion. This conclusion is also qualitatively supported by the rigorous analyses by Aboustit and Reddy[1] of a similar problem.

18.7   DISPLACEMENT BOUNDING THEOREMS IN SHAKEDOWN OF FOUNDATIONS

It was indicated in Section 18.2 that the deformations associated with the elastic shakedown load on a structure may be inadmissible from the serviceability point of view. In other words, if a structure shakes down under a given set of loads and their range, it does not necessarily mean that the displacements occurring before it reaches the shakedown state would not be intolerably large. It has been shown[10] that in case of collapse overall displacements are bounded for all loads less than collapse. In a similar manner, the boundedness of the overall displacements can be demonstrated for all loads below the shakedown limit. However, for the realistic assessment of the safety of a foundation subjected to fluctuating loads, an *a priori* determination of the order of magnitude of local quantities such as displacement and plastic strains is of great importance. Upper bounds for plastic strains in the shakedown of structures have been formulated by Vitiello[31] and Maier.[17] Ponter[25] has derived a general principle which allows the evaluation of upper bounds of local displacement. Another bounding theorem for local displacements has been formulated by Capruso.[3] Both Ponter and Capruso conclude that bound theorems proposed by them could provide a useful guide to the order of displacements. Capruso's theorem[3] is relatively simple and holds promise of application in conjunction with finite element techniques. It is discussed with a numerical application in the next section.

18.7.1 Upper bounds on residual displacements (Capruso's theorem)

The general theorem states that the residual displacement occurring in any loading programme starting from an undisturbed state at $t = 0$ up to any time $t = \tau$ may be bounded by the inequality

$$\int_S p_i^f u_i^R(\tau)\, dS + \int_V x_i^f u_i^R(\tau)\, dV \le \frac{1}{2} \Bigg[\int_V A_{ijhk} \sigma_{ij}^{R*} \sigma_{hk}^{R*}\, dV$$

$$+ \frac{1}{\lambda - 1} \int_V A_{ijhk} \bar\sigma_{ij}^R \bar\sigma_{hk}^R\, dV \Bigg] \quad (18.17)$$

where p_i^f and x_i^f are some fictitious surface tractions and fictitious body
     forces respectively, and the elastic stresses field due to
     these is represented by σ_{ij}^f;
  $u_i^R(t)$ represents the residual displacement vector at any instant
     of time;
    σ_{ij}^{R*} is a distribution of self-equilibrating stresses such that

$$F(\sigma_{ij}^{R*} + \sigma_{ij}^f) \le 0 \quad (18.18)$$

     for any t in the interval $0 \le t \le \tau$
    $\bar\sigma_{ij}^R$ is another distribution of self-equilibrating stresses such
     that

$$F(\bar\sigma_{ij}^R + \lambda \sigma_{ij}^E(t)) \le 0 \quad (18.19)$$

     for any t in the interval $0 \le t \le \tau$, in which $\sigma_{ij}^E(t)$ is the stress
     field due to varying loads assuming the material behaviour
     to be linear elastic and $\lambda > 1$ is the factor of safety with
     respect to shakedown load, A_{ijhk} is the tensor of elastic
     coefficients.

The proof of equation (18.17) relies on Drucker's stability postulate. It may be noted that p_i^f and x_i^f are arbitary values and can be chosen. Equation (18.17) in general can be reduced to a mathematical programming problem. However, simple bounds can be easily found from equation (18.17). If we assume $x_i^f = 0$ and $p_i^f =$ static collapse load (p^c), equation (18.17) may be written as

$$u_a^R \le \frac{1}{2p^c} \left\{ \int_V A_{ijhk} \sigma_{ij}^{RC} \sigma_{hk}^{RC}\, dV + \frac{1}{\lambda - 1} \int_V A_{ijkl} \bar\sigma_{ij}^{RS} \bar\sigma_{hk}^{RS}\, dV \right\}. \quad (18.20)$$

Here

u_a^R represents the residual displacement in the direction a;

σ_{ij}^{RC} is the self-equilibrating stress field associated with monotonic collapse load (p^c) as the fictitious load;

λ, σ_{ij}^{RS} are shakedown safety factor and self-equilibrating stress fields respectively

Another way of writing the bound is to assume $x_i^f = 0$ and $p_i^f = $ load associated with initial yield (p^E). Now

$$\int A_{ijhk}\sigma_{ij}^{R*}\sigma_{hk}^{R*}\,dV = 0$$

in equation (18.17) and this gives

$$u_a^R \leqslant \frac{1}{2(\lambda-1)p^E}\int_V A_{ijhk}\bar{\sigma}_{ij}^{RS}\bar{\sigma}_{hk}^{RS}\,dV. \qquad (18.21)$$

Equations (18.20) and (18.21) thus represent two estimates of the upper bound.

18.7.2 Numerical example of estimation of upper bound of displacement

The numerical example chosen is that of a rigid plane strain footing of width 'b' subjected to repeated vertical load resting on an elastoplastic foundation. For this problem the shakedown load is the same as the collapse load. The deformations at collapse or shakedown load are unbounded. A finite element discretization having 192 degrees of freedom in quarter space and the Tresca criterion of yield are adopted. The computed collapse load is $5.30c$ and initial yield takes place at $2.70c$. Integrals involved in equations (18.20) and (18.21) can be easily computed for various values of λ. Figure 18.11 shows variation of non-dimensionalized residual vertical settlement (δ_R/b) with different upper bounds of fluctuating vertical load represented by ($\lambda - 1$). Computations are made according to equation (18.20) as well as equation (18.21). To assess the accuracy of the bounds, computations were made for up to 50 cycles of applications. Residual displacements at the end of the 50th cycle are also plotted on Figure 18.11. It is seen that displacement bounds, though not very good, can be useful guides to the designer.

18.8 SHAKEDOWN AND STRAIN-HARDENING MODELS OF SOILS

The discussion on shakedown so far has been with reference to ideal elastoplastic behaviour of soils. Although shakedown theorems have been extended to strain-hardening materials,[22] computational difficulties appear

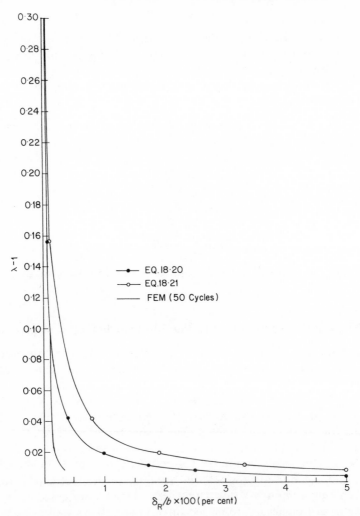

Figure 18.11 Variation of computed non-dimensionalized upper bound residual vertical settlement on repeated loadings with $(\lambda - 1)$

to be formidable at the present stage. An alternative appears to be the development of practical rules. As far as isotropic hardening models like critical state model are concerned, it appears straightforward to adopt similar practical rules as discussed in this chapter for ideal plasticity. The procedure would be similar to that adopted in the application of collapse theorems for strain-hardening materials. Once yield parameters corresponding to hardened yield surface have been estimated, ideal plasticity can be assumed. This would lead to material inhomogeneity in the boundary value problem but should not pose any special problems in finite element context. For kinematic hardening models elastic shakedown criterion would be too restrictive and practical rules for plastic shakedown will have to be developed if the full potential of these sophisticated models of soil behaviour is to be exploited in future. Such rules may also indicate as to what extent sophistication in the soil models is fruitful.

18.9 CONCLUSIONS

A brief survey of developments in the area of shakedown analysis has been presented with a view to stimulating research on this topic in the context of the problems of soils subjected to cyclic loading. With any non-linear constitutive law for soil, it is virtually impossible to carry out explicit analysis for more than a few cycles due to constraints on computing costs. Determination of shakedown loads could help our understanding in such problems. Unfortunately, precise determination of shakedown loads leads to mathematical programming problems of great magnitude, which again involve very excessive computing costs. As an alternative, simplified practical rules have been suggested for shakedown analysis for situations of interest to geotechnical engineers, assuming the material model to be linear elastic and ideally plastic. An attempt has also been made to obtain displacement/settlement bounds. It is seen that the first-cycle shakedown load for a specified load path and the displacement bound for cyclic loading along that path can be obtained as a by-product from an incremental elastoplastic collapse analysis, with little extra computation. The lower bound on shakedown and the upper bound on the local displacement so obtained may serve as a useful guide to the designer of geotechnical structure.

ACKNOWLEDGEMENT

The author is thankful to Mr. W. S. Abdullah and Mr. K. G. Sharma, his colleagues at the Department of Civil Engineering, University College of Swansea for help on computing in many problems. He is also indebted to the late Professor E. H. Davis of the University of Sydney and to Professor O. C. Zienkiewicz, for encouragement and support in the past few years.

REFERENCES

1. Aboustit, B. L., and Reddy, D. V. Finite element linear programming approach to foundation shakedown. In Pande, G. N. and Zienkiewicz, O. C. (eds.), *Soils Under Cyclic and Transient Loading*, Balkema, Rotterdam, 1980.

2. Belytschko, T., and Hodge, P. G., Jr., Plane stress shakedown analysis by finite elements *Intl. J. Mech. Sci.* **14,** 619–25, 1970.
3. Capruso, M. A displacement bounding principle in shakedown of structures subjected to cyclic loads, *Intl. J. Solids Struct.* **10,** 77–92, 1974.
4. Chen, W. F. *Limit Analysis and Soil Plasticity,* Elsevier, Amsterdam, Oxford and New York, 1975.
5. Corradi, L., and Zavelani, A. A linear programming approach to shakedown of structures, *Computer Methods in Appl. Mech. Eng.* **3,** 37–53, 1974.
6. Dafalias, Y. F. and Herrmann, L. R. Chapter 10, this book.
7. Hoeg, K. 'Foundation engineering for fixed offshore structures' *Behaviour of Offshore Structures, Proc.* Trondheim, **I,** 39–69, 1976.
8. Hung, N. D., and Konig, J. A. A finite element formulation for shakedown problems using a yield criterion of the mean, *Comp. Meth. Appl. Mech. Eng.* **8,** 179–92, 1976.
9. Janbu, N., Grande, L., and Eggereide, K. Effective stress stability analysis of gravity structures, *Behaviour of Offshore Structures, Proc.* Trondheim, **I,** 449–66, 1976.
10. Koiter, W. T. A new general theorem on shakedown of elastic plastic structures *Proc. Kon. Ned. Ak. Wet.* **B59,** 24–34, 1956.
11. Koiter, W. T. General theorems for elastoplastic solids, *Progress in Soild Mechanics,* Vol. 1, North Holland, Amsterdam, pp. 166–221, 1960.
12. Leckie, F. A. Shakedown pressures for flush cylinder-sphere shell interactions, *J. Mech. Eng. Science.* **7**(4), 1965.
13. Leckie, F. A. A review of bounding techniques in shakedown and ratchetting at elevated temperature, *W.R.C. Bull.* 195, 1974.
14. Lee, K. L. End restraint effects on undrained static tri-axial strength of sand. *J. Geotech. Eng. Dn. ASCE.,* **GT6,** 687–719, 1978.
15. Maier, G. Shakedown theory in perfect elasto/platicity with associated and non-associated flow laws: A finite element linear programming approach, *Mecanica,* **4,** 259–60, 1969.
16. Maier, G. A matrix structural theory with interacting yield planes, *Mecanica,* **5,** 54–66, 1970.
17. Maier, G. A shakedown theory allowing for work hardening and second order geometric effects *Intl. Symp. on Foundations of Plasticity,* Warsaw, 417–33, 1972.
18. Melan, E. Theorie statisch unbestimmter systemme, *Pel. Pub. 2nd Congress Intn. Assoc. Bridge and Structural Eng.* Berlin, 43–64, 1936.
19. Melan, E. Theory statisch unbestimmter System aus ideal plastischem Baustoff, *Sitzber, Akad. Wiss. Wien, Abt* IIa, **145,** 195–218, 1936.
20. Melan, E. Der Spannungzustand eines 'Mises–Henckyschou' Kontinuums bie verandlicher Belastung, *Sitzber, Akad. Wiss. Wien, Abt.* IIa, **147,** 73–87, 1938.
21. Mroz, Z., Norris, V., and Zienkiewicz, O. C. An anisotropic hardening model for soils and its application to cyclic loading, *Intl. J. Num. and Anal. Meth. in Geomech.* **2,** 203–221, 1978.
22. Neal, B. G. Plastic collapse and shakedown theorems for structures of strain hardening materials. *J. Aero. Science,* **17,** 297–306, 1950.
23. Pande, G. N., Davis, E. H., and Abdullah, W. S. Shakedown of elasto-plastic continua with special reference to soil-rock structures. In Pande, G. N., and Zienkiewicz, O. C. (eds.), *Soils Under Cyclic and Transient Loading,* Balkema, Rotterdam, 1980.
24. Penny, R. K., and Leckie, F. A. Solutions for strains at nozzles in pressure vessels, *W.R.C. Bull.* No. 90, 1963.

25. Pointer, A. R. S. An upper bound on small displacements of elastic, perfectly plastic structures, *J. Appl. Mech.* **39**, 959, 1972.
26. Rowe, P. W. Displacements and failure modes of model offshore gravity platforms founded on clay, *Conf. Offshore Europe*, Spearhead Publications, 1975.
27. Rowe, P. W., Craig, W. H., and Proctor, D. C. Model studies of gravity studies on clay *Proc. Conf. Behaviour of Offshore Structures Proc.* Trondheim, **I,** 439–48, 1976.
28. Salençon, J. *Théorie de la Plasticité pour les Applications à la Mécanique des Sols,* Eyrolles, Paris.
29. Symonds, P. S. Shakedown in continuous media, *J. App. Mech.* **28,** 85, 1951.
30. Terzaghi, K., *Theoretical Soil Mechanics* Wiley, New York, 1943.
31. Vitiello, E. Upper bounds to plastic strain in shakedown of structures subjected to cyclic loads, *Meccanica* **7**(3), 205, 1972.
32. Zarka, J., and Casier, J. *Cyclic Loading on an Elastic Plastic Structure: Practical Rules* Report of Laboratoire de Mechanique des Solides, Ecole Polytechnique, Palaiseau, 1978.
33. Zienkiewicz, O. C., Lewis, R. W., Norris, V. A., and Humpheson, C. *Numerical Analysis for Foundations of Offshore Structures with Special Reference to Progressive Deformation,* Society of Petroleum Engrs. of AIME, SPE 5760, 1976.
34. Zienkiewicz, O. C., Norris, V. A., Winnicki, L. A., Naylor, D. J., and Lewis R. W. A unified approach to the Soil Mechanics problems of Offshore Foundations, *Numerical Methods in Offshore Engineering,* eds. Zienkiewicz, Lewis and Stagg, Wiley, 1978.

Soil Mechanics—Transient and Cyclic Loads
Edited by G. N. Pande and O. C. Zienkiewicz
© 1982 John Wiley & Sons Ltd

Chapter 19

Comments on the use of Physical and Analytical Models

P. W. Rowe and I. M. Smith

A   PHYSICAL MODELS (P. W. ROWE)

19.1   INTRODUCTION

The degree of acceptance of any analytical model depends on its power to reproduce physical field and model measurements. In the absence of field data a model begs prediction irrespective of any deviations it may incur from field similitude.

Computational facilities and expertise have progressed ahead of the establishment of constitutive equations for effective stress–strain–volume change laws under continually varying stress paths. Under unidirectional stress paths, developments continue[8] but for cyclic loading under fixed, let alone rotational, principal stress directions, there are no foreseeable solutions. The fundamental reason for this is that particulate matter degrades and rearranges under each change in stress state, creating an infinite number of new soil 'materials' for which empirical relations are required. Partly for this reason, simple soil element models are to be preferred as long as they reflect the major mechanisms, and especially if this permits three-dimensional analysis, for the purpose of optimization of dimensions rather than precise prediction of displacements.

19.2   ANALYSIS WITHOUT PHYSICAL MODELS

The simplifications inherent in some but not all theoretical treatments of sand were put to the test during 1974–5, when several attempts were made to predict the lateral displacement of a surface caisson subjected to cyclic loading at Neeltje Jans. For the case of a densified sand foundation where the initial soil state could be judged fairly reliably, the displacement predictions varied quite widely from each other. The accuracy of 'predictions' tends to improve as the order of the expected behaviour becomes more

491

certain following comparison with physical modelling, since this leads to re-examination of the basic assumptions. One conclusion was that increase in complexity of the analysis did not necessarily result in improved accuracy.[11]

19.3 EXAMPLE OF OPTIMIZATION FROM ANALYTICAL AND PHYSICAL MODEL AGREEMENT

A particular value of analytical models lies in their potential for parametric studies. In this respect a simple 'spring constant' analysis was sufficient to illustrate that there was an optimum depth of caissons buried in sand and subjected to cyclic and static lateral forces above the sea bed, which leads to a minimum lateral displacement at sea bed/sill level, a finding supported by physical model studies in the centrifuge (Figure 19.1). Thus, even if there remained uncertainty regarding the size of the displacement, one could more readily take a design decision based on the agreed optimum depth. The

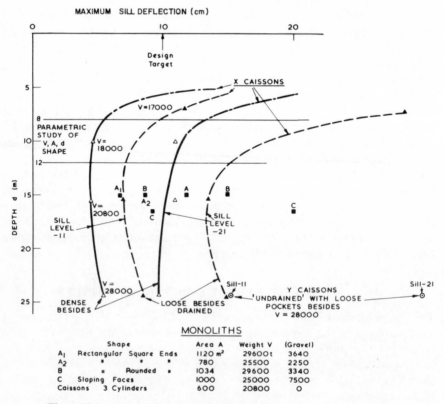

Figure 19.1 Optimum depth for founding cyclically loaded caissons in sand

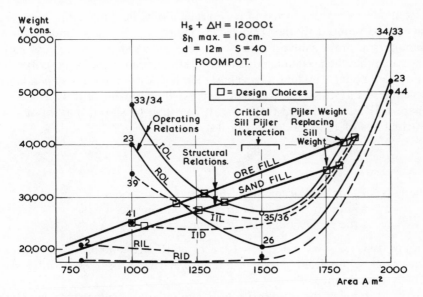

Figure 19.2   Optimum area of caissons at given spacing in sand

same analytical method ultimately produced good numerical agreement with centrifuge model displacements for specific cases,[21] notwithstanding that the analysis suffers at the outset from the fact that the stiffness parameters are a function of boundary geometry, and no full account could be taken of the three-dimensional features of the problem. As it happened, centrifuge models also indicated the optimum area of caissons at a given spacing taking into account the three-dimensional soil–structure interaction including the sill, before theory had been developed[19] (Figure 19.2).

A clear finding, from the Oosterschelde storm-surge barrier studies, consistent with the history of soil mechanics, is that no theory coupled with element test data can be relied upon until it has been checked against a range of physical field/model studies. The most convincing data come from full-scale field observations which always take precedence because they include the surprises inherent in the natural strata. Where such observations are difficult to obtain offshore, or are impossible to obtain at the design stage in the case of novel projects, a theory should be checked against the physical model.

19.4   EXAMPLE OF ANALYTICAL SIMPLIFICATIONS ON DISPLACEMENTS

A limited number of field model tests were run (at considerable cost), e.g. the $\frac{1}{10}$th scale caisson tests on sand in Holland.[17] Despite the attractions of a large model, a feature is that in a 1g gravitational field the applied stresses have to be reduced by the scale factor. However, the strains induced in a granular soil are sensitive to the stress level at all stages of a stress path. Assuming that the accumulated strain to a stress point σ can be described by an overall Young's molulus (E) value, then it is known that the dependence of stiffness on stress level can be expressed approximately by $E = K\sigma^y$, where it is often assumed that $y = \frac{1}{2}$. It can then be shown that for similar distributions of K in model and prototype,

$$\frac{\delta_p}{\delta_m} = \frac{1}{N_s}\left(\frac{\sigma_p}{\sigma_m}\right)^{1-y},$$

where δ_p, δ_m are displacements of prototype and model
      σ_p, σ_m are stress levels of prototype and model
      N_s is the ratio model size/prototype size.

On this basis for $y = \frac{1}{2}$ and $1/N_s = 10$, $\delta_p/\delta_m = 10\sqrt{(10)}$ and prototype performance can be predicted. (Analytical models have also been used with $y = 0$, namely E constant.)

Notwithstanding the complexity of computational accuracy working from soil elements, one might expect it to be possible at least to scale from a large model to a prototype. However, at Manchester a carefully conducted series of triaxial tests subject to cyclic loading, with corrections for boundary effects, showed that the power y was not constant.[7] Figure 8 in Ref. 7 shows that at small stress ratios y tends to, but does not reach, zero where E would be constant as for a true elastic solid; whereas towards failure at high stress ratio y tends to, but does not reach, unity as for a purely plastic material. The finding has a sound physical base and the variation of y throughout a cyclic loading programme illustrates the danger of the use of gross simplifications in the treatment of granular materials. Over the range of Oosterschelde designs the centrifugal models, which offer a true multi-element stress path test, gave $y \simeq 0.15$ for tilt, which reflected elastic states at depth and $y = 0.1–0.65$ for horizontal sliding, depending on the factor of safety and depending on combinations of tilt and horizontal strain. For the $\frac{1}{10}$th field model type design $y \simeq 0.2$ and the scale factor was 60 compared with 31 for $y = 0.5$. No analytical model exists which can resolve this order of discrepancy, especially when coupled with variations in storm sequences.

19.5 MECHANISMS FROM PHYSICAL MODELS

Physical models help to formulate input to analytical models.

Figure 19.3 illustrates gravity platform model performance towards failure on saturated uniform overconsolidated clay. The lateral displacement swings, for a purely oscillatory lateral force ΔH having no static component,

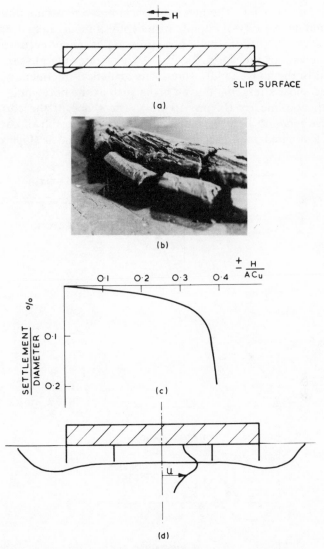

Figure 19.3   Progress to failure of cyclically loaded gravity platform model on clay

increase with ΔH at a gradually increasing rate until the lateral displacement swing accelerates to an unacceptably large value. At this stage a slip surface of weakened clay is found just below the platform (Figure 19.3(a)), and clay has been progressively displaced to the outer leading edges (see the leaf-like formation of the displaced clay layers; Figure 19.3(b)). The clay displacement causes settlement (Figure 19.3(c)), the shape of this curve being similar to that for lateral swing. The maximum pore pressure occurs directly below the base and decreases very rapidly with depth. If moderately deep skirts are added (Figure 19.3(d)), the peak pore pressure at failure occurs at the level of the underside of the skirts. Figure 19.4 illustrates the physical processes which lead to the above failure state. Any analytical treatment will result in the applied shear stress, on a given plane such as the horizontal, decreasing with depth non-linearly (Figure 19.4(a)). The shape of the corresponding strain curve tends to be even more non-linear. The result of each cycle of stress on a drained sand is a net reduction in volume v (Figure 19.4(b)),

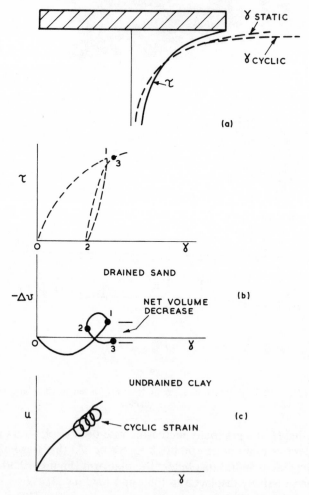

Figure 19.4 Physical processes underlying gravity platform performance

increasing in size with that of the stress reversal. This is a direct result of strain rather than stress, for any treatment of particle mechanics proves a parameter $dv/d\gamma$. For an undrained clay there is a net increase in pore water pressure (Figure 19.4(c)) if the cycle of stress exceeds a certain minimum, and this also increases in size with time. Consequently the pore pressure generation just below the platform must develop ahead of that lower down. The reduction in effective stress then induces more strain, which accelerates generation of pore pressure, and by progressive action the maximum degree of weakening ultimately develops at the plane of maximum stress which is at the soil-structure boundary. This means that any finite element analysis of this problem requires the use of a series of fine nets moved progressively very local to the base. Only the work of Smith[24] so far simulates the model observations.

More especially the high pore pressure gradient in the vertical direction implies that for clays consolidation can occur between storms and for sands it can occur during generation of pore pressure throughout the storm. Denoting time factor/cycle

$$T_c = \frac{k}{\gamma_w m_v} \frac{1}{H_z D^2}$$

it has been found that the soil foundation just below the base remains 'undrained' for $T_c \leqslant 3 \times 10^{-6}$ and remains fully drained for $T_c \geqslant 10^{-4}$. At $T_c = 3 \times 10^{-5}$ pore pressure development just begins to be significant. This means that for a 100 m platform on very dense sand with $m_v = 0.01$ m²/MN, pore pressures would only be expected for $k \leqslant 3 \times 10^{-6}$ m/s and for the Oosterschelde caisson 25 m wide for $k \leqslant 2 \times 10^{-7}$ m/s. No pore pressure generation was found at Neeltje Jans on dense sand with a caisson 15 m wide. No significant pore pressure generation has been reported from the North Sea.

In contrast, in the case of very loose sands a model with $m_v = 0.6$ m²/MN subject to 1 Hz frequency led to liquefaction. A 25 m base on such a loose sand with $k = 10^{-5}$ m/s at 0.1 Hz would begin to generate pore pressures.

It follows that laboratory liquefaction tests of undrained sands has direct application either to platforms on fine very loose sands which no one would contemplate, or to thin sand layers intercalated with silt or clay, a potentially dangerous situation.

In this respect it will be appreciated that there is a complete contrast between the mechanisms of earthquake and storm loading. An earthquake shakes all the foundation. Pore pressure gradients must be much smaller. The frequency is high. Even dense sands can develop almost instantaneous pore pressures, increasing with depth, so that the surface pressures can actually rise with time due to the flow of water from below. It has not always been clear that these mechanisms differed.

19.6 CONCLUSIONS

Analytical treatments should explore the effect of alternative soil element models in the light of observed physical soil–structure model mechanisms and performance. Undrained states can not always be adopted even for simplification in the presence of the generation of high pore pressure gradients. For drained or partially drained sand, improved account has to be taken of the non-linear effect of scale.

B   ANALYTICAL MODELS (I. M. SMITH)

19.7 INTRODUCTION

In principle, rather complicated analytical studies of the behaviour of soils under transient and cyclic loading can be conducted. In practice this is rarely done, due to a combination of cost and of uncertainty as to the validity of the results. It will inevitably remain the case that analytical models will not prevail until it has been shown that they are capable of reproducing the results of laboratory or field scale model tests (in the absence of prototype observations, particularly of hopefully rare failures). The present contribution begins with remarks on some analytical predictions of field and model behaviour of cyclically loaded soils which appear in the literature. Subsequent sections go on to make much more general comments as to what can be learned qualitatively from such analytical work.

19.8 ANALYSES OF PROTOTYPES AND MODELS

19.8.1 Dams subjected to earthquakes

Seed[20] has described the state of the art in this field. The failures of the San Fernando dams in the earthquake of 1971 permitted back-analyses to be conducted. In Seed's opinion the 'pseudo-static' method of analysis[15], is adequate for soils with a well-defined yield stress level: for example, 'clays and clayey soils, dry or moist cohesionless soils or extremely dense cohesionless soils.' This method does not even involve a finite element analysis of the progress of deformation to collapse (Smith and Hobbs[23] for $\phi = 0$ soils or Zienkiewicz, Humpheson and Lewis[27] for frictional soils).

In the case of loose to medium-dense cohesionless soils which 'do not exhibit any clearly defined yield strength and whose behaviour is complicated by the possibility of large pore pressure build-ups and redistribution during and following an earthquake' Seed proposed an analysis procedure combining finite element computations to predict cyclic stress levels, with laboratory testing of 'sufficient' representative soil elements to obtain undrained excess pore water pressure distributions at the end of the earthquake. Possibility of collapse would then be checked by a conventional 'limit equilibrium' (slip circle) analysis. In fact in the case of the lower San Fernando Dam failure this procedure did not lead to a prediction of collapse. Additional assumptions regarding possible pore water pressure migrations had to be made.

It is clear that in this type of problem, although finite element computations which combine dynamic analysis with generation/redistribution of pore water pressure and the progress of deformations to collapse are in principle possible,[3] they have not yet proved feasible in practical cases.

19.8.2 Gravity offshore oil/gas platforms

Rowe has referred to his tests on model gravity platforms bearing on saturated, uniform, overconsolidated clay. Smith[24] first applied finite element methods in an attempt to simulate the test results. A simple elastic–von Mises plastic constitutive model was assumed to represent the clay and failure under various combinations of horizontal and vertical load was computed. These solutions compared well with limit equilibrium methods for the cases in which such solutions existed (uniform foundations of homogeneous clay). The power of the finite element computations became apparent when non-homogeneous sites were studied. In the case of strong soil overlying weak it could be shown that the stronger material dominated the sliding type of failure mode. Conversely it was shown that if weakening under cyclic loading was predominantly a near-surface phenomenon, sliding failure would occur very close to the platform base, and required a fine mesh in that region. This much at least confirmed the model data on failure.

However, turning to cyclic deformations, just a few (3) cycles of alternating load were applied and only showed in principle that vertical plastic movements occurred under cyclic horizontal forces sufficient to cause yield. In contrast, the centrifuge fatigue tests ran for some 5000 cycles.

Using an improved bilinear work hardening elastic–plastic model, Smith and Molenkamp[25] were able to increase the number of alternating load cycles to about 25. A much clearer picture of the progress of plastic settlement of the platform emerged (Figure 19.5). It was also possible to establish a correlation between vertical settlement and horizontal force level broadly in agreement with Rowe's data, although the predicted magnitudes of settlement were too large (Figure 19.6). Finally, although the number of cycles computed was two orders of magnitude less than in the tests, a plot of

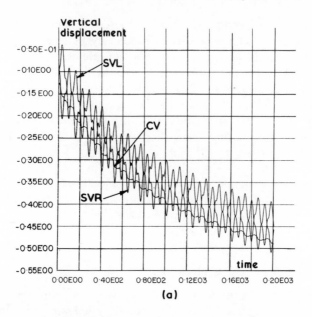

(a)

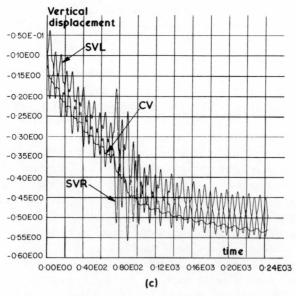

(c)

Figure 19.5 Computed plastic settlement of cyclically loaded gravity

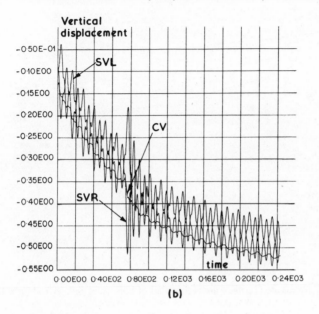

(b)

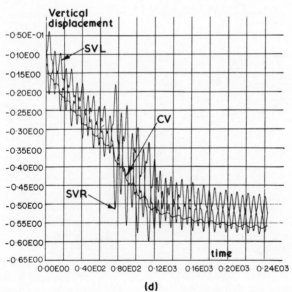

(d)

platform on clay subject to increasing numbers of large waves

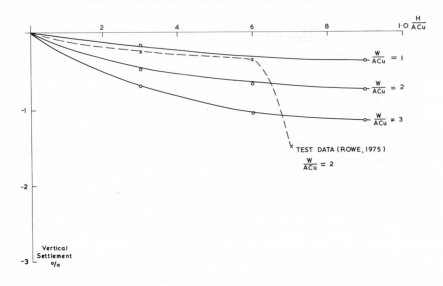

Figure 19.6   Computed settlements as functions of horizontal and vertical force levels

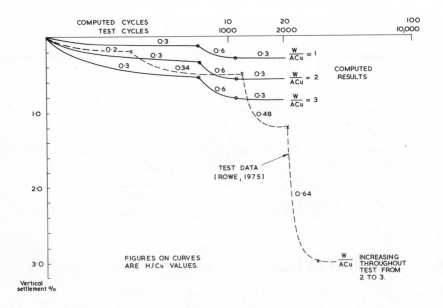

Figure 19.7   Computed settlements as functions of number of load cycles

vertical settlement against number of cycles at various vertical force levels was in qualitative agreement with the test data (Figure 19.7).

Prevost and Hughes[16] attempted to reproduce Rowe's model test data assuming the clay to be anisotropic and elastic–kinematically hardening plastic. They say that 'all the analytical results are consistent with the experimental results obtained by Rowe in his centrifuge tests.' It is difficult to see how this can be so when only five cycles of horizontal load at one vertical force level could be analysed. Many more exhaustive comparisons are needed before such a complicated constitutive model gains general acceptance.

Dumas and Lee[5] compared results from finite element analyses with Rowe's data and with data from their own model tests at 1g. Their analyses apparently involved elastic soil with modulus reducing according to cyclic strain level and a very coarse mesh beneath the platform base. At any rate they were able to apply some 2500 cycles of load in their computations. For Rowe's tests they overpredicted the movements while for their own tests they under-predicted them by factors up to 5.

What emerges from the above comparisions is that it is doubtful at present whether improved constitutive models have anything to offer the practising engineer. The simpler models have enough adjustable parameters to fit such model or field data as may be available, but of course this means that the analytical methods cannot at present be used in a truly predictive role.

19.9 QUALITATIVE OBSERVATIONS FROM ANALYSES

19.9.1 The nature of cyclic pore pressures

Analysis shows that for surface excitation, high pore water pressure gradients exist near ground surface. In addition, pore water pressures are in general out of phase with applied total stresses with consequent weakening potential in terms of effective stresses.[13,24]

19.9.2 Detailed analysis of behaviour throughout a cycle, over many cycles, and of hysteresis

Figure 19.5 shows that rather detailed information emerges from analyses as to the gradual progress of plastic settlement of a cyclically loaded gravity platform on clay. By monitoring the stress distributions immediately beneath the platform it is possible to show that plastic deformations continue until such time as the initial shear stresses, which are self-equilibrating and directed towards the centre of the base, have attained their final plastic

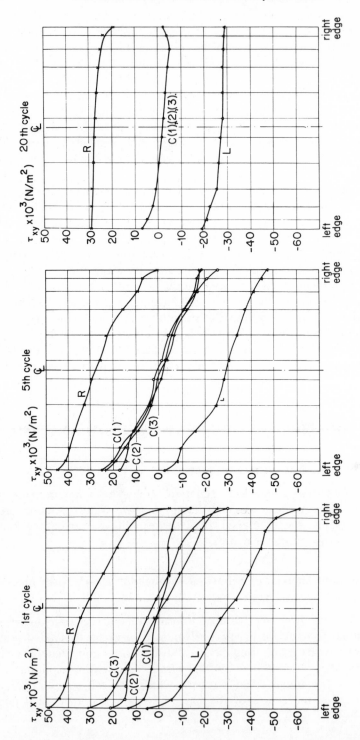

Figure 19.8 Redistribution of shear stress beneath gravity platform as cycling proceeds

state. Figure 19.8, taken from Ref. 25, shows the shear stress under the base during the first, fifth and twentieth load cycles. In the Figure, L refers to the state in which the cyclic force is pushing at its maximum to the left, similarly R to the right while $C(1)$, $C(2)$ and $C(3)$ refer to the three states during a cycle when there is no horizontal force acting. The liquidation of the initial shear stress due to cycling is quite vividly demonstrated.

Figure 19.9 shows typical hysteresis loops of horizontal force and moment against horizontal displacement and rotation respectively. Due to the

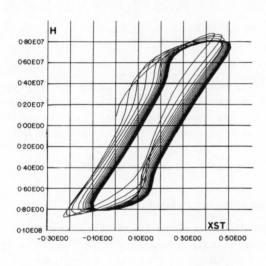

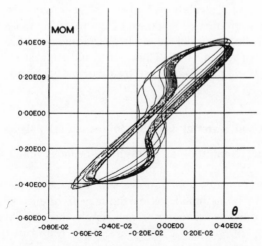

Figure 19.9 Hysteresis of horizontal force and moment

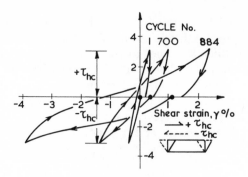

Figure 19.10 Hysteresis loops in 'simple' shear tests after Andersen et al.[1]

bilinear elastic–plastic constitutive model assumed, the loops all have a simple 'S' shape. No data of this type from physical tests have been presented. However, data from element tests on clays, e.g. Andersen et al.[1] (see Figure 19.10), indicate a stiffening of the material in the region of load reversal. To account for this physical observation some viscous resistance could be incorporated in the mathematical model.

19.9.3 The nature of dynamic amplification

Physical models have not yet delivered any information as to the order of dynamic amplification of cyclic motions. At centrifuge scales the frequencies that would have to be used are quite unrealistic given changing soil response at high rates of loading. Calculations by two different methods[25] give typical amplifications of low-frequency response of about 10%. However, calculations using the non-linear model show ultraharmonic amplification (i.e. at $\frac{1}{3}$ natural frequency) of about 50%. Thus one could suggest suitable adjustment of cyclic test wave parcels at these frequencies, based on the analytical results.

19.9.4 Nature of soil material damping in linearized analyses

It is recognized that computations of cyclic non-linear material behaviour are so expensive that linearized equivalents must be adopted whenever possible. Even then, various assumptions can be made as to the nature of material damping to be assumed. At earthquake frequencies this is not so important, but in the low-frequency range, applicable to wave loading of offshore structures, differences of 15% have been found in the hysteretic damping ratios according to the assumptions of Lysmer et al.[12] and of Kausel, Roesset, and Christian[9] (Figure 19.11). The latter's assumption of

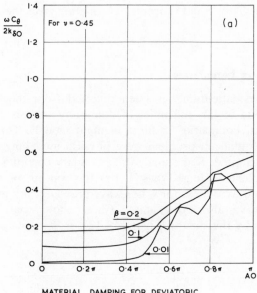

MATERIAL DAMPING FOR DEVIATORIC
DEFORMATION ONLY.

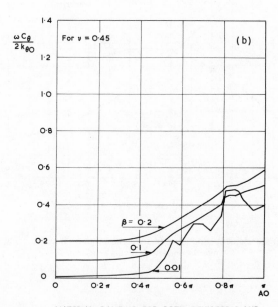

MATERIAL DAMPING FOR BOTH DEVIATORIC AND
ISOTROPIC DEFORMATION.

Figure 19.11 Material damping in linearized analyses

damping for both deviatoric and isotropic deformation gives the less conservative results.[25]

19.9.5 Simulation of boundaries

In dynamic analyses, attention has been diverted from the constitutive behaviour of soil due to the difficulties of doing any analysis at all in the presence of artificial boundaries in finite element models. This of course would be an equally difficult problem were physical modelling of dynamic processes to be attempted. Some analytical boundary conditions are quite complicated to implement, (e.g. Waas[26]), but the concept of infinite elements (Bettess[2]) used by Bettess for water wave problems, has proved to be effective for solid waves also. For example, Chow[4] shows accurate dynamic stiffness predictions by this means (Figure 19.12).

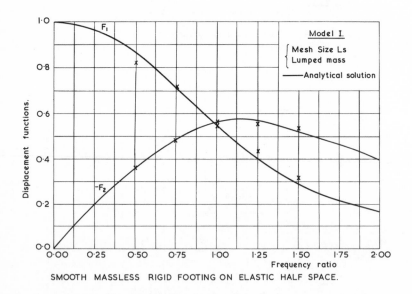

SMOOTH MASSLESS RIGID FOOTING ON ELASTIC HALF SPACE.

Figure 19.12 Finite element modelling of dynamic footing vibration using 'infinite' boundary elements

19.9.6 Computation of soil behaviour

The contractive/dilative behaviour of soil under monotonic loading, has been well understood at least since the work of Rowe.[18] Computations of

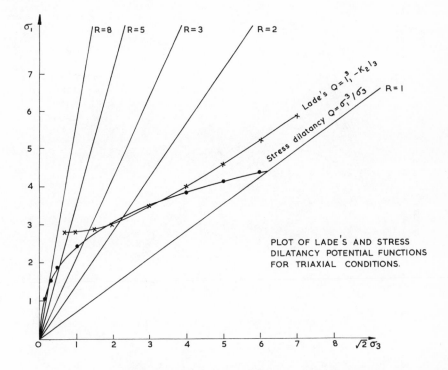

Figure 19.13 Plastic potential functions for sand

simple boundary value problems have also been done (e.g. Smith[22]). The type of plastic potential function derived by Rowe is shown in Figure 19.13. However, it should be remembered that even for monotonic loading these calculations are expensive, and because of the extreme non-linearity of the problem very small load or displacement steps are necessary. Figure 19.14 shows computations of a simple triaxial test for a stress-dilatancy-type constitutive model (Griffiths[6]). If large load steps are taken it is quite possible to predict a completely wrong response—for example, the volume decrease stage can be missed altogether. This simple calculation taxes the power of modern computers and so the prospect of applying realistic constitutive assumptions in cyclic analyses remains a daunting one.

Lade *et al.*[10] have presented alternative plastic potentials to Rowe's (see also Figure 19.13). These have quite the wrong shape except in the strongly dilatant region and it is no surprise that a second 'cap' potential has been added subsequently. Since Rowe's potential has essentially the correct shape it seems sensible to employ it in future work.

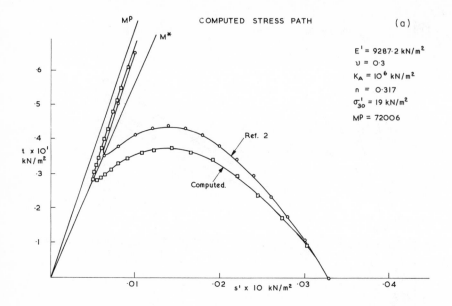

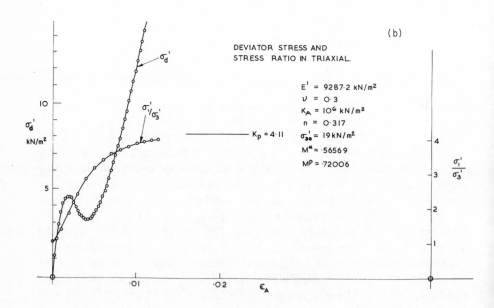

Figure 19.14 Computed undrained stress paths for dense sand

REFERENCES

1. Andersen, K. H., Hansteen, O. E., Hoeg, K., and Prevost, J. H. Soil deformations due to cyclic loads on offshore structures, Chapter 13 in *Numerical Methods in Offshore Engineering*, Wiley, 1977.
2. Bettess, P. Infinite elements, *Int. J. Num. Meth. Eng.*, **11**, p. 53–64 1977.
3. Chang, C. T. *Nonlinear Response of Earth Dams and Foundations in Earthquakes*, Ph.D. Thesis, University of Wales, 1979.
4. Chow, Y. K. *Dynamic Behaviour of Piles*, Ph.D. Thesis, University of Manchester, 1980.
5. Dumas, F., and Lee, K. L. Cyclic movements of offshore structures on clay, *Proc. Conf. Soil Mechanics in the Marine Environment*, ASCE, 1979.
6. Griffiths, D. V. *Applications of Viscoplasticity in Soil Mechanics*, Ph.D. Thesis, University of Manchester, 1980.
7. Hain, S. J. An application of cyclic triaxial testing to field model tests, *Soils Under Cyclic and Transient Loading*, **I**, 23–31, Balkema, Rotterdam, 1980.
8. Hardin, B. O. The nature of stress–strain behaviour for soils, *Earthquake Engineering and Soil Dynamics*, ASCE Geot. Eng. Division Speciality Conference, 1978.
9. Kausel, E., Roesset, J. M., and Christian, J. T. Non linear behaviour in soil-structure interaction, *J. Geotech. Eng. Div.*, ASCE, **102** (GT11), Proc. Paper 12579, 1159–70, 1976.
10. Lade, P. V., Ozawa, K., Duncan, J. M., and Booker, J. R. *Specialty Session on Constitutive Models for Soil*, ISSMFE Conf., Tokyo, 1977.
11. Lambe, T. W. *et al.* Caisson tests at Neeltje Jans, the Netherlands, *Symposium on Foundation Aspects of Coastal Structures*, Delft, 1978.
12. Lysmer, J., Udaka, T., Tsai, C. F., and Seed, H. B. FLUSH—A computer program for approximate 3-D analysis of soil-structure interaction problems, *Earthquake Engineering Research Centre*, Report No. EERC 75–30, College of Engineering, University of California, Berkeley, 1975.
13. Madsen, O. S. Wave-induced pore pressures and effective stresses in a porous bed, *Geotechnique*, **28**(4), 377–94, 1978.
14. Molenkamp, F., and Smith, I. M. Hysteretic and viscous material damping, *Int. J. Num. Anal. Meth. Geomech.*, **4**, 1980.
15. Newmark, N. M. Effect of earthquakes on dams and embankments, *Géotechnique*, **15**(2), 139–60, 1965.
16. Prevost, J. H., and Hughes, T. J. R. Analysis of offshore structure foundations subjected to cyclic wave loading, *Offshore Technology Conference*, Houston, 1978.
17. de Quelerij, L., and Broeze, J. J. Model tests on piers scale 1 : 10, *Symposium on Foundation Aspects of Coastal Structures*, Delft, 1978.
18. Rowe, P. W. The stress dilatancy relation for static equilibrium of an assembly of particles in contact, *Proc. Royal Soc.*, **269**, 500–27, 1962.
19. Rowe, P. W., and Craig, W. H. Prediction of caisson and pier performance by dynamically loaded centrifugal models, *Symposium on Foundation Aspects of Coastal Structures*, Delft, 1978.
20. Seed, H. B. Considerations in the earthquake-resistant design of earth and rockfill dams, *Géotechnique*, **29**(3), 215–63, 1979.
21. Sellmeyer, J. B. Simple numerical methods to determine displacements and stability of piers, *Symposium on Foundation Aspects of Coastal Structures*, Delft, 1978.

22. Smith, I. M. Plane plastic deformation of soil, *Proc. Roscoe Memorial Symposium*, Cambridge, 548–63, 1971.
23. Smith, I. M., and Hobbs, R. Finite element analyses centrifuged and built-up slopes, *Géotechnique*, **24**(4) 531–59, 1974.
24. Smith, I. M. Aspects of the analysis of gravity offshore structures, *Proc. 2nd Int. Conf. on Numerical Methods in Geomechanics*, Blacksburg, Virginia, V.I, 419–24, ASCE Publications, 1976.
25. Smith, I. M., and Molenkamp, F. Dynamic displacements of offshore foundations due to low frequency sinusoidal loading, *Géotechnique* **30**, No. 2, 179–205, 1980.
26. Waas, G. *Analysis Method for Footing Vibrations through Layered Media*, Tech. Rep. S-71-14, U.S. Waterways Station, 1972.
27. Zienkiewicz, O. C., Humpheson, C., and Lewis, R. W. Associated and non-associated viscoplasticity and plasticity in soil mechanics, *Géotechnique* **25**(4), 671–89, 1975.

Soil Mechanics—Transient and Cyclic Loads
Edited by G. N. Pande and O. C. Zienkiewicz
© 1982 John Wiley & Sons Ltd

Chapter 20

Laboratory Investigations of the Behaviour of Soils under Cyclic Loading: A Review

D. M. Wood

20.1 INTRODUCTION

Soil mechanics is a field of engineering study where it is rather easy for two camps to form: those who believe they can describe mathematically how soil ought to behave, and those who observe the behaviour of soil. Opportunities for détente between these camps should, of course, be grasped, but the area of cyclic loading forms one part of the field where the first camp has presently very much established for itself a position of hegemony.

This dominance has arisen because of the contrast between the ever fertile imaginations of those seeking new ideas for construction of mathematical models, and the paucity of good quality experimental data against which these models may be matched. This contrast exists also with static or monotonic testing and modelling of soil response, but the results of cyclic tests are much more dominated by the vicissitudes of the testing procedures.

Any consideration of the results of laboratory investigations of the behaviour of soils under cyclic loading must begin, then, by considering the apparatus and test procedures that are to be used to investigate this behaviour—an area where further research is still required to develop a better understanding.[131]

20.2 PRELIMINARY IDEAS

In order to define completely the stress state for an element of any continuous material six independent components of stress need to be specified. These six components may be three normal stresses and three shear stresses acting on the mutually orthogonal faces of a cube of material (Figure 20.1(a)). Alternatively we may choose to orient our cube of material in such a way that we have no shear stresses, but only three normal stresses—the three principal stresses—acting on the faces of the cube

513

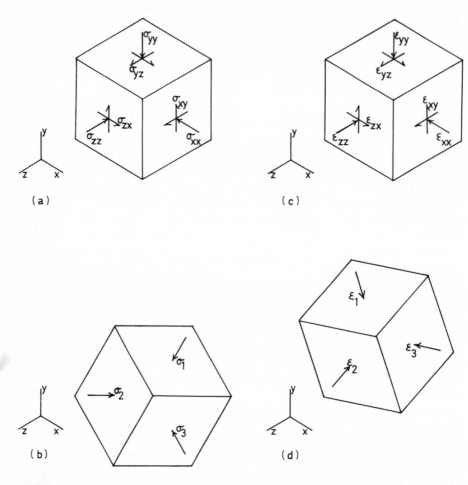

Figure 20.1 (a)   Stress state specified by normal and shear stresses referred to fixed axes. (b) Stress state specified by principal stresses. (c) Strain state specified by direct and shear strains referred to fixed axes. (d) Strain state specified by principal strains.

(Figure 20.1(b)). Three further parameters are then required to specify the orientation of the cube—the directions of the three mutually perpendicular principal axes of stress. For any change of loading, strains will develop in the element of material and six parameters will be required to define the strain state. These may, similarly, be three direct strains and three shear strains (Figure 20.1(c)) or three principal strains and their directions (Figure 20.1(d)). There is in general no *a priori* reason why principal axes of stress and strain should be coincident.

In certain cases the initial state for an element of soil, or the incremental state imposed by some external loading from above or below the element, may degenerate. For any structure or foundation which is long in comparison to its width, plane strain conditions may be relevant with $\varepsilon_{zz} = \varepsilon_{xz} = \varepsilon_{yz} = 0$, $\sigma_{xz} = \sigma_{yz} = 0$. Three independent stress or strain components remain: the stress σ_{zz}, which is a principal stress, is not an independent component; it will automatically adjust to maintain the plane strain condition and can, for example, be considered as dependent on the normal stresses and shear stress in the plane of shear $(\sigma_{xx}, \sigma_{yy}, \sigma_{xy})$.

Under the centreline of a circular load placed on level uniform ground of large lateral extent, conditions of axial symmetry will apply with $\varepsilon_{yz} = \varepsilon_{zx} = \varepsilon_{xy} = 0$ and $\varepsilon_{xx} = \varepsilon_{zz}$ (or alternatively $\sigma_{yz} = \sigma_{zx} = \sigma_{xy} = 0$ and $\sigma_{xx} = \sigma_{zz}$).

In general, however, simultaneous variation of all six components of stress or strain will occur and any attempt to model mathematically the behaviour of a mass of soil under general loading requires a soil model which can cope with simultaneous variation of all three principal stresses and rotation of all three principal axes—and such a soil model needs to be built upon test data obtained under similarly general conditions, resulting from experimental explorations of six-dimensional stress or strain space.

Such explorations are perhaps easy to describe mathematically but are difficult to visualize, and difficult also to apply in a controlled fashion to laboratory test specimens. Most laboratory test apparatus have two or at best three degrees of freedom: the no-man's-land of stress space is thus rather extensive. It is important, however, to appreciate that different apparatus will be restricted to travelling in different sections of stress or strain space; soil is a highly non-linear, path-dependent material and although it may be possible to describe current states or changes in state in different apparatus in terms of a common set of parameters (for example, principal stresses and strains, or stress and strain invariants) there is no particular reason why the response of the soil elements, investigated separately, should under this common frame of reference turn out to be identical. The only way that soil response under different test paths can be compared is by constructing a soil model with a set of basic hypotheses and testing this model against observed behaviour. On the other hand, however, it is valuable to have complete information about the components of stress and strain for a test element so that differences in response *can* be observed in a common frame of reference.

Obviously tests need to be conducted in as wide a range of apparatus as possible so that, restricted though each apparatus may be, overall, coverage of stress or strain space may be as extensive as possible. Principal stress and strain spaces provide a useful vehicle for comparing capabilities of certain available apparatus. Since these are three-dimensional spaces, paths in each can be displayed as two orthogonal views as shown in Figures 20.2(a,b,c,d):

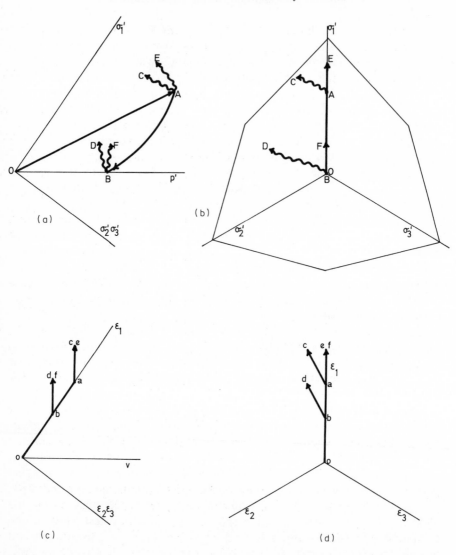

Figure 20.2 (a) and (b)   Orthogonal views of principal stress space: (a) is a view containing true view projections of the σ_1' axis and the p' axis or space diagonal $\sigma_1' = \sigma_2' = \sigma_3'$; (b) is the π-plane view down the space diagonal, a constant p' section through the Mohr–Coulomb failure criterion is also shown. (c) and (d) Corresponding orthogonal views of principal strain space. Strain path *oa* and stress path *OA* are for one-dimensional loading; strain paths *ac* and *ae* are constant volume plane strain and axisymmetric paths from *a* resulting in stress paths such as *AC* and *AE*. Strain path *ab* and stress path *AB* are for one-dimensional unloading; strain paths *bd* and *bf* are constant volume plane strain and axisymmetric paths from *b* resulting in stress paths such as *BD* and *BF*. (e) Division of π-plane into sectors with different relative magnitudes of σ_1', σ_2' and σ_3'

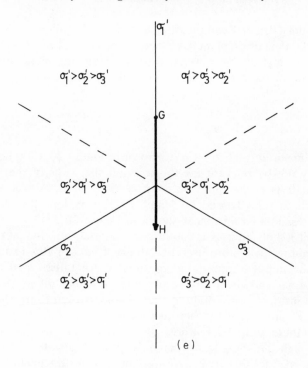

(e)

in each case a deviatoric view down the space diagonal ($\sigma_1 = \sigma_2 = \sigma_3$, Figure 20.2(b); or $\varepsilon_1 = \varepsilon_2 = \varepsilon_3$, Figure 20.2(d)) and an orthogonal view in which the space diagonal and one axis appear in true view (Figures 20.2(a,c)).

An element of soil beneath level ground of large lateral extent will have been one-dimensionally compressed in its geological past. It may currently be in a normally consolidated state (*OA* and *Oa*) or an overconsolidated state (*OAB* and *Oab*). In the ground beneath a long embankment, or behind a long retaining wall, or (approximately) under an offshore gravity platform subjected to lateral wave loading, the element may be subjected to undrained plane strain shearing—a strain-controlled path specified in strain space (*ac* or *bd*) and having a result in stress space (*AC* or *BD*). This path can be matched by a plane strain device such as the biaxial apparatus of Hambly[31,32] but in spite of their obvious simple relevance, plane strain apparatus forcing fixed principal axes have been infrequently used for monotonic tests and apparently never used for cyclic tests. An ideal constant volume simple shear test also follows this same path in principal strain space—but an extra dimension is really required in order that the rotation of principal axes in that apparatus may be displayed.

An undrained triaxial compression test can only follow paths which are vertical in the deviatoric plane (*AE* and *ae* or *BF* and *bf*)—and is clearly a quite different test of the material response from the plane strain test. A typical section of the Mohr–Coulomb failure criterion

$$\frac{\sigma'_i}{\sigma'_j} = \frac{1 \pm \sin \phi'}{1 \mp \sin \phi'} \qquad (i = 1, 2, 3; j = 2, 3, 1)$$

at constant mean normal stress p' is shown in Figure 20.2(b). This criterion may or may not be relevant for the soil, but the paths of the triaxial and plane strain tests strike quite different sections of it. A true triaxial apparatus such as that devised by Hambly[30] can explore the whole of the principal stress and strain spaces shown in Figure 20.2.[126]

While soil is behaving as a continuous material, and has not developed planes of failure, it is more useful to think of stress paths in terms of the nature of the implied exploration of stress space rather than in terms of resolved normal and shear stresses on any particular plane, (for example, Lee[48]). A transition from a state of triaxial compression to triaxial extension (*G* to *H* in Figure 20.2(e)) is then just seen as a smooth path in stress space and, while there may be features of material response associated with passing through the stress space diagonal, through a state of purely hydrostatic stress, the fact that σ'_1 is no longer the major principal stress is not of itself significant: changes in the relative magnitudes of the three principal stresses occur for any path crossing any of the solid or dashed radial lines in Figure 20.2(e). It is not helpful to talk of sudden rotations of principal axes occurring every time one of these lines is crossed.

An element of soil which has been subjected to the one-dimensional compression history (*OA* and *Oa* or *OAB* and *Oab* in Figure 20.2) has an anisotropic structure, and responds anisotropically to changes of stress or strain[71,85] even if its initial stress state is by chance isotropic, as at *B*. Beneath level ground such a soil is cross-anisotropic with anisotropy axially symmetric about a vertical axis.[36,75] The need to be able to accommodate this anisotropy in any test that is performed on such a material has been well illustrated by Saada[84] and Saada and Bianchini.[87] Unless the principal axes imposed by the testing device coincide with the axes of anisotropy, the sample is restrained from deforming as it wishes (with principal axes of strain increment and of stress not coincident) and responds by deforming non-uniformly (Figure 20.3). A triaxial test on an inclined sample of cross-anisotropic soil may give us some indication of the anisotropy of *strength* properties—if the sample fails on a discrete failure plane—but does not provide useful information on the stress–strain behaviour of the soil before failure. Even a conventional triaxial test on a horizontal sample

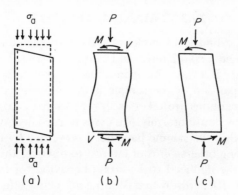

Figure 20.3 Exaggerated deformation of inclined specimen of anisotropic soil tested in the triaxial apparatus: (a) uniform state of stress; (b) end platens are horizontal, rough and rigid; (c) displacement when end platens are horizontal, smooth and rigid (after Saada and Bianchini[87])

needs to be interpreted with care: in general the initial circular cross-section is not retained during shearing but tends to become elliptical—two separate components of radial strain should strictly be measured.

Under sloping ground, axes of anisotropy are not vertical and attempts to match loading paths for elements of soil from such conditions with simple laboratory tests are unlikely to be successful.

Saada and Bianchini[87] conclude that the only valid test on a sample of a cross-anisotropic material, with the axes of the sample not coincident with the axes of anisotropy, is a test on a hollow cylindrical sample (of appropriate proportions[132]) in which torque and axial load (and equal internal and external pressures) are applied together in such a way that the axes of principal stress are maintained coincident with the axes of anisotropy of the sample throughout the test.

It has been tacitly assumed in sketching Figures 20.2(a,b) that it is on the effective stresses (that is the differences between the total applied stresses and the pore pressure) that the response of a soil element depends. This has been often demonstrated for the behaviour of soils under monotonic loading, and, as will be seen later, it applies equally to the behaviour of soils under cyclic loading. Investigation of the behaviour of soils in terms only of total stresses may be of some interest, but understanding of the changes in response that are seen needs to be related to the effective stress changes that are occurring.

20.3 TRIAXIAL TESTING

In the triaxial test cylindrical samples of soil are subjected to controlled axial strain, through rigid end platens, and controlled radial stress, through a rubber membrane. Because the membrane is usually sealed to the end platens lateral expansion of the sample is often restricted at the ends and unless special precautions are taken (see, for example, Rowe and Barden[83]) the sample deforms non-uniformly. Analyses of the effect of end restraint on the stress field within the sample have been presented by Balla[8] and Perloff and Pombo[74] among others—it is of interest to discover the consequences of end restraint for the deduced stress–strain behaviour of the soil sample.

Experimentally, Richardson and Whitman[79] show effective stress paths deduced in 'fast' and 'slow' undrained tests on normally consolidated clay (Figure 20.4(a)), with pore pressures measured at the ends of the sample. Evidently much smaller pore pressures are registered in the fast test than in the slow test. The restraint at the ends concentrates the deformation into the central part of the specimen. For this normally consolidated clay larger positive pore pressures develop at the centre, and in the slow tests pore water is able to flow down the pore pressure gradient, increasing the water content at the ends of the sample at the expense of the centre (Figure 20.4(b)). An opposite effect is observed in tests on heavily overconsolidated samples where there is a tendency for negative pore pressures to develop in the central deforming region.

Carter (unpublished) has performed finite element analyses of tri-axial samples in which the effect of finite permeability of the soil has been considered. Pore water flow influences the results of drained tests through the applied strain rate. Fast tests produce a response observed externally similar to that in ideal undrained tests. Slow tests behave as

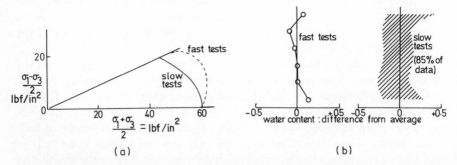

Figure 20.4 (a)   Effect of speed of testing on effective stress path observed in undrained triaxial tests on normally consolidated clay (after Richardson and Whitman.[79]) Note this is not $q : p'$ space (b) Distribution of water content over height of sample of normally consolidated clay after slow undrained triaxial shearing (after Richardson and Whitman[79])

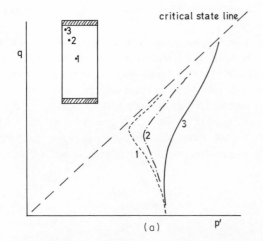

Figure 20.5 (a) Stress paths computed by finite element method for three elements within a triaxial specimen with smooth ends during a fast drained compression test (after Carter, unpublished) (b) Regions of expansion and compression predicted by finite element method within a triaxial specimen with rough ends during a slow undrained compression test (after Carter, unpublished)

should an ideal drained test. Different elements of soil within the sample follow quite different stress paths (Figure 20.5(a)). In an undrained test, although no overall change in volume may occur, individual elements of soil within the sample may expand or contract (Figure 20.5(b)), and different stress paths are again followed.

Clearly for good quality triaxial testing measurements of pore pressure and of deformation must be made in the central deforming region of the sample. Deformation measurements in the central region should include both axial and radial deformations: it cannot be assumed that zero overall volume change implies no volume change in the region of interest.

A further effect which may permit volume changes to occur in nominally undrained triaxial tests is that of membrane penetration. Membrane penetration is significant particularly in coarse grained soils—Lade and Hernandez[46] suggest that the effect is small for particle sizes less than 0.1–0.2 mm. The magnitude of the effect depends on the difference between the pore pressure and the cell pressure—and hence on the radial effective stress. As noted by Sangrey,[89] 'the membrane penetration which takes place under a given change in pore pressure (due to shear) is the same direction of volume change as that which would take place if the shear had been conducted as a drained test.' Thus the effect of membrane penetration is to reduce the measured pore pressure, and Lade and Hernandex[46] observe, for example, that resistance to undrained liquefaction may consequently be overestimated.

Sangrey[89] found that an effect analogous to membrane penetration could produce pore pressure errors in undrained tests on samples of clay with enlarged frictionless end platens—because of small pockets of water trapped between the membrane and the platens at re-entrant angles.

The ideas behind stress path testing tell us that we should strive as far as possible to apply in our laboratory tests paths which are relevant to field elements under consideration. As we have already seen, this may not be feasible with the simple testing apparatus that are available to us. In these

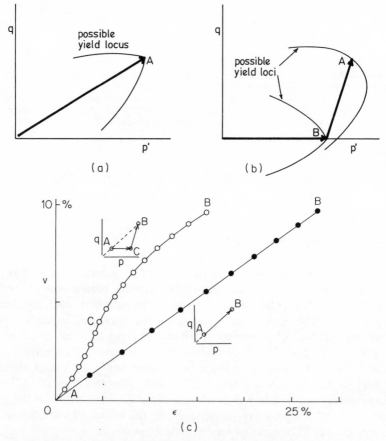

Figure 20.6 (a) and (b) Possible effect on final yield locus of stress path to reach point A; (a) correct anisotropic consolidation OA; (b) isotropic consolidation OB followed by constant cell pressure shearing BA. (c) Effect of consolidation stress path on strains predicted for anisotropic consolidation with correct (AB) and stepped (ACB) paths (after Wood[125])

circumstances testing takes on a slightly different role: there is no need to try to justify each stage of each test by comparison with some complex field stress or strain path—each test provides information on one area of soil response and no test may in practice be specially relevant for field predictions. Nevertheless when trying to match behaviour of elements of soil beneath level ground of large lateral extent the initial consolidation of the soil will have been anisotropic. Anisotropic soils behave quite differently from isotropic ones and anisotropic consolidation in laboratory tests is important.

Saada[86] has indicated how difficult it is to change the initial consolidation structure or orientation of soil particles. Lewin[51] has shown the anisotropic plastic potentials that are deduced for anisotropically consolidated samples—and suggests that this anisotropy is very persistent even with a substantial rise in the stress level. Anisotropic consolidation aiming for a constant stress ratio path such as *OA* in Figure 20.6(a) should apply stress changes close to this path. Arrival at the anisotropic stress state *A via* the isotropic stress state *B* (Figure 20.6(b)) may not only produce quite different strains during the consolidation phase (Figure 20.6(c) shows predicted strains for supposed anisotropic consolidation stress paths used for Newfield clay by Namy[64]; these have been predicted[125] using a simple isotropic Cam clay model of soil behaviour which may not necessarily be correct in its detailed predictions but certainly indicates the possible effect) but also lead to different subsequent response: compare putative yield loci at the end of consolidation to *A* in Figures 20.6(a) and 20.6(b).

20.4 SIMPLE SHEAR TESTING

In a simple shear apparatus soil samples are subjected to shear stresses and strains under conditions of plane strain, and of no direct strain in the direction of shearing. Thus $\varepsilon_{zz} = \varepsilon_{xx} = 0$ (Figure 20.7(a)) and the two remaining degrees of freedom are ε_{yx} and ε_{yy}. (Arthur *et al.*[4] have discussed the simple shear apparatus in comparison with other test apparatus in which controlled rotation of principal axes of stress and strain is possible.) Ever since the elastic analyses of Roscoe[80] it has been recognised that the stress and strain conditions within the simple shear sample are non-uniform largely because of the absence of complementary shear stresses on the ends of the sample. This implies a non-uniform distribution of shear stress over the top and bottom faces (Figure 20.7(b)) and a corresponding non-uniform distribution of normal stress over these faces in order to maintain moment equilibrium of the sample (Figure 20.7(c)). No statement of material properties or of magnitude of strains is needed in proposing these distributions: complete or partial absence of the shear stresses on the ends of the sample implies some similar variation for even infinitesimal elastic deformations.

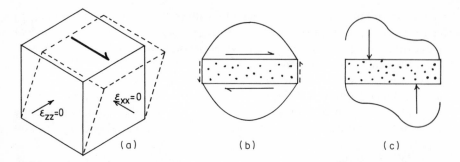

Figure 20.7 (a) Simple shear state of strain. (b) Non-uniform distribution of shear stress from absence of complementary shear stresses on ends of simple shear sample. (c) Non-uniform distribution of normal stress to preserve moment equilibrium of the simple shear sample

There are essentially two types of simple shear apparatus in use today. One, developed by stages at Cambridge University, encloses a sample, rectangular in plan, between rigid boundaries which are hinged or linked together in such a way that simple shear boundary deformation can freely occur[112] (Figure 20.8(a)). The other, developed at the Norwegian Geotechnical Institute[10] encloses a sample, circular in plan, within a rubber membrane with a spiral wire binding (Figure 20.8(b)).

The stress state in the rectangular apparatus has been analysed for an elastic material by Roscoe[80] and Prévost and Höeg[76]. Duncan and Dunlop[20] have introduced a more complicated material model allowing for progressive failure and anisotropic soil modulus values. The philosophy at Cambridge has been to try to come to terms with the known non-uniformities of stress

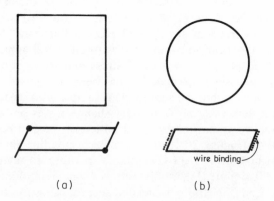

Figure 20.8 (a) Rectangular simple shear apparatus
(b) Circular simple shear apparatus

by measuring the boundary stress distribution with an array of load cells, and using these measurements to deduce the stress state in the central part of the sample, the part least influenced by the end effects.[130] Because the boundary stresses are not uniform, it is unlikely that the internal deformations will follow uniformly the imposed boundary deformations. Roscoe's[80] analysis showed that the deformations were considerably more uniform than the stresses, but radiographic observations such as those produced by Wood and Budhu[129] show that sand samples are certainly not deterred from developing major variations of density.

The stress state in the circular apparatus is analysed by Lucks *et al.*[53] who conclude that 70% of the sample has a very uniform stress condition. This conclusion is questioned by Wright, Gilbert, and Saada[132] both from a theoretical point of view and also in the light of photoelastic experiments that they conducted to obtain direct measurements of the internal stress state: they show that the circular section cannot maintain a plane strain condition and that shear stresses normal to the plane of shear (σ_{yz}) and variations of shear stress in the plane of shear (σ_{xy}) are too large to be neglected. An analytical treatment is also provided by Shen *et al.*[103,104] who include in a finite element analysis both soil elements and elements for the reinforced membrane, and illustrate the effect on uniformity of conditions, and on apparent material properties, of different height-to-diameter ratios and different ratios of soil properties to membrane properties. It is clear that the longer the sample in comparison with its height the more uniform conditions are—but some of their other results need to be treated with caution since statements about non-uniformities after significant strains are very significantly influenced by the actual properties of the material being tested.

Experimental verification of the non-uniformity of the internal volumetric strains in a sand sample in a circular simple shear apparatus is provided by Casagrande.[16] He reports tests in which a saturated sample was frozen and cut up so that the water content of each portion could be determined (Figure 20.9). Evidently maintaining the overall volume of a simple shear test specimen constant does not prevent redistribution of pore water or pore air within the sample.

If the primary problems in the simple shear test come from the effect of the ends then it may be possible to improve matters by escaping from the ends. Seed[100] shows a configuration for an ideal simple shear test (Figure 20.10(a)). With a device of this sort, with unconfined ends, but with load cells in the top, bottom, and lateral boundaries, the stress state in the centre can be readily deduced. A long simple shear device is used by Kovacs[42] testing clay samples with no lateral restraint at all. He concludes that a length-to-height ratio of at least 6 is desirable in order to obtain consistent results in terms of average observed behaviour. A shaking table can also be

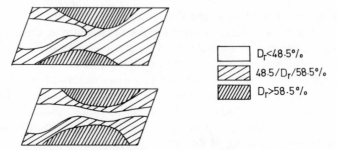

Figure 20.9 Redistribution of water content in vertical slices parallel to direction of shearing after 25 cycles of simple shear of medium-loose specimen of sand (after Casagrande[16])

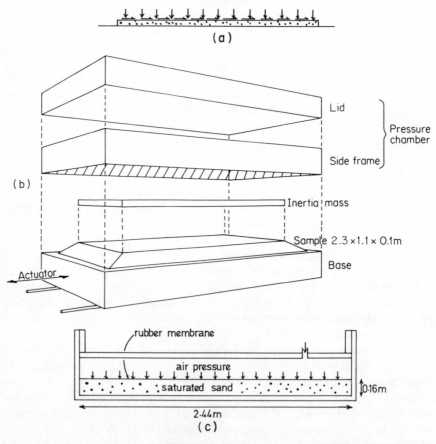

Figure 20.10 (a) Desirable form of simple shear test with no complementary shear stresses (after Seed[100]) (b) Schematic view of shaking table device used by De Alba, Seed, and Chan.[18] (c) Shaking table model container used by Finn, Emery, and Gupta[24]

used as a device of this type: De Alba, Seed and Chan[18] report the use of a shaking table with a $2.3 \times 1.1 \times 0.1$ m sand sample to investigate the liquefaction resistance of the sand. Their loading was applied by means of an inertial mass placed on top of the sand sample (Figure 20.10(b))—so the actual shearing path applied was dependent on the dynamic properties of the whole system. They deduce that so far as number of cycles to produce liquefaction is concerned, corrected results from shaking table tests agree well with results of carefully conducted small-scale simple shear tests. This is a somewhat curious result. Liquefaction testing will be discussed further below: it is very clear that liquefaction response is particularly imperfection-sensitive and it may be that some similar imperfections build up in both test devices.

Our intention here is to concentrate on simple laboratory tests, but it is clear that shaking tables can be used for more extensive investigation of soil behaviour under cyclic loading. A device such as that used by Finn, Emery, and Gupta[24] (Figure 20.10(c)) may appear to be a rather simple shearing test: a sand sample loaded by air pressure and contained in a rigid vessel shaken according to some prescribed programme. In fact, however, for anything at all to happen within this sample non-uniformities must develop within a rather complicated boundary value problem.

Use of results of simple shear tests and attempts to relate results of simple shear tests to results of other tests require knowledge of all the components of stress acting on the soil sample. In other words, we require not only the normal and shear stresses on the top and bottom boundaries (σ_{yy} and σ_{yx}) (Figure 20.11(a)) but also the lateral normal stress σ_{xx} in the plane of deformation, and the lateral normal stress σ_{zz} (a principal stress) normal to the plane of deformation (Figure 20.11(a)). These last two lateral stresses will not in general be equal: measurements obtained by Wood and Budhu[129] show much more rapid build up of σ_{xx} than of σ_{zz} for a simple shear test on dense sand in a rectangular apparatus (Figure 20.11(b)). In a circular apparatus, binding the membrane with resistance wire may permit an estimate of lateral stress to be made[61,135] but this is an average lateral stress which is also influenced by other factors in the shearing of the circular sample.[129]

The lateral stresses change in response to the applied shearing of the sample under the imposed conditions of no direct strain in either of the horizontal directions ($\varepsilon_{xx} = \varepsilon_{zz} = 0$). Control of lateral stress around the sample, for example by placing a circular simple shear apparatus in a cell under fluid pressure, may provide an interesting test configuration, but it is no longer simple shear—and the forcing of all stresses normal to the boundary to be equal cannot be compatible with the conditions of plane strain ($\varepsilon_{zz} = 0$) or of zero horizontal strain ($\varepsilon_{xx} = 0$). Pyke[78] describes an apparatus with articulated circular platens which will permit the complete

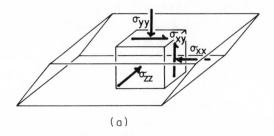

(a)

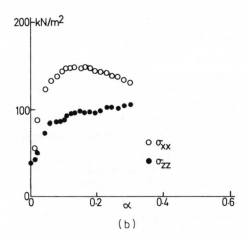

(b)

Figure 20.11 (a) Stress components required to specify stress state in simple shear apparatus. (b) Variation of lateral stress σ_{xx} (in plane of shearing) and σ_{zz} (normal to plane of shearing) in monotonic simple shear test on dense Leighton Buzzard sand (after Wood and Budhu[129])

stress ellipse to be applied to the boundaries of an initially circular sample—with deformation of this circular section to an ellipse permitted. This may also provide some information about conditions in another part of stress space but does not sound much like simple shear.

Finally, where sand samples are contained within membranes as in the circular apparatus, or the shaking table tests of De Alba, Seed, and Chan,[18] and are tested undrained so that pore pressures are able to build up, then membrane penetration effects need again to be considered.

20.5 TEST AND PREPARATION PROBLEMS

We have already discussed the problems associated with the apparatus commonly used for cyclic testing of soils. The principal effects of these problems are that samples cannot be treated accurately as single elements of soil, and that the manner in which tests are performed and interpreted should take this into account. Analysis of each sample becomes a rather elaborate boundary value problem and a simple interpretation may give misleading results. In neither the triaxial nor the simple shear test does prevention of drainage from or volume change of the sample guarantee that no changes of volume or pore water movements occur within the sample.

In order to avoid the effects of the compliance and response of the pore water pressure measuring system on observed behaviour in undrained tests (for example, Martin, Finn, and Seed,[56]) constant volume drained tests have sometimes been preferred.[22,26,60,117] Changes in total stress (radial stress in the triaxial, vertical stress in the simple shear test) to keep the volume constant are then interpreted as changes in pore pressure that would be seen in undrained tests. There is no guarantee, however, that the specimen in either the triaxial or the simple shear test would actually choose to generate the uniform pore pressure that this implies.

Radiographic evidence reported by Wood and Budhu[129] shows that dense sand tested in the simple shear apparatus can readily develop zones of preferential dilation within the test specimen. Arthur *et al.*[4] discuss the various directions, relative to the directions of major principal strain increment and major principal stress, in which failure zones may tend to form within a sand specimen. Even if these have not been observed it is very likely that such zones will develop within specimens of sand in most testing apparatus. The formation of such zones is a bifurcation problem (as discussed for heavily overconsolidated clay by Schofield and Wroth[95])—the sand very happily exploits any pre-existing weaknesses or imperfections within its structure. Observed response for tests on sand samples can be expected to be very dependent on details of sample preparation and testing procedure.

There have been a number of papers recently[44,45,62,110] in which the dramatic effect of method of sample preparation on strength or on number of cycles to produce liquefaction at any given stress level has been shown. Liquefaction testing seems to be very largely a phenomenological study of imperfection sensitivity, so the finding is not surprising. Finn, Pickering, and Bransby[25] show the importance in the simple shear test of being able to generate adequate friction on the horizontal boundaries—increasing the boundary roughness increases the liquefaction resistance. This must be some rather complex apparatus effect since their preferred rough boundary— sharp ribs 0.5 mm high at 2 mm centres—can hardly have failed to produce local disturbance in the samples with grain size range 0.15–0.6 mm.

These results can all be related back to the studies of fabric variation in shear tests on sand samples conducted by Oda.[66–8] Oda defines fabric as the spatial arrangement of solid particles and associated voids: the particular fabric controlling material response at any moment depends on the soil particles—their grading, angularity, and proportions—and on the history of the deposit, which includes, of course, the manner by which it came to be deposited. Oda's investigations of fabric required impregnation of sand samples with polyester resin which was allowed to set so that the samples could be cut into thin sections. Though this technique may well not have left the fabric undisturbed his conclusions on the significance of fabric cannot be faulted.

Stress and strain histories will also influence future response through their effect on fabric and this will inevitably lead to perplexity in trying to compare results from different test devices. There is little reason to suppose that the stress or strain history imposed by a simple shear test produces the same changes in fabric as the history imposed by a triaxial test. It is not surprising then that Seed and Peacock[99] find that none of their proposed criteria for linking liquefaction tests in the simple shear apparatus and the triaxial apparatus is particularly successful.

Dynamic prestraining, or seismic history, will also influence future response—for example, resistance to liquefaction.[23,101] The changes in density produced may be negligible but the particle rearrangements[136] or the possible wear on particle contacts[19] may have an important effect on the soil fabric.

In view of the problems associated with reconstituting sand samples, Finn, Bransby, and Pickering[23] conclude that resistance to liquefaction can only be determined on undisturbed samples. Seed[102] takes a more pragmatic view of the sampling difficulties and settles rather for correlations between penetration resistance and liquefaction resistance of sand deposits determined *in situ.*

20.6 SPECIFICATION FOR CYCLIC TESTING

From what has been discussed so far it is fairly easy to establish certain specifications for judging reported results of cyclic testing. Evidently a prime consideration must be that some awareness is shown of the problems associated with the various testing apparatus, and that some attempt should be made to come to terms with these problems. Obviously the material being tested must be carefully described with details of its origin, and, if possible, the particle size distribution. Where relative densities are quoted for cohesionless soils it is vital that the methods by which the maximum and minimum void ratios were obtained, and the values of void ratio themselves, should be specified. The method of preparation of the samples needs also to be described—for example, Mulilis *et al.*[62] conclude that pluviation of sand, possibly with low-frequency vibration, produced very uniform samples: other methods of preparation produced very non-uniform samples.

The methods of measuring pore pressure and deformations in the deforming region of the sample must be described. Devices for measuring axial and radial deformations at the centre of a triaxial sample are described by Pappin,[69] for example. Pore pressure measurements at the ends of a triaxial specimen are unlikely to be useful unless the test is conducted sufficiently slowly for equalisation of pore pressures to occur: Sangrey, Pollard, and Egan[92] note that no rapid test can be meaningful when applied to *in situ* conditions if the non-uniformities of pore pressure within the test specimen are large. If results are to be interpreted in terms of effective stresses—and it seems that this approach is the most hopeful for understanding soil behaviour—then slow tests are to be preferred to rapid ones. This will, of course, ignore the possible viscous effects that different rates of loading have on the material response. Slow cyclic tests in which one can be certain of the reliability of the measured pore water pressures are to be preferred to fast tests, and are more likely to provide the necessary clues for modelling cyclic behaviour of soils.

Most cohesive soils show some creep or relaxation effects—continuing deformation under constant load, or continuing build-up of pore pressure when volumetric deformation is prevented. Since build-up of pore pressure changes the effective stresses it will also change the material response. It is necessary, as discussed by Sangrey, Pollard, and Egan[92] to perform creep or relaxation tests in parallel with cyclic loading tests so that the separate effects of time and of cyclic loading itself can be clearly identified.

20.7 BEHAVIOUR OF SANDS UNDER CYCLIC LOADING

A summary of some published work on behaviour of sands under cyclic loading is given in Table 20.1. There has been a large volume of material published on the resistance of sands to liquefaction under cyclic loading. Some of the results have already been referred to. Summaries of liquefaction philosophies are given by, *inter alia*, Casagrande[16] and Seed.[102] Tests to discover resistance of soils to liquefaction do not usually produce much information on the stress–strain behaviour of sand under cyclic loading. The definition of liquefaction is problematical, and the selection of a number of cycles to produce liquefaction under particular cyclic stress conditions may be useful as an index test[77] more than as a fundamental aspect of material behaviour.

Most of the results listed in Table 20.1 come from tests in the triaxial apparatus. Castro[17] is clearly aware of the non-uniformities that can develop

Table 20.1   Summary of published work on

Sand	d_{10} (mm)	d_{50} (mm)	Coefficient of uniformity	e_{max}	e_{min}	Preparation	Dry (D) Saturated (S)	Apparatus	Pore pressure measurement	Deformation measurement
Banding sand	0.097	—	1.8	0.84	0.50	Tamping	S	Triaxial	Ends	Boundaries
Fontainebleau sand	0.16	0.26	—	(0.96)	(0.57)	?	S	Triaxial	?	Boundaries
Crystal silica No. 20	—	—	—	0.97	0.64	Pouring and vibration	D	Triaxial	—	Boundaries
Crystal silica No. 20	0.4	0.55	—	0.973	0.636	'Screened' and vibrated	D	Circular simple shear	—	Boundaries
North Sea sand	0.08 (0.1 with fines removed)	0.13	—	0.92	0.57	Moist tamping, vibration	S	Triaxial	Ends	Boundaries
Fuji River sand	0.22	—	2.21	1.08	0.53	Wet spooning ($e_0 \simeq 0.75$) tamping and vibration ($e_0 \simeq 0.53$)	S	Triaxial	?	Boundaries
Monterey 20–30	—	0.62	1.2	0.85	0.53	Wet raining and vibration ($e_0 \simeq 0.69$)	S	Cubic test apparatus	Corner	Approximately from flexible boundaries
Leighton Buzzard 14/25	0.72	0.85	1.2	0.78	0.53	Pluviation	D	Rectangular simple shear	—	Boundaries (and internal)

within triaxial samples, but none of the listed investigations makes any attempt to measure pore pressures or deformations at the centre of the sample away from the effects of the boundaries.

Similarly the results of the simple shear tests reported by Silver and Seed[107,108] take no account of the non-uniformities of boundary stresses and internal deformations in the simple shear apparatus.

It seems likely that the results reported by Wolfe, Annaki, and Lee[124] (looking anyway only at liquefaction resistance, and with rather suspect deformation measurements) must be discounted on grounds of membrane penetration problems. However, this paper is interesting in showing an attempt to break out of the narrow confinement of simple shear and triaxial

behaviour of sands under cyclic loading

Membrane penetration correction	Pulse shape	Frequency (Hz)	Number of cycles	Tests	References
?	~Sinusoidal	Various	~20	Liquefaction in static and cyclic loading, cyclic σ_a (compression–extension) apparatus effects	Castro[17]
Yes	Constant strain rate		~20	Cyclic σ_r; cyclic σ_a; various amplitudes, various mean values drained/undrained stress–strain response	Habib and Luong,[29] Luong[54,55]
—	Sinusoidal	0.5	~300	Various strain amplitudes shear moduli, damping (comparisons with simple shear)	Park and Silver[70]
—	Triangular	1	~300	Volume changes; shear moduli, damping; various σ_v, e_0 constant strain amplitudes	Silver and Seed[107,108]
?	Sinusoidal	0.2	~10000	Cyclic σ_a (compression–extension) liquefaction	Procter[77]
?	Constant axial strain rate 0.24%/min		10–30	Cyclic σ_a (compression–extension); drained/undrained; constant q/p' amplitude; stress–strain response	Tatsuoka[114] Tatsuoka and Ishihara[115,116]
?	?	0.5	~1000	Liquefaction reconsolidation and reliquefaction	Wolfe, Annaki and Lee[124]
—	Constant strain rate		~200	Constant strain amplitude stress–strain response apparatus effects	Wood and Budhu[129]

testing. The one series of tests which could not be performed in the triaxial apparatus is their series D—so-called cyclic plane stress tests—in which one principal stress (σ_1) is kept constant while the other two are varied by equal and opposite amounts.

For an initially isotropic stress state the test paths can most usefully be studied in the π-plane view of principal stress space (Figure 20.12). Whereas the octahedral stress ratios (q/p) on the Mohr–Coulomb failure criterion towards which the path of a cyclic triaxial test tends at its compression and extension extremities are quite different—so that for a constant amplitude of stress ratio the mobilised strength is greater in extension than in compression—the mobilised strengths at the two ends of

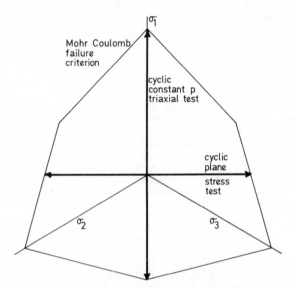

Figure 20.12 Exploration of π-plane view of stress space implied in cyclic constant p-triaxial compression–extension test and cyclic plane stress test, relative to Mohr–Coulomb failure criterion

the 'cyclic plane stress' cycle are identical. One might expect, then, that in octahedral terms these latter tests might appear stronger. Observed response is, however, also influenced by the initial structural anisotropy of the sand resulting from its initial deposition and formation in the 1 direction of the apparatus.

Sand, being a relatively permeable material, relying on no obviously time-dependent bonding, might be expected to show little sensitivity to rate of testing. Peacock and Seed[72] show that varying the frequency of stress cycles from $\frac{1}{6}$ to 4 Hz produces little effect on the number of cycles to cause liquefaction in the simple shear apparatus.

There appears to be some disagreement as to whether dry and saturated (drained) samples of sand will show the same stress–strain response. Finn and Vaid[26] state that, so far as stress changes in constant volume tests were concerned (and prediction of liquefaction resistance from these), the results from the two kinds of sample seemed identical, but do not actually present any comparisons. Moussa[60] shows the results of two parallel monotonic constant-volume simple shear tests on dry and saturated samples, giving broadly similar results (though with slightly different initial void ratios). Feda[22] compares the stress paths for constant-volume triaxial tests on dry and saturated samples. He concludes that the response is very close, but shows no points in the initial parts of the tests, where the major changes in stress state occur, so the validity of his assertion cannot be tested. Silver and Park[109] report results obtained in cyclic triaxial tests on Crystal silica No. 20 sand (sinusoidal axial loading waveform with frequency 0.5 Hz). Unfortunately, they do not perform constant-volume drained tests on their dry and saturated samples, so that their conclusion that the damping values for saturated specimens were higher than for dry specimens should be treated with caution. They do conclude that equivalent moduli are determined from the two types of test provided they are compared at the same effective stress.

Typical results of a cyclic triaxial compression test on loose sand reported by Castro[17] are shown in Figure 20.13. Residual pore pressures are left after the first few cycles, causing the effective stress path to migrate towards the origin, until liquefaction failure occurs in the sixth cycle. The strains in the first five cycles are very small.

An interesting attempt to link description of the behaviour of sand in cyclic loading with that in monotonic loading is made by Ishihara, Tatsuoka, and Yasuda[39] and by Ishihara, Lysmer, Yasuda, and Hirao[40] working largely with the results of a good series of tests on Fuji River sand performed by Tatsuoka[114] and reported by Tatsuoka and Ishihara.[115,116]

The tests performed include both drained and undrained cyclic triaxial tests, with a concentration on tests on loose samples. The volumetric strains developed in the drained tests correspond to the pore pressure development in the undrained tests: typical results from compression tests on loose sand are shown in Figures 20.14 and 20.15. It is interesting to note that in each of the pairs of diagrams in each figure, the sand more or less reverts to the virgin loading behaviour as soon as a cycle of loading passes the previous maximum stress ratio—which operates quite closely as a yield condition. This is particularly striking in Figures 20.14(b) and 20.15(a), which show up the volumetric or pore pressure response of the sand.

It is not the intention here to describe the details of the model that is used by Ishihara, Tatsuoka, and Yasuda[39] to make their predictions. Two ingredients of the model may, however, perhaps be mentioned. Introducing a set of yield loci, which indicate the threshold between elastic and plastic response, it is found that the sand has independent recall for its histories of loading in compression and extension. Thus yielding in compression is governed by the maximum stress ratio that has previously been reached in compression: similarly in extension. It is not readily apparent how this sort of directional yield condition should best be extended to more general three- (or six-) dimensional states of stress.

A second concept that is introduced is that of an angle of phase transformation, which defines critical stress ratios in compression and extension at which 'the behaviour of sand as a solid is lost and transformed into that of a liquefied state.'

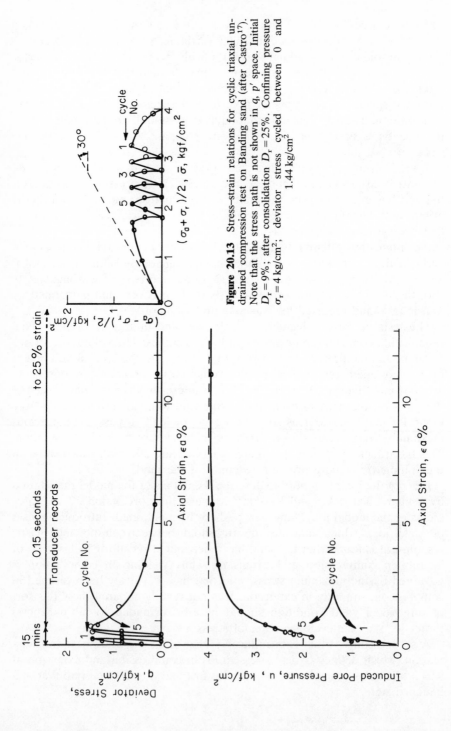

Figure 20.13 Stress–strain relations for cyclic triaxial undrained compression test on Banding sand (after Castro[17]). Note that the stress path is not shown in q, p' space. Initial $D_r = 9\%$; after consolidation $D_r = 25\%$. Confining pressure $\sigma_r = 4 \, \text{kgf/cm}^2$; deviator stress cycled between 0 and 1.44 kgf/cm^2

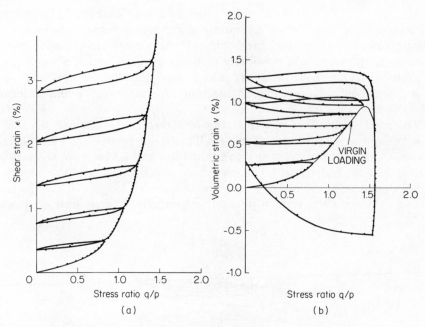

Figure 20.14 Cyclic triaxial drained compression test on loose Fuji River sand (after Tatsuoka[114]). Initial $e = 0.723$; confining pressure $\sigma_r = 0.5$ kg/cm^2

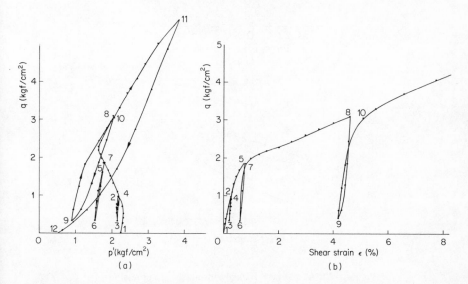

Figure 20.15 Cyclic triaxial undrained compression test on loose Fuji River sand (after Tatsuoka[114]). Initial $e = 0.749$; confining pressure $\sigma_r = 2.25$ kg/cm^2

The concept of an 'angle of phase transformation' is supported by results of a number of tests. Below this threshold removal of load produces finite, elastic strains. Above this threshold, unloading produces excessive pore pressures, and there is a possibility of liquefaction occurring if loading is repeated. This threshold, the stress ratio at which the effective stress paths in q, p' space show a sharp change of direction, separates two different patterns of behaviour.

The implication is that once this stress ratio threshold is attained, liquefaction failure is possible, though complete liquefaction is not always inevitable. Evidently the loose samples in Figures 20.14(b) and 20.15(a) are showing a strong dilative tendency at high stress ratios which would tend to resist the development of a collapse flow structure.

The behaviour observed in drained and undrained cyclic tests on loose

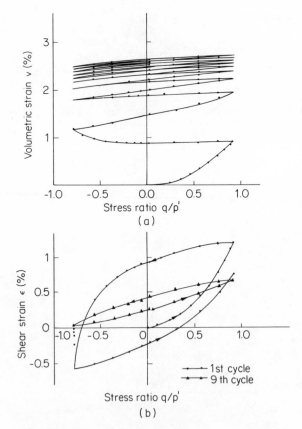

Figure 20.16 Cyclic triaxial drained compression-extension test on loose Fuji River sand (after Tatsuoka[114]). Initial $e = 0.750$; confining pressure $\sigma_r = 2$ kg/cm^2

sand, in which the stress amplitude is kept constant, is shown in Figures 20.16 and 20.17. These are compression-extension tests. The drained test shows volumetric compression occurring at a progressively decreasing rate as the sand densifies: one may suppose that the densification is associated with some continuing plastic hardening. The undrained test shows the expected progressive migration as the pore pressure builds up (volumetric compression being prevented) until eventually the phase transformation line is reached and a state of cyclic mobility—liquefaction with limited though large deformations—is reached.

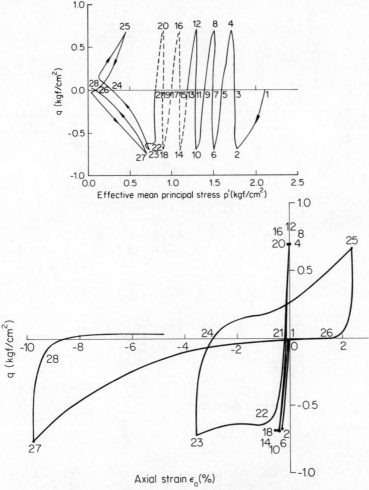

Figure 20.17 Cyclic triaxial undrained compression-extension test on loose Fuji River sand (after Tatsuoka[114]). Initial $e = 0.737$; confining pressure $\sigma_r = 2.1 \, \text{kg/cm}^2$

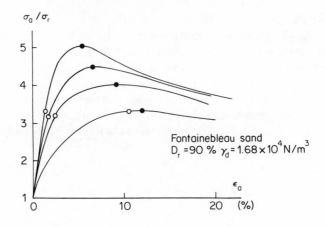

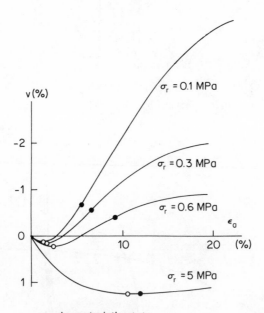

o characteristic state

● peak stress ratio

Figure 20.18 Observation of 'characteristic state' in triaxial drained compression tests on dense Fontainebleau sand (after Habib and Luong[29]). Initial $D_r = 90\%$. Characteristic state is attained when incremental volumetric strains change from compressive to expansive

The 'phase transformation' line of Tatsuoka and Ishihara seems to have some similarities with the 'concept of characteristic state' of Habib and Luong[29] and Luong.[54,55] This characteristic state is defined as the stress level corresponding to the passage from volumetric compression to volumetric dilation—the finite strain stress level at which the rate of volume change is zero (Figure 20.18). This characteristic state concept has certain similarities with the critical state concept of Roscoe, Schofield, and Wroth[81] and Schofield and Wroth[95]—and indeed a simple flow rule, describing the way in which energy input is distributed, whether of the Cam clay type[82] or of the Taylor[117] type, leads one to expect that the stress ratio at which the volumetric strain rate is zero will be the same whether this is a low or finite strain pre-peak transition from compression to dilation, or a large strain, ultimate constant-volume state of flow. Even an elaborate model such as that of Nova and Wood[65] retains a single stress ratio for these two conditions. Habib and Luong make a point of the fact that so far as experimental determination of the characteristic state is concerned it is '*directe donc facile*' whereas for the critical state it is '*par extrapolation donc délicate*'.

The selection of the characteristic state is illustrated in Figure 20.18 for conventional drained triaxial compression tests on Fontainebleau sand. It is suggested that the characteristic state provides also a good asymptote for the stress path in constant-volume tests whether drained or undrained (Figure 20.19). Beyond the characteristic state large deformations are possible

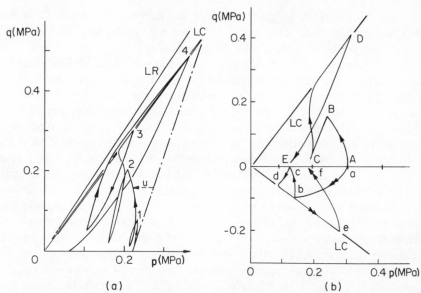

Figure 20.19 'Characteristic state' (LC) as asymptote to stress paths in (a) undrained compression and (b) constant volume compression and extension tests on medium dense Fontainebleau sand (after Luong[54]) (LR = failure line)

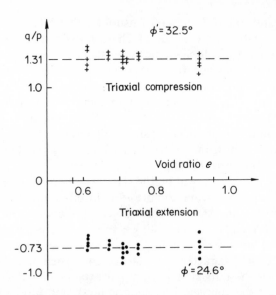

Figure 20.20 Independence from void ratio of 'characteristic states' observed in triaxial compression and extension tests on Fontainebleau sand (after Habib and Luong[29])

because the interlocking grain structure has become destroyed. It appears that the characteristic state stress ratio, like the critical state stress ratio, is largely independent of void ratio (Figure 20.20).*

The influence of the characteristic state on behaviour of Fontainebleau sand is shown for constant amplitude drained triaxial compression cycles in Figure 20.21: the stress ratios in compression and extension at which the volumetric strain rate is zero are not affected by the cycling. Figure 20.22 shows results obtained for small drained cycles of stress applied about various stress levels within a larger compression–extension cycle. The overall volumetric response, contraction or dilation, depends on the mean stress level: below the characteristic state the sand contracts; above, it expands.

This then leads to the summary of behaviour in Figure 20.23 for drained or undrained tests. In drained tests, cycling above the characteristic state leads to dilation of the sand and hence build up of strains: below the characteristic state compression occurs, with progressive stiffening. The characteristic state marks the boundary between incremental collapse and shakedown, to draw an analogy with plastic theory of structures. A similar threshold is provided for undrained loading: below the characteristic state

---

* But note that very different characteristic angles of friction seem to apply in compression and extension.

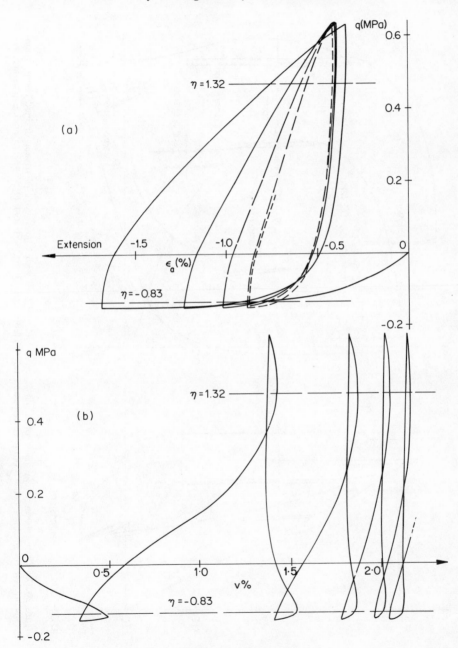

Figure 20.21 Cyclic triaxial drained compression-extension test on medium dense Fontaineb-leau sand (after Habib and Luong[29]). Initial $D_r = 62\%$; confining pressure $\sigma_r = 200$ kN/m^2

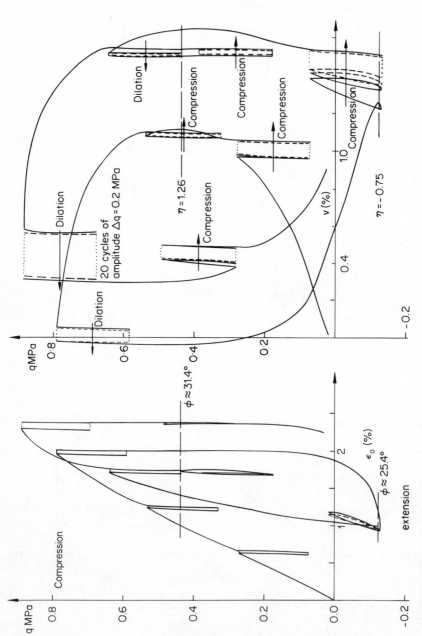

Figure 20.22 Compression or expansion observed in small cycles of load within large cycle of triaxial drained compression–extension test on medium dense Fontainebleau sand (after Habib and Luong[29]). Initial $D_r = 64\%$; confining pressure $\sigma_r = 200 \ kN/m^2$

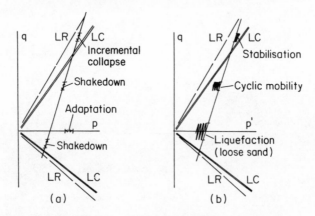

Figure 20.23 Summary of different phenomena observed in (a) drained cyclic loading and (b) undrained cyclic loading on sand (after Luong[54])

positive pore pressures and eventual liquefaction can develop; above, prevented dilation leads to a tendency to develop negative pore pressures which rapidly stabilise the deformation.

20.8 BEHAVIOUR OF CLAYS UNDER CYCLIC LOADING

A summary of some published work on behaviour of clays under cyclic loading is given in Table 20.2. This area has received rather more coverage than has the cyclic loading of sands, but the assessment of the amount of reliable data is not easy. Few of the investigations show any awareness of the testing problems discussed in earlier sections. Discussion of test results in terms of total stresses is in general not particularly helpful for understanding soil behaviour, but very few of the published test series have been conducted sufficiently slowly and with such pore pressure measuring devices that reliable pore pressures could be obtained. Lee and Focht[49] have produced a survey, essentially in terms of total stress response, of the strength of clay subjected to cyclic loading.

Bishop and Henkel[9] observe the effects of applying stress pulses to undrained samples of Weald clay. They observe that residual pore pressures were left after the removal of the load, and that the magnitude and sign of these pore pressures was dependent on the consolidation history of the clay and on the magnitude of the transient stress. A typical result is shown in Figure 20.24(a), where loading of an undisturbed sample of Weald clay leaves a residual negative pore pressure. If dissipation of this pore pressure

Table 20.2 Summary of published work

Clay	LL	PL	% < 2 μ	Preparation	Apparatus	Pore pressure measurement	Deformation measurement	Creep correction	Volumetric compression results	Pulse shape	Frequency Hz
Drammen clay	55	28	45–55	Undisturbed	Triaxial	Ends sides (some)	Boundaries	No	Yes	Sinusoidal trapezoidal	0.1
					Circular simple shear	(Constant volume)	Boundaries	No	Yes	Trapezoidal sinusoidal	(0.05) & 0.1
Drammen clay	?	?	?	?	Triaxial	Ends	Boundaries	No	No	Sinusoidal	0.1
Kaolinite	65	30	92	Reconstituted	Triaxial	Ends	Boundaries	(No?)	Yes	Rectangular	0.01 –4
Kaolinite (Hydrite UF)	82	42	100	Reconstituted	Triaxial	Ends	Boundaries	Yes	No	Sinusoidal	0.05 & 0.1
Kaolinite (Spestone)	72	40	73	Reconstituted	True triaxial	—	Boundaries	No	Yes	Controlled strain rates	
Kaolinite (Spestone)	72	40	73	Reconstituted	Triaxial	Ends	Boundaries	No	Yes	Controlled strain rates	
Keuper marl	32	18	18	Reconstituted	Triaxial	Ends sides	Boundaries	(No?)	Yes	Sinusoidal	10
Holocene marine clays	~66 ~34	~28 ~17	? ~70	Undisturbed	Triaxial	Ends	Boundaries	No	No	Sinusoidal	1/15 & 1
Illite	57	26	64	Reconstituted	Triaxial	Ends	Boundaries	No	Yes	Slow cycles	
Kaolinite	70	40	64	Reconstituted	Triaxial	Ends	Boundaries	Yes	No	Sinusoidal	0.1
Bentonite	105	51	69								
Champlain Sea clay	34–41	18–20	?	Undisturbed	Triaxial	Ends	Boundaries	No	No	Rectangular and curved/ rectangular	1/30 & 0.25
Edgar plastic kaolin	?	?	?	Reconstituted	Hollow cylinder	Ends	Boundaries	No	No	Triangular	1/30
Newfield clay	27–31	17–19	32	Undisturbed	Triaxial	Ends	Boundaries	No	Yes	Controlled strain rates	
San Francisco Bay mud	88	43	?	Undisturbed	Triaxial	None	Boundaries	No	No	Trapezoidal triangular	2
					Triaxial Circular simple shear	None	Boundaries	No	No	Rectangular triangular	1 & 2
Seattle clay	52	26	?	Reconstituted	Triaxial	Ends	Boundaries	No	No	Sinusoidal	1 2
Happisburgh till	25	12	15	Reconstituted	Triaxial	Sides	Boundaries	No	No	Sinusoidal triangular controlled strain rate	?
Halloysite	34	20	?	Reconstituted	Triaxial	Ends	Boundaries	No	Yes	Sinusoidal	0.2
Anchorage silty clay	35	21	?	Undisturbed	Triaxial Circular simple shear	None	Boundaries	No	No	Rectangular triangular	1
				Remoulded	Circular simple shear	None	Boundaries	No	Yes	Rectangular	1
Hamilton, Ontario lacustrine clay	34	20	8	Undisturbed	Triaxial	Ends	Boundaries	No	No	Rectangular	1/120

on behaviour of clays under cyclic loading

Number of cycles	Tests (see key)	References
≯8000		Andersen[1,2]
	ABCDEFGH(±3%)	Andersen, et al.[3,138]
	IJK(L?)MNOPQRSU	Brown, Andersen, and McElvaney[15]
≯2000		
1250	BCFGH(±3%)IJ	Kvalstad and Dahlberg[43]
	K(L)NTU	
~1000	ACEH(30%)IJKOTU	Brewer[12]
500	ACEFIKNOU	Motherwell and Wright[59]
~6	Drained tests; constant p; exploration of stress space	Wood[125]
~4	Undrained cycles compared with static tests on overconsolidated samples	Wroth and Loudon[133] Loudon[52]
10^6	ACEIJKNQRU	Brown, Lashine, and Hyde[13] Hyde and Brown[38], Lashine[47]
5000	BCEJKNOQRU	Fischer, Koutsoftas, and Lu[27] Koutsoftas[41]
~20	ACEFIJKLNQRSU	France and Sangrey[28]
~1000	ACEGIJKNQRSU	Meimon and Hicher[57]
~3500	ACEH(plane)IJKU	Mitchell and King[58]
~1000	ABCEFIJ(K)(L)U combined torsion and axial stress	Saada, Bianchini, and Shook[88]
~10	ADEFIJKLNQRSU	Sangrey[89]; Sangrey and France[93] Sangrey, Henkel, and Esrig[90]
~1000	ABCEFH(?)IPU	Seed[96]
200–500	BCDEH(?)KNOPU inclined samples	Seed and Chan[97] Thiers[120] Thiers and Seed[121,122]
3600	BCEIJKLNOQRU	Sherif and Wu[105]
7200		Sherif, Wu, and Bostrom[106]
~50	BCDEIJ(K)LQOU	Takahashi, Hight, and Vaughan[113]
100	BDEJKLMNQ	Taylor and Bacchus[118]
30–50	BCEH(?)HU inclined samples	
		Thiers[120] Thiers and Seed[121,122]
20	BCH(?)U	
~1000	ACEIJKLU	Wilson and Greenwood[123]

Key to tests
A triaxial compression only
B triaxial compression and extension
C stress controlled cycles
D strain controlled cycles
E isotropic consolidation
F anisotropic consolidation
G "storm" loading
H number of cycles to failure (failure criterion in parentheses: permanent shear strain; cyclic shear strain amplitude; formation of failure plane).
I permanent strains
J permanent pore pressures
K cyclic strains (moduli)
L cyclic pore pressures
M damping ratio
N post cycling monotonic strength
O effect of frequency
P effect of pulse shape
Q effect of overconsolidation ratio
R monotonic tests at different overconsolidation ratios
S effect of drainage within cyclic loading
T variation of σ_r and σ_a
U undrained cycles
V drained cycles

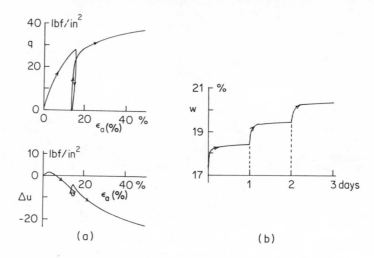

Figure 20.24 (a) Stress–strain response in undrained test on undisturbed Weald clay (after Bishop and Henkel[9]) (b) Increase in water content as consequence of drainage following daily pulses of deviator stress (amplitude ~0.4×undrained shear strength). Undisturbed Weald clay (after Bishop and Henkel[9])

is permitted, then the sample sucks in water, its water content increases (Figure 20.24(b)), and the clay softens.

A systematic study of the effect of applying slow undrained cycles of triaxial compression loading with constant amplitude q to samples of Newfield clay is reported by Sangrey[89] and Sangrey, Henkel, and Esrig.[90] It was found that the development of pore pressure caused the effective stress cycles to migrate either to failure (Figure 20.25), or to equilibrium without failure (Figure 20.26). When failure was reached, the effective stress ratio was the same as that for monotonic loading—the same value of q/p' or friction angle ϕ' was relevant. When samples reached equilibrium, subsequent tests to failure also ended at the same effective stress ratio. Failure was never attained in another series of tests reported by Sangrey[89] in which a constant peak effective stress ratio was applied.

The results of these tests can be summarised in the form shown in Figure 20.27 where equilibrium lines are shown for samples starting with different overconsolidation ratios but the same maximum consolidation stress. For each overconsolidation ratio there appears to exist a critical level of repeated loading below which failure will not occur. Equilibrium points then lie on a line joining the initial state to this critical level on the failure line.

The direction of migration of the effective stress path, as evinced by the slope of the equilibrium line, is dependent on the overconsolidation ratio. Positive pore pressures build up in the normally consolidated sample,

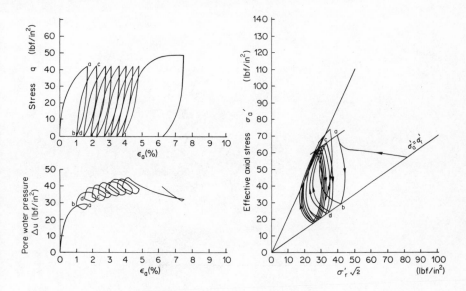

Figure 20.25   Cyclic triaxial undrained compression test on Newfield clay (after Sangrey[89]). Confining pressure $\sigma_r = 57$ lbf/in^2; cyclic deviator stress $q = 42.1$ lbf/in^2

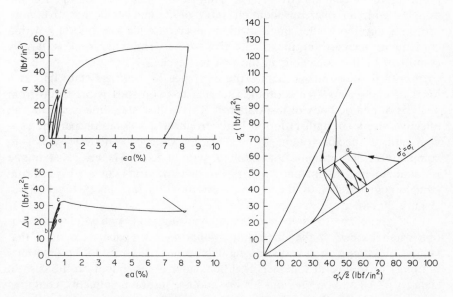

Figure 20.26   Cyclic triaxial undrained compression test on Newfield clay (after Sangrey[89]). Confining pressure $\sigma_r = 57$ lbf/in^2; cyclic deviator stress $q = 26.2$ lbf/in^2

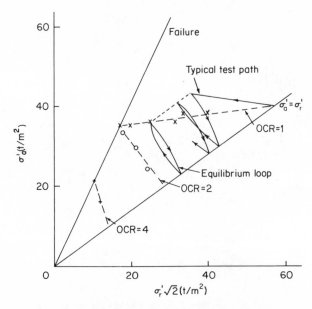

Figure 20.27 End points for cyclic triaxial undrained compression tests on Newfield clay reaching equilibrium (after Sangrey, Henkel and Esrig[90])

reducing the mean effective stress. Almost no migration occurs for the samples with an overconsolidation ratio of 2, and small negative pore pressures develop in the samples with an overconsolidation ratio of 4 with a consequent increase in mean effective stress. The migration is largely controlled by the behaviour in the first few cycles.

In critical state soil mechanics the existence of a critical state line as a locus of end points of monotonic shearing tests on soils is proposed from analysis of a large body of test data.[81,95] This critical state line is not just an effective stress strength criterion but introduces a volumetric parameter—void ratio e, specific volume V $(=1+e)$, or water content w $(=e/Gs)$—to complete the description of the ultimate state of a soil. The critical state line is found to be straight and to pass through the origin in q, p' space and straight and parallel to the normal consolidation line in V, $\log p'$ space (Figure 20.28).

This approach to soil mechanics, viewing soil behaviour in two-dimensional views of the three-dimensional q, p', V space, recognizes that volumetric compression and shearing behaviour of soils cannot be separated but must be looked at together. Soil behaviour is dependent on the current state (q, p', V) and on previous history and not just on an effective confining pressure before shearing.

The concept of a critical state line linking ultimate states, towards which

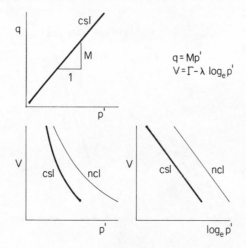

Figure 20.28 Critical state in q, p'; V, p'; V, $\log p'$ spaces (csl = critical state line; ncl = normal consolidation line). At critical state shearing can continue with

$$\frac{\partial q}{\partial \varepsilon} = \frac{\partial p'}{\partial \varepsilon} = \frac{\partial V}{\partial \varepsilon} = 0$$

the state of a soil sample will tend to move in a monotonic test, permits the division of soil response into broad classes—so-called 'wet' and 'dry' of critical (Figure 20.29). Samples 'wet' of critical (A and B), that is with states in V, p' space initially between the critical state line and the isotropic normal compression line, when tested in conventional compression tests are expected to contract in volume in drained tests, or develop positive pore pressures (with consequent reduction in mean effective stress) in undrained tests. Conversely, samples initially 'dry' of critical (C and D) are expected to expand in volume or develop negative pore pressures.

The precise boundary between 'wet' and 'dry' patterns of behaviour depends on the type of test that is used to detect the boundary. Typically, transitions of response can be expected to occur for overconsolidation ratios around 2. Evidently, the results reported by Sangrey[89] fit in broadly with this qualitative description of soil behaviour—with the transition from migration towards lower or higher mean effective stresses occurring for the samples with an overconsolidation ratio of 2.

Three-dimensional paths in q, p', V space may not be easy to visualise. Provided it is always borne in mind that there are three parameters which may be varying simultaneously, a two-dimensional normalised view of stress space may be more convenient. A convenient normalisation parameter is the equivalent pressure, p_e, which is defined as the pressure on the normal

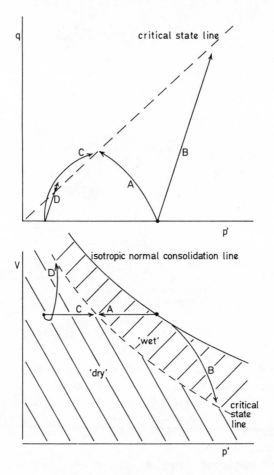

Figure 20.29 Division of soil response into broad
classes: initial states 'wet' or 'dry' of critical

consolidation line at the current water content or specific volume (Figure
20.30). (The concept of equivalent pressure is used by Terzaghi[119] and
Hvorslev[37] to allow comparison in a stress space of results of tests having
different water contents.) We may now plot results of triaxial tests, for
example, in a non-dimensional stress space q/p_e, p'/p_e.*

---

* Exploration of V, p' space is much harder for samples of sand than for samples of clay.
Hendron,[35] for example, shows that in one-dimensional compression tests on samples of
Pennsylvania sand pressures in excess of $20 \, MN/m^2$ are needed to eliminate the effect of
different initial densities—and at these pressures considerable particle crushing has occurred,
and the material has consequently changed from its initial condition. It is difficult to change the
structure or packing of a sand without performing some sort of shear test—under monotonic or
oscillatory loading.

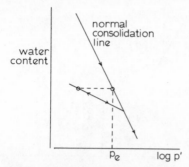

Figure 20.30 Definition of equivalent
pressure p_e

Wroth and Loudon[133] show migration of effective stress state in a cyclic
undrained triaxial compression test on kaolin (Figure 20.31(a)). They show
that the effect of loading and unloading, leaving a residual pore pressure, is
to make the clay behave as an overconsolidated soil—its isotropic effective
stress level at the beginning of the second cycle is lower than it was at the
beginning of the first cycle. This can be observed by comparison with Figure

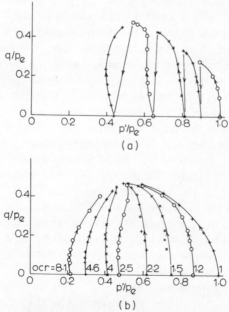

Figure 20.31 (a) Cyclic undrained triaxial compression test on normally consolidated kaolin; (b)
undrained tests on overconsolidated samples of
kaolin (ocr = overconsolidation ratio) (after
Wroth and Loudon[133])

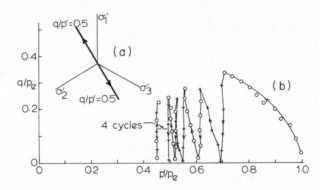

Figure 20.32 Cyclic drained true triaxial test on normally con-solidated kaolin with constant $p' = 150 \text{ kN/m}^2$ (after Wood[125])

20.31(b) where the results of a series of monotonic tests on samples with different overconsolidation ratios are shown (and in which the non-dimensionalisation with p_e is needed because the samples did not have the same water content at the start of undrained shearing). Wroth and Loudon continue by showing that not only are the stress paths similar for individual cycles in Figure 20.31(a) and for individual tests in Figure 20.31(b), but also the strains developed are similar. This notion that the behaviour of soil in cyclic loading is governed by its effective stress state (however that may have been arrived at) is central to models, such as that of van Eekelen and Potts[21] for the behaviour of Drammen clay under cyclic loading.

Development of pore pressure in an undrained test is an indication of prevented volume change. Normalisation with p_e enables us to follow the path of a drained test in the same two-dimensional stress plot. Figure 20.32 shows the path of a drained test at constant mean effective stress performed in the true triaxial apparatus by Wood[125] on an initially isotropically consolidated sample of kaolin. The path of this test is shown in the π-plane in Figure 20.32(a)—like the so-called plane stress test of Wolfe, Annaki, and Lee[124] discussed above, it is symmetrical relative to the Mohr–Coulomb failure criterion: with the same peak stress ratio q/p' at each end of the cycle it mobilises the same angle of friction at each end of the cycle. Volumetric migration has taken the place of pore pressure migration, but the general effect in Figures 20.32(b) and 20.31(a) is the same.

The true triaxial apparatus used by Wood[125] can very simply be program-med to perform tests at constant mean stress. Cyclic loading with constant mean total stress seems rarely to have been achieved in the conventional triaxial apparatus where problems exist of ensuring that cyclic changes of the correct amplitude and at the correct phase are applied. The disadvantage of tests in which the total mean normal stress is not kept constant is that a cyclic pore pressure change in phase with the applied mean normal stress change is generated (as for example in Figure 20.33, taken from Kvalstad and Dahlberg[43]) which can only tend to confuse the interpretation of observed pore pressure developments.

Migration occurs also in undrained creep tests such as those on San Francisco Bay mud reported by Arulanandan, Shen, and Young[6] (Figure 20.34). Their results show that in terms of effective stresses the failure condition is not altered by creep loading—which corresponds with Sangrey's[89] finding for cyclic loading. They also show that the development of pore pressure due to creep (prevented volumetric relaxation) is largely independent of shear stress q.

The need to take account of the effects of creep before attributing experimental observations to cyclic loading has been mentioned by Wood[128] and by Sangrey, Pollard, and Egan.[92] The separation of effects seems in

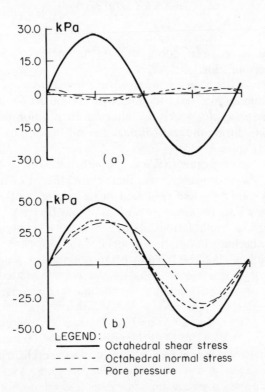

Figure 20.33 Pore pressure response in triaxial sample of Drammen clay to variation in p', (a) $p' \approx$ constant; (b) $\sigma_r =$ constant (after Kvalstad and Dahlberg[43])

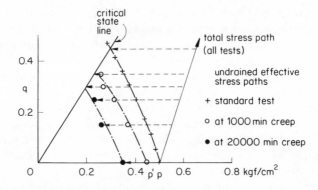

Figure 20.34 Migration through stress space as result of undrained creep (prevented volumetric relaxation) under constant deviator stress. Normally consolidated San Francisco Bay mud, confining pressure $\sigma_r = 0.5$ kg/cm² (after Arulanandan, Shen and Young[6])

general not to have been done, though the necessary creep data are available for certain soils.

That failure can occur in cyclic tests or creep tests at a lower deviator stress than in monotonic tests to failure is an indication that though the critical state line in q, p' space is not affected, its position in $V, \log p'$ space must be changing during the test—it must be shifting or flattening as a result of the loading (Figure 20.35).

Parallels can here be drawn (Wood[128]) with the work of Schmertmann[94] and Yudhbir.[137] Schmertmann shows that remoulding or disturbance flattens the normal consolidation line observed in oedometer tests (Figure 20.36). Yudhbir[137] shows data obtained by Simons[111] and Bjerrum[11] and notes that remoulding flattens the relationship between water content and strength (Figure 20.37). Perhaps in building models of soil response the flattening of the critical state line could be linked with the cumulative length of the strain path, or with some measure of cumulative energy dissipated.

The degree of any degradation is obviously dependent on the clay type—in particular its mineralogy and plasticity. The kaolin whose behaviour is shown in Figures 20.31 and 20.32 has a clay content of about 73% and a plasticity index of 32. The Happisburgh till, for which Takahashi, Hight, and Vaughan[113] report results of careful slow cyclic triaxial tests has a clay content of 15% and a plasticity index of about 13. Its response to undrained compression–extension cycling is shown in Figure 20.38 and appears to have some of the characteristics of the cyclic behaviour of sand discussed above (compare Figure 20.17). Takahashi, Hight, and Vaughan are very aware of the need to obtain accurate and reliable pore pressure

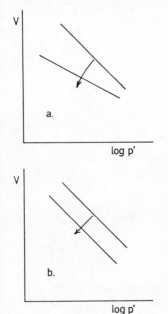

a.

log p'

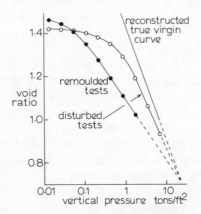

b.

log p'

Figure 20.35 (a) Tilting; or (b) translation of critical state line in V, log p' space implied by experimental observation that failure can be reached at lower deviator stress in cyclic or creep tests than in monotonic tests

measurements and suggest that 'effective stress paths based on a global measurement of pore pressure in fast monotonic or cyclic loading tests may travel outside a failure envelope defined by slow monotonic loading tests'.

There are two effects here: one is the response of the complex combination of sample, apparatus and transducer; the other is the possible rate dependency of the fundamental response of the soil. Undrained strengths

Figure 20.36 Effect of disturbance on slope of one-dimensional normal consolidation line for marine organic silty clay (after Schmertmann[94])

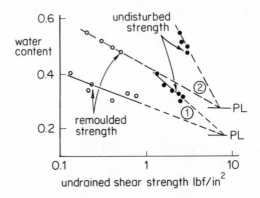

Figure 20.37 Effect of remoulding on strength of Drammen clay. Data from (1) Simons[111]; (2) Bjerrum[11]; (after Yudhbir[137])

measured in slow monotonic tests may not provide the most useful stress parameter for comparing amplitudes of applied fast cyclic loading. Motherwell and Wright[59] attempt to make some allowance for the usual stiffer and stronger behaviour in fast tests by extrapolating from the first quarter cycle of their cyclic tests in order to estimate a hypothetical strength.

We have so far deliberately concentrated on slow cyclic tests because the results of these tests are more likely to provide a reliable guide to the behaviour of clay than are fast tests. Fast tests may illustrate a general trend, and may have a direct relevance for some particular prototype situation. In cyclic stress controlled tests on Keuper marl, conducted with a frequency of 10 Hz, Brown, Lashine, and Hyde[13] show the expected 'wet' and 'dry' tendencies for pore pressure generation in samples with overconsolidation ratios of 2 and 20 (Figure 20.39).

Taylor and Bacchus[118] show migration of effective stress path in constant strain amplitude triaxial tests on halloysite at 0.2 Hz (Figure 20.40(a)). Applying a constant strain amplitude, as the soil weakens the stress amplitude that can be attained declines. The strain amplitude contours in this q, p' plot form a rather similar pattern to that found by Wroth and Loudon[133] from tests on kaolin at different overconsolidation ratios.

The paths of static tests to failure after cyclic loading reported by Taylor and Bacchus[118] resemble the expected paths of tests on overconsolidated samples (Figure 20.40(b)), but results of such monotonic tests are unfortunately not provided for comparison. The effective stress failure criterion appears to remain unaltered by this fast cyclic loading, while the undrained strength has been reduced—findings similar to those discussed above for slow cyclic tests.

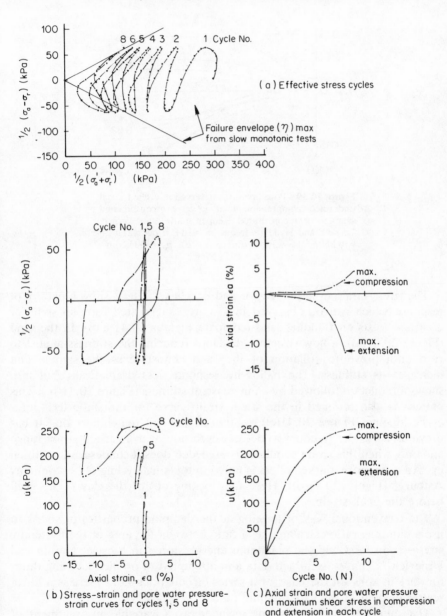

Figure 20.38 Cyclic triaxial undrained compression–extension test on normally consolidated Happisburgh till (after Takahashi, Hight, and Vaughan[113]). Confining pressure $\sigma_r = 300$ kN/m². Note that the stress path is not shown in q, p' space

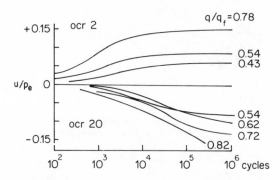

Figure 20.39   Pore pressures observed in cyclic triaxial undrained compression loading of overconsolidated samples of reconstituted Keuper marl (after Brown, Lashine and Hyde[13]): frequency = 10 Hz; ocr = 2: p_e = 650 kN/m^2, q_f = 360 kN/m^2; ocr = 20: p_e = 400 kN/m^2, q_f = 195 kN/m^2

The stress–strain cycles of Taylor and Bacchus's[118] tests show a pattern of response which can be observed in a more exaggerated form in some of Castro's[17] tests on Banding sand (compare Figures 20.41(a,b)). In the sand (Figure 20.41(a)), a flow structure develops reducing the stiffness almost to zero until prevented dilation of the sand almost locks the sample and increases its stiffness. The halloysite responds less dramatically but also shows a reduction followed by an increase of stiffness (Figure 20.41(b)). This pattern is also retained in the stress–strain curve for the static tests after cyclic loading (Figure 20.41(c)), as though the soil remembers that it has previously had the freedom to deform cyclically over a certain strain range and only when this strain condition is exceeded does it condescend to stiffen up. An even more marked effect is noted in torsional cycling of bentonite by Astbury[7] (Figure 20.41(d)). Here the 'locking up' of the clay occurs well before the peak strain.

The conventional way of looking at the damping properties of soil is to use a damping ratio calculated as a ratio between the area of the hysteretic stress–strain loop and the maximum energy stored in a cycle. Hardin and Drnevich[33,34] have assembled data and attempted to predict damping ratios for various soils on the basis of a series of equations—but there is a lot of scatter evident in the experimental data.

The number of cycles is not included as a variable in treatment of damping ratio by Hardin and Drnevich[34]. Independence of number of cycles may be plausible for very small strain amplitudes for which soil behaviour can be reasonably approximated as viscoelastic, with ideal elliptical hysteresis loops. At larger strains, where the loops tend to collapse in shape to the *S*-shapes shown in Figure 20.41 as the number of cycles is increased, it

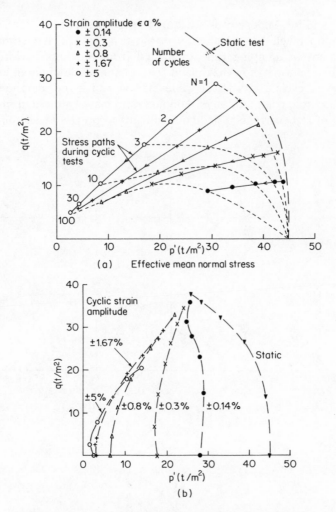

Figure 20.40 (a) Migration of effective stress state in cyclic strain controlled triaxial undrained compression–extension tests on normally consolidated halloysite; (b) stress paths for monotonic tests after 100 cycles of loading on normally consolidated halloysite (after Taylor and Bacchus[118]) Confining pressure $\sigma_r = 64$ lbf/in². Frequency of cyclic loading 0.2 Hz

is clear that the soil is not behaving in the assumed ideal fashion, and treatment simply in terms of shear modulus and damping ratio is masking fundamental aspects of the behaviour of the soil.

Hardin and Drnevich[33,34] and Seed and Idriss[98] show that for small strain amplitudes (less than about 0.1%) damping ratio tends in general to increase with strain amplitude. The deviation from this pattern found by Taylor and

Bacchus[118] and by Andersen[1] is attributed to the S-shaped hysteresis loops which develop much more markedly in the tests with higher strain amplitudes. Values of damping ratio calculated from the tests performed by Taylor and Bacchus[118] show maxima at a strain amplitude of about $\pm 1\%$ for any number of cycles, with lower values at lower or higher amplitudes.

Clearly accurate computation of damping ratio may require determination of the area of rather irregular geometrical figures—precise measurements of

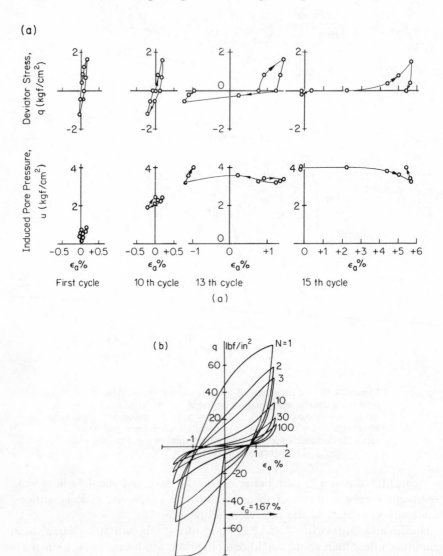

(a)

First cycle   10th cycle   13th cycle   15th cycle

(a)

(b)

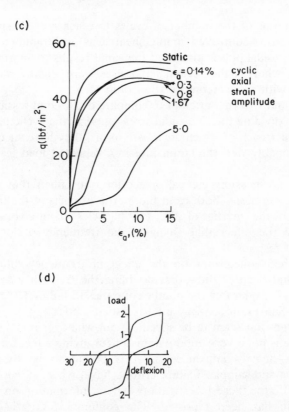

Figure 20.41 (a)   Stress–strain curves observed in cyclic triaxial undrained compression–extension tests on Banding sand (after Castro[17]). Initial $D_r = 60\%$; after consolidation with $\sigma_r = 4 \, \text{kg/cm}^2$, $D_r = 65\%$. (b) Stress–strain curves observed in cyclic strain controlled triaxial undrained compression–extension test on normally consolidated halloysite (after Taylor and Bacchus[118]). Frequency 0.2 Hz; confining pressure $\sigma_r = 64 \, \text{lbf/in}^2$. (c) Stress–strain curves for monotonic triaxial undrained compression tests on normally consolidated halloysite following cyclic loading (after Taylor and Bacchus[118]). Confining pressure $\sigma_r = 64 \, \text{lbf/in}^2$. (d) Torsional hysteresis loop for bentonite (after Astbury[7]). 1 unit of load = 1.2 kN/m²; 1 unit of deflexion = 2×10^{-4} strain

stress and strain around the cycle are required, and there must be no phase errors between the measuring systems for stress and strain.

Table 20.2 includes a note of the pulse shape used by the various investigators in their cyclic tests. Rectangular, trapezoidal, triangular and sinusoidal have all been used. The relative effect of different pulse shapes can be expected to depend on the proportion of time that the soil spends at the higher loads. Thus, for example, Seed and Chan[97] and Thiers[120] show that rectangular stress pulses have a much more serious effect than triangular pulses; and Murayama and Shibata[63] seem to show that rectangular stress pulses are also more damaging than sinusoidal pulses. Andersen[1] reports that, in terms of the number of cycles to reach a cyclic strain amplitude of ±3% in stress-controlled simple shear tests on Drammen clay, rectangular and trapezoidal pulses produced the same results for normally consolidated samples, whereas the rectangular pulses were slightly more damaging for samples with an overconsolidation ratio of 4.

Hyde and Brown[38] attempt to compare cyclic compression tests and creep tests. By dividing the sinusoidal stress pulse in the cyclic test into a series of steps under each of which creep will occur, they have succeeded in predicting reasonably well the strain rate in a repeated load test from creep test results.

There are possibly inertial or acceleration effects that need to be taken into consideration—both from the inertia of parts of the loading apparatus, and from the inertia of the soil particles themselves. These will also influence the effect that changing the frequency of loading has on the recorded response.

Published conclusions on the effect of frequency differ. Thiers[120] and Thiers and Seed[122] show that doubling the frequency from 1 Hz to 2 Hz increases considerably the number of cycles to failure for cyclic simple shear tests on San Francisco Bay mud (plasticity index ~45).* Their criterion of failure does not seem to be specified, but whatever it is, a number of cycles to failure is not a very good parameter for giving a feel for the behaviour of the soil—heavily influenced as it is bound to be by imperfections of apparatus and sample. Failure criteria have, where possible, been noted in Table 20.2 for those investigations where information on number of cycles to failure has been reported. The commonest criteria are cyclic strain amplitude criteria, but these cannot distinguish between a steady build-up of deformation and a sudden loss of strength (Figure 20.42).

Sherif and Wu[105] show that in general increasing the frequency from 1 Hz to 2 Hz has little effect on the permanent and cyclic strains and pore pressures, for a given cyclic stress level, for cyclic triaxial tests on Seattle clay (plasticity index ~26).

Andersen[1] reports cyclic stress-controlled simple shear tests on Drammen clay (plasticity index ~26) and also concludes that changing the frequency of loading from 0.1 Hz to 0.05 Hz has no significant effect on the number of cycles required to reach a cyclic strain amplitude of ±3% at any overconsolidation ratio. The data are, however, rather limited and scattered.

Brewer[12] shows that for the frequency range 0.01–4 Hz, the lower the frequency, in general the higher the strain, for cyclic triaxial tests on kaolinite (plasticity index ~35). However, Brown, Lashine, and Hyde[13] state

---

* Thiers[120] attempts to correlate results of cyclic simple shear and cyclic triaxial tests. He concentrates, in his triaxial tests, on tests on inclined samples. The fundamental objections to the use of inclined samples have been discussed above.

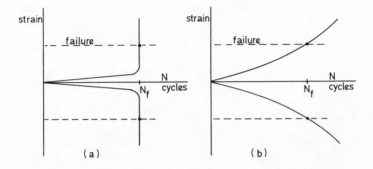

Figure 20.42 The limitation of number of cycles to failure as a material parameter. (a) and (b) show putative plots of strain against number of cycles for cyclic stress controlled tests. Each test reaches the cyclic strain failure criterion at the same number of cycles, but the characters of the response are quite different

that varying the frequency in the range 0.01–10 Hz has no significant effect, for cyclic triaxial tests on Keuper marl (plasticity index ~14).

It seems likely that the effect of frequency is greater for more plastic clays, and intuitively it seems reasonable that slower cycles give the clay, as a viscous material, more time to follow the applied load and are likely to produce greater strains and greater pore pressures.

The development of permanent axial strain in relatively slow cyclic triaxial compression tests on a lacustrine clay from Hamilton, Ontario, is reported by Wilson and Greenwood[123] and shown in Figure 20.43. The amplitudes are expressed in terms of the peak deviator stress q as a proportion of the value of q to produce failure in monotonic loading. As might be expected, the larger the amplitude, the larger the resulting deformation, and, indeed, for the two higher amplitudes the test has to be halted when the continuing deformation of the sample can no longer be accommodated and failure is deemed to have occurred.

By contrast, cyclic tests in which the amplitude of strain is controlled can be taken to as many cycles as is desired. The soil may develop no strength, but control need never be lost. Typical results, obtained with halloysite by Taylor and Bacchus[118] are shown in Figure 20.44—their results being reinterpreted in the form of variation of stress amplitude with number of cycles. Taylor and Bacchus applied strain cycles passing from compression to extension and, perhaps surprisingly, obtain rather symmetric stress–strain loops—with the same stress amplitudes being reached in both compression and extension. They show stress–strain loops only for amplitudes up to ±1.67% (axial strain) compared with a strain to failure in monotonic loading of about 10%. The results of true triaxial tests of Pearce[73] show that at low strains kaolin behaves isotropically (in the sense of Lensky[50]) and is unaware

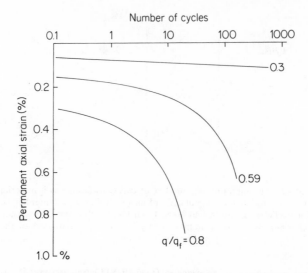

Figure 20.43 Development of permanent axial strain in cyclic triaxial undrained compression tests on lacustrine clay from Hamilton, Ontario (after Wilson and Greenwood[123]). Frequency 1/120 Hz

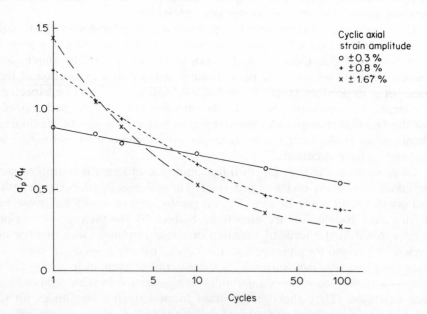

Figure 20.44 Decay of stress amplitude in cyclic strain controlled triaxial undrained compression–extension tests on normally consolidated halloysite (after Taylor and Bacchus[118]). Confining pressure $\sigma_r = 64$ lbf/in². Frequency 0.2 Hz

of the impending anisotropy of the Mohr–Coulomb failure criterion. Perhaps the halloysite of Taylor and Bacchus[118] is also initially in this happy state.

The results in Figure 20.44 are again non-dimensionalised with reference to a static strength. Evidently the increased rate of straining in the cyclic test (0.2 Hz in comparison with 1% per minute axial strain rate—in other words 80 times as fast for the amplitude ±1.67%) increases the initial stiffness and sends the deviator stress in the first cycle well above the static strength. Larger strain amplitudes have a more dramatic effect on stiffness as the tests proceed—the order of the curves reverses at larger numbers of cycles.

Strain-controlled and stress-controlled tests explore different areas of the overall cyclic stress–strain response of a soil—a good soil model should be able to link the two together and predict either set from the other. The problem in predicting strain-controlled tests from stress-controlled tests is one of being able to cope with a varying amplitude of stress. The problem is similar to that involved in trying to predict the response in a storm of an element of soil under an offshore structure: the storm will cause to be applied to the structure, and hence to the soil, a whole spectrum of different amplitudes (and frequencies) of loading. Various simple techniques have been devised for achieving this. We may, in particular, mention the method described by Andersen[2] and based on test results obtained from cyclic tests on Drammen clay. Predicted and measured paths for strain-controlled simple shear tests on Drammen clay are shown in a plot of stress amplitude against number of cycles in Figure 20.45.

These tests on Drammen clay form part of a large programme of triaxial and simple shear tests reported in detail by Andersen[1] and summarised from various aspects by Andersen,[2] Andersen et al.[3,138] and Brown, Andersen, and McElvaney.[15] This vast range of tests investigated many areas of the behaviour of Drammen clay under cyclic loading. It suffers somewhat from having been conducted, inevitably, as a joint venture of four testing laboratories—and there remain some uncertainties and differences between the laboratories. It is also unfortunate that no samples were tested with overconsolidation ratios between 1 (normally consolidated) and 4—precisely the area in which, as we have seen, we may expect major transitions of behaviour to occur. That said, it remains a major source of experimental data which cannot all be briefly summarised here.

Some success seems to be found by Andersen[1,2] in using a non-dimensional cyclic effective stress level to compare different sets of stress-controlled cyclic tests. The stress level (SL) is defined very simply as (Figure 20.46):

$$SL = \frac{\tan \phi'_{mob}}{\tan \phi'}$$

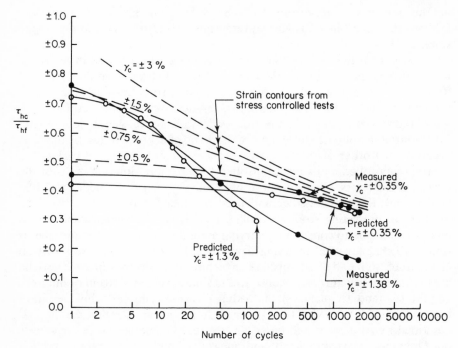

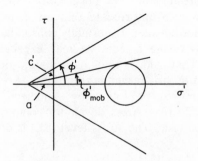

Figure 20.45 Measured and predicted paths of strain controlled cyclic simple shear tests on Drammen clay with overconsolidation ratio = 4 (after Andersen *et al.*[138]). τ_{hf} and τ_{hc} are shear strength and cyclic shear stress in simple shear tests

Figure 20.46 Definition of effective stress level: SL = $(\tan \phi'_{mob})/(\tan \phi')$

where ϕ'_{mob} is the mobilised angle of friction determined from the effective stress state, and ϕ' is the angle of friction of the material determined from monotonic tests. Where monotonic tests show the existence of an attraction, a (or cohesion intercept $c' = a \tan \phi'$) the mobilised angle of friction is calculated with reference to the same attraction.* Evidently, in order to calculate the mobilised friction in the simple shear test, an assumption is required about the stress state in the simple shear sample—a value of the lateral stress in the plane of shearing is required.

With this definition of stress level it is clear that the development of pore pressure will lead to a change in stress level even if the deviator or shear stress amplitude is kept constant. Variations of stress level with number of cycles are shown by Andersen[1] for (a) cyclic compression triaxial tests; (b) cyclic compression-extension triaxial tests; and (c) cyclic two-way simple shear tests on Drammen clay at three different overconsolidation ratios. Typical results, for the compression–extension triaxial tests, at overconsolidation ratios of 1 and 4 are shown in Figure 20.47.

Contours of cyclic strain are shown on these figures and to a first approximation it can be suggested that these contours are nearly horizontal—in other words, that the cyclic strain: cyclic stress level relationship is independent of the number of cycles. Data from these figures are replotted in Figure 20.48 and the spread of all results shown in Figure 20.49 where the range of data from the three sets is indicated.†

All sorts of assumptions have been made on the route to the generation of Figure 20.49, but treatment in terms of effective stresses does appear to assist in unifying the results of different types of test on samples of Drammen clay with different overconsolidation ratios. Of course, in order to be able to make use of this unified relationship knowledge of effective stress changes occurring in any particular cyclic test is required.

An aspect of cyclic loading behaviour of clays which is studied by Andersen[1] and also by Andersen *et al.*,[3] Brown, Andersen, and McElvaney,[15] France and Sangrey[28] and Meimon and Hicher[57] (*inter alia*) is the effect of allowing drainage to occur between periods of cyclic loading.

---

* Curiously, in determining values of cyclic stress level for compression–extension triaxial tests, the effective stress level appears to be calculated only for the extension phase of the cycle. Logically, if a higher angle of friction is mobilised in extension than in compression for each cycle, then the tests should be interpreted as having a permanent extension contribution to the applied effective stress level—and the greater damage associated with compression–extension cycling than with compression only cycling can be readily understood.

† It is tacitly assumed by Andersen that shear strain—presumably defined as $\varepsilon_a - \varepsilon_r$ for the triaxial apparatus, and defined as the ratio of horizontal shear displacement to height for the simple shear apparatus—provides the correct parameter for comparing results of triaxial and simple shear tests. As mentioned above this is not necessarily the case, since the strain conditions in the two types of apparatus are so completely different. An octahedral shear strain (such as our deviator strain ε) might be a more rational, but ultimately no more defensible, strain parameter to choose.

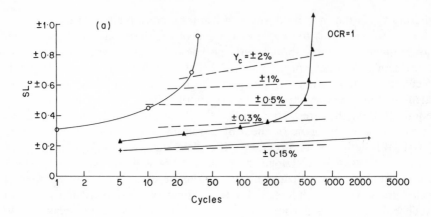

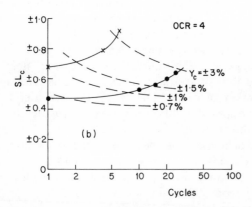

Figure 20.47 Development of effective cyclic stress level (SL$_c$) with number of cycles in cyclic triaxial undrained compression–extension tests on Drammen clay (after Andersen[1]). Contours of cyclic shear strain shown ($\gamma = 1.5\varepsilon$ for triaxial conditions)

Obviously as soon as there is a question of changes in water content occurring during a test it is very important to ensure that the changing void ratio is accounted for in presentation of the data: the appropriate reference undrained strength or equivalent pressure, p_e, will also be changing during the test. Data concerning the volumetric compression and unloading of the soil are needed (as they are also where results of tests on samples at different initial water contents have to be compared).

The general pattern of response is clear: where residual positive pore pressures are left dissipation of these must lead to decrease in water content and stiffening of the soil, with the converse result where residual negative

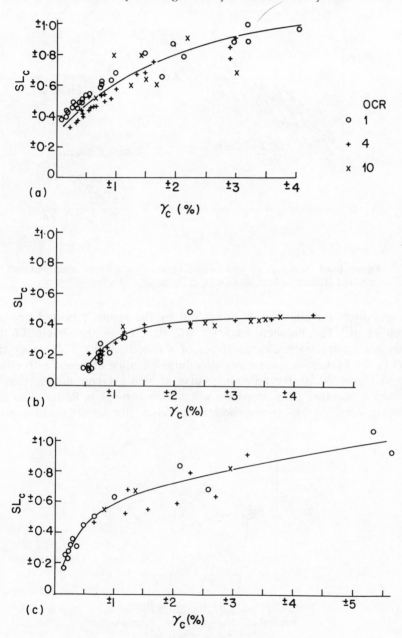

Figure 20.48 Relationships between cyclic shear strain and effective cyclic stress level for cyclic undrained tests on Drammen clay (after Andersen[1]): (a) simple shear; (b) triaxial compression (one-way); (c) triaxial compression–extension (two-way)

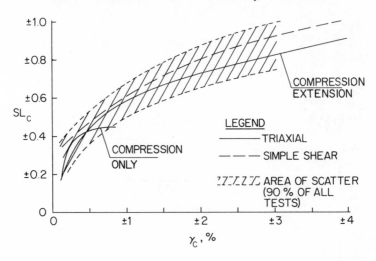

Figure 20.49 Summary of relationships between cyclic shear strain and effective cyclic stress level for all tests on Drammen clay (after Andersen[2])

pore pressures remain. This is borne out by the results reported by Andersen *et al.*[3] The detailed response will depend on the details of the volumetric compression characteristics of the soil. Brown[14] discusses the effect of the hysteretic volumetric unloading–reloading response with reference to Figure 20.50: the unloading curve at high overconsolidation ratios (for which negative pore pressures will be dissipating) is flatter than the reloading curve at low overconsolidation ratios (for which positive pore

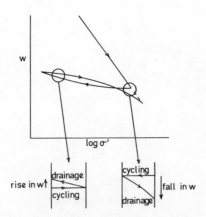

Figure 20.50 Paths during undrained cyclic loading and subsequent drainage in *w*, log *p'* space (after Brown[14])

pressures will be dissipating). The change in water content (and in anticipated soil response) for a given residual pore pressure will be larger in the latter case—a result confirmed by Brown, Andersen and McElvaney.[15]

20.9   CONCLUSION

We have discussed separately the cyclic loading behaviour of sands and clays. In practice this categorisation is too simplistic—the term 'clay' covers materials of great mineralogical differences and with very different plasticities. It is unreasonable to expect that any one model will be able in detail to match results from all different soil types.

Analysis of cyclic test results should produce some indication of the mechanisms of behaviour that are relevant whether at the particulate level or at the macroscopic level. Thus, for example, Youd[134,136] has discussed factors contributing to compaction of sands in terms of energy associated with displacement of sand particles. On the other hand, Sangrey *et al.*[91] describe a qualitative model for the behaviour under cyclic loading of sands, silts and clays: the treatment is firmly against the background established in critical state soil mechanics, where study of the paths of tests in the complete state space of void ratio and stresses is considered vital.

An interesting correlation (Figure 20.51) is produced by Sangrey *et al.*[91] between the critical level of repeated loading below which failure (however defined) will not be reached, and the bulk compliance ($\kappa/V_0 = 1/$bulk modulus) of the soil—for undrained tests on isotropically normally consolidated soils, of the contractive or 'wet' type. The greater susceptibility of sands to cyclic loading is linked with their very much stiffer unloading–reloading response. A very much greater pore pressure is required to transport the sand across V, p' space to the ultimate state line—which Sangrey *et al.* term

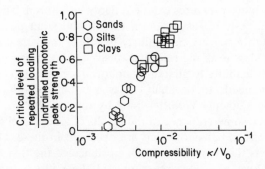

Figure 20.51 Relationship between critical level of repeated loading and bulk compliance κ/V (κ = slope of elastic swelling line) (after Sangrey *et al.*[91])

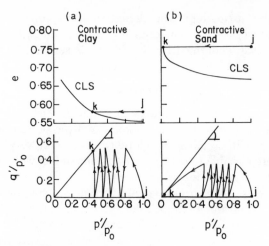

Figure 20.52 Comparison of state paths of contractive specimens of (a) clay (after Sangrey et al.[90]) and (b) sand (after Castro[17]) when subjected to cyclic undrained loading. CLS = cyclic limit state (after Sangrey et al.[91])

the 'cyclic limit state line' (Figure 20.52). A similar correlation could no doubt be drawn between critical level of repeated loading and plasticity index.

The two approaches are not mutually exclusive—a particulate cause may be sought for the continuing build-up of pore pressure. As the pore pressure builds up the chance of the interparticle stress at any contact reducing to zero also increases, and the likelihood of the sand being able to develop large strains with minimal stiffness increases.

Mechanisms of behaviour provide an observational guide to soil model building. We have seen that effective stresses and pore pressures provide the key controlling the behaviour of soils under cyclic loading. Test procedures need to be such that pore pressures can be reliably monitored. This leads us to prefer slow cyclic tests to fast ones on the grounds that pore pressure measurements in the former are more likely to reflect the actual intentions of the soil. Even in slow tests the relevance of particular apparatus may be limited through their inability to guarantee uniformity of conditions within a sample—which then needs to be analysed as a boundary value problem rather than a single element. Wood[127] ends by noting 'We are therefore more inclined to place some credence in the results of slower cyclic tests and to rely on the patterns of behaviour that are observed in these. One of the principal observations that is made is the importance of the first few cycles on the changes in both the pore pressure and the stiffness, and a final unanswered question concerning the more rapid tests is: to what extent are effects beyond the first few cycles the results of doubtful experimental procedure and testing inadequacies?' These remarks need to be emphasised. The reason for beginning here by concentrating on the philosophy behind and the details of soil testing is that study of test results cannot be divorced from consideration of the way in which they were obtained. Poor testing techniques cannot produce results which it is worthwhile spending time trying to match with complicated models.

NOTATION

a	attraction $= c' \cot \phi'$
c'	effective cohesion
$d_{10} d_{50}$	grain diameters for which 10% and 50% of the soil are finer
D_r	relative density $= (e_{max} - e)/(e_{max} - e_{min})$
e	void ratio
$e_{max} \; e_{min}$	maximum and minimum void ratios
G_s	specific gravity of soil particles
LL PL	liquid limit and plastic limit
p'	mean effective stress $= (\sigma'_1 + \sigma'_2 + \sigma'_3)/3$
p_e	equivalent pressure $=$ value of p' on normal consolidation line at current water content or specific volume.
q	deviator stress $= \{[(\sigma_2 - \sigma_3)^2 + (\sigma_3 - \sigma_1)^2 + (\sigma_1 - \sigma_2)^2]/2\}^{1/2}$
SL	effective stress level $= \tan \phi'_{mob}/\tan \phi'$
u	pore water pressure
v	volumetric strain $= \varepsilon_1 + \varepsilon_2 + \varepsilon_3$
V	specific volume $= 1 + e$
w	water content $= e/G_s$
xyz	Cartesian coordinates
α	shear strain—ratio of horizontal displacement to sample height—in simple shear apparatus
γ	shear strain—diameter of Mohr circle of strain
Γ	critical state parameter defining position of critical state line in $V : \ln p'$ space
ε	deviator strain $= \frac{1}{3}\{2[(\varepsilon_2 - \varepsilon_3)^2 + (\varepsilon_3 - \varepsilon_1)^2 + (\varepsilon_1 - \varepsilon_2)^2]\}^{1/2}$
$\varepsilon_a \; \varepsilon_r$	axial and radial strain in triaxial test
$\varepsilon_{xx} \varepsilon_{yy} \varepsilon_{zz}$ $\varepsilon_{yz} \varepsilon_{zx} \varepsilon_{xy}$	components of strain tensor
$\varepsilon_1 \; \varepsilon_2 \; \varepsilon_3$	principal strains
η	stress ratio $= q/p'$
$\kappa \; \lambda$	slope of unloading and virgin loading lines in $V, \ln p'$ space
M	critical state parameter: slope of critical state line in q, p' space.
$\sigma_a \; \sigma_r$	axial and radial stresses in triaxial test
$\sigma_{xx}\sigma_{yy}\sigma_{zz}$ $\sigma_{yz}\sigma_{zx}\sigma_{xy}$	components of stress tensor
$\sigma_1\sigma_2\sigma_3$ $\sigma_i (i = 1, 2, 3)$	principal stresses
ϕ'	angle of friction
ϕ'_{mob}	mobilised angle of friction

Compressive stresses and strains are taken as positive. A prime denotes an effective stress.

REFERENCES

1. Andersen, K. H. Research project, *Repeated Loading on Clay*. Summary and interpretation of test results. NGI internal report 74037–9, 1975.
2. Andersen, K. H. Behaviour of clay subjected to undrained cyclic loading. *Proc. Int. Conf. on Behaviour of Off-Shore Structures*, Trondheim, **1**, 392–403, 1976.
3. Andersen, K. H., Brown, S. F., Foss, I., Pool, J. H., and Rosenbrand, W. F. Effect of cyclic loading on clay behaviour. *Design and Construction of Off-Shore Structures*, Institution of Civil Engineers, London, 75–9, 1976.
4. Arthur, J. R. F., Dunstan, T., Al-Ani, Q. A. J. L., and Assadi, A. Plastic deformation and failure in granular media. *Géotechnique*, **27**(1), 53–74, 1977.
5. Arthur, J. R. F., Chua, K. S., Dunstan, T., and Rodriguez del C., J. I. *Imitating Changes in Principal Stress Directions*. Preprint 3603. US/British Developments in Off-Shore Platforms, A.S.C.E. Convention, Boston, 1979.
6. Arulanandan, K., Shen, C. K., and Young, R. B. Undrained creep behaviour of a coastal organic silty clay. *Géotechnique*, **21**(4), 359–75, 1971.
7. Astbury, N. F. Science in the ceramic industry. *Proc. Roy. Soc. of London*, Series A, **258**, 27–46, 1960.
8. Balla, A. Stress conditions in the triaxial compression test. *Proc. 4th Int. Conf. Soil Mech.*, London, **1**, 140–3, 1957.
9. Bishop, A. W., and Henkel, D. J. Pore pressure changes during shear in two undisturbed clays. *Proc. 3rd Int. Conf. Soil Mech.*, Zurich, **1**, 94–9, 1953.
10. Bjerrum, L. and Landva, A. Direct simple-shear tests on a Norwegian quick clay. *Géotechnique*, **16**(1), 1–20, 1966.
11. Bjerrum, L. Engineering geology of Norwegian normally-consolidated marine clays as related to settlements of buildings. 7th Rankine Lecture, *Géotechnique*, **17**(2), 81–118, 1967.
12. Brewer, J. H. *The Response of Cyclic Stress in a Normally Consolidated Saturated Clay*, Ph.D. thesis, North Carolina State University, Raleigh, 1972.
13. Brown, S. F., Lashine, A. K. F., and Hyde, A. F. L. Repeated load triaxial testing of a silty clay. *Géotechnique*, **25**(1), 95–114, 1975.
14. Brown, S. F. Discussion: Effect of cyclic loading on clay behaviour. *Design and Construction of Off-Shore Structures*, Institution of Civil Engineers, London, 81, 1976.
15. Brown, S. F., Andersen, K. H., and McElvaney, J. The effect of drainage on cyclic loading of clay. *Proc. 9th Int. Conf. on Soil Mechanics and Foundation Engineering*, Tokyo, **2**, 195–200, 1977.
16. Casagrande, A. *Liquefaction and Cyclic Deformation of Sands: a Critical Review*. Harvard Soil Mechanics Series No. 88, January 1976.
17. Castro, G. *Liquefaction of Sands*. Ph.D. thesis, Harvard University. Harvard Soil Mechanics Series No. 81, 1969.
18. De Alba, P., Seed, H. B., and Chan, C. K. Sand liquefaction in large-scale simple shear tests. *Proc. A.S.C.E.*, **102** (GT9), 909–27, 1976.
19. Drnevich, V. P., and Richart, F. E. Dynamic prestraining of dry sand. *Proc. A.S.C.E.*, **96** (SM2), 453–69, 1970.
20. Duncan, J. M., and Dunlop, P. Behaviour of soils in simple shear tests. *Proc. 7th Int. Conf. Soil Mechs.*, Mexico, **1**, 101–9, 1969.
21. Eekelen, H. A. M. van and Potts, D. M. The behaviour of Drammen clay under cyclic loading. *Géotechnique*, **28**(2), 173–96, 1978.
22. Feda, J. Constant volume shear tests of saturated sand. *Archiwum Hydrotechniki*, **18**(3), 349–67, 1971.

23. Finn, W. D. L., Bransby, P. L., and Pickering, D. J. Effect of strain history on liquefaction of sand. *Proc. A.S.C.E.*, **96** (SM6), 1917–34, 1970.
24. Finn, W. D. L., Emery, J. J., and Gupta, Y. P. A shaking table study of the liquefaction of saturated sands during earthquakes. *Proc. 3rd European Symp. on Earthquake Engineering*, Sofia, Bulgaria, 253–62, 1970.
25. Finn, W. D. L., Pickering, D. J., and Bransby, P. L. Sand liquefaction in triaxial and simple shear tests. *Proc. A.S.C.E.*, **97** (SM4), 639–59, 1971.
26. Finn, W. D. L., and Vaid, Y. P. Liquefaction potential from drained constant volume cyclic simple shear tests. *Proc. 6th World Conf. on Earthquake Engineering*, New Delhi, **6**, 2157–62, 1977.
27. Fischer, J. A., Koutsoftas, D. C., and Lu, T. D. The behaviour of marine soils under cyclic loading. *Proc. Int. Conf. on Behaviour of Off-Shore Structures*, Trondheim, **2**, 407–17, 1976.
28. France, J. W., and Sangrey, D. A. Effects of drainage in repeated loading of clays. *Proc. A.S.C.E.*, **103** (GT7), 769–85, 1977.
29. Habib, P., and Luong, M. P., Sols pulvérulents sous chargement cyclique. *Matériaux et Structures Sous Chargement Cyclique*, Association Amicale des Ingénieurs Anciens Elèves de l'Ecole Nationale des Ponts et Chaussées (Palaiseau, 28–29 Sept.) 49–79, 1978.
30. Hambly, E. C. A new true triaxial apparatus. *Géotechnique*, **19**(2), 307–9, 1969.
31. Hambly, E. C., *Plane Strain Behaviour of Soft Clay*. Ph.D. thesis, Cambridge University, 1969.
32. Hambly, E. C. Plane strain behaviour of remoulded normally consolidated kaolin. *Géotechnique*, **22**(2), 301–17, 1972.
33. Hardin, B. O., and Drnevich, V. P., Shear modulus and damping in soils: measurement and parameter effects. *Proc. A.S.C.E.*, **98** (SM6), 603–24, 1972.
34. Hardin, B. O., and Drnevich, V. P. Shear modulus and damping in soils: design equations and curves. *Proc. A.S.C.E.*, **98** (SM7), 667–92, 1972.
35. Hendron, A. J. *The Behaviour of Sand in One-dimensional Compression*. Ph. D. thesis, University of Illinois, Urbana, 1963.
36. Henkel, D. J. The relevance of laboratory-measured parameters in field studies. *Stress–strain behaviour of soils. Proc. Roscoe Memorial Symposium*, G. T. Foulis & Co., 669–75, 1971.
37. Hvorslev, M. J., *Über die Festigkeitseigenschaften gestörter bindiger Böden*. Ingeniørvidenskabelige Skrifter A No. 45, København. English translation No. 69–5. U.S. Waterways Experimental Station, Vicksburg, Miss., 1969. (*Physical properties of remoulded cohesive soils*) 1937.
38. Hyde, A. F. L., and Brown, S. F. The plastic deformation of a silty clay under creep and repeated loading. *Géotechnique*, **26**(1), 173–84, 1976.
39. Ishihara, K., Tatsuoka, F., and Yasuda, S. Undrained deformation and liquefaction of sand under cyclic stresses. *Soils and Foundations* **15**(1), 29–44, 1975.
40. Ishihara, K., Lysmer, J., Yasuda, S., and Hirao, H. Prediction of liquefaction in sand deposits during earthquakes. *Soils and Foundations* **16**(1), 1–16, 1976.
41. Koutsoftas, D. C. Effect of cyclic loads on undrained strength of two marine clays. *Proc. A.S.C.E.*, **104** (GT5), 609–20, 1978.
42. Kovacs, W. D. Effect of sample configuration in simple shear testing. *Proc. Symp. on Behaviour of Earth and Earth Structures Subjected to Earthquakes and Other Dynamic Loads*, Roorkee, India, 82–6, 1973.
43. Kvalstad, T. J., and Dahlberg, R. Cyclic behaviour of clay as measured in

laboratory. *Proc. Int. Symp. on Soils under Cyclic and Transient Loading,* Swansea (eds. G. N. Pande and O. C. Zienkiewicz), Balkema, Rotterdam, **1,** 157–67, 1980.

44. Ladd, R. S. Specimen preparation and liquefaction of sands. *Proc. A.S.C.E.,* **100** (GT10), 1180–4, 1974.

45. Ladd, R. S. Specimen preparation and cyclic stability of sands. *Proc. A.S.C.E.,* **103** (GT6), 535–47, 1977.

46. Lade, P. V., and Hernandez, S. B. Membrane penetration effects in undrained tests, *Proc. A.S.C.E.* **103** (GT2), 109–25, 1977.

47. Lashine, A. K. F. Deformation characteristics of a silty clay under repeated loading. *Proc. 8th Int. Conf. Soil Mech.,* Moscow **1**(1), 237–44, 1973.

48. Lee, K. L. Fundamental considerations for cyclic triaxial tests on saturated sand. *Proc. Int. Conf. on Behaviour of Off-Shore Structures,* Trondheim, **1,** 355–73, 1976.

49. Lee, K. L., and Focht, J. A. Strength of clay subjected to cyclic loading. *Marine Geotechnology* **1**(3), 165–85, 1976.

50. Lensky, V. S. Analysis of plastic behaviour of metals under complex loading. *Proc. 2nd Symp. on Naval Structural Mechanics,* Brown University. In Lee, E. H., and Symonds, P. S. (eds.), *Plasticity,* Pergamon Press, New York, 259–78, 1960.

51. Lewin, P. I. The influence of stress history on the plastic potential. *Proc. Symp. on Role of Plasticity in Soil Mechanics,* Cambridge, 96–105, 1973.

52. Loudon, P. A. *Some Deformation Characteristics of Kaolin.* Ph.D. thesis, Cambridge University, 1967.

53. Lucks, A. S., Christian, J. T., Brandow, G. E., and Höeg, K. Stress conditions in NGI simple shear test. *Proc. A.S.C.E.,* **98** (SM1), 155–60, 1972.

54. Luong, M. P. Les phénomènes cycliques dans les sables. *Journée de Rhéologie: Cycles dans les Sols—Rupture—instabilités.* Publication no. 2 de l'Ecole Nationale des Travaux Publics de l'Etat, Vaulx-en-Velin, France (16 pages + figures), 1979.

55. Luong, M. P. Stress–strain aspects of cohesionless soils under cyclic and transient loading. *Proc. Int. Symp. on Soils under Cyclic and Transient Loading,* Swansea (eds. G. N. Pande and O. C. Zienkiewicz), Balkema, Rotterdam, **1,** 315–24, 1980.

56. Martin, G. R., Finn, W. D. L., and Seed, H. B. Effects of system compliance on liquefaction tests. *Proc. A.S.C.E.,* **104** (GT4), 463–79, 1978.

57. Meimon, Y., and Hicher, P. Y. Mechanical behaviour of clays under cyclic loading. *Proc. Int. Symp. on Soils under Cyclic and Transient Loading,* Swansea (eds. G. N. Pande and O. C. Zienkiewicz), Balkema, Rotterdam, **1,** 77–87, 1980.

58. Mitchell, R. J., and King, R. D. Cyclic loading of an Ottawa area Champlain Sea clay. *Canadian Geotechnical Journal,* **14**(1), 52–63, 1977.

59. Motherwell, J. T., and Wright, S. G. Ocean wave load effects on soft clay behaviour. *Proc. A.S.C.E. Specialty Conf., Earthquake Engineering and Soil Dynamics,* Pasadena, **2,** 620–35, 1978.

60. Moussa, A. *Constant Volume Simple Shear Tests on Frigg Sand.* Norwegian Geotechnical Institute, Internal Report 51505-2, 1973:

61. Moussa, A. A. *Radial Stresses in Sand in Constant Volume Static and Cyclic Simple Shear Tests.* Norwegian Geotechnical Institute, Internal Report 51505-10, 1974.

62. Mulilis, J. P., Seed, H. B., Chan, C. K., Mitchell, J. K., and Arulanandan, K. Effects of sample preparation on sand liquefaction. *Proc. A.S.C.E.,* **103** (GT2), 91–108, 1977.

63. Murayama, S., and Shibata, T. On the dynamic properties of clay. *Proc. 2nd World Conf. on Earthquake Engineering*, Tokyo, **1,** 297–310, 1960.
64. Namy, D. L. *An Investigation of Certain Aspects of Stress–Strain Relationships for Clay Soils.* Ph.D. thesis, Cornell University, Ithaca, 1970.
65. Nova, R. and Wood, D. M. A constitutive model for sand. *Int. J. Num. Anal. Meth. Geomechs*, **3**(3), 255–78, 1979.
66. Oda, M. Initial fabrics and their relations to mechanical properties of granular material. *Soils and Foundations*, **12**(1), 17–36, 1972.
67. Oda, M. The mechanism of fabric changes during compressional deformation of sand, *Soils and Foundations*, **12**(2), 1–18, 1972.
68. Oda, M. Deformation mechanism of sand in triaxial compression tests. *Soils and Foundations*, **12**(4), 45–63, 1972.
69. Pappin, J. W. *Characteristics of a Granular Material for Pavement Analysis.* Ph.D. thesis, University of Nottingham, 1979.
70. Park, T. K. and Silver, M. L. Dynamic triaxial and simple shear behaviour of sand. *Proc. A.S.C.E.*, **101** (GT6), 513–29, 1975.
71. Parry, R. H. G., and Nadarajah, V. Anisotropy in a natural soft clayey silt. *Eng. Geol.*, **8,** 287–309, 1974.
72. Peacock, W. H., and Seed H. B. Sand liquefaction under cyclic loading simple shear conditions. *Proc. A.S.C.E.*, **94** (SM3), 689–708, 1968.
73. Pearce, J. A. A new true triaxial apparatus. *Stress–strain behaviour of soils. Proc. Roscoe Memorial Symp.*, G. T. Foulis & Co., 330–9, 1971.
74. Perloff, W. H., and Pombo, L. E. End restraint effects in the triaxial test. *Proc. 7th Int. Conf. SM & FE*, Mexico, **1,** 327–33, 1969.
75. Pickering, D. J. Anisotropic elastic parameters for soil. *Géotechnique*, **20**(3), 271–6, 1970.
76. Prévost, J.-H. and Höeg, K. Reanalysis of simple shear soil testing. *Canadian Geotechnical Journal*, **13**(4), 418–29, 1976.
77. Procter, D. C. Requirements of soil sampling for laboratory testing. In George, P., and Wood, D. (eds.), *Off-Shore Soil Mechanics*, Lloyd's Register of Shipping and Cambridge University Engineering Dept., 131–53, 1976.
78. Pyke, R., Discussion: Measurement of dynamic soil properties. *Proc. A.S.C.E. Specialty Conf., Earthquake Engineering and Soil Dynamics*, Pasadena, **3,** 1474–7, 1978.
79. Richardson, A. M., and Whitman, R. V. Effect of strain-rate upon undrained shear resistance of a saturated remoulded fat clay. *Géotechnique*, **13**(4), 310–24, 1963.
80. Roscoe, K. H. An apparatus for the application of simple shear to soil samples. *Proc. 3rd Int. Conf. Soil Mech.* Zurich, **1,** 186–91, 1953.
81. Roscoe, K. H., Schofield, A. N., and Wroth, C. P. On the yielding of soils. *Géotechnique*, **8**(1), 22–52, 1958.
82. Roscoe, K. H. and Schofield, A. N. Mechanical behaviour of an idealised 'wet' clay. *Proc. 3nd Eur. Conf. Soil Mech.*, Wiesbaden, **1,** 47–54, 1963.
83. Rowe, P. W., and Barden, L. Importance of free ends in triaxial testing. *Proc. A.S.C.E.*, **90** (SM1), 1–27, 1964.
84. Saada, A. S. Testing of anisotropic clay soils. *Proc. A.S.C.E.*, **96** (SM5), 1847–52, 1970.
85. Saada, A. S., and Bianchini, G. F. Strength of one-dimensionally consolidated clays. *Proc. A.S.C.E.*, **101** (GT11), 1151–64, 1975.
86. Saada, A. S. Discussion: Anisotropy in heavily overconsolidated kaolin. *Proc. A.S.C.E.*, **102** (GT7), 823–4, 1976.
87. Saada, A. S., and Bianchini, G. F. Discussion closure: Strength of one-dimensionally consolidated clays. *Proc. A.S.C.E.*, **103** (GT6), 655–60, 1977.

88. Saada, A. S., Bianchini, G. F., and Shook, L. P. The dynamic response of anisotropic clay. *Proc. A.S.C.E. Specialty Conf., Earthquake Engineering and Soil Dynamics*, Pasadena, **2**, 777–801, 1978.
89. Sangrey, D. A. *The Behaviour of Soils Subjected to Repeated Loading*. Ph.D. thesis, Cornell University, 1968.
90. Sangrey, D. A., Henkel, D. J., and Esrig, M. I., The effective stress response of a saturated clay soil to repeated loading. *Canadian Geotechnical Journal*, **6**(3), 241–52, 1969.
91. Sangrey, D. A., Castro, G., Poulos, S. J., and France, J. W. Cyclic loading of sands, silts and clays. *Proc. A.S.C.E. Specialty Conf., Earthquake Engineering and Soil Dynamics*, Pasadena, **2**, 836–51, 1978.
92. Sangrey, D. A., Pollard, W. S. and Egan, J. A. Errors associated with rate of undrained cyclic testing of clay soils. *Dynamic Geotechnical Testing*, ASTM: STP 654, 280–94, 1978.
93. Sangrey, D. A., and France, J. W. Peak strength of clay soils after a repeated loading history. *Proc. Int. Symp. on Soils under Cyclic and Transient Loading*, Swansea (eds. G. N. Pande and O. C. Zienkiewicz), Balkema, Rotterdam, **1**, 421–30, 1980.
94. Schmertmann, J. H. The undisturbed consolidation behaviour of clay. *Trans. A.S.C.E.*, **120**, 1201–33, 1955.
95. Schofield, A. N., and Wroth, C. P. *Critical State Soil Mechanics*. McGraw-Hill, London, pp. 310, 1968.
96. Seed, H. B. Soil strength during earthquakes. *Proc. 2nd World Conf. on Earthquake Engineering*, Tokyo, **1**, 183–94, 1960.
97. Seed, H. B., and Chan, C. K. Clay strength under earthquake loading conditions. *Proc. A.S.C.E.*, **92** (SM2), 53–78, 1966.
98. Seed, H. B., and Idriss, I. M. *Soil Moduli and Damping Factors for Dynamic Response Analysis*. Earthquake Engineering Research Centre, University of California, Berkeley, Report EERC 70–10 (December), 1970.
99. Seed, H. B., and Peacock, W. H. Test procedures for measuring soil liquefaction characteristics. *Proc. A.S.C.E.*, **97** (SM8), 1099–1119, 1971.
100. Seed, H. B. Evaluation of soil liquefaction effects on level ground during earthquakes. *Liquefaction problems in geotechnical engineering*. A.S.C.E. Annual Convention and Exposition, Philadelphia, 27 September–1 October, Preprint 2752, 1–104, 1976.
101. Seed, H. B., Mori, K., and Chan, C. K. Influence of seismic history on liquefaction of sands. *Proc. A.S.C.E.*, **103** (GT4), 257–70, 1977.
102. Seed, H. B. Soil liquefaction and cyclic mobility evaluation for level ground during earthquakes, *Proc. A.S.C.E.*, **105** (GT2), 201–55, 1979.
103. Shen, C. K., Herrmann, L. R., and Sadigh, K. Analysis of cyclic simple shear test data. *Proc. A.S.C.E. Specialty Conf., Earthquake Engineering and Soil Dynamics*, Pasadena, **2**, 864–74, 1978.
104. Shen, C. K., Sadigh, K., and Herrmann, L. R. An analysis of NGI simple shear apparatus for cyclic soil testing. *Dynamic Geotechnical Testing*, ASTM: STP 654, 148–62, 1978.
105. Sherif, M. A., and Wu, M.-J. *The Dynamic Behaviour of Seattle Clays*. Dept. of Civil Engineering, University of Washington, Seattle. Contract no. GC-2023, Soil Engineering Report 6, January 15, 1971.
106. Sherif, M. A., Wu, M.-J., and Bostrom, R. C. Reduction in soil strength due to dynamic loading. *Proc. Int. Conf. on Microzonation for Safer Construction Research and Application*, Seattle, Washington, (NSF), **2**, 439–54, 1972.
107. Silver, M. L., and Seed, H. B. Deformation characteristics of sands under cyclic loading. *Proc. A.S.C.E.*, **97** (SM8), 1081–98, 1971.

108. Silver, M. L., and Seed, H. B. Volume changes in sands during cyclic loading. *Proc. A.S.C.E.*, **97** (SM9), 1171–82, 1971.
109. Silver, M. L., and Park, T. K. Testing procedure effects on dynamic soil behaviour. *Proc. A.S.C.E.*, **101** (GT10), 1061–83, 1975.
110. Silver, M. L., Chan, C. K., Ladd, R. S., Lee, K. L., Tiedemann, D. A., Townsend, F. C., Valera, J. E., and Wilson, J. H. Cyclic triaxial strength of standard test sand. *Proc. A.S.C.E.*, **102** (GT5), 511–23, 1976.
111. Simons, N. E. Comprehensive investigations of the shear strength of an undisturbed Drammen clay. *Proc. A.S.C.E. Research Conf. on Shear Strength of Cohesive Soils*, Boulder, Colorado, 727–45, 1960.
112. Stroud, M. A. *The Behaviour of Sand at Low Stress Levels in the Simple Shear Apparatus*. Ph.D. thesis, Cambridge University, 1971.
113. Takahashi, M., Hight, D. W., and Vaughan, P. R. Effective stress changes observed during undrained cyclic triaxial tests on clay. *Proc. Int. Symp. on Soils under Cyclic and Transient Loading*, Swansea (eds. G. N. Pande and O. C. Zienkiewicz), Balkema, Rotterdam, **1**, 201–9, 1980.
114. Tatsuoka, F. *Shear Tests in a Triaxial Apparatus—a Fundamental Study of the Deformation of Sand* (in Japanese). Thesis, Tokyo University, 1972.
115. Tatsuoka, F., and Ishihara, K. Yielding of sand in triaxial compression. *Soils and Foundations*, **14**(2), 63–76, 1974.
116. Tatsuoka, F., and Ishihara, K., Drained deformation of sand under cyclic stresses reversing direction. *Soils and Foundations*, **14**(3), 51–65, 1974.
117. Taylor, D. W. *Fundamentals of Soil Mechanics*. Wiley, New York, pp. 700, 1948.
118. Taylor, P. W., and Bacchus, D. R., Dynamic cyclic strain tests on a clay. *Proc. 7th Int. Conf. Soil Mech.*, Mexico, **1**, 401–9, 1969.
119. Terzaghi, K. Festigkeitseigenschaften der Schüttungen, Sedimente und Gele, in Auerbach, F., and Hort, W. (eds.), *Handbuch der physikalischen und technischen Mechanik*, Leipzig, Barth **4**(2), 513–78, 1931.
120. Thiers, G. R. *The Behaviour of Saturated Clay under Seismic Loading Conditions*. Ph.D. thesis, University of California, Berkeley, 1965.
121. Thiers, G. R., and Seed, H. B. Cyclic stress–strain characteristics of clay. *Proc. A.S.C.E.*, **94** (SM2), 555–69, 1968.
122. Thiers, G. R., and Seed, H. B. Strength and stress–strain characteristics of clays subjected to seismic loading conditions. *Vibration Effects of Earthquakes on Soils and Foundations*, ASTM: STP 450, 3–56, 1969.
123. Wilson, N. E., and Greenwood, J. R. Pore pressures and strains after repeated loading of saturated clay. *Canadian Geotechnical J.*, **11**(2), 269–77, 1974.
124. Wolfe, W. E., Annaki, M., and Lee, K. L. Soil liquefaction in cyclic cubic test apparatus. *Proc. 6th World Conf. on Earthquake Engineering*, New Delhi **6**, 2151–6, 1977.
125. Wood, D. M. *Some Aspects of the Mechanical Behaviour of Kaolin under Truly Triaxial Conditions of Stress and Strain*. Ph.D. thesis, Cambridge University, 1974.
126. Wood, D. M. Explorations of principal stress space with kaolin in a true triaxial apparatus. *Géotechnique*, **25**(4), 783–97, 1975.
127. Wood, D. M. Cyclic loading of soil samples. In George, P. J., and Wood, D. M. (eds.), *Off-Shore Soil Mechanics*, Lloyd's Register of Shipping and Cambridge University Engineering Department, 363–87, 1976.
128. Wood, D. M. Comments on cyclic loading of clay. *Proc. Int. Conf. on Behaviour of Off-Shore Structures*, Trondheim, **2**, 418–24, 1976.
129. Wood, D. M., and Budhu, M. The behaviour of Leighton Buzzard sand in cyclic simple shear tests. *Proc. Int. Symp. on Soils under Cyclic and Transient*

Loading, Swansea (eds. G. N. Pande and O. C. Zienkiewicz), Balkema, Rotterdam, **1**, 9–21, 1980.

130. Wood, D. M., Drescher, A., and Budhu, M. On the determination of the stress–state in the simple shear apparatus. *ASTM Geotechnical Testing Journal*, **2**(4), 211–222, 1980.

131. Woods, R. D. Measurement of dynamic soil properties. *Proc. A.S.C.E. Specialty Conf., Earthquake Engineering and Soil Dynamics*, Pasadena, **1**, 91–178, 1978.

132. Wright, D. K., Gilbert, P. A., and Saada, A. S. Shear devices for determining dynamic soil properties. *Proc. A.S.C.E. Specialty Conf., Earthquake Engineering and Soil Dynamics*, Pasadena, **2**, 1056–75, 1978.

133. Wroth, C. P., and Loudon, P. A. The correlation of strains within a family of triaxial tests on overconsolidated samples of kaolin. *Proc. Geotechnical Conf.*, Oslo, **1**, 159–63, 1967.

134. Youd, T. L. Densification and shear of sand during vibration. *Proc. A.S.C.E.*, **96** (SM3), 863–80, 1970.

135. Youd, T. L. and Craven, T. N. Lateral stress in sands during cyclic loading. *Proc. A.S.C.E.*, **101** (GT2), 217–21, 1975.

136. Youd, T. L. Packing changes and liquefaction susceptibility. *Proc. A.S.C.E.*, **103** (GT8), 918–22, 1977.

137. Yudhbir, Field compressibility of soft sensitive normally consolidated clays. *Geotechnical Engineering*, **4**(1), 31–40, 1973.

138. Andersen, K. H., Pool, J. H., Brown, S. F. and Rosenbrand, W. F. Cyclic and static laboratory tests on Drammen clay. *Proc. A.S.C.E.*, **106** (GT5), 499–529, 1980.

Soil Mechanics—Transient and Cyclic Loads
Edited by G. N. Pande and O. C. Zienkiewicz
© 1982 John Wiley & Sons Ltd

Chapter 21

The Cyclic Simple Shear Test

W. D. L. Finn, S. K. Bhatia, and D. J. Pickering

21.1 INTRODUCTION

The cyclic simple shear test is the foundation on which rests most of the practical applications of theoretical soil dynamics. The fundamental importance of this test evolved from attempts to explain in a quantitative way the extensive liquefaction of saturated sands that occurred in 1964 during the earthquakes in Alaska and Niigata, Japan. The liquefaction failures occurred primarily on level ground, and in formulating a procedure for analysis, Seed and Idriss[21] made two basic assumptions:

(1) seismic excitation is primarily due to shear waves propagating vertically and
(2) level ground conditions may be approximated by horizontal layers with uniform properties.

Under these conditions the ground deforms in shear only and may be analysed by treating a vertical column of soil as a shear beam. An element of the shear beam is shown in Figure 21.1 before and during the earthquake. The initial effective vertical stress is σ'_{v0} and the initial effective horizontal stress is $K_0\sigma'_{v0}$ in which K_0 is a stress-history dependent soil parameter. During the earthquake, because of the development of pore water pressure, these initial effective stresses change to σ'_v and $K\sigma'_v$. Both total stress[20] and effective stress dynamic response analyses[6] based on these assumptions have proven effective in assessing liquefaction potential and estimating the acceleration response to various seismic inputs.

To support analyses based on shear beam behaviour, it was necessary to develop a laboratory test which would simulate the postulated field behaviour and allow the measurement of the parameters needed for the analytical models. The simple shear test developed by Roscoe[19] for static testing of soils simulated the required field conditions, in theory. This test was adapted to cyclic loading conditions by Peacock and Seed[16] in 1968 and they produced the first laboratory data on the behaviour of saturated sands in cyclic simple shear.

In 1970, Finn, Bransby, and Pickering[4] and Finn, Pickering, and Bransby[5]

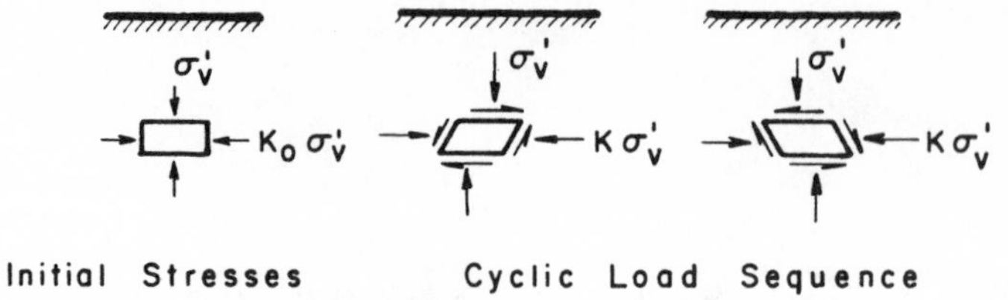

Initial Stresses   Cyclic Load Sequence

Figure 21.1 Idealized field loading conditions

reported experimental investigations of the effects of various boundary conditions on the stress and deformation conditions within the test sample. They established the experimental requirements for ensuring as uniform deformations as possible within the sample. The investigations were carried out in a Roscoe-type apparatus which incorporated a number of innovations to improve the capability of the apparatus to approach the ideal simple shear state.[5,17]

The Roscoe-type of apparatus using a sample of square cross-section and generating simple shear conditions in plane strain most closely approximates the ideal field conditions assumed for the shear beam analogy. However, other types of apparatus have been developed, most notably that developed by Bjerrum and Landva.[1] This apparatus enclosed a circular sample in a wire-reinforced rubber membrane. This apparatus does not generate plane strain conditions and analysis by Lucks *et al.*[12] indicates that under ideal conditions approximately 70% of the sample is uniformly stressed. In practice, the region would be smaller. Some of the problems with samples enclosed in wire-reinforced membranes and comparisons between sand behaviour in Bjerrum-type and Roscoe-type apparatuses have recently been investigated by Wood and Budhu.[25] For research into fundamentals of sand behaviour under cyclic loading the Roscoe-type apparatus is superior but the Bjerrum-type apparatus, because of its greater simplicity, appears more suitable for use in engineering practice.

Recently, there has been a revival of interest from the experimental point of view in the basic mechanics of the simple shear test and renewed interest in the stress and deformation conditions induced by the Roscoe-type apparatus characterized by a sample container with rigid walls. Fundamental studies on the drained behaviour of sand at large strains in the Cambridge simple shear apparatus have been reported by Wood and Budhu[25] and Wood, Drescher, and Budhu.[24] In these studies the distribution of boundary stresses were measured by an array of load cells around the sample and internal deformations were investigated by X-rays.

Studies on the small strain behaviour of sands which is of primary interest in seismic loading problems in both drained and undrained cyclic simple shear tests have been conducted over the last three years at the University of British Columbia. These studies, which will be reported below, resulted in the development of new test equipment and new testing procedures. Before discussing the recent studies, some earlier experimental work on the effects of boundary conditions will be reviewed and previously unpublished stress and deformation analyses of samples in the simple shear test will be given.

21.2 EXPLORATION OF EFFECTS OF BOUNDARY CONDITIONS

Simple shear in the x, y plane is defined as pure shear in which $\partial u/\partial y$ or $\partial v/\partial x = 0$, where u and v are displacements in the x- and y-directions. A sample of length L and height $2H$ shown in Figure 21.2, undergoes simple shear strain, γ, in the x, y plane with $\partial v/\partial x = 0$. In the case of the ideal undrained test (constant volume is maintained) no vertical displacements of the upper and lower boundary surfaces occur and $v = 0$. However, the sloping lengths of the sides increase as shear progresses. In a simple shear apparatus with rigid container walls (see the deformed sample in Figure 21.7) the increasing length of the sloping sides is accommodated by having the side walls of the sample container in the x, y plane long enough for the maximum shear strain condition and having a slip contact at the top or bottom to allow the changes in length during shearing deformation. The slipping required to accommodate the change in length violates the ideal boundary conditions for simple shear at the ends. The shear stresses that occur during slip are indeterminate so to ensure controlled conditions during a test these frictional shear stresses, τ_{xy}, should be reduced as much as possible, ideally to zero. Therefore, instead of the ideal boundary conditions, a 'practical' set of boundary conditions is adopted as shown in Table 21.1.

A very clear picture of the effect of various boundary conditions is given by testing plasticine samples with a coloured grid as shown in Figure

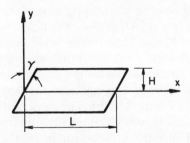

Figure 21.2 Simple shear deformation

Table 21.1   Ideal and 'practical' boundary conditions in simple shear
apparatus

	Ideal	'Practical'
Top and bottom	$u = kh$ and $-kh$ $v = 0$	$u = kh$ and $-kh$ $v = 0$
Ends	$u = ky$ $v = 0$	$u = ky$ $\tau_{xy} = 0$

21.3(a).[17] Although small strain behaviour is of primary interest, this 'phenomenological' study of boundary conditions was conducted at large strains (up to $\gamma = 30°$) so that the grid deformations would show up clearly. The rough metal boundary plates at the top and bottom of the sample were placed in direct contact with the plasticine and oiled paper was placed between the plasticine and the ends of the box to approximate as closely as possible the zero shear stress condition. After the test, the top and bottom plates adhered tightly to the plasticine sample and had to be cut away with a wire saw. It seemed clear that no slip occurred along these boundaries during the large strain cyclic shearing. In Figure 21.3(b) it may be seen that even for large strains under the 'practical' boundary conditions, the strain field consists of essentially uniform shear over most of the sample and the magnitude of that shear is roughly comparable to the prescribed boundary deformations. Around the boundaries and especially near the ends are narrow zones of lesser shear strain, but even after three cycles of $\gamma = 30°$ no localized failure zone has developed nor are the shear strains anywhere markedly higher than the average.

The effect of slip between the top and bottom plates in addition to slip on the ends is explored by inserting oiled paper between these plates and the sample surfaces. Results of this kind of test are shown in Figure 21.3(c). In this case there is a significant variation in shear strain throughout the sample and only a relatively small zone near the centre approaches a shear strain distribution corresponding to the boundary deformations. For comparison with the experimental pattern of shearing deformation, an elastic solution for an isotropic sample subject to the 'practical' boundary conditions in Table 21.1 is shown in Figure 21.3(d). The solution is for a length to height ratio $L/2H = 1.778$, the ratio for the UBC apparatus. Note the marked similarity in the shear strain pattern for the elastic solution in Figure 21.3(d) and the experimental pattern corresponding to approximately the same boundary conditions in Figure 21.3(b). Two very tentative conclusions may be drawn from this kind of study; essentially uniform simple shear conditions can be induced in most of the sample by the 'practical' boundary conditions and elastic solutions may be useful in predicting the effects of changes in sample geometry and boundary conditions for small shear strains.

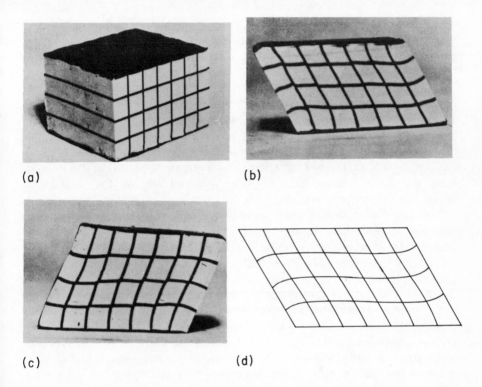

(a)   (b)

(c)   (d)

Figure 21.3 Deformation fields by experiments and theory

21.3 THEORETICAL ANALYSIS OF THE CYCLIC SIMPLE SHEAR TEST

For the analysis of small strain deformations, $\gamma \leqslant 0.5\%$ the sand is assumed to be an anisotropic elastic solid. Other studies by Finn[3] of soil response to imposed boundary deformation conditions have shown that for small strain behaviour elastic response gives a useful indication of stress and deformation patterns. The results of the tests with plasticine models tend to confirm this finding. The analysis was carried out analytically using stress functions and the resulting series solutions evaluated by computer. Figures 21.4–21.6, showing some of the important results, are photographs of the computer plotter output. To give as detailed a picture as possible of the effects of sample geometry and soil properties on stress and strain distributions throughout the sample a large number of photographs are shown. This has necessitated producing each to a very small scale. The stresses and deformations are all presented in non-dimensional form. The computer notation for the stresses in the figures is as follows; $\text{SIGX} = \sigma_x/\tau_0$, $\text{SIGY} = \sigma_y/\tau_0$,

TAUXY $= \tau/\tau_0$ in which σ_x and σ_y are normal stresses in the x- and y-directions, τ_0 is the average applied cyclic shear stress and τ is the computed value of the shear stress. All analyses are for the 'practical' boundary conditions only.

The stress and deformation analyses were formulated for an elastic homogeneous anisotropic material. Solutions for specified isotropic and anisotropic materials were obtained by inserting the appropriate relations between the elastic constants in the general solution. Therefore, for the examples quoted here, the isotropic material is defined by $E_x/E_y = 1$, $\mu_{xy} = 0.50$, $\mu_{xx} = 0.5$, $E_x/G_{xy} = 3.00$ where E is Young's Modulus, μ the Poisson ratio and G the shear modulus; the subscripts define the domain of application of the constants.

The stress and deformation conditions for an isotropic sample with dimensions approximating those in the UBC apparatus is shown in Figure 21.4. Under ideal boundary conditions, SIGX $=$ SIGY $= 0$ and TAUXY $= 1$. Deviations from these conditions indicate the distortions introduced in the ideal simple shear stress and deformation field by the loss of complementary shears on the ends. It can be seen that high stress concentrations occur on the sloping ends near the corners (Figure 21.4(a)) and on both horizontal surfaces (Figure 21.4(b)). Although these concentrations die out very rapidly, nevertheless they are felt throughout the body of the sample. The shear stress is fairly uniform over most of the sample though higher than unity in the middle half of the horizontal surfaces where TAUXY $= 1.1$ (Figure 21.4(c)). This value corresponds to $Y = 0$. Despite the local stress concentrations the deformation field is very uniform. The horizontal displacement field shows distortion mainly near the ends (Figure 21.4(d)) and except near the corners, the y-displacement field is sensibly zero (Figure 21.4(e)).

Similar results are shown in Figure 21.5 in which the ratio of length to height of the sample $L/2H = 6$. Note that, in this case, the stress and deformation conditions in the sample are remarkably uniform except near the corners and ends. Especially notable are the extensive regions with negligible normal stresses in the x- (Figure 21.5(a)) and y-directions (Figure 21.5(b)) and the greatly improved uniformity in the distribution of shear stresses (Figure 21.5(c)). The horizontal and vertical deformation fields shown in Figure 21.5(d) and Figure 21.5(e), respectively, indicate conditions remarkably close to ideal simple shear deformations. The principal deviation from the ideal conditions is at the sloping ends. These theoretical results give strong support to the validity of the data from the large scale cyclic simple shear tests conducted by De Alba, Chan, and Seed[2] in which high length to height ratios were used.

Finally, some results are shown for an anisotropic material in Figure 21.6. The length to height ratio is $L/2H = 2$. The material is specified by $E_x/E_y =$

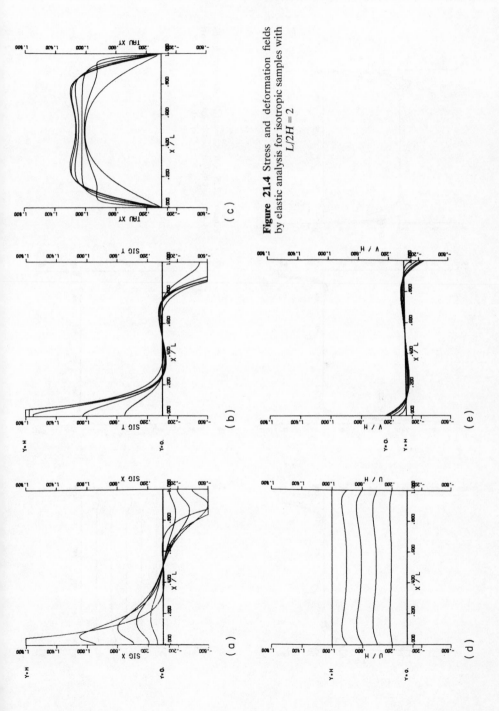

Figure 21.4 Stress and deformation fields by elastic analysis for isotropic samples with $L/2H = 2$

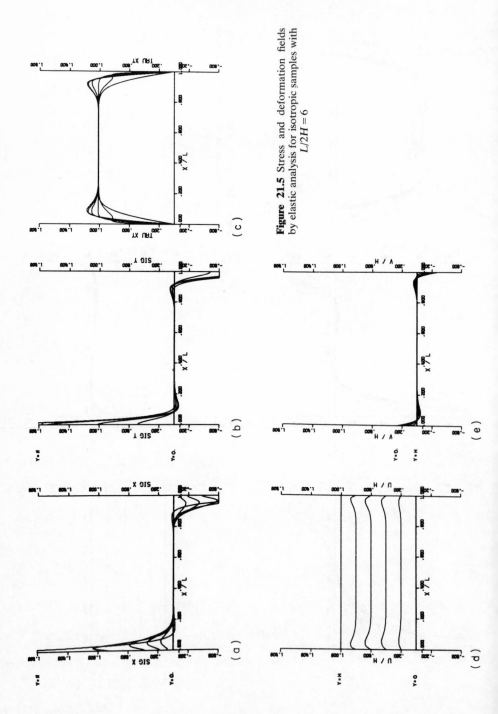

Figure 21.5 Stress and deformation fields by elastic analysis for isotropic samples with *L/2H = 6*

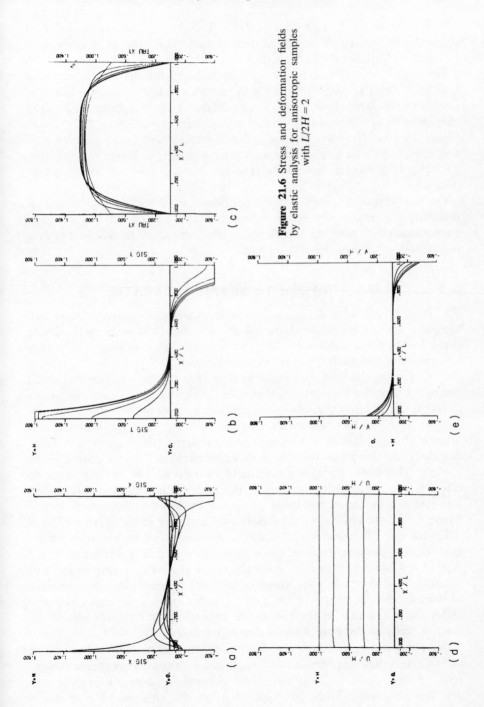

Figure 21.6 Stress and deformation fields by elastic analysis for anisotropic samples with $L/2H = 2$

1.5, $\mu_{xy} = 0.20$, $\mu_{xx} = 0.20$, and $E_x/G_{xy} = 6.00$. It is cross-anisotropic with rotational symmetry about the vertical y-axis. This type of anisotropy is common in horizontally bedded soils.

Horizontal normal stresses are greatly reduced for the anisotropic case (Figure 21.6(a)) in comparison with the stresses computed for the isotropic sample of the same dimensions (Figure 21.4(a)). There appears to be little change in the distribution of vertical stresses; compare Figure 21.6(b) with Figure 21.4(b). The shear stresses are remarkably uniform (Figure 21.6(c)) and approach those in a sample with a large length to height ratio (Figure 21.5(c)). The deformation fields (Figures 21.6(d), (e)) are exceptionally uniform.

On the basis of the elastic analyses it seems clear that for the 'practical' boundary conditions attainable in existing test equipment a very good approximation to uniform simple shear conditions may be attained over a substantial volume of the test sample.

21.4 NEW SIMPLE SHEAR APPARATUS

The most recent advance in cyclic simple shear testing using a rigid-wall sample container is the development of the constant volume cyclic simple shear test by Finn and Vaid.[8] This test was developed to reduce to a very low value the errors due to system compliance.

In evaluating the results of cyclic loading tests on saturated sands carried out under undrained conditions, it is assumed that no volume changes occur during the test. However, compliance in test systems allows volume expansion to occur in the supposedly constant volume saturated sample. This volume change, having the same effect as partial drainage would have, decreases the tendency for the pore water pressure to rise during cyclic loading. Therefore, undrained tests tend to overestimate the resistance to liquefaction. A theoretical analysis of the errors in liquefaction data caused by system compliance has been presented by Martin, Finn, and Seed.[14] These errors are always on the unsafe side and may range up to 100%.

Errors, due to system compliance, are common to all undrained liquefaction testing. The magnitude of the compliance is primarily a function of the type of test equipment and the grain size of the sand. In the various kinds of triaxial and shake table tests, compliance arises primarily as a consequence of membrane penetration effects and in simple shear tests with rigid container walls it is due to the membrane expanding into the corners of the sample cavity as the pore water pressure increases.

An alternative procedure for determining the undrained behaviour of sand without the complications of compliance is to conduct constant volume tests on dry sands. In these tests, the changes in confining pressures to maintain constant volume are equivalent to the changes in pore water pressure in the corresponding undrained test. Constant volume triaxial testing of sand as an equivalent method of obtaining its behaviour under undrained loading was first suggested by Taylor.[22] Application of the constant volume test to simple shear conditions was proposed by Pickering.[18] Constant volume testing using the Bjerrum-type apparatus has been developed by Moussa.[15]

21.5 DESCRIPTION OF TEST APPARATUS

The University of British Columbia (UBC) simple shear apparatus, described by Finn, Pickering, and Bransby[5] has been modified to permit cyclic shear testing at constant sample volume. The modified apparatus is shown in Figure 21.7. The two components of linear horizontal strain are identically

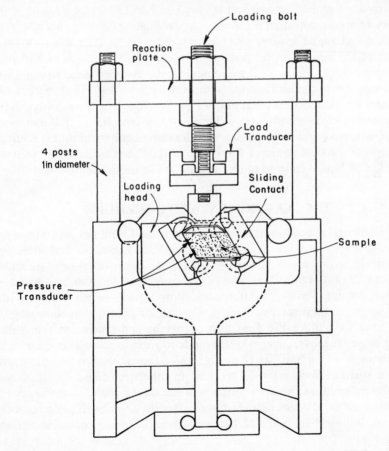

Figure 21.7 Constant volume cyclic simple shear apparatus at UBC

zero in this simple shear apparatus. Thus, a constant volume condition is achieved by clamping the loading head to prevent vertical strain. A horizontal reaction plate is clamped to four vertical posts which are threaded into the body of the simple shear apparatus. A stiff vertical load transducer is attached to the sample loading head and carries on its upper side a heavy loading bolt which passes through a central hole in the reaction plate. The desired vertical pressure on the sample is applied by tightening the loading bolt nut on the underside of the reaction plate. Simultaneously, the loading head is clamped in position by tightening the loading bolt nut on the top side of the reaction plate. Another important innovation was the incorporation of two small stiff pressure transducers (350 kPa capacity and full-scale deflection of 0.0015 cm) on one of the movable lateral boundaries in order to monitor the lateral stresses during cyclic loading.

Maximum gross volume change introduced at the onset of liquefaction in this so-called constant-volume test is very small and arises as a result of the recovery of elastic deformation in the vertical loading components when the load on the clamped loading head is reduced to zero. The use of a thick reaction plate, heavy vertical posts and loading bolt, and a very stiff load transducer reduces the vertical movement of the clamped head to a negligible amount. For liquefaction tests with initial $\sigma'_{v0} = 2$ kg/cm^2 (196 kPa) this movement amounted to a maximum of 5×10^{-4} cm which was only 5% of the movement of the floating head due to system compliance in liquefaction tests on saturated undrained samples in the same equipment and is equivalent to a total vertical strain of the order of 0.02%. Thus, a more accurate evaluation of liquefaction potential can be made using the new test.

21.6 EXPERIMENTAL PROCEDURE

An experimental procedure for preparing saturated sand samples within the simple shear apparatus has been described by Finn, Pickering, and Bransby.[5] Evolution in testing methods has since led to an improved procedure which gives samples of more uniform density and, therefore, yields more consistent and reproducible results. In the earlier procedure, densification of the sample to the desired relative density was carried out prior to siphoning off excess sand to achieve the final desired height and seating of the ribbed loading plate. A looser layer of sand tends to form at the top due partly to the siphoning action and partly to digging of the ribbed loading plate into the sand surface. Such a looser layer in an otherwise dense sample would lower the overall liquefaction resistance of the sand sample. In the improved procedure, the sand is not densified until after the top ribbed plate has been seated on the sand surface and the membrane has been sealed to the top loading cap.

A constant volume test can be performed on either dry or saturated sands.

Dry sand is deposited within the membrane in the apparatus through a funnel tip permitting a free fall. The funnel is traversed across the plan area of the sample so that the sand surface remains sensibly level. Pouring is continued until an excess of sand over that required for the sample has been deposited. Excess sand over the final elevation is then siphoned off using a small vacuum. Saturated samples are prepared to the desired height by deposition under water in the manner described in Ref. 5. The top ribbed plate is now placed on the sand surface, the membrane closed over it, and sealed to the loading head. The desired final density for either kind of test is then obtained by vertical vibrations while the sample is kept under a seating pressure of approximately 0.1 kg/cm². The top ribbed plate thus follows the settlement of the sand surface and assumes a proper seating, while the entire sample gets uniformly densified without development of looser zones under the top plate. The vertical pressure is then increased to the required value of the overburden pressure. The cyclic shear load may be applied using either square, ramp or sine wave forms.

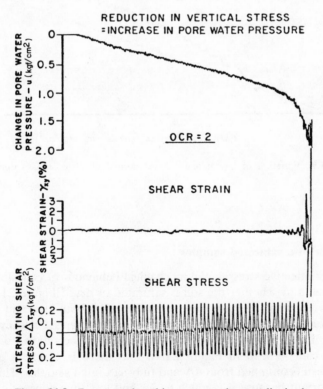

Figure 21.8 Response of sand in constant volume cyclic simple shear test

21.7 CONSTANT VOLUME CYCLIC SIMPLE SHEAR TESTS

A typical record of a constant-volume simple shear cyclic loading test on medium dense dry sand is shown in Figure 21.8. In the constant-volume test, the pore water pressure remains constant (usually zero) and the normal effective stress may decrease to zero. The simultaneous reductions in horizontal and vertical effective stresses during cyclic loading are shown in Fig. 21.9 for a typical test.

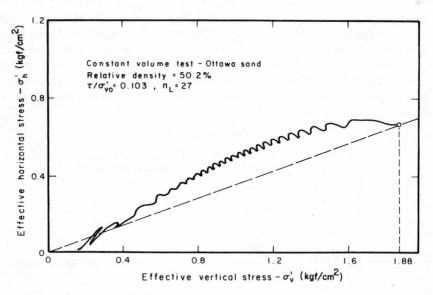

Figure 21.9 Variation of lateral and vertical normal effective stresses during cyclic loading

21.7.1 Dry vs. saturated samples

In terms of effective stresses, the mechanical behaviour of cohesionless soils is not affected by whether the soil is saturated or dry.[10] Figure 21.10 shows data from constant-volume liquefaction tests of both dry and saturated samples. Identical values of initial confining stresses, σ'_{vo}, and cyclic shear stress amplitude, τ, were used in all tests and only the relative density was varied. It can be seen in Figure 21.10 that there is excellent agreement between results obtained from dry and fully saturated samples. This result is quite important because it is much easier and less time consuming to work with dry sands than with saturated sands.

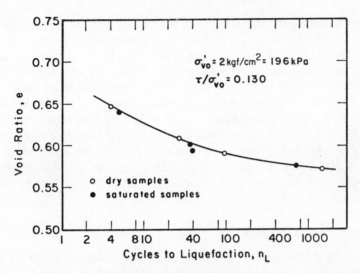

Figure 21.10 Behaviour of dry and saturated sands in constant volume
cyclic simple shear

21.7.2 Effect of system compliance

A series of conventional liquefaction tests was carried out on saturated
undrained samples using the new procedure for sample formation. Constant
values of initial confining stress and cyclic shear stress ($\tau/\sigma'_{v0} = 0.13$) were
used and only the sample density was varied. The results from this series of
tests are shown in Figure 21.11. Corresponding results obtained by constant

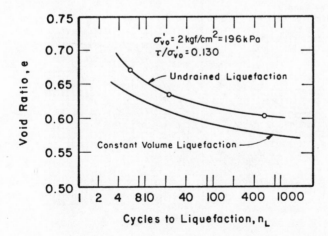

Figure 21.11 Effect of system compliance on measured resistance to liquefaction

volume liquefaction tests are also shown for comparison. It is clear from Figure 21.11 that for identical methods of sample preparation the resistance to liquefaction measured in conventional cyclic simple shear tests on undrained saturated samples is consistently larger than that measured in the constant volume tests. Since the test samples were prepared in an identical manner, the increased resistance to liquefaction in undrained tests is directly attributable to the system compliance in the undrained test. The apparent volume change at the onset of liquefaction in undrained tests amounted to a movement of loading head which was 20 times as large as the corresponding movement in the constant-volume tests. Thus, the use of conventional undrained test data may result in an overestimation of the liquefaction resistance of sand.

21.7.3 Stress–strain behaviour

Typical stress–strain data from constant stress and constant strain cyclic loading tests at constant volume are shown in Figures 21.12 and 21.13 respectively. The softening effect of increasing pore water pressure in the constant-stress test is shown by the increasing strain amplitudes and larger and flatter hysteresis loops (Figure 21.12). In the constant-strain test, the softening shows up in the decreasing shear stresses required to generate the required strain (Figure 21.13). Figure 21.13 shows clearly the effect of friction in the cyclic shear apparatus. The energy loss due to friction is represented by the almost rectangular hysteresis loop. All test data are corrected for this effect.

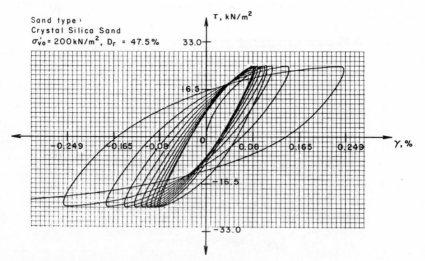

Figure 21.12 Stress–strain loops in constant-stress–constant-volume cyclic simple shear test

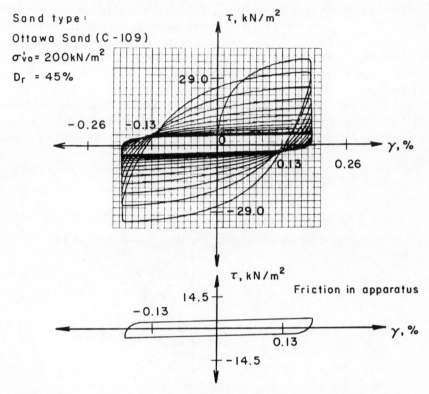

Figure 21.13 Stress–strain loops in constant-strain–constant-volume cyclic simple shear test (showing apparatus friction)

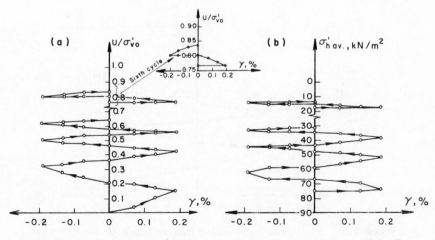

Figure 21.14 Pore water pressures and effective lateral stresses during cyclic loading

The pore water pressure response during the loading history shown in Figure 21.13 is shown in Figure 21.14(a). Note that as the number of load cycles increases the development of pore water pressure becomes limited to the unloading part of the strain cycle. The decrease in lateral effective stress accompanying the increase in pore water pressure is shown in Figure 21.14(b).

Stress–strain behaviour during a drained test is illustrated in Figure 21.15(a) for a constant cyclic strain test. Note that, in this case, the sand strengthens during cyclic loading and each successive cycle requires a greater

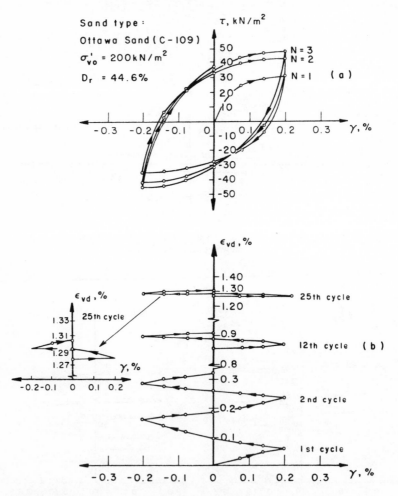

Figure 21.15 Stress–strain loops and volumetric strains in constant–strain drained cyclic loading test

shear stress to achieve the prescribed strain level. This strain-hardening is accompanied by volumetric strain, ε_{vd}. The development of volumetric strain with loading cycles is shown in Figure 21.15(b). Note the similarity in form between the volumetric strains (Figure 21.15(b)) and pore water pressures in Figure 21.14(a). Note also that as loading progresses more of the volumetric strain is accumulated during the unloading portions of the load cycles. Martin, Finn, and Seed[13] based a pore pressure generation model on this close affinity between volumetric strains and pore water pressure.

The behaviour of the lateral stresses during drained cyclic loading is shown in Figure 21.16. The average lateral stress (Figure 21.16(a)) approaches an asymptotic value after a few cycles and the stress–strain loops are almost closed. The individual lateral stresses in the top and bottom pressure transducers from which the average is computed are shown in Figures 21.16(b) and 21.16(c). The transducers are located at the third

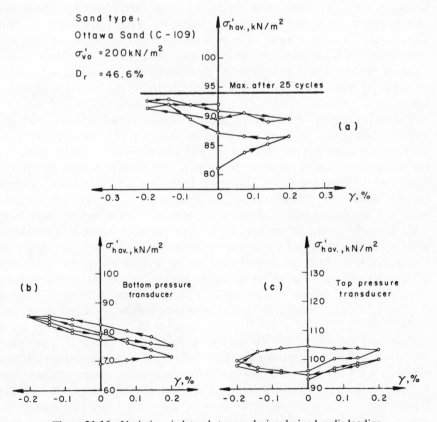

Figure 21.16 Variations in lateral stresses during drained cyclic loading

points of the end pieces and reflect the stress concentrations at the ends of the stress cycles predicted by the theoretical elastic analyses.

21.7.4 Extension to more general stress fields

The cyclic simple shear test is designed to simulate field conditions in which, prior to an earthquake, there are no initial shear stresses acting on the planes which will be subjected to cyclic shear stresses by the earthquake loading. This condition is appropriate for horizontally layered soil deposits with a horizontal ground surface subjected to shear waves propagating vertically. However, there are many cases of practical interest in which initial static shear stresses will act on these planes. For example, significant initial static shear stresses exist on horizontal planes below the surface of a natural slope. Even in the case of horizontal ground, the presence of a structure will introduce initial static shear stresses on horizontal planes. These stress conditions may be modelled in the simple shear apparatus by imposing an initial static shear stress, τ_s, under undrained conditions and then applying the cyclic shear stress, τ, about the static equilibrium position. The pattern of development of pore water pressure in this kind of test is quite different from that without the initial shear stress. A comprehensive study of the undrained behaviour of sand under these conditions has been reported by Vaid and Finn[23] from which the following brief description is taken.

Typical examples of the cyclic loading behaviour of samples with and without shear stress reversal are shown in Figure 21.17. The value of τ_s was identical for each sample except one. For comparison, the behaviour of a sample with no initial static shear stress is also shown. It may be seen in Figure 21.17 that sample 91, in which there was no shear stress reversal $(\tau < \tau_s)$, underwent progressive straining and pore water pressure build-up (decrease in vertical confining stress in constant volume simple shear tests) with successive cycles of loading. Pore water pressure increased very slowly after about 20 cycles and attained a maximum value of $0.8\sigma'_{v0}$ after about 80 cycles, beyond which no further increase occurred with cycles of loading. Thus, a state of initial liquefaction $(u_{max} = \sigma'_{v0})$ did not occur, although shear strain γ kept on increasing (Figure 21.17(b)), reaching 10% at 280 cycles. Sample 16 showed behaviour similar to that of sample 91, even though the cyclic loading imposed a shear stress reversal $(\tau > \tau_s)$. A state of initial liquefaction did not occur; the maximum pore water pressure approached the value $0.9\sigma'_{v0}$ asymptotically. This behaviour is at variance with that reported by Lee and Seed.[11] Their data showed that the slightest stress reversal during cyclic loading triaxial tests gave rise to a condition of initial liquefaction. A condition of initial liquefaction did develop in sample 67 in which the degree of stress reversal (ratio of minimum to maximum peak

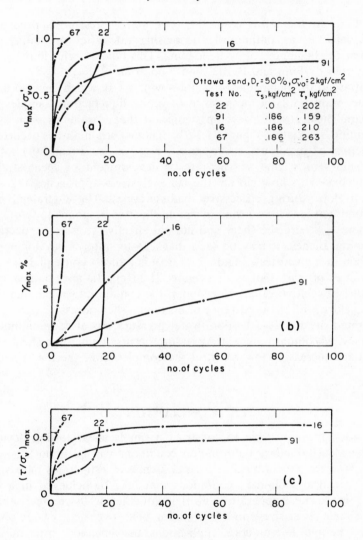

Figure 21.17 Cyclic loading behaviour of loose Ottawa sand with and without initial static shear stresses

shear stress amplitudes) was more severe than in sample 16. Sample 22, with $\tau_s = 0$ (with the highest degree of shear stress reversal = 1) developed the condition of initial liquefaction.

The characteristic difference in the shape of pore water pressure curves in tests with and without initial static shear stress may be noted. The accelerated rate of increase in pore water pressure shortly before the incidence of initial liquefaction in tests with $\tau_s = 0$ is not seen in the tests with non-zero values of τ_s. In the latter case, the pore water pressure approaches a terminal value at a continuously decreasing rate after an initial rapid build-up at the beginning of cyclic loading. This type of behaviour has also been reported by Finn and Byrne[7] and Finn *et al.*[9]

In contrast to the behaviour of samples with $\tau_s = 0$, which develop hardly any shear strain until the occurrence of initial liquefaction, samples with initial static shear stresses develop significant shear strains as soon as the cyclic loading is initiated, Figure 21.17(b). The net strain always occurred in the direction of maximum shear stress $(\tau_{max} = \tau_s + \tau)$ irrespective of the relative values of τ_s and τ. The rate of development of shear strain is intimately linked to how closely the state of stress approaches the static failure envelope during successive loading cycles. The undrained static failure condition for Ottawa sand in simple shear corresponds to $(\tau/\sigma'_v)_{max} = 0.59$, where τ, σ'_v are the shear and normal effective stresses, respectively, on horizontal planes. It may be seen that a faster rate of development of shear strain in a given test (Figure 21.17(b)) is a direct consequence of the higher values of τ/σ'_v mobilized (Figure 21.17(c)). In any loading cycle, irrespective of relative values of τ_s and τ, the condition of maximum shear strain coincided with the occurrence of maximum shear stress, $(\tau_{max} = \tau_s + \tau)$. At maximum shear stress, the pore water pressure was at its minimum due to dilatancy. Maximum pore water pressure occurred when the shear stress was zero and increasing towards τ_s at the completion of the load cycle.

21.8   CONCLUSION

It does not appear possible to construct a simple shear testing device in which the ideal boundary deformation conditions will be imposed on test samples. However, theoretical and experimental investigations have shown that the 'practical' boundary conditions that may be achieved in a well-designed simple shear apparatus with rigid container walls create essentially uniform simple shear stress and deformation fields over a very large portion of the test sample. For practical applications, the deviations from the ideal may be ignored. From the research point of view, by sophisticated instrumentation techniques the stress and strain state in the uniform segment of the sample can be measured and the response of the soil to ideal simple shear conditions deduced.

The cyclic simple shear test may be extended to model more complex seismic loading conditions in the field by incorporating an appropriate initial static shear stress in the test procedure. This modelling will be useful as long as the primary excitation source in the field is shear waves propagating vertically.

Since Roscoe's pioneering work in the 1950s, there has been a steady evolution in the development of simple shear testing and a greater understanding of the mechanics of the test. This evolution is likely to continue.

ACKNOWLEDGEMENTS

The authors express their appreciation to Ms. Desiree Cheung who typed the manuscript and Mr. Richard Brun who drafted the figures. The financial support of the National Research Council of Canada under grant no. 1498 is gratefully acknowledged. Mrs. Shobha K. Bhatia was supported by a Commonwealth Scholarship from Canada.

NOTATION

D_r	relative density
e	void ratio
EX, E_x	Young's modulus in the x-direction
EY, E_y	Young's modulus in the y-direction
G_{xy}	shear modulus in an x, y plane
$2H$	height of simple shear sample
K_0	coefficient of earth pressure at rest
K	ratio of horizontal to vertical effective stress
L	length of simple shear sample
$L/2H$	length to height ratio
MUXX, μ_{xx}	Poisson's ratio in a plane at right angles to the y-axis
MUXY, μ_{xy}	Poisson's ratio in an x, y plane
n_L	cycles to liquefaction
$\text{SIGX} = \sigma_x/\tau_0$	normal stress in x-direction/applied cyclic shear stress
$\text{SIGY} = \sigma_y/\tau_0$	normal stress in y-direction/applied cyclic shear stress
$\text{TAUXY} = \tau/\tau_0$	computed shear stress/applied cyclic shear stress
u_{max}	maximum pore water pressure
u	pore water pressure
U	elastic displacement in x-direction
V	elastic displacement in y-direction
x, y	direction of co-ordinate axes in space
ε_{vd}	volumetric strain
σ'_{v0}	vertical effective confining stress
σ'_v	effective vertical stress
σ'_h	effective horizontal stress
$\sigma'_{h\,ave}$	average effective horizontal stress
σ_x	normal stress in x-direction
σ_y	normal stress in y-direction
τ_{xy}	shear stress
τ_0	average applied cyclic shear stress
τ	computed shear stress on x, y plane, cyclic shear stress
τ/σ'_{v0}	cyclic shear stress/vertical effective confining stress
τ_s	initial static shear stress
τ_{max}	maximum shear stress

REFERENCES

1. Bjerrum, L., and Landva, A. Direct simple-shear tests on a Norwegian quick clay, *Geotechnique*, **16**(1), 1–20, 1966.
2. De Alba, P., Chan, C. K., and Seed, H. B. *Determination of Soil Liquefaction Characteristics by Large-scale Laboratory Tests*, Report No. EERC 75-14, Earthquake Engineering Research Center, University of California, Berkeley, May 1975.
3. Finn, W. D. L. Boundary value problems of soil mechanics, *J. Soil Mechanics and Foundations Div.*, *ASCE*, **89** (SM5), 39–72, September 1963.
4. Finn, W. D. L., Bransby, P. L., and Pickering, D. J. Effects of strain history on liquefaction of sand, *J. Soil Mechanics and Foundations Div.*, *ASCE*, **97** (SM6), Proc. Paper 7670, 1917–34, November 1970.
5. Finn, W. D. L., Pickering, J., and Bransby, P. L. Sand liquefaction in triaxial and simple shear test, *J. Soil Mechanics and Foundations Div.*, *ASCE*, **97** (SM4), Proc. Paper 8039, 639–59, April 1971.
6. Finn, W. D. L., Lee, K. W., and Martin, G. R. An effective stress model for liquefaction, *J. Geotech. Engineering Div.*, *ASCE*, **101** (GT6), Proc. Paper 13008, 517–33, June 1975.
7. Finn, W. D. L., and Byrne, P. M. Liquefaction potential of mine tailing dams, *Proc. 12th International Conference on Large Dams*, Mexico City, **1**, 153–76, 1976.
8. Finn, W. D. L., and Vaid, Y. P. Liquefaction potential from drained constant volume cyclic simple shear tests, *Proc. 6th World Conference on Earthquake Engineering*, New Delhi, 1977.
9. Finn, W. D. L., Lee, K. W., Maartman, C. H., and Lo, R. Cyclic pore pressures under anisotropic conditions, Preprint, *Specialty Conference on Earthquake Engineering and Soil Dynamics*, Pasadena, 15 pp., June 19–21, 1978.
10. Finn, W. D. L., Vaid, Y. P., and Bhatia, S. Constant volume cyclic simple shear testing, *2nd International Conference on Microzonation*, San Francisco, Calif., 26 November–1 December 1978.
11. Lee, K. L., and Seed, H. B. Dynamic strength of anisotropically consolidated sand, *J. Soil Mechanics and Foundation Div.*, *ASCE*, **93** (SM5), Proc. Paper 5451, 169–90, September 1967.
12. Lucks, A. S., Christian, J. T., Brandow, G. E., and Höeg, K. Stress conditions in NGI simple shear test, *Proc. ASCE*, **98** (SM1), 155–60, 1972.
13. Martin, G. R., Finn, W. D. L., and Seed, H. B. Fundamentals of liquefaction under cyclic loading, *J. Geotech. Engineering Div.*, *ASCE*, **101** (GT5), Proc. Paper 11284, 423–38, May 1975.
14. Martin, G. R., Finn, W. D. L., and Seed, H. B. Effects of system compliance on liquefaction tests, *J. Geotech. Engineering Div.*, *ASCE*, **104** (GT4), Proc. Paper 13667, 463–79, 1978.
15. Moussa, A. A. Equivalent drained–undrained shearing resistance of sand to cyclic simple shear loading, *Géotechnique*, **25**(3), 485–94, 1975.
16. Peacock, W. H., and Seed, H. B. Sand liquefaction under cyclic loading simple shear conditions, *J. Soil Mechanics and Foundations Div.*, *ASCE*, **94** (SM3), 689–708, 1968.
17. Pickering, D. J. *A Simple Shear Machine for Soil*, Ph.D. Dissertation, University of British Columbia, Vancouver, 1969.
18. Pickering, D. J. Drained liquefaction testing in simple shear, *J. Soil Mechanics and Foundation Engineering Div.*, *ASCE*, **99** (SM12), 1179–84, 1973.

19. Roscoe, K. H. An apparatus for the application of simple shear to soil samples, *Proc. 3rd Int. Conf. on Soil Mechanics*, Zurich, **1**(2), 186–91, 1953.
20. Schnabel, P. B., Lysmer, J., and Seed, H. B. *SHAKE: A Computer Program for Earthquake Response Analysis of Horizontally Layered Sites*, Report No. EERC 72-12, University of California, Berkeley, 1972.
21. Seed, H. B., and Idriss, I. M. Analysis of soil liquefaction: Niigata earthquake, *J. Soil Mechanics and Foundations Div.*, ASCE, **93** (SM3), 83–108, 1967.
22. Taylor, D. W. *Fundamentals of Soil Mechanics*, Wiley, New York, N.Y., 7th printing, pp. 130–3, 1954.
23. Vaid, Y. P., and Finn, W. D. L. Effect of static shear on liquefaction potential, *J. Geotech. Engineering Div.*, ASCE, **105** (GT10), Proc. Paper 14909, 1233–46, 1979.
24. Wood, D. M., Drescher, A., and Budhu, M. On the determination of the stress state in the simple shear apparatus, *ASTM Geotechnical Testing Journal*, **2**, No. 4, 1980.
25. Wood, D. M., and Budhu, M. The behaviour of Leighton Buzzard sand in cyclic simple shear test, *International Symposium on Soils under Cyclic and Transient Loading*, Swansea, **1**, 9–21, 7–11 January 1980.

Soil Mechanics—Transient and Cyclic Loads
Edited by G. N. Pande and O. C. Zienkiewicz
© 1982 John Wiley & Sons Ltd

Chapter 22

Soil Liquefaction Studies in the People's Republic of China

W. D. L. Finn

22.1 INTRODUCTION

A delegation from the People's Republic of China attended the International Symposium on Soils under Cyclic and Transient Loading held at University College, Swansea, Wales, 7–11 January, 1980. In addition to a formal presentation by Professor Huang Wen-Xi, included in the proceedings of the conference, Dr. Wang Wen-shao presented a very interesting description of experimental, laboratory and field studies of liquefaction of saturated sandy and silty soils during earthquakes carried out by the Research Institute of Water Conservancy and Hydroelectric Power, over the past 20 years. It seemed to members of the Advisory Committee of the Conference that Dr. Wang's contribution was an important chapter in the history of soil liquefaction studies and a summary should be included in the state-of-the-art volume. At the request of the Chairman, Professor O. C. Zienkiewicz, F.R.S., the author undertook the task. Since Dr. Wang's presentation traced the development of soil liquefaction studies in China, it seemed appropriate to supplement it with material from other Chinese research institutions, where such material clarified and extended the picture of development. The author was greatly assisted in this respect by discussions with the Chinese delegation of geotechnical engineers which visited Vancouver in October 1979. This delegation was headed by Vice-President He Xiang of the Academy of Building Research, China. Reports by members of that delegation on liquefaction during the Tang-shan earthquake of 1976 were especially helpful as well as the report of the Chinese delegation to the 2nd U.S. National Conference on Earthquake Engineering, Stanford, 1979. Deng En-Cheng and Wang Zhi-Liang of the General Research Institute of Building and Construction, Ministry of the Metallurgical Industry, Beijing provided information on the procedures for non-linear dynamic analysis. All contributions are acknowledged appropriately in the text; individuals who contributed ideas in discussions are identified in the acknowledgements section at the end of the paper. It is the author's hope that what follows is a reasonably accurate picture of soil liquefaction research in the People's Republic of China and that the contributions of the various researchers are fairly presented.

22.2 EARLY LABORATORY INVESTIGATIONS

The first dynamic triaxial test apparatus, based on the suggestions of Professor Huang Wen-Xi of Qinghua University, was developed in 1959.[7] The triaxial apparatus was mounted on a vibrator and carried a heavy piston which transferred a vertical load to the sample. The vertical inertia forces generated during vertical vibration applied cyclic deviator stresses to the triaxial specimen which was maintained under constant confining pressure.

Test data obtained using this equipment was published by Wang in 1962.[16] These results showed the effects of applied static and dynamic stresses on the development of pore water pressure in the triaxial sample. The presentation of data was somewhat different from that used to-day. The non-dimensional rise in pore water pressure, u/σ_3, was plotted against either σ_1/σ_3 or against $\Delta\sigma_1/\sigma_1$. In these ratios, u = pore water pressure, σ_1 = maximum axial stress, $\Delta\sigma_1$ = cyclic increase in axial stress, and σ_3 = confining pressure. Current procedures for the presentation of liquefaction data were adopted later by Wang Wen-shao in 1974,[8] using the concept of the cyclic stress ratio τ/σ_0' (τ = cyclic shear stress, σ_0' = initial confining pressure) and cycles of loading N.

Some of the earliest known studies on the effect of pre-shearing or pre-vibration on liquefaction potential and pore pressure rise were carried out in the triaxial equipment. The saturated sand samples were pre-vibrated for various periods of time and then the excess pore water pressure which had developed was drained off. The samples were then vibrated again with the same intensity of vibration as previously under undrained conditions and the pore water pressures recorded. The results showed that pre-vibration or pre-shearing, over the range of strains generated in the tests, increased the resistance to liquefaction. The rate of increase in pore water pressure and its ultimate magnitude for a given excitation were reduced by pre-vibration. These results were reported by Wang Wen-shao in 1962.[16] The beneficial effect of pre-vibration was attributed to the generation of a more stable structure in the sand. Similar conclusions were arrived at independently by Finn, Bransby, and Pickering in 1970.[2]

One of the surprising aspects of these results is the early date at which they were derived. It is clear that the study of cyclic loading of saturated sands under controlled conditions was well under way in China by 1966 when rapid development of the field began in the western world stimulated by the large number of liquefaction failures that occurred during the Niigata and Alaskan earthquakes of 1964.[14]

22.3 METHODS OF DYNAMIC ANALYSIS

Fundamental to any study of liquefaction is a method of dynamic response analysis for estimating with sufficient accuracy the cyclic shear stresses induced in the ground during an earthquake. Methods of dynamic analysis used in China are in the mainstream of recent developments in the field. Equivalent linear wave-propagation methods, similar to that developed by Schnabel, Lysmer, and Seed[13] are used, as well as equivalent linear methods based on mode superposition solutions. More recently, a non-linear method of analysis has been developed by Wang Zhi-Liang, Han Qing-yu and Zhou Gen-shou[20] which treats the soil as a non-linear, hysteretic material. In this method the initial loading curve, as derived in laboratory tests, is treated as a skeleton or backbone curve and subsequent unloading and reloading curves are defined by the Masing criterion.[3,12] Analyses with this model are as yet restricted to total stress analyses. Changes in stiffness during excitation are not taken into account. A particularly ingenious procedure has been developed for modifying the hysteresis loops to maintain comparability between computed and experimentally determined measures of damping. A damping ratio degradation factor $k(\gamma_{max})$ is introduced which is dependent on the maximum shear strain. The factor is incorporated in the equation of the hysteresis loop in such a way that Masing behaviour is preserved.

The defining relations may be analytically represented as follows:

$$\text{loading curve:} \qquad \tau = f(\gamma) \qquad (22.1)$$

$$\text{unloading–reloading:} \quad \frac{\tau - \tau_m}{2} = f\frac{(\gamma - \gamma_m)}{2} \qquad (22.2)$$

generalized Masing curve with damping correction:

$$\tau - \tau_m = k(\gamma_m)\left[2f\left(\frac{\gamma - \gamma_m}{2}\right) - G(\gamma_m)(\gamma - \gamma_m)\right] + G(\gamma_m)(\gamma - \gamma_m) \quad (22.3)$$

Some interesting studies have been carried out by Wang Zhi-Liang[20] comparing the results of non-linear and equivalent linear analyses which show that, generally, the equivalent linear analysis results in significantly larger stresses. Results of analyses for 10 sites in Beijing are shown in Table 22.1. The results for shear stresses are comparable to those obtained by Finn, Martin, and Lee[4] from similar analyses.

A method of dynamic effective stress analysis has been presented by Wang Wen-shao[17-19] in which changes in porosity during drained cyclic loading are related to pore water pressures during undrained cyclic loading. In structure the method is similar to the model of pore water pressure proposed by Martin, Finn, and Seed[11] and incorporated by Finn, Lee, and Martin[3] in a procedure for dynamic effective stress analysis. Wang assumes that the porosity-consolidation pressure lines for sand are parallel as

Table 22.1   Computed results of ten profiles in Beijing

Profile No.	Depth (m)	Ground maximum acceleration Base maximum acceleration			Maximum shear stress (kgf/cm²)		
		1	2	3	1	2	3
B25	27.1	2.24	2.74	2.37	0.46	0.62	0.57
B18	32	1.53	2.17	2.47	0.41	0.53	0.52
B17	48.3	1.42	1.99	1.87	0.63	0.68	0.65
B16	142.3	0.75	1.12	1.22	1.07		1.04
A22	58.9	1.11	1.22	1.50	0.77	0.96	0.91
A26	60	1.22	1.22	1.59	0.83	1.13	0.98
A24	70	1.24	1.19	1.58	0.92	1.17	1.14
A13	98.7	0.84	1.21	1.20	0.97	1.31	1.16
A10	114.2	0.87	1.28	1.19	1.31	1.70	1.19
A7	136.9	0.60	0.65	1.12	0.98	1.50	1.24

Notes: 1. Visco-elastoplastic method with Masing behaviour (Wang Zhi-Liang).
2. Equivalent linear wave propagation method (Wang Zhi-Liang).
3. Equivalent linear mode superposition method (Fu Sheng-Cong).
Type A: Max.Acc. 125 gal. predominant period 0.3 s.
Type B: Max.Acc. 100 gal. predominant period 0.3 s.

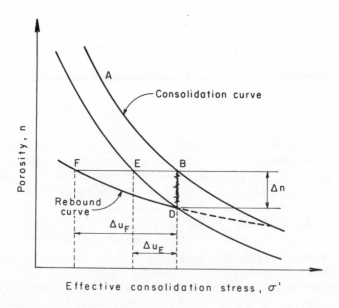

Figure 22.1 Model for pressure development during undrained cyclic loading

in Figure 22.1 and that during drained cyclic loading a point such as *B* will move to point *D* with a porosity change Δn. During undrained cyclic loading the porosity does not change and *B* translates to *E* giving a pore water pressure Δu_E. This model of soil behaviour is based on the assumption that during vibration the consolidation lines shift position parallel to themselves so that point *B* initially on the line *AB* stays on that line during vibration or cyclic loading as it translates to *ED*. The associated changes in porosity and pore water pressure are determined by the assumption that in either drained or undrained vibration, *B* will end up on a consolidation line at either *E* giving the increment in pore water pressure or at *D* giving the change in porosity.

There is an interesting difference between the fundamental assumption of the Wang model and the model proposed by Martin, Finn, and Seed.[11] The latter assume that point *B* will move to the rebound curve through *D* during vibration and not to the virgin consolidation line. They argue that if a porosity change *BD* is prevented by undrained conditions, this porosity change must be absorbed by rebound to maintain zero volume change or, at most, to limit volume change to the compression in the water. Hence, they consider that in undrained cyclic loading the point *B* moves to *F* on the rebound curve through *D*. Since the slope of the rebound curve is flatter than that of the virgin consolidation line, their assumption results in larger computed pore-water pressure increments, Δu_F.

22.4 CASE HISTORIES

22.4.1 Liquefaction damage to Baihe Dam[10]

The Baihe main dam is a sloping core earthdam with a maximum height of 66 m and a crest length of 960 m. The dam is situated at the confluence of the Chaohe and Baihe Rivers about 90 km from Beijing and retains the Miyun reservoir which has a capacity of 4300 million m^3. The sloping core of the earthdam is composed of medium to heavy silty loams. The design dry density of the impervious core was 1.70 t/m^3. Both the downstream dam shell and the 3–5 m thick upstream protective layer were placed with sand and gravel materials. The design dry densities of sand and gravels were 1.71–2.07 t/m^3, depending on the gravel content of the materials. The actual dry densities had reached or exceeded the design standard. The dam foundation is an alluvium of sand and gravel, 44 m deep. A combination of concrete diaphram wall and grout curtain is used for controlling the underground seepage. A typical cross-section of the earthdam is shown in Figure 22.2. Since the filling of the reservoir in 1960, the dam, which was once subjected to 93% of the design water head, has given good performance.

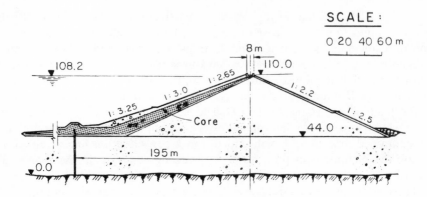

Figure 22.2   Cross-section of Baihe dam[10]

During the Tangshan earthquake on 28 July 1976, a large slide occurred in the upstream protective layer of sand and gravel which covered a wide area of about 60 000 m², with a slide length of about 900 m, and a slide volume of about 150 000 m³. As described by Liu Ling-yao, Li Kueifei and Bing Dong-ping,[10] at the moment of sliding, the reservoir water level was about 21 m below the crest, while the slide mass was almost entirely located below the water level. Almost all of the damaged portions were within the protective layer below the water surface, and only a localized damage occurred in the sloping core. The condition after failure is shown in Figures 22.3-5. The flowing soils were deposited over a long distance, the greater part at a distance of more than 40 m from the upstream toe of the dam and a small part more than 100 m away. The soil particles of the failure mass became smaller and smaller with increasing distance away from the upstream toe. Cobbles and coarse gravels were deposited at the upper part of the dam slope, while the fine materials were deposited beyond the toe of the upstream slope. From the evidence, a liquefaction failure was indicated.

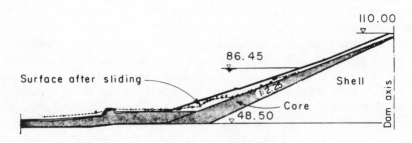

Figure 22.3   Failure slope in dam after Tangshan earthquake[10]

Figure 22.4 Top of failure surface shows above water level[10]

Figure 22.5 Slide surface exposed after lowering of reservoir level[10]

Table 22.2   Acceleration and period of the earthquake on 28 July at Baihe Dam

Observation point	Location	Direction	Maximum acceleration	Period of major pulses (s)
a	Dam crest	Parallel to dam axis	160	0.57
a	Dam crest	Normal to dam axis	128	0.60
a	Dam crest	Vertical	66	0.25, 0.11,
b	Downstream dam toe	Parallel to dam axis	39	0.2, 0.1–0.3, 0.45–0.70
b	Downstream dam toe	Normal to dam axis	53	0.2, 0.11, 0.4–0.8
b	Downstream dam toe	Vertical	50	0.13, 0.25

Just after the earthquake, measurements of the settlement and displacement of the earthdam were made. The maximum settlement was 59 mm and the maximum horizontal displacement was 28 mm towards downstream. No other earthquake damage such as cracks or abnormality of seepage were found.

According to the seismograms recorded at the dam site, the duration of the earthquake was 114 s. The maximum acceleration, the period, the duration and other characteristics are listed in Table 22.2 and Table 22.3. It can be seen from the tables that the seismic motions were of long duration and had a significant vertical component of acceleration and a large fraction of long-period pulses. The number of strong pulses (corresponding to 65% of maximum acceleration) exceeded 60, which is much greater than the statistical number of strong pulses (approximately 30) for other earthquakes of magnitude 8. The seismic intensity at the site was estimated to be 6° on the Chinese scale. The Chinese intensity scale contains 12° of intensity and corresponds approximately to the modified Mercalli scale. However, since

Table 22.3   Time duration of the earthquake on 28 July at Baihe Dam

Observation point	Direction	$t(s)$ ($>\frac{2}{3}a_{max}$)	$t(s)$ ($>\frac{1}{2}a_{max}$)	$t(s)$ $\left(>\frac{1}{e}a_{max}\right)$*
a	Parallel to dam axis	13.7	23.1	35.0
a	Normal to dam axis	16.4	23.0	31.0
a	Vertical			
b	Parallel to dam axis	0.1	16.7	20.6
b	Normal to dam axis			
b	Vertical	4.4	30.6	40.0

* e—base of natural logarithm
a—seismic acceleration

construction in China is quite different to ours direct comparisons of intensity cannot be made.

At first glance it seems surprising that a flow slide obviously triggered by high pore water pressures could occur in such coarse material with short drainage paths. In fact, the drainage was poor because the gravelly material was sandwiched between the impermeable core and a rubble revetment set in cement mortar. When the beneficial effects of high permeability are negated by no provisions for drainage, gravelly materials are susceptible to liquefaction as shown by Wong, Seed, and Chan[21] in laboratory tests and by Finn, Yong, and Lee[6] and Finn and Yong[5] in studies of the liquefaction of unfrozen gravel layers sealed by ice or frozen ground during the Alaskan earthquake of 1964.

An experimental investigation of the liquefaction potential of the sand–gravel materials was carried out to determine if sufficient porewater pressures could be developed to generate a flow slide. A cylindrical container of diameter 40 cm was used in the tests, which were performed on a shaking table. The peak acceleration was 0.2 g (amplitude: 36 mm, frequency: 100 cycles/min). Specimens prepared to the design dry density were laterally confined by the container walls and allowed to drain in the vertical direction. Therefore, neither stress nor drainage conditions were controlled. Results, though significant in interpreting gravel behaviour, are qualitative. The degree of liquefaction was calculated by the following equation:

$$K_t = \frac{U_t}{(\bar{\sigma}_1)_0} \times 100\%,$$

where K_t = degree of liquefaction, %;

U_t = pore-water pressure due to vibration at time t, kg/cm^2;

$(\bar{\sigma}_1)_0$ = initial effective vertical pressure at the point of measurement, kg/cm^2.

It can be noticed from the grain size distribution curves, as shown in Figure 22.6, that the sand and gravel materials comprising the protective layer of Baihe earthdam have an average gravel content of about 60%, being discontinuous in size distribution and lacking the intermediate particles. The fine materials (grain size less than 5 mm) from the sand and gravel materials contain on the average about 76% of medium to fine sands.

The investigation showed that the sand and gravel materials could be subject to liquefaction, and that the degree of liquefaction depended on the gravel content of the materials. Figure 22.7 illustrates that the degree of liquefaction is 74% for fine materials comprising only particles less than 5 mm, increases to 90.5% for sand and gravel materials with a gravel content of 50% and then decreases progressively to very low values for sand and gravel materials with a gravel content of 70%. This may be expected in

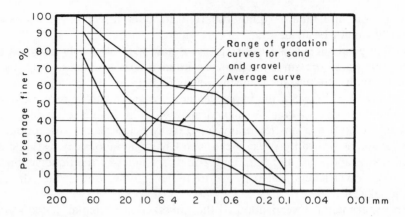

Figure 22.6 Grain-size distribution curves for materials in the upstream protective layer of Baihe dam[10]

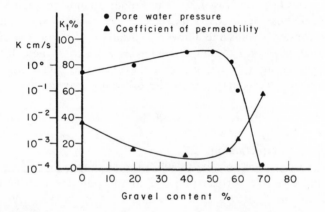

Figure 22.7 Effect of gravel content on permeability, K, and degree of liquefaction, K_f[10]

these tests because the permeability appears to increase quite rapidly for gravel contents above about 60%.

The protective layer of Baihe Dam had an average gravel content between 50–60%. Therefore, on the basis of the laboratory tests, pore pressure ratios of up to 90% are possible during excitation. However, it must be pointed out again that both stress and drainage conditions were not controlled during the laboratory tests and the results, therefore, to some extent must be considered qualitative. Nevertheless, the findings are very significant in elucidating the mechanism of failure and are in keeping with those of Wong, Seed, and Chan.[21]

22.4.2   Ground liquefaction during Tangshan earthquake

Widespread liquefaction occurred during the Tangshan earthquake and the effects of liquefaction were especially severe between Tangshan and the coast and along the old courses of rivers.[9] Damage, due to liquefaction, resulted in the embedment of zones of intensity IX within a broad general zone of intensity VII. Settlements up to 3 m were reported. Despite the widespread manifestations of liquefaction such as sand boils, slumping of embankments and banks of rivers and canals, the Tangshan earthquake resulted in little new quantitative information on liquefaction because strong motion records were not available in most regions where liquefaction occurred.

A highly interesting novel approach to the study of field liquefaction caused by the Tangshan earthquake was developed by Fang Hong-qi, Wang Zhung-qi and Zhao Shu-dong of the Geotechnical Investigation Institute of the Chinese Academy of Building Research.[1] They related the macrofeatures of liquefaction and the damage modes to the local microgeomorphic conditions. The study resulted in the allocation of damage parameters to various types of geomorphic conditions in liquefiable areas.

One of the basic tools used in the study was air-photo interpretation. Air-photos were taken 3 days after the Tangshan earthquake when the field patterns of liquefaction such as sand boils, sand ejections from fissures, etc. were still preserved. On the basis of the air-photo evidence, the macrofeatures of liquefaction were divided into three categories. Each category was associated with particular geomorphic and topographic features and deduced pattern of wave motion. The details of categorization are shown in Table 22.4 which, as a matter of interest, is reproduced directly from Fang's report.[1] Three examples of the categories are shown in Figure 22.8 (densely scattered stars), Figure 22.9 (arteries) and Figure 22.10 (vortex).

In the 'scattered stars' category, the stars are the lightly coloured sand boils. Their spacing and size reflect the embedded depth of the liquefied layer or the overburden pressure on it (Figure 22.8). The network pattern is typical of the convex bank side of a river bend where ground motion may be intensified by the focusing of surface waves from the curvilinear boundary of the bank. Such ground movement was considered to have a definite direction of propagation parallel to the axis of the river bend (Figure 22.9). The tortile category is the most novel and unusual. It is assumed to result from the intersection of reflected surface waves from a number of boundaries such as a series of river bends. Near the intersection points the waves moving in different directions set up strong torsional vibrations in the ground which result in strongly curved fissures through which the liquefied sand is ejected. The sand ejecta may show up in air-photos as vortex patterns (Figure 22.10).

Table 22.4   Macroscopic features of liquefaction and its environmental characteristics (reproduced directly from Ref. 1)

Categories of Patterns	Subdivided pattern	Legend	Geomorphic and Topographic Feat	Principal Mode of Surface Wave	plate No.
Scattered stars (s.s)	sparsely s.s		plain in topog., simple in geom., away from river bank.	Evenly transmitted surface travelling waves.	1
	densely s.s				2
	cloudily s.s				3
Network	branches		On the convex bank of rive bend, topographically low, near by the river	Stationary surface waves reflected and focused within the river bend.	4
	ateries				5
	radiation				6
Tortile	broom		inside the river bends skirt, topographically plain and low.	The resultant wave front vectors from multi-directional reflection and intersected to form torsional movement of ground.	7
	vortex				8
	Curly hair				9

Figure 22.8   Scattered stars liquefaction pattern[1]

Figure 22.9   Network liquefaction pattern[1]

Figure 22.10   Tortile liquefaction pattern[1]

The different macrofeatures to be seen on air-photos are assumed to represent different mechanisms of generating liquefaction and on the basis of the work done to date, it is believed that these macrofeatures can be reliably related to geomorphic and topographic features. From the damage to structures which occurred in regions with various liquefaction macrofeatures a damage index was developed giving the relative damage in the various regions. For the categories illustrated in Figures 22.8–10, the damage indices were as follows: scattered stars 0.4, arteries 0.6, and vortex 0.9.

Fang, Wang, and Zhao[1] suggest that the use of geomorphic and topographic features together with the damage indices may be a useful procedure for zoning large areas for potential damage.

22.5 EMPIRICAL CRITERION FOR LIQUEFACTION

Field studies of liquefaction by Chinese engineers, primarily from 1966 to 1970, resulted in the development of an empirical criterion for assessing the liquefaction potential of sand in the field, based on the results of the standard penetration test. The criterion was incorporated in the Chinese seismic code (TJ11-74) in 1974. The development of the criterion has been described in detail by Xie Junfei[22] and useful additional information is given by Jennings.[9]

The criterion is given by the equation

$$N_{crit} = \bar{N}\{1 + 0.125(d_s - 3) + 0.05(d_w - 2)\},$$

where d_s = depth to sand layer under consideration, m
   d_w = depth of water table below ground surface, m
   $\bar{N}$ = function of the shaking intensity or acceleration in gravity units.

Intensity	Acceleration, g	$\bar{N}$ (blows/30 cm)
VII	0.075	6
VIII	0.15	10
IX	0.30	16

$\bar{N}$ was evaluated from field data for 12 cases of foundation failure due to sand liquefaction and 58 cases having macroscopic surface phenomena which made it possible to distinguish between liquefied and non-liquefied sites. The ground water levels in these cases varied from 1 to 3 m with an average of 2 m and the depths of liquefaction varied from 2 to 4 m with an average of 3 m. These data are shown plotted in Figure 22.11 from which the critical $\bar{N}$ values of 6, 10, 16 at intnsities VII, VIII and IX are obtained. The extension to other depths of liquefiable layers, d_s, and other depths to the water table, d_w, was carried out using the simplified liquefaction analysis developed by Seed and Idriss.[15] This extension is achieved by means of the

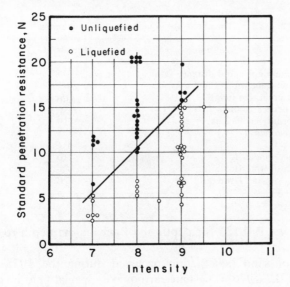

Figure 22.11 Field data on liquefaction used to define critical $\bar{N}$ values[9]

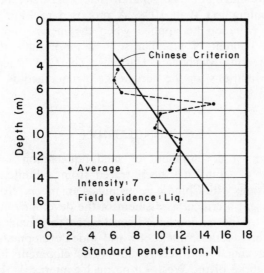

Figure 22.12 Correlation of code criterion and field data on liquefaction[22]

Table 22.5   Thresholds for liquefaction (After Jennings[9])

Condition	Threshold
Maximum epicentral distance (km)	$\log D_{max} = 0.87M - 4.5$
Minimum intensity (Chinese scale)	6
Mean grain size (mm)	$0.02 < D50 < 1.0$
Clay particle content (%)	<10
Uniformity coefficient	<10
Relative density (%)	<75
Void ratio	>0.8
Plasticity index I_p	<10
Depth to water table (m)	<5
Depth of sand layer (m)	<20

correlation factors of 0.125 for depth and 0.05 for the depth to the water table.

Many cases of sand liquefaction occurred during the 1975 Haicheng earthquake and the 1976 Tangshan earthquake. Special field investigations were made to check the reliability of the criterion. A typical correlation is shown in Figure 22.12, which shows that the criterion gives a reasonable estimate of liquefaction potential. In general, however, the criterion appeared to be somewhat conservative when compared with the field data from the Tangshan earthquake. It is considered that the Tangshan materials contained considerable fines which increased the liquefaction resistance. It was found that soils which liquefied generally had a plasticity index less than 10 and a clay content less than 10%. A proposed revision to the code suggests that the equation be applied when the mean grain size $D_{50} > 0.05$ mm and the content of granular soils is more than 40%. At the Institute of Engineering Mechanics, Harbin, a chart has been developed giving threshold conditions for the occurrence of liquefaction. The chart is described in Table 22.5.[9]

22.6   CONCLUSION

The broad pattern of development of sand liquefaction studies in the People's Republic of China has been traced from a limited number of reports and discussions with Chinese geotechnical engineers. No attempt has been made to present a critical assessment of the development. Comments and interpolation by the author have been limited to placing some of the developments in the broader context of world-wide developments. Although many details are lacking, it does seem clear that developments in China have largely paralleled those of the West. Of particular interest is the early date at which cyclic triaxial testing for liquefaction potential was introduced by Chinese researchers. Recent techniques using air-photo interpretation for evaluating field data on liquefaction are novel and ingenious and may develop into useful tools for assessing the likelihood and consequences of liquefaction in particular geomorphic and topographic environments.

ACKNOWLEDGEMENTS

Discussions with the following were very useful in the preparation of this paper: He Xiang, Vice-president, Academy of Building Research of China; Li Mingqing, Director, Geotechnical Investigation Division, Design Bureau, State Administration of Building Construction; Wang Zhongqi, Deputy Chief Engineer, Geotechnical Investigation Institute, Academy of Building Research of China; Lin Zaiguan, Deputy Chief Engineer, Geotechical Investigation Institute of Shanxi Province; Yan Renjue, Deputy Chief Engineer, Soil and Foundation Section, General Research Institute of Building and Construction, Ministry of Metallurgical Industry; Professor Huang Wen-Xi, Qinghua University, Beijing, and the following members of the Research Institute of Water Conservancy and Hydroelectric Power: Dr. Wang Wen-shao, Deputy Head, Earthquake Engineering Department, and Liu Ling-yao, Li Kueifen and Bing Dong-ping.

REFERENCES

1. Fang, Hong-qi, Wang, Zhung-qi, Zhao, Shu-dong, *Macroscopic Mechanism of Soil Liquefaction and its Influence on Earthquake Damage of Ground*, Report, Geotechnical Investigation Institute, Chinese Academy of Building Research, Beijing, 1979, 1–17; also contributed to 26th Session of International Geological Congress, 1979.
2. Finn, W. D. L., Bransby, P. L., and Pickering, D. J., Effect of strain history on liquefaction of sand, *J. Soil Mechanics and Foundations Division, ASCE,* **97** (SM6) Proc. Paper 7670, 1917–34, 1970.
3. Finn, W. D. L., Lee, K. W., and Martin, G. R., An effective stress model for liquefaction, *J. Geotech. Engineering Division, ASCE,* **103** (GT6), Proc. Paper 13008, 517–33, June 1977.
4. Finn, W. D. L., Martin, G. R., and Lee, M. K. W., Comparison of dynamic analyses for saturated sands, *Proc. ASCE Geotechnical Engineering Division,* Speciality Conference, Earthquake Engineering and Soil Dynamics, Pasadena, Calif., **1,** 472–91, 19–21 June 1978.
5. Finn, W. D. L., and Yong, R. N., Seismic response of frozen soils, *J. Geotech. Engineering Division, ASCE,* **104** (GT10) 1225–41, October 1978.
6. Finn, W. D. L., Yong, R. N., and Lee, K. W., Liquefaction of thawed layers in frozen soil, *J. Geotech. Engineering Division, ASCE,* **104** (GT10), Proc. Paper 14107, 1243–55, October 1978.
7. Huang, Wen-Xi, Investigation on stability of saturated soil foundations and slopes against liquefaction, *Proc. 5th International Conference on Soil Mechanics and Foundation Engineering,* **2,** 1961.
8. Institute of Hydrotechnical Research, *Research Report on Seismic Stability of*

Saturated Soil of Strata II_2 *in Foundation of Shentou Power Plant, Shuoxian, Shanxi*, Beijing (in Chinese), August 1974.

9. Jennings, P. C. (ed.), *Earthquake Engineering and Hazards Reduction in China*, Committee on Scholarly Communication with the People's Republic of China, CSCPRC Report No. 8, National Academy of Sciences, Washington, D.C., 1–189, 1978.

10. Liu, Ling-yao, Li, Kueifen, Bing, Dong-ping, *Earthquake Damage of Baihe Earth Dam and Liquefaction Characteristics of Sand and Gravel Materials*, Report, Research Institute of Water Conservancy and Hydroelectric Power, Beijing, 1–11, August 1979.

11. Martin, G. R., Finn, W. D. L., and Seed, H. B., Fundamentals of Liquefaction Under Cyclic Loading, *J. Geotech. Engineering Division, ASCE*, **101** (GT5), Proc. Paper 11284, 423–38, May 1975.

12. Masing, G., Eigenspannungen und Verfestigung beim Mesing, *Proc. 2nd International Congress of Applied Mechanics*, Zurich, Switzerland, 1926.

13. Schnabel, P. B., Lysmer, J., and Seed, H. B., *SHAKE: A Computer Program for Earthquake Response Analysis of Horizontally Layered Sites*, Report No. EERC 72–12, Earthquake Engineering Research Center, University of California, Berkeley, California, December 1972.

14. Seed, H. B., and Lee, K. L., Liquefaction of saturated sands during cyclic loading, *J. Soil Mechanics and Foundations Division, ASCE*, **92** (SM6), Proc. Paper 4972, 105–34, November 1966.

15. Seed, H. B., and Idriss, I. M., Simplified procedure for evaluating soil liquefaction potential, *J. Soil and Mechanics and Foundations Division, ASCE*, **97** (SM9), Proc. Paper 8371, 1249–73, September 1971.

16. Wang, Wen-shao, Study on pore water pressures of saturated sands due to vibration, *J. Hydraulic Engineering*, No. 2, Beijing (in Chinese), May 1962.

17. Wang, Wen-shao, Generation, dispersion and dissipation of pore water pressures of saturated sands during vibration, *Proc. 1st National Conference on Soil Mechanics and Foundation Engineering*, China Civil Engineering Society, Beijing (in Chinese), December 1962.

18. Wang, Wen-shao, *Some Findings in Soil Liquefaction*, Report, Research Institute of Water Conservancy and Hydroelectric Power, Beijing, 1–17, August 1979.

19. Wang, Wen-shao, Strength, liquefaction and failure of saturated sands under cyclic loading, *J. Hydraulic Engineering*, No. 1, Beijing (in Chinese), 1980.

20. Wang, Zhi-Liang, Han, Qing-yu, and Zhou, Gen-shou, *Wave Propagation Method of Site Seismic Response by Visco-elastic Model*, Report, General Research Institute of Building and Construction, Ministry of Metallurgical Industry, PRC, July 1979.

21. Wong, R. T., Seed, H. B., and Chan, C. K., Cyclic loading liquefaction of gravelly soils, *J. Geotech. Engineering Division, ASCE*, **101** (GT6), Proc. Paper 11396, 571–83, June 1975.

22. Xie, Junfei, *Empirical Criteria of Sand Liquefaction*, Special Session on Earthquake Engineering in China, 2nd U.S. National Conference on Earthquake Engineering, Stanford University, 22–24 August 1979.

Subject Index

anisotropy, 518

Biot's equations, 2, 12, 18, 88
boundary layer, 12, 22, 23
bounding surface, 257
bounding theorems, 483

consolidation equations, 7, 25, 48, 58
consistency condition, 181, 441
creep, 414, 531, 555
critical state, 219, 285, 541, 550
cyclic hardening index, 301
cyclic mobility, 539

damage parameter, 125
damping ratio, 113, 296, 506, 560
densification 80, 408
degradation of elastic modulus, 134, 324, 459, 461, 556
dilatancy, 325, 346, 441, 456, 508
dimensional analysis, 9, 448
drained behaviour, 8, 77, 326, 542
dynamic amplification, 506
dynamic analysis 95, 106, 155, 431, 498

Endochronic theory, 74, 112, 123, 377, 419, 423
equivalent cycle, 121
equivalent viscous damping ratio, 297
equations of motion, 4, 148

failure surface, 314
  see also yield surface
fatigue
  model, 461
  parameter, 460
flow rule, 75, 261, 319, 442

hardening rules/functions, 76, 180, 185, 300, 318, 390, 420, 441, 485
hysteresis, 238, 313, 473, 503, 598
Hooke's law, 20, 22

image stress point, 258

Ins model, 203
intrinsic time, 378

Koiter's theorem 474

liquefaction, 54, 66, 73, 115, 123, 165, 360, 434, 447, 531, 598, 609, 613

Masing rule, 108, 145, 611
mass balance of flow, 5, 84
Melan's theorem, 474
membrane penetration, 328, 521
Miner's rule, 461
modified Cam clay model, 220
multi-surface model, 187
multi-axial response, 128

phase transformation
  line, 135
  angle, 139
physical models, 491
plastic potential, 74, 346, 509
pore pressure
  dissipation, 53, 64, 57
  generation, 55, 61, 64, 108, 497
  model, 110
  ratio, 54, 126

ratcheting, 73, 370, 472

San Fernando dam, 95, 499
simple shear, 523, 583
shakedown, 73, 469, 474, 545
shear travel, 412
skeleton curve, 145

time integration, 84, 156
two surface model, 190

undrained behaviour, 7, 77, 87, 304, 328, 335, 542

visco-plasticity, 74, 174, 207

yield surface, 74, 135, 204, 222, 316

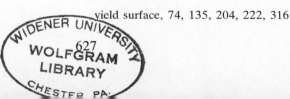